TELECOMMUNICATIONS PRIMER

Feher/Prentice Hall Digital and Wireless Communication Series

Carne, *Telecommunications Primer: Signal, Building Blocks, and Networks*

Feher, *Wireless Digital Communications: Modulation and Spread Spectrum Applications*

Garg / Wilkes, *Wireless and Personal Communications Systems*

Pelton, *Wireless Satellite Telecommunications: The Technology, the Market and the Regulations*

IEEE Telecommunications Handbook Series

Books in the Series

Bhatnagar, *Network Synchronization*

Boucher, *Traffic System Design Handbook*

Carne, *Telecommunications Primer: Signal, Building Blocks, and Networks**

Garg / Wilkes, *Wireless and Personal Communications Systems**

Nellist, *Understanding Telecommunications and Lightwave Systems, Second Edition*

Reeve, *Subscriber Loop Signaling and Transmission Handbook: Analog*

Reeve, *Subscriber Loop Signaling and Transmission Handbook: Digital*

Thomas, *Cable Television Proof-of-Performance: A Practical Guide to Cable TV Compliance Measurements Using a Spectrum Analyzer**

If you are interested in becoming an author, contributor, or reviewer of a book in this series, or if you would like additional information about forthcoming titles, contact:

Whitham D. Reeve
Series Editor, IEEE Telecommunications Handbook Series
P.O. Box 190225
Anchorage, Alaska 99519-0225
w.reeve@ieee.org

*A Prentice Hall PTR/IEEE PRESS copublication

TELECOMMUNICATIONS PRIMER

Signals, Building Blocks, and Networks

E. BRYAN CARNE

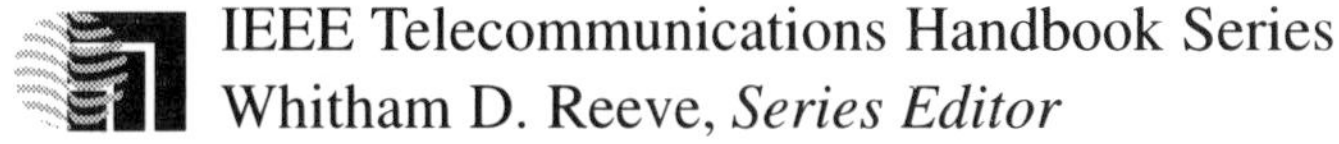

The Institute of Electrical and Electronics Engineers, Inc., New York

Prentice Hall PTR
Upper Saddle River, New Jersey 07458

Library of Congress Cataloging-in-Publication Data for Original Edition
Carne, E. Bryan
Telecommunications primer: signals, building blocks, and networks
/E. Bryan Carne.
p. cm.
Includes bibliographical references.
ISBN 0-13-206129-5
1. Telecommunication. I. Title.
TK5101.C29863 1995
621.382—dc20 95-16021
CIP

Printed in the United States of America

10 9 8 7 6 5

ISBN 0-13-490426-5
IEEE Order Number PC5670

Prentice-Hall, Inc.
A Simon & Schuster Company
Upper Saddle River, New Jersey 07458

This edition is a copublication of the IEEE PRESS in association with Prentice Hall PTR.

The publisher offers discounts on this book when ordered in bulk quantities. For more information, contact:

IEEE PRESS Marketing
Attn: Special Sales
P.O. Box 1331 / 445 Hoes Lane
Piscataway, NJ 08855-1331
1-800-678-IEEE

Prentice-Hall International (UK) Limited, *London*
Prentice-Hall of Australia Pty. Limited, *Sydney*
Prentice-Hall Canada Inc., *Toronto*
Prentice-Hall Hispanoamericana, S.A., *Mexico*
Prentice-Hall of India Private Limited, *New Delhi*
Prentice-Hall of Japan, Inc., *Tokyo*
Simon & Schuster Asia Pte. Ltd., *Singapore*
Editora Prentice-Hall do Brasil, Ltda., *Rio de Janeiro*

To

JOAN

SPECIAL THANKS

From 1989 until 1992, I was BellSouth Distinguished Visiting Professor of Telecommunications and Information Systems Management at Christian Brothers University, Memphis, Tennessee. It is a pleasure to thank BellSouth Foundation for funding the Chair. Thanks are also due to the following people:

- Brother Theodore Drahmann, Brother Peter Drake, and the community of Brothers of the Christian Schools, for welcoming me and introducing me to the LaSallian tradition of education
- Dr. Ray House, Dean of the School of Business, for his support and for providing me with a wide spectrum of teaching opportunities
- Craig Blackman, Director of Graduate Engineering Programs, School of Engineering, for his encouragement and friendship
- The many students who, through their questions, unknowingly contributed to the organization and contents of this book.

Thank you all.

Contents

PART ONE: SIGNALS

Preface

It used to be that *communication-at-a-distance* was limited to employing the telegraph and telephone for important messages. The technology of implementation was devoted to pole-climbing, stringing wires, cleaning metal contacts, and maintaining wet batteries. As such, it was an avocation pursued by craftsmen employed by entrepreneurs eager to seize the opportunity to make a profit. With the coming of solid state electronics, and the ability to travel freely anywhere in the world, all that has changed. High-performance integrated circuit chips, reduced instruction set processors, and optical fibers are commonplace in a network that carries data, image, voice, and video messages of increasing complexity at higher and higher speeds. The result is a telecommunications environment filled with specialized equipment that is being incorporated in the day-to-day operations of corporations and the daily activities of individuals. Global corporations depend on expanding streams of information to run their businesses and to compete in a world economy. Through the use of telecommunications-linked, information systems, corporations control operations, improve productivity, and create economic networks. Furthermore, with voice, video, and document communications, they facilitate interactive communication among their staffs, and with their customers and suppliers. In increasing numbers, individuals rely on pervasive telecommunication facilities to organize their lives, to shop, to pay their bills, and make their living. They even want to communicate from the car, the airplane, the beach, and the shopping mall—from anywhere except the point at which traditional facilities terminate.

Today, communication-at-a-distance is provided by a range of facilities that incorporate the latest technologies and permeate commerce and society. As they transform the world into an information-dependent global village, their capabilities must be understood and their potential appreciated. *Telecommunications Primer* is offered as an aid to achieving such a state. In choosing the title I was guided by the literal definitions of these words:

- **telecommunications:** the technology of communication-at-a-distance that permits information to be created anywhere and used everywhere, without delay.
- **primer**: a book of elementary principles.

They accurately reflect my objective—in as simple a way as possible, describe contemporary facilities that support communication-at-a-distance, particularly those that contribute to the operation of automated enterprises. My arena is the information age, the period that began in 1991 when, according to the United States Department of Commerce, capital expenditures on computers and communications equipment first exceeded expenditures on other industrial equipment and machinery. My purpose is to provide a comprehensive reference for those persons who seek to apply telecommunications-based services to further the economic well-being of geographically-dispersed enterprises. They are the *knowledge workers*, part business persons, part computer scientists and part engineers, who need an up-to-date understanding of the spectrum of telecommunications so as to be able to implement electronic highways that will allow information to be created anywhere, and used everywhere.

Probably, it is every author's dream to write a book in which themes follow logically on one another and the subjects can be covered in a continuous stream of information without detours or repetitions. Not many subjects lend themselves to this model—and the more complex they are, the greater is the confusion. So it is with telecommunications. There are many paths to be followed, and dissimilar applications to be considered. Nevertheless, a defensible organization is a prerequisite to a successful book. With that in mind, the main divisions are as follows:

Introduction. Because telecommunication is a national activity, it is subject to government regulation. It is also an international activity, and subject to standards adopted by supernational organizations. Without them, it would be impossible to conduct communications with many parts of the world. Telecommunications is also a commercial activity that is used by the global corporation to link facilities together, support strategic plans, and increase productivity. In the introduction we touch on all these things to define the scope of the applications assumed in the book.

Signals. In Part One, we introduce the reader to the many kinds of signals present in communication systems. They are the manifestations of messages exchanged-at-a-distance. Their generation, transmission, and reception are the fundamental tasks performed by communication systems. To understand the divisions of telecommunication we must understand the difference between analog and digital signals, digital voice and digital data, low-speed data and high-speed data, video and compressed video, noise and message signals, etc. In Chapter 2, we identify the parameters used to describe signals and signal-related topics, and discuss some of the properties of noise signals. In Chapter 3 we discuss voice, music, and television signals, and in Chapter 4, we discuss the properties of data signals.

Building Blocks. In Part Two, we introduce the reader to devices and procedures that perform in specialized ways to generate, transport, and receive signals. To create a communication system, we must ensure that these building blocks are compatible and appropriate to the task. In data communication, the sequences of bits must be preserved, some form of error control is likely to be invoked, and the processors must follow common procedures if the activity is to be successful. In person-and-person telecommunication, the communication process is simplified by the intelligence of the users—but it is complicated by their psychological nature. So too, is communication between person and machine. Mediated by a terminal, it requires that the user adapt to the requirements of the machine. In Chapter 5, we describe the hardware connections and software functions used to implement communication between data processing machines. In Chapter 6, we describe the headers and trailer associated with the procedures employed to achieve communication between cooperating systems. In Chapter 7, we describe the special considerations that apply to personal telecommunication. They include psychological factors that are not present in machine communication. In Chapter 8, we describe the types of, and functions performed by multiplexers, and in Chapter 9, we describe the principles of operation of circuit switches and packet switches.

Networks. In Part Three, we examine the various kinds of networks that are employed to support the transfer of information between persons, machines, and persons and machines. In Chapter 10, we describe the public switched telephone network. It provides pervasive voice and low-speed data communications, connecting anyone to anyone else with compatible equipment. Evolving from an analog system that provided simple voice communication, it is in transition to a wideband, digital format designed to deliver digital voice, high-speed data, and digital television to all who wish to subscribe. In Chapter 11, we address local-area networks. They make distributed data processing a reality. Using them, it is possible to collect data, create information and distribute it within offices, over floors of buildings, in entire buildings, across campuses and throughout urban cores. In Chapter 12, we focus on wide-area, data-only networks. We start with packet networks that provide fully-featured services to several million users. They are quite reliable, but relatively slow. For faster data communication services, we describe frame relay, and switched multimegabit data service. Finally, we describe what may be the most used of all wide-area network techniques, IBM's System Network Architecture. In Chapter 13, we describe enterprise networks—networks maintained by individual corporations—that are important elements in the way they conduct their businesses. It is these networks that support knowledge workers in their quest to improve the information operations of their corporations. In Chapter 14, we describe ways in which radio is used to create wireless networks and remove location restrictions from communication. Paced by improvements in portable telephones and personal computers, these networks provide flexible connections to a wide variety of mobile stations. They present what may be the ultimate opportunity in telecommunication; universal personal telecommunication—the ability to communicate between stations located anywhere, over channels that are not constricted by geographic constraints. In Chapter 15, we describe some specialized topics involved in the management of telecommunication networks. They include a framework for network management that provides a common set of concepts for managers, users, and owners, maintaining message integrity, and ensuring information privacy. Finally, in Chapter 16, we present a short discussion of traffic engineering and the influence traffic patterns have on the performance of telecommunication facilities.

To round out the presentation of the spectrum of telecommunications, and to help the reader navigate through what is an embarrassing, but necessary, array of specialized words and phrases, we include a substantial listing of acronyms and definitions. For those that have the desire to go further, we include a short list of books and special issues of magazines that might help.

While it is possible to read straight through from cover to cover, it is unlikely that many readers will have the time and fortitude to do so. I suggest that the reader refers to topics of immediate interest and works from there to discover what is pertinent to the enquiry at hand in other Chapters. To this end, I have included frequent references to related Sections throughout the text. In an undertaking of this magnitude there are bound to be errors. It is my hope that they have been reduced to an acceptable level by the constant re-reading and re-working that I have employed.

E. Bryan Carne
Peterborough, NH

Telecommunications Primer

1

Introduction

By Wednesday

- **Telegraphy**: in 1843, by interrupting and restoring the current flowing in a wire strung between Washington, DC and Baltimore, MD, Samuel Morse and Alfred Vail ushered in the age of instant communication. Their electric telegraph was the first, large-scale application of the newly organized sciences of electricity and magnetism, and the first, practical method of providing real-time communication beyond line-of-sight. Using it, trained operators began to exchange messages over the copper wires that marched across the countryside in the company of the railroads. With the formation of the Associated Wire Service in 1846, and Western Union in 1856, even towns of modest size were able to send and receive messages and to know of important events shortly after they occurred. Moreover, these developments were not restricted to the United States—in 1865, the International Telegraph Union (ITU) was formed by twenty nations. Its charter promoted worldwide communication by attempting to standardize equipment and operating procedures.
- **Telephony**: in 1876, Alexander Graham Bell received a patent for the telephone. In the same year, at the Centennial Exhibition in New York City, he demonstrated the transmission of voice signals over a wire. This feat stimulated people to buy telephones for use over private circuits, and led to the formation of local telephone companies that operated under a Bell franchise. In 1885, the American Telephone and Telegraph Company (AT&T) was formed to build and operate long-lines facilities that could interconnect local Bell companies. In 1893, when Bell's original patents expired, independent companies entered the market and opened up the less populated areas of America. In 1900, there were some 1.6 million telephones in service in the United

States, and, by 1913, there were some 20,000 telephone companies providing connections within local communities. Long-distance messages continued to be carried by telegraph. In 1932, the ITU became the International Telecommunication Union with responsibility for radio, telegraph, and telephone regulations. Today, with more than 500 million worldwide, the vast majority of telephone subscribers have ready access to each other, and, in the developed world, hardly anyone sends telegrams. That so many telephones can participate in this global voice network is due to the determination of the world's telecommunication carriers and their support of the recommendations of the ITU.

1.1 COMMUNICATION AND TELECOMMUNICATION

In this Section, we examine the general nature of communication and the meaning of some common terms.

1.1.1 Communication

Derived from the Latin *communicare*, meaning to make common, to share, to impart, or to transmit, communication is generally understood to be the *activity associated with distributing or exchanging information.* As shown in **Figure 1.1**, communication may be *one-way* in the sense of an announcement, or *two-ways.* If it is two-ways, the exchange may be *interactive*, as in a conversation between persons in which information is transmitted one-way at a time, or *simultaneous*, as in an exchange between machines in which data are transmitted in two directions at once. These relationships can be formalized using the idea of a communication *channel*, i.e., a one-way communication path. The combination of two channels, one in each direction between communicating entities, is called a *circuit.* Shown in **Figure 1.2,** the ways in which information can be exchanged are:

- **Simplex** (SX): information flows in only one direction (a single channel is employed).

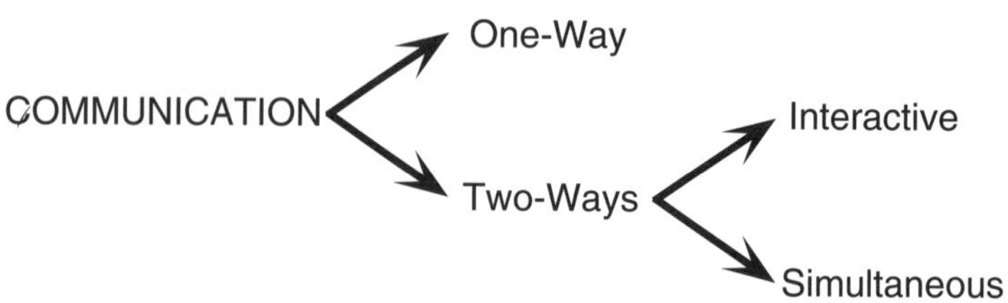

Figure 1.1 Communication Can Be One-Way or Two-Ways

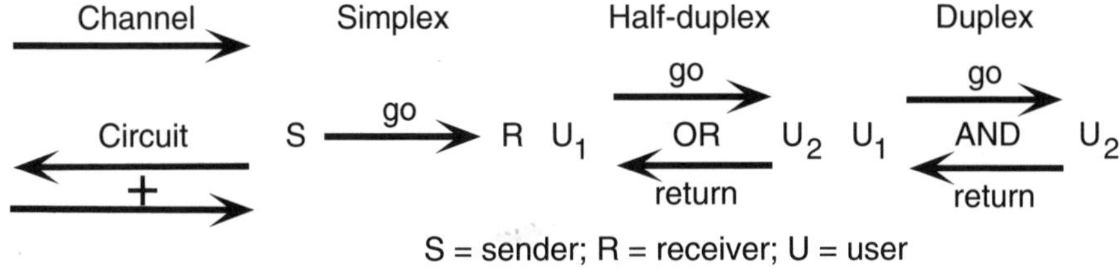

Figure 1.2 Channels, Circuits, and Styles of Message Flows

- **Half-duplex** (HDX): information flows in two directions, but only in one direction at a time (go or return). If a single channel is used, it must be *turned around* to allow information flow in the reverse direction. If a circuit is used, only one channel is employed at a time.
- **Duplex** (DX), also known as **full-duplex** (FDX): information can flow in two directions at the same time. A circuit is employed.

Information is distributed over a variety of connections. Shown in **Figure 1.3**, the basic topologies are:

- **One-to-one**: connects one location to one other location (e.g., a telephone connection); an alternative nomenclature is *point-to-point*

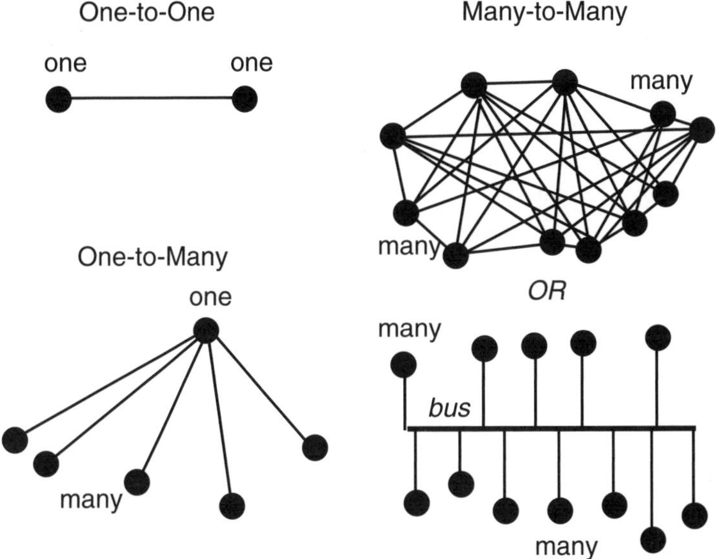

Figure 1.3 Styles of Connections

- **One-to-many**: connects one location to many other locations (e.g., a cable television connection that distributes programs from a head-end to thousands of homes); an alternative nomenclature is *point-to-multipoint.* When the number of receiving locations becomes very large, we may use the description **one-to-all**. It is a connection that links one location to all possible locations (e.g., a broadcasting arrangement). An alternative nomenclature is *point-to-all.*
- **Many-to-many**: connects many locations to many locations (e.g., a conferencing arrangement, or a local area network); an alternative nomenclature is *multipoint-to-multipoint.*

In all, nine different combinations of styles of flow and types of connection are possible. However, not all of them produce practical arrangements. For instance, if several stations send at once, many-to-many, duplex connections can be very confusing. A case-in-point is provided by a local-area network (LAN) in which all the data devices are connected to a common bus. If more than one unit sends a data burst at the same time, the messages collide, and neither message is received correctly. Rules built into the electronics must control the operation of the LAN so that only one station transmits at a time. This situation is diagrammed in **Figure 1.4**. Another example is provided by the many-to-many, duplex connection that is employed in a conference among remote parties held over telephone facilities. If all parties speak at once, no one will be understood. The chairperson must control the meeting in such a way that only one person speaks to all the others at one time.

From this discussion, we conclude that the existence of a connection is a *necessary,* but not *sufficient*, condition for communication. In addition, there must be procedures that organize and discipline the users so that messages flow in acceptable ways. In *machine*-oriented communication, specific rules (known as protocols) are required to guarantee a successful exchange—the structure of the datastream must be defined exactly, and data transfer must be implemented under procedures that assure recovery from all possible error conditions. In *person*-oriented communications, we rely on the intelligence of the users to ensure the proper flow of information.

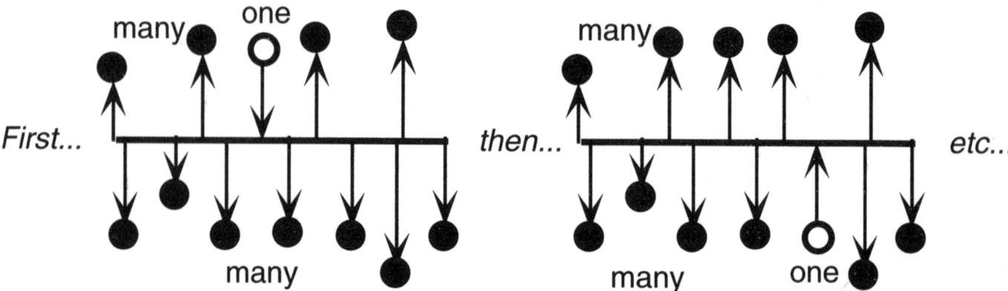

Figure 1.4 Control of Many-To-Many Connection

1.1.2 Information

Information—that which is distributed or exchanged by communication—is created by, or used in, virtually every human activity and machine operation. When associated with machines, it may be defined as *organized or processed data*—the output that results from processing data according to a given algorithm. It exists in the pattern of bits produced from data presented to an application program that was created to fulfill a (productive) purpose. When associated with persons, it may be defined as the *substance of messages*—that which is known about, or may be inferred from, particular facts or circumstances. It is created in the mind of the receiver by the sender's message. That these definitions depend on context, underscores the important differences that exist between data communication (i.e., machine-oriented communication) and personal communication (i.e., person-oriented communication).

1.1.3 Telecommunication

Literally, telecommunication means communication from afar (*tele*—from the Greek τηλε—far off); it is *the action of communicating-at-a-distance*. In the broadest sense, it can include several ways of communicating (letters, telegraph, telephone, etc.); however, it is customary to associate it only with electronic communication systems such as telephone, data communication, radio, and television.

1.1.4 Telecommunications

Telecommunications is the term used to mean the technology of communication-at-a-distance. It is a *transporting* technology that transfers messages among information users, a *linking* technology that connects islands of information users, and an *enabling* technology that makes it possible for information that is created *anywhere* to be used *everywhere* without delay. For instance, exploiting the capabilities of telecommunications to span continents and oceans to communicate across several time-zones, managers of modern enterprises see to it that their colleagues in remote locations can employ the same sophisticated information technologies as are used at headquarters, and all of them can have access to the same information on a timely basis. Also, they connect to their customers and to their suppliers, making it possible for all of them to share product and service information. Such activities depend on the availability of telecommunication services. They can be divided into two classes:

- **Data services**: services that collect data for processing in equipment that executes applications (programs), and distribute the resulting information to persons and machines that must act on it.

- **Personal services**: voice, video, and document services used for the exchange of information among persons, and between persons and machines such as:
  - one-to-one, real-time, interactive communication (principally voice telephone), or delayed communication (voice, text, or image electronic mail), or data communication (information retrieval)
  - one-to-many, real-time, interactive communication (voice or video seminar), or delayed communication (voice, text, or image electronic mail)
  - many-to-many, real-time, interactive communication (voice or video conference), or delayed communication (voice, text or image electronic mail).

Stimulated by competition, such services are available throughout most of the developed world. The combination of service, equipment, and associated support persons (if any) that implements a specific capability for communicating-at-a-distance is known as a telecommunication *facility*.

1.1.5 Intercommunication Services

Telecommunication that takes place between two sites is known as *intercommunication*. When machines are involved, intercommunication takes place on-demand, between units that are authorized to communicate. A common example of machine-centered intercommunication is the transmission of data files between computers. It may occur over the public telephone network or over special connections reserved for the purpose. When persons are involved, intercommunication takes place by agreement at a mutually convenient time, and is controlled by the participants—as their conversation develops, they vary the format and content of the exchange. The most common example of person-centered intercommunication is the use of the telephone network to make a telephone call—but person-centered intercommunication also includes text (as in electronic mail), image (as in facsimile), and video (as in videoconferencing) services. So as not to interfere with others, each pair of users of an intercommunication service requires a separate, exclusive connection. This sets a limit to the number of parties that can be served simultaneously by a particular complement of equipment (a network). As a result, at times of peak use, some of those who wish to communicate may not be able to do so immediately. Intercommunication services are likely to become congested during periods of peak demand.

1.1.6 Mass Communication Services

Telecommunication in which information flows simultaneously from a single (transmitting) site to a large number of (receiving) sites, is known as *mass communication*. The contents of the message, and the times at which transmission takes place, are controlled by the originator, and there is no response from the receiving sites. Among persons, mass communication consists of broadcasting television or radio programs from central, terrestrial stations, or over cable television facilities, or from geostationary satellites, directly to large audiences. As these examples suggest, mass communication is associated principally with the delivery of entertainment and information services. Among machines, mass communication may be employed to update a database or datafile that is possessed by many secondary stations. Sometimes, this is accomplished using a direct-broadcasting satellite. Unlike intercommunication, mass communication does not suffer capacity limitations in the sense of a limit on the number of sites that can receive a service; however, there is a limit to the number of originators that can operate simultaneously. For instance, in the case of radio broadcasting, only a certain number of frequency assignments are possible; and, in the case of cable television, the technical characteristics of the cable used to distribute the signals set a limit to the number of channels that can be provided.

1.2 REGULATION AND STANDARDS

Providing communications facilities to a nation is an activity that invites government participation. Raising questions of public financing, monopoly operations, and economies of scale, in many countries, telecommunication is the responsibility of a specific agency or department of government, or a quasi-independent organization created and chartered for the purpose. Often, the communications function is organized to embrace all aspects of intercommunication—mail, telegraph, and telephone—and is administered by an entity that is referred to as the PTT (Post, Telegraph, and Telephone Administration). Responsible for their national networks, these organizations administer standards within the country and cooperate with the recommendations of ITU to achieve international communications. In the United States, Bell System practices and procedures became *de facto*, the practices and procedures followed by the entire domestic industry, and U.S. telecommunications facilities developed independently from European telecommunications facilities. Because ITU is an arrangement among governments, relations between it and the United States are managed by the Department of State; it is supported by other government agencies and the telecommunications industry.

1.2.1 Regulation of U. S. Telecommunications Industry

In 1910, Congress placed the activities of all telephone companies engaged in interstate business under the control of the Interstate Commerce Commission (ICC). In 1912, ICC and the Department of Justice (DoJ) asserted that, to the detriment of the independents, AT&T was engaging in anti-competitive practices. As a result of this action, AT&T agreed to dispose of its stock in Western Union, to obtain approval of ICC before acquiring additional independents, and to allow independents to connect to AT&T facilities. Known as the Kingsbury Commitment, the agreement laid the basis for later actions that produced a government-regulated, AT&T-dominated,

national telecommunications network. In 1921, in the Graham Act, Congress granted telecommunications companies an exemption from the Sherman Antitrust Act that permitted them to consolidate competing properties. As a result, telecommunications companies established monopoly positions in their serving territories. In return, they accepted the principle of *universal service*—within their franchise areas, the companies agreed to serve all customers on an equal basis. In 1934, in the Communications Act, Congress created a separate organization—the Federal Communications Commission (FCC)—to regulate the emerging telecommunications industry. Consisting of a board of seven Commissioners appointed by the President and a relatively small staff, the FCC is empowered to regulate all interstate telecommunications facilities and services, as well as international traffic within the United States. An important objective is to prevent the companies it regulates from using their monopoly powers to generate unreasonable profits. Today, it oversees the activities of providers of telegraph, telephone, radio, television, and cable TV services.

1.2.2 Data Communication

The introduction of electronic computers in the late 1940s, the realization in the 1950s that they were capable of performing data processing, and the development of a vigorous, data processing community, produced demands for communication between remote machines. Ignored by the telephone companies, users began experimenting with various techniques. Led by the Advanced Research Projects Agency (ARPA) of the Department of Defense (DoD), government-backed researchers pioneered packet switched networks and procedures. Led by IBM, industrial and commercial users adopted networks and procedures based on Systems Network Architecture (SNA). In this environment, existing facilities were pressed to do much more than provide telephone service—their ability to carry voices over a distance became only part of the story—in addition, they slowly learned how to facilitate the remote use of technologies that store, retrieve, and analyze information. In the middle 1960s, encouraged by the success of these activities, Specialized Common Carriers challenged the Bell System for the right to carry data signals over private facilities between major cities. Since the middle of the 20th century, industry, commerce, and government have been caught up in a computer-based, telecommunications-assisted, *information revolution* in which telecommunication facilities provide the infrastructure that transports data and information among a rapidly growing, data processing population.

1.2.3 Deregulation of U.S. Telecommunications Industry

In 1984, after a complicated series of events that included the settlement of a federal anti-trust suit by AT&T, the provision of all services, except local exchange services, was deregulated and opened for competition. To achieve this, AT&T Corporation was divided into seven independent regional companies (called

RBOCs—regional Bell operating companies) that provide local exchange services within specific geographical areas, and an independent, long-distance carrier and manufacturing organization that retained the AT&T name. Practices and procedures are coordinated by Bellcore (Bell Communications Research), an organization funded by the seven RBOCs. The breakup of AT&T promoted competition among a myriad of long-distance carriers, and opened the way for the creation of the most sophisticated telecommunications environment in the world. Except for local services regulated by State Commissions, and radio-based services that must comply with the frequency assignment procedures of the FCC, telecommunications services are largely unregulated, and are provided by competing companies. Furthermore, recent developments have introduced the threat of competition between local exchange companies (LECs) and cable television (CATV) companies. LECs are preparing to provide television as well as telephone services, and CATV companies are preparing to provide telephone services as well as television programs.

The success of deregulation in the United States caused the idea to spread to other nations, and telecommunication administrations are slowly opening their national markets to the presence of independent carriers and to equipment from competing companies. Since 1984, competition among providers in the United States has increased the sophistication of available facilities and produced a frenzy of activities in standards-making bodies. Working together in national, regional, and international bodies, telecommunications administrations, manufacturers, and users are designing architectures to accommodate the anticipated requirements of a global information-age. Mindful of the experience of product developers that standardization is the prerequisite for low-cost implementation and market development, they are promulgating architectural and physical standards so that manufacturers can produce the equipment needed to implement these networks, and users can pursue applications that build demand for the services they can provide. Today, despite its reputation as being dull and boring, standards are a hotbed of activity as organizations challenge each other to design and build tomorrow's networks.

1.2.4 Principal International Standards Organizations

From the middle of the 19th century, the standardization of operating procedures and equipment performance has been the concern of organizations established by governments to ensure international compatibility of communications networks. It is an essential, continuing activity if the world's public networks are to support multinational corporations and satisfy their needs for sophisticated facilities to serve world markets. Within the last few years, all the major standards-setting bodies have gone through a reorganization to speed up their work. At the global level, standards are produced by three organizations:

- **International Telecommunication Union**: concerned with the creation of standards that facilitate international telecommunication. Information on its objectives and organization are given later in this section.

- **International Standards Organization** (ISO): concerned with improving international collaboration and communication, and promoting smooth and equitable growth of international trade throughout the world. ISO is the leader in standardizing communications between cooperating, data processing systems.
- **International Electrotechnical Commission** (IEC): concerned with promoting safety, compatibility, interchangeability, and acceptability of international electrical standards.

Recently, ISO and IEC have recognized the need for a single body to pursue the development of standards that support the needs of the information technology industry:

- **ISO/IEC Joint Technical Committee 1** (JTC1): formed in 1987 by the merger of information technology committees of ISO and IEC, JTC1 is responsible for consensus standards in the area of information processing systems.

(1) International Telecommunication Union

Founded in 1865 as the International Telegraph Union, since 1947, the International Telecommunication Union has been a *specialized* agency of the United Nations. Its objectives are to:

- Maintain and extend international cooperation for the improvement and rational use of telecommunications
- Promote development of technical facilities and their operation in order to improve the efficiency of telecommunication services and make them available to the public
- Harmonize the actions of nations in pursuit of these goals.

Reduced to practice, they are achieved by the *regulation* of radio communications, the *standardization* of international telecommunications, and the *development* of global networks. In 1993, ITU was reorganized into three sectors that reflect these functions:

- **Radiocommunications Sector** (ITU-R): formed by merging most of the activities of the *International Radio Consultative Committee* (CCIR: Comité Consultatif International Radio), and the *International Frequency Registration Board* (IFRB), the sector is concerned with radio issues and the allocation of the electromagnetic spectrum. Work is accomplished through world and regional conferences and study groups, and regulations are promulgated by the Radio Regulations Board.

- **Telecommunications Standardization Sector** (ITU-T): formed principally from the *International Consultative Committee on Telegraphy and Telephony* (CCITT—Comité Consultatif International Télégraphique et Téléphonique), the sector develops and adopts *ITU-T Recommendations* designed to facilitate global telecommunication. Work is accomplished through world conferences and study groups.
- **Telecommunications Development Sector** (ITU-D): formerly the *Bureau of Telecommunication Development* (BDT), the sector is responsible for promoting the development of technical facilities and services likely to improve the operation of public networks. Work is accomplished through world and regional conferences and study groups.

Administratively, each Sector is headed by a Director who is supported by a staff Bureau and an Advisory Board. The latter consists of professionals drawn from PTTs and non-PTTs who manage the standardization process to ensure that the work plans of the study groups reflect the increasing number of market-driven services and the needs of the non-PTT providers created by deregulation.

(2) Regional Standards Organizations

While recognizing the lead role of ITU in global standardization, in various regions of the world telecommunications organizations have banded together to form groups that develop regional standards when no international standards exist, or influence international standards so that they take account of regional requirements. Among such organizations are:

- **European Telecommunications Standards Institute** (ETSI): created by the Committee for Harmonization of the European Conference for Post and Telecommunication (CEPT), and originally intended to serve the needs of the twelve states of the European Community (EC), ETSI now counts Eastern and Central European countries in its membership. Its objective is to produce the technical standards that are necessary to achieve a large, unified, European telecommunications market.
- **Telecommunication Technology Committee** (TTC): established in 1985 to develop and disseminate Japanese domestic standards for deregulated technical items and protocols, TTC is composed of representatives of Japanese carriers, equipment manufacturers and users.
- **Committee T1—Telecommunications**: established in the United States in 1984, Committee T1 is sponsored by the Exchange Carriers Standards Association (ECSA), accredited by ANSI and approved by FCC as an appropriate forum to address telecommunications networks standards issues. Pledged to work towards consistent, worldwide, telecommunications standards, but principally concerned with matters pertaining to the operation of

domestic public networks, Committee T1 develops consensus standards and technical positions related to interfaces for United States and associated North American telecommunications networks.

(3) Other Standards Organizations in the United States

In the United States, telecommunications-related standards are developed by several other organizations:

- **American National Standards Institute** (ANSI): manages the voluntary development of national standards. Coordinates the activities of government units, industry groups, consumer organizations, and professional societies and associations in forming consensus. Provides verification of due process and the supervision, designation, publication, and maintenance of approved standards. Identifies areas in which standards are needed.
- **Electronic Industries Association** (EIA): a non-profit organization that represents the interests of manufacturers. Focuses on electronics, national defense, telecommunications, education and entertainment, and technical developments. Provides a public forum for the discussion of national laws and policies. The Association is well known for its work in standardizing connections between data terminals and other equipment.
- **Institute of Electrical and Electronic Engineers** (IEEE): a professional society, IEEE develops technical performance standards. The Institute pioneered in standards for local-area networks.
- **National Institute for Standards and Testing** (NIST): a government unit (formerly the National Bureau of Standards) charged with providing an orderly basis for the conduct of business by maintaining measurement standards. NIST coordinates voluntary product standards and develops computer software and data communication standards for the federal government [Federal Processing Information Standards (FIPS)]. With industrial organizations, NIST sponsors the North American ISDN Users Forum (NAIUF) that provides users opportunities to influence developing ISDN concepts, identifies and develops ISDN application profiles, develops standards-based ISDN implementation agreements, and solicits the participation of users, product providers, and service providers. NAIUF is neither a standards-making body, nor a testing and certification body. It holds three plenary meetings each year.

1.2.5 Networks for the 21st Century

The universal goal of these organizations is a network that provides a full range of data, voice, video, and document services to support global economic activities. So that the physical facilities can evolve as technologies develop, standards are stated in terms of performance, not technical configuration. These standards enable suppliers to design equipment that performs as the user expects, and works with equipment designed by another supplier. In this way, interoperability is assured without specifying the precise method of implementation, and equipment can be upgraded as new technology makes improved performance possible.

1.3 OPERATING ENVIRONMENT

In the United States, a mixture of public and private networks carry data and voice traffic. These networks are operated by several categories of carriers:

- Voice and low-speed data are carried by the public switched telephone network operated by common carriers and other common carriers
- Data is carried by public data networks operated by value-added carriers
- Data and voice are carried by private networks operated by businesses in their own interest.

By no means are these networks separate physical entities. When expedient, they lease transport facilities and rights of way from each other, and compete to provide switching and relaying services.

1.3.1 Public Switched Telephone Network

Telephone service is provided over the public switched telephone network (PSTN) by local exchange carriers (LECs), inter-exchange carriers (IXCs), and other common carriers (OCCs). Facilities belonging to these companies are interconnected to create a network that transports messages between users located anywhere within the United States, and connects to the networks of the rest of the countries of the world.

(1) Local Exchange Carriers

- Approximately 80% of the nation's telephones are served by 22 local exchange companies that are grouped into seven independent holding companies called Regional Bell Operating Companies (RBOCs)
- Approximately 10% of the nation's telephones are served by the local exchange companies of GTE Corporation
- The remainder are served by more than 1300 other local exchange companies.

Local exchange companies controlled by RBOCs provide public, switched telephone services within limited areas. Called LATAs (local access and transport areas), in most cases, they are the area defined by users that share a common area code. Within each LATA, the LEC provides the following:

- Connections between users located within the same exchange area

- Connections between users located in different exchange areas that are in the same LATA
- Delivery of traffic addressed to an exchange area outside the LATA to inter-exchange carriers (IXCs). When the message reaches the destination LATA, it is handed-off by the IXC to the LEC facilities that serve the called exchange.

Technically, local exchange companies that are not controlled by RBOCs, are not subject to this discipline; they may operate in whatever manner federal and state regulators permit. In practice, however, they operate in a manner that is compatible with the RBOCs.

*(2) **Inter-Exchange Carriers***

Inter-Exchange Carriers provide long-distance transport services between points-of-presence (POPs) that are established within each LATA; they serve as the connecting point between IXC networks and the LEC network. With approximately two-thirds of the market, AT&T is the dominant IXC. As such, it is regulated by the FCC. AT&T's nearest competitors are MCI, with greater than 10% of the market, and Sprint, with something less than 10% of the market. They operate full domestic networks and interconnect with international carriers over facilities they own, and facilities they lease from other carriers. MCI, Sprint, and the other carriers that operate facilities in the long-distance market in competition with AT&T are known as *other common carriers* (OCCs).

1.3.2 Public Data Networks

Public data networks are operated by *value-added carriers*. Largely, by leasing lines from common carriers, they have constructed packet networks (known as VANs, value-added networks) that serve all parts of the United States and connect with international carriers. Access is obtained over dial-up or leased lines from the customer's premises to the nearest VAN node.

1.3.3 Satellite Carriers

Several companies offer satellite services. Through earth stations on the customer's premises, or at the carrier's location, they transport data, voice, or video to the fifty states and to overseas locations.

1.3.4 International Carriers

Known formally as *international record carriers* (IRCs), several companies provide voice and data services to all parts of the world. In many cases, they have joined together to finance the procurement and installation of undersea cables or satellite facilities.

1.3.5 Private Networks

Most large businesses operate private networks that interconnect their facilities. Originally, individual leased lines provided the data connections. Today they have grown into switched networks that support a full range of data, voice, and video services. As requirements become more sophisticated, they are targets of opportunity for carriers who would provide *virtual* networks—networks based on carrier facilities that appear to the customer as if they were private facilities.

1.4 TELECOMMUNICATIONS IN THE INFORMATION AGE

1.4.1 Areas of Application

In a study by the Office of Technology Assessment of the United States Congress (see the *Further Reading* list included at the back of this book), technologies for collecting, storing, manipulating, and communicating information are identified as having the potential to revolutionize the economy of the United States. Two areas of application singled out in the report are:

- **Productivity**: information technologies can increase the productivity of operations where real changes may never have been considered seriously
- **Economic networks**: information technologies can link production systems together in ways that improve the performance of entire economic networks.

One (productivity) is concerned with improving the internal efficiency of economic units; the other (economic networks) is concerned with improving the ways in which economic units cooperate to produce finished goods. What the authors of the OTA study had in mind is that the information technologies have a decisive role to play in stimulating and maintaining a strong, expanding, national economy in the information age of the 21st century. Their statements imply that telecommunications (for collecting and communicating information), databases (for storing information), and data processors and information systems (for manipulating information) will have a significant impact on the future success of business enterprises. As individual entities, and in combination with one another, they can contribute to increased margins and improved performance.

(1) Telecommunications and Automation

Modern business enterprises employ computers and computer-controlled machinery to sustain their competitive edge. They use them to control operations, increase productivity, incorporate custom features in their products, maintain quality standards, and provide accurate, timely responses to the demands of their markets. To continue to reduce the cost of units of output, and to improve the efficiency of operations, they must constantly increase the level of automation. Moreover, as the overall life-cycle of each product is reduced by technical changes and competitive offerings, there is a need to reduce development, engineering, procurement, and deployment times. To promote timely decisions that have a high likelihood of being correct, a growing amount of data must be collected from many locations and processed into information that is distributed in a timely fashion. Achieving this performance depends on speedy collection and distribution facilities, information systems that organize the data in useful ways, and the free flow of comments and instructions among employees, suppliers, and customers. By providing voice, data, and video services over networks of increasing sophistication, telecommunication facilities contribute to the successful attainment of these goals, and to the cost-effective operation of today's enterprises.

(2) Information Systems

An information system is a software entity that organizes data to produce information for the benefit of an enterprise. More specifically, information systems process data to create useful information for strategic, management control and operating purposes. Information systems can be classified according to the sorts of functions they perform, as follows:

- **Evaluative**: one that provides information with which to plan future activities
- **Informative**: one that provides information with which to guide (control) present activities
- **Supportive**: one that provides information with which to operate the enterprise.

The types of systems needed to support each product or service depend on current market share, and position in the life-cycle. Thus, for a new product or service, evaluative systems support can be used to identify markets and develop a positioning strategy; for a mature product or service, informative and supportive systems are used to improve manufacturing efficiency and match customer requirements with optional capabilities.

(3) Telecommunications and Information Systems

What sets one enterprise apart from another is the use it makes of the combination of telecommunication facilities and information systems. Some of the applications that are improving the way in which enterprises conduct their business include the following:

- **Manufacturers** link together the computer-based tools they use to design, engineer, and manufacture products so as to create computer-integrated-manufacturing (CIM). Today, CIM is an essential capability for organizations that must match the quality and turnaround time of global competitors. Tomorrow, more powerful capabilities will link them and their suppliers, distributors, and customers so as to make the producing process as efficient as it can be.

- **Financial** institutions broker global financial markets, serve multinational businesses, buy and sell currencies around the world, and sell financial services based on proprietary information packages. Today, these activities are pursued by those who have access to a real-time network that integrates many islands of information and expertise. Tomorrow, to retain the attention of a dynamic, global customer, it will be necessary to cover the world with instant information, acceptances, and analysis.
- **Passenger airlines** depend on filling seats; today, their strategy is three-fold:
  - provide attractive fares and attractive schedules, and provide them to travel agents ahead of competing airlines
  - provide travel agents with special terminals linked to the airline's database so that the agents can sell and confirm tickets on the airline
  - build customer loyalties through frequent flier, senior citizen, and other programs that are based on records stored in reservation systems.

  Tomorrow, new customer services will be required to stay ahead of aggressive followers.
- **Property** and **casualty insurance** companies link agents with proprietary programs and databases so that they can quote the best available terms to clients—immediately. Today, independent agents have the same access through special terminals. Tomorrow, all will have universal access to information from anywhere. In this environment, underwriters must provide additional services to agents if they expect to retain their customers.

While the success of these applications is due in large measure to the information systems that they employ, the results would be diminished or negated without the immediacy of capable telecommunication facilities. Furthermore, the benefits they provide only produce temporary market leadership. To remain the leader, a stream of new services must be added to offset aggressive attempts by followers to regain market share. With the right kind of telecommunications facilities in place, this can be done relatively easily.

1.4.2 Linking Production Systems Together

(1) Manufacturing Chain

In **Figure 1.5**, an electronic goods manufacturing chain is shown; goods are manufactured in a sequence of operations beginning with raw materials and ending with the finished article. At each step, the input is transformed into what is needed for the next step. Thus, raw materials are processed into industrial grade materials. In

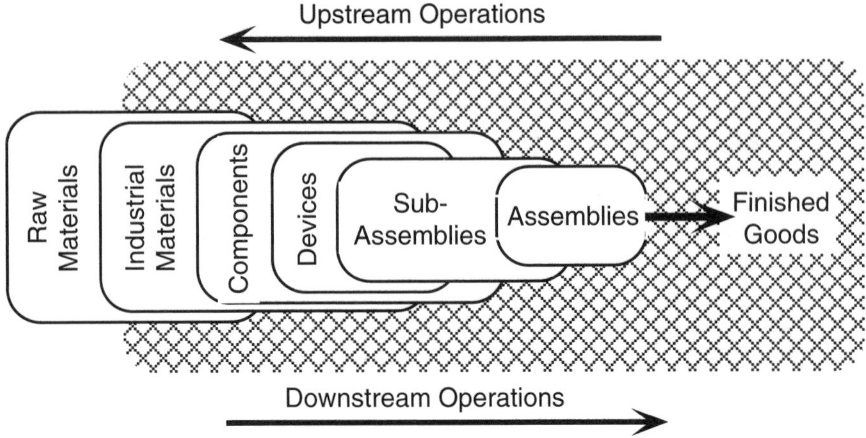

Figure 1.5 Manufacturing Chain

turn, they are made into components, then devices, sub-assemblies, and assemblies to produce the final product. The steps may be performed one-at-a-time by separate companies, or a single company may perform several steps. At each step, the product may be dispatched to other manufacturing chains; for this reason, the shading denoting finished goods extends back up the chain to industrial materials. So as to complete the goods in a timely and cost-effective way, information is passed along the chain.

(2) Inter-Enterprise Uses of Telecommunications

In **Figure 1.6**, part of a manufacturing chain that produces electronic goods for consumers is shown. Company A, a low-cost producer, links backward to two levels of suppliers to manage inventories and acquire confidence in the suppliers' abilities to deliver their products as required. Company B, an aggressive supplier of sub-assemblies, links backward to suppliers D, E, and F to assure their ability to support just-in-time manufacturing, and forward to Company A to provide information support that will differentiate the product from others. The link from B to A represents a switching cost (to A) that may prevent A from changing suppliers in times of poor performance by B. At the component level, company D links forward to company B for the same reasons. Company C, less-aggressive than company B, only links back to supplier G to manage inventories.

(3) Linking to Suppliers

The path to finished goods may lead through several firms, each of which is responsible for a portion of production. To achieve the timely delivery of high-quality goods, the final, downstream firm in the chain has to depend on the performance and skill of its upstream suppliers; in turn, they depend on their suppliers.

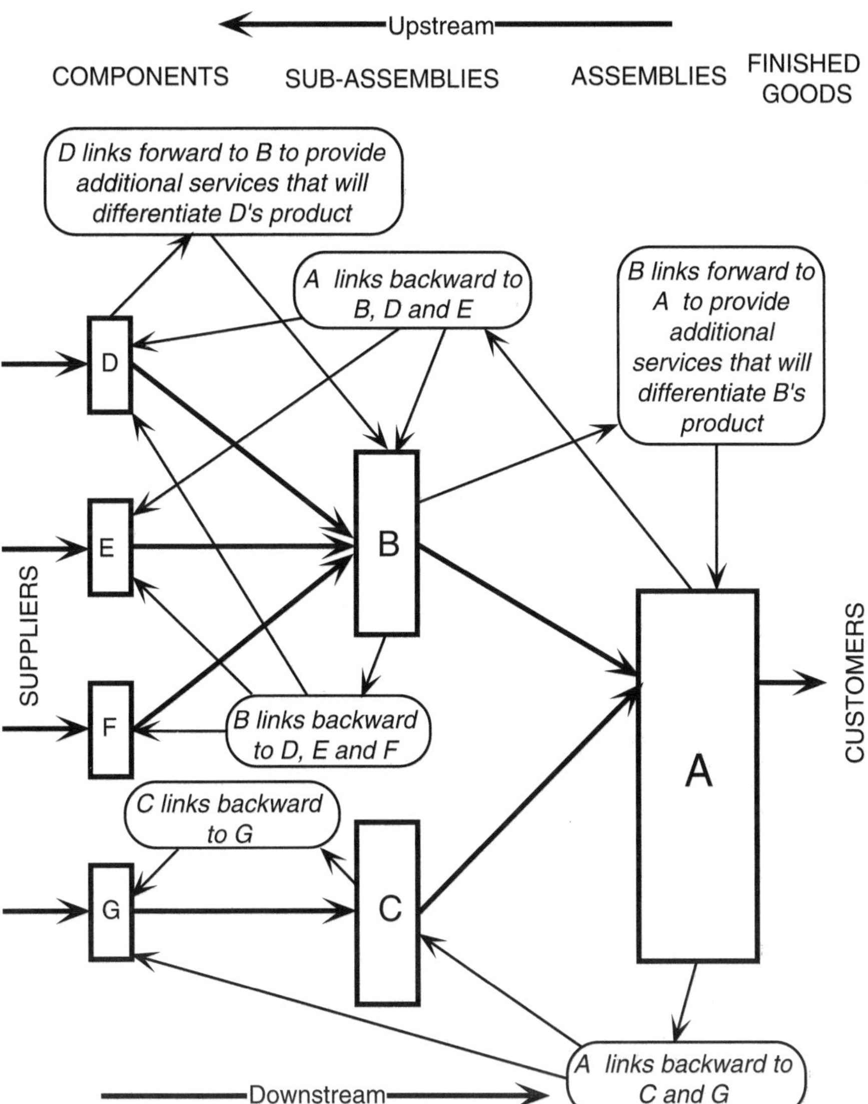

Figure 1.6 Use of Inter-Enterprise, Telecommunications-Based Systems to Manage a Manufacturing Network

To gather information with which to build confidence in their performance and the strengths of their commitments, each firm can extend telecommunications links to information nodes in their suppliers' internal networks—if they agree. The data they provide will permit the customer to better evaluate progress on its orders, and to estimate the probability of on-time deliveries. Moreover, once the link is established, the customer can send information to the supplier concerning future orders and schedules, and, once suppliers and customer gain confidence in each other—and the system—the customer can take advantage of just-in-time operation as a way of reducing the cost of money for goods that might otherwise be held in the customer's inventory. In fact, through the use of several telecommunications-based information systems, inventory costs can be reduced all the way upstream. Furthermore, with customer-to-supplier linkages between the firms in the chain, the individual supplier-to-customer segments can operate more efficiently, and the entire chain is likely to achieve a higher percentage of on-time deliveries of higher-quality, less costly, finished goods.

(4) Linking to Customers

In Figure 1.6, company B has links backward to its own suppliers, and forward to its customer. Over the link, company B provides information services to company A that are in addition to the data that A collects over its link to B. By providing additional services to A, Company B, an aggressive supplier, is attempting to build a relationship with A that will provide B with a competitive advantage. At worst, the additional services will differentiate B from other suppliers, and at best they will create a relationship that prevents A from going elsewhere for B's products.

(5) Advantage of Occupancy

One of the ways to create links forward is to place a terminal in the customer's office and support uses that the customer considers important. Not all of them need to be related to the supplier's product. For instance, if the product is a common use item, inventory management and automatic re-ordering might be included. Also, the system could provide inventory services for related products—including those supplied by competitors—making it more difficult for the competition to persuade the customer to replace the terminal with one they supply, and encouraging the notion that the convenience of automatic re-ordering could be extended to more products if they came from the same supplier. Fully exploiting the advantage of occupancy means finding creative ways to assist the customer and benefit the supplier.

(6) Integrating the Chain

If the final supplier of finished goods is a significant customer of the other firms in the manufacturing process, it can use this fact to put links in place that connect it to the other firms so as to be able to integrate data from the entire manufacturing chain—and communication does not have to be between manufacturing entities only. Engineering may collect data and exchange information with other engineering organizations in the chain; marketing may share information with the other marketing organizations, etc. Given cooperation on both sides, the more activities that can be coordinated, the greater the levels of efficiency that are likely to be achieved. Such networks will build relationships among suppliers and customers so that the entire manufacturing chain for a particular set of finished goods becomes the single, integrated, economic network referenced in the OTA study.

(7) Electronic Data Interchange

In industries such as transportation, automobiles, groceries, pharmaceuticals and personal care products, dominant firms have built economic networks of the sort described and have standardized message formats and messaging procedures for themselves and their suppliers. The development of broader, general-purpose standards for electronic data interchange (EDI) is now the task of Industry Associations coordinated by the ANSI Accredited Standards Committee X12. For buyer and seller transactions, this group has adopted a message architecture that accommodates a broad range of practices; it includes a feature set that can be tailored to different industries, and it is becoming the architecture of choice for many applications. EDI is most beneficial for industries with regular, repetitive buying, selling, and distributing functions.

1.4.3 Strategic Uses of Telecommunications

(1) Support of Marketing Strategy

Telecommunications networks can be used to pursue different, information-driven, marketing strategies; for instance, if the firm seeks to:

- **Lower product costs** so as to become the low-cost producer (and market leader), the telecommunication network will emphasize linkages backward to suppliers, and suppliers' suppliers so as to increase the efficiency of the manufacturing chain
- **Enhance product differentiation** so as to be the preferred source of supply, the telecommunication network will include linkages forward to the customer so as to provide information services that are not provided by competitors
- **Produce specialized products** that are tailored to particular market segments and to particular customers, the network will emphasize integrated operations so that custom parts and sub-assemblies can be produced with a minimum of difficulty.

(2) Building Barriers

By investing in electronic networks to link together all of the activities associated with a particular product, its markets, and supporting materials, producers can manage the integrated community of firms that support them. Given the right tools for analysis and decision making, they can modify production schedules and adjust incoming supplies and outgoing distribution in an optimum manner to respond to changes in the marketplace, and fluctuations in the world environment. Under the control of a dominant producer, the combination of a pervasive telecommunications network and powerful information systems tools can bind groups of firms together in a dependent way. If other producers wish to compete, they must establish a similar combination of telecommunications-based information systems capabilities—not an easy task. Thus, in building an integrated network, the producer has thrown up a significant barrier to would-be followers.

(3) Over-integration and Loss of Flexibility

Participation in these sophisticated arrangements entails more than just connecting to the network. Suppliers and distributors must share some of their proprietary information with the producer, and the producer must reveal proprietary plans and strategies. Thus, firms doing business as part of an integrated network become beholden to each other, and, after a period of confidence-building interaction, they become dependent on each other. Continuing success will bind them even closer together. Eventually, they may lose the flexibility to pursue new implementations of existing products because the community they have created does not have the necessary expertise or incentives to shift from the mode they are in. In this case they are in danger of losing their dominant position to a less well-established network that has greater flexibility to go after new opportunities.

(4) Retaining Customers

When a producer builds a telecommunications-based information system forward to a customer, it can offer information concerning product characteristics and applications that differentiate the supplier from its competitors. Properly thought out, these services can be important to the customer, yet inexpensive for the producer. A case in point is an original equipment manufacturer (OEM) that supplies a computer system for incorporation in a customer's product. On-line programming help, on-line maintenance guides, on-line applications news, and similar information, can enhance the OEM's position with the buyer. The strategy can be particularly effective when the goods involved are complex, and are likely to go through an applications evolution in their lifetime. With a linkage in-place that is filled with useful information, the buyer is likely to find that switching to another supplier will have a cost—in time, in convenience, or in training. At times of stress in their relationship, the buyer may find that the services to be a persuasive reason not to switch to another OEM's products. At times of price competition from other OEMs, the supplier may find the services provide some protection against the buyer substituting another's product.

(5) Pre-Emptive Strikes

The first implementations of some of the examples cited earlier have been called pre-emptive strikes. The term describes the aggressive use of new technologies, or old technologies in new ways, to the advantage of the innovator, such that:

- The move surprised the innovator's competitors

- The move was well received by the innovator's customers, and attracted customers served by competitors
- The move reshaped the markets to the advantage of the innovator
- Competitors required significant time, and had to make substantial investments, to catch up.

Through the use of the combination of telecommunications facilities and information systems, many companies have succeeded in attaining these ends. We shall briefly describe one of the best-known examples.

(6) American Airlines

Through the use of telecommunications-based information systems, American Airlines created a new field on which other airlines were forced to play. In the 1970s, they built an electronic reservation and ticketing system known as SABRE. Once the computers and communication links were operational, and SABRE was available at the ticketing desks in airports, terminals were moved into travel agents' offices—to create links forward to customers—thereby assisting them and differentiating American from other airlines. At the same time, having terminals in agents' offices allowed American to provide promotional materials to them, and, unless the potential traveler had a strong preference for another line, encouraged the agents to *book American.* In 1981, American announced a frequent flier bonus program. Designed to foster brand loyalty, it was a linkage forward to the ultimate customers—the travelers. Building a record of flights taken by full-fare passengers (typically business travelers) for which tickets were dispensed through SABRE, American awarded points for miles flown, kept track of the total, and converted them into additional tickets at the request of the flier. Their action met with acclaim from travelers and expanded the fraction of all frequent fliers whose carrier of choice was American Airlines. Caught at a disadvantage, American's competitors were forced to respond on American's terms. Even those airlines with proprietary reservation systems had to spend a great deal of time and money adapting them to function as a system that could total the distance flown by individual fliers—a record that has little to do with selling a ticket to fill a seat—but which presented a sizeable barrier to entry into the new marketplace of electronic ticketing and bonus programs. American continues to exploit the advantages of occupancy given by a terminal in the travel agent's office, and the records of frequent fliers stored in SABRE's processors. The combination of telecommunications and data processing capabilities they put together has helped American survive twenty turbulent years in the airline industry.

(7) Managements' Role

Telecommunications is one of many resources that are employed in running an enterprise. Normally, the level of management review of, and concern for, telecommunications activities will depend on the fraction of the corporation's total budget that is allocated to running expenses and capital acquisitions. When they become significant expenditures, top managers will want to understand the processes well enough to be able to appreciate the risks and the rewards. When strategic results are expected, members of the top-management team must get involved. Certainly, the pre-emptive strike described above could not have happened without this level of support.

1.4.4 Increasing the Productivity of Operations

Internally, the ready availability of telecommunications resources has changed the way in which business is conducted. What is more, they have altered work requirements so that the skills possessed by persons presently on the job may not be adequate for the future, and they are changing the way in which work is organized. Intra-enterprise use of telecommunications has aided in the acceptance of automated enterprises, automated offices, and automated factories.

(1) Person-Centered Applications

Telecommunications-based systems are used within the enterprise to improve the efficiency of operations through the use of the following:

- **Materials**: a materials specialist who keys in a part number that defines a manufactured sub-assembly, the quantity required, and the need date, may cause:
  - the display of current inventory
  - the ordering of critical parts from a supplier
  - the issuing of a revised manufacturing schedule for the sub-assembly to a facility somewhere else
- **Integrated circuits**: an engineer who inserts a description of the functions to be performed by an integrated circuit may cause:
  - the design and simulation of a circuit
  - the generation of the layout information for an integrated circuit chip to perform the task
  - the development of the software instructions to adapt it to a specific set of operating circumstances
  - the development of a test suite that ensures the chip and software perform as specified.

Each of these examples reduces labor, speeds up activities, and increases the expectation of customer satisfaction. They substitute information processing, and the linking and transporting abilities of telecommunications facilities, for the labor of several persons. More than likely, the information systems in these examples will reside in different locations and the tasks cited are enabled by the availability of telecommunication facilities.

(2) Machine-Centered Applications

Not all activities need be in direct response to actions by persons. They may be initiated between machines operating under the control of programs designed to transfer information and perform specific functions at certain times, or when certain conditions prevail:

- **Daily activity consolidation**: during the night, a machine at the headquarters of a supermarket chain will call for information on the day's transactions from machines at branch locations, balance the total activity in the company, return consolidated records to each branch for use the next day, and provide a summary of the day's activities to the operating managers when they arrive in the morning
- **Process control**: periodically, a machine at an operations center that supervises a pipeline will receive status reports from machines at important nodes in the network. Should equipment failure or unusual operating conditions occur, it may transmit instructions to the controllers so as to mitigate the difficulty.

These examples reduce the amount of labor involved and increase the expectation of a higher quality output. They are enabled by the availability of telecommunication facilities.

(3) Scope of Intra-Enterprise Telecommunications

In **Figure 1.7**, a manufacturing enterprise is represented as a value-producing process surrounded by R&D, engineering, and marketing resources. It is coupled to the manufacturing chain by distribution channels, and to executive and administrative functions by telecommunications-assisted, information systems. The company's telecommunications-based networks collect data from all parts of the enterprise and distribute information to control and action points, as required. The pervasive nature of their applications is suggested by the ellipse that extends over all resources and the distribution channels. For firms whose operations are located in a limited area, the telecommunication facilities supply a full-range of services over short pathways between information nodes. For firms whose operations are distributed across a continent, the telecommunication facilities supply a full-range of services over local and long-distance pathways between information nodes. For firms that do not manufacture goods (hardware or software), the diagram can be modified to reflect the generation of value-added services.

1.5 AUTOMATED ENTERPRISE

Pervasive use of telecommunications-based systems, and other information technology, creates an environment in which more and more tasks become candidates

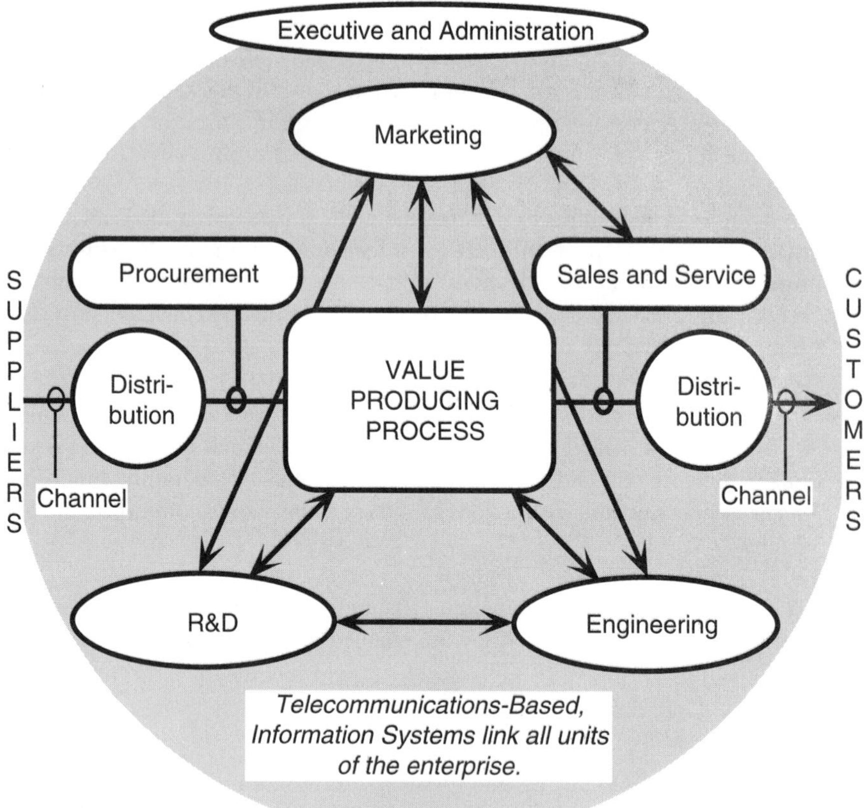

Figure 1.7 Manufacturing Enterprise

for automation. As each information systems-using company substitutes information technologies for human labor, it creates more information and further opportunities for automation. As each information systems-developing company perfects new capabilities, they create further opportunities for information technologies to substitute for labor. As a consequence, in the modern enterprise, the levels of automation, and dependence on information, increase inexorably. Companies that are caught up in information revolution in this manner make extensive use of telecommunications-based systems; they are called *automated enterprises*. We divide their activities into two categories; they are known as automated office and automated factory.

1.5.1 Automated Office

In an automated enterprise, the use of information technologies to improve the effectiveness of persons performing office-based tasks has come to be known as *office automation*, and offices that adopt these services are known as *automated offices*. The specific functions performed in an automated office depend on the work to be done and on the capabilities of the equipment installed. Besides pervasive voice telephone services and desktop computing capabilities, automated offices are likely to incorporate telecommunications-based services that:

- Achieve desk-to-desk, local, or remote transfer of text, data, and images as easily as voice information is transferred over the telephone
- Provide the ability to reach local or remote databanks, making central storage feasible and ensuring that all available, relevant data can be assembled and reviewed when needed
- Support electronic mail (text, voice, or graphics) and teleconferencing (computer, voice, or video) so that ideas can be exchanged and problems discussed with associates anywhere, thereby providing a higher level of expertise everywhere.

By fostering these activities, telecommunications-based systems merge the individual offices of the geographically distributed corporation into an extended electronic workspace in which to attend to the running of the enterprise.

1.5.2 Automated Factory

A set of persons, machines, data processors, and other facilities that work together in a timely and cost-effective manner to produce goods is called a factory. In an automated enterprise, the use of information technologies to improve the effectiveness of factory processes is known as *factory automation*, and factories that adopt these services are known as *automated factories*. Under the control of software systems supervised by operators, automated factories that manufacture and distribute hard goods employ computer-integrated manufacturing (CIM) facilities. They consist of numerically-controlled machines, robots, and other tools organized into flexible machining cells that may be served by automated, materials handling devices. In addition, supervisory functions such as order entry, inventory control, scheduling, testing, etc. are integrated into the operation of the factory by information systems that collect data and feed information directly to the appropriate equipment for processing or procurement. Automated factories that manufacture and distribute software products and services employ expert systems, high-level languages, and software engineering procedures. The term automated factory identifies operations that manufacture goods; they may be hard goods—hardware, or soft goods—software or services.

1.5.3 Knowledge Workers

In many firms, virtually all the physical labor components of traditional jobs are performed by automated devices—leaving only the intellectual components to be performed by workers. As more of the remaining activities are assumed by information systems, the result is:

- A steady increase in the technical skills demanded of a decreasing number of workers
- A decrease in the number of routine tasks performed by all employees
- A rapid decrease in the number of management persons engaged in collecting, analyzing, summarizing, and distributing information.

Overall, the total number of workers per million dollars of revenue is declining, and for those jobs that remain, higher skills and/or different skills, are required. Automated enterprise is creating a group of persons that does not fit within the traditional working environment. Known as *knowledge workers*, they perform information-based tasks and communicate knowledge and know-how directly to manufacturing, operations, and other segments of the enterprise. They are the group of talented workers that understands the potential of new telecommunication facilities and information systems, and can apply them to achieve the objectives of the enterprise. Their tools are terminals and their workspaces are combinations of computers, telecommunication networks, and software-directed devices distributed throughout the enterprise.

1.5.4 Changing Organizational Structure

The classical, hierarchical management structure is of little value when the enterprise is dependent on knowledge workers and automation. Knowledge is power, and in many situations, the understanding of how to manipulate it makes the knowledge worker indispensable. In contrast, the manager who does not have this understanding is powerless—and expendable. The traditional hierarchical structure in which level is sacrosanct is giving way to loose-knit, *ad-hoc* groups of specialists that form teams (task forces) to suit the requirements of the moment. They bypass regular management channels as they solve problems and perform other tasks. The result is a reduction in the number of manager levels and the elimination of many jobs that existed solely to summarize and interpret information and pass on directives. **Figure 1.8** conceptualizes these changes:

- **Contemporary firm**: in diagram A, a traditional organization is shown as a pyramid with layers occupied by officers, managers, and workers. Information flows upwards through several layers to get to the chief executive, and instructions flow down to workers that produce the company's products. An important characteristic of this organization is that the person-in-charge (manager) is the person-in-control of an operation (group of workers). The corporation is under overall pressure from global competitors to increase

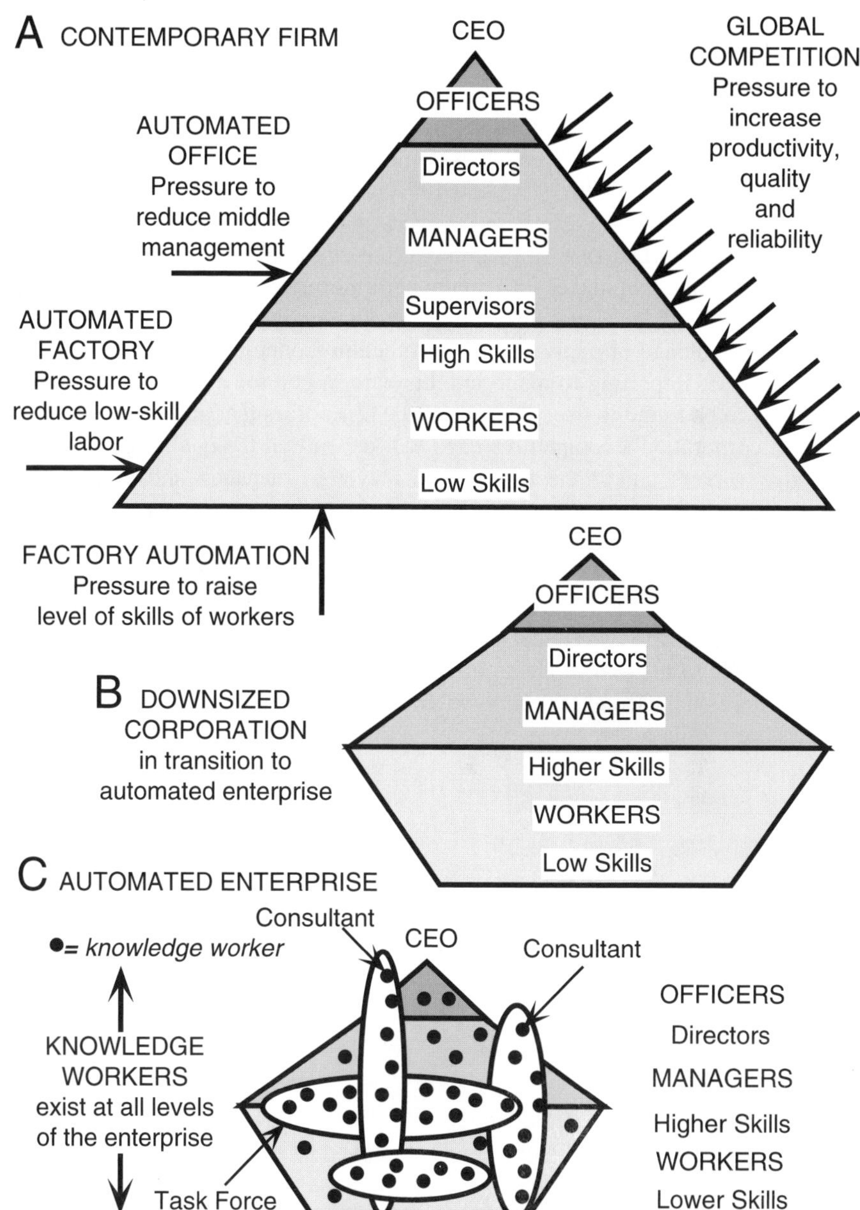

Figure 1.8 Changes in the Composition of a Firm as it Becomes an Automated Enterprise

productivity, product quality, and reliability. In addition, the corporation is under internal pressures from automation to reduce the number of middle managers, to reduce the number of low skilled workers, and to increase the skill levels of the workers that remain.

- **Downsized corporation**: in diagram B, the corporation is responding to these pressures. There are fewer managers, and fewer workers in an organization that requires higher levels of skills in all persons in order to exploit higher levels of automation.
- **Automated enterprise**: in diagram C, the corporation has reached the level of automation that makes it an automated enterprise. There are still fewer persons involved, and at all levels there are knowledge workers who have risen to the challenge of applying new automation techniques for the good of the enterprise. Reporting to managers, they are in control of operations, and are likely to be found in one, or more, of the task forces that guide the enterprise. The corporation's competitive edge will depend on the quality of its knowledge workers, and on the teamwork displayed by managers and workers. This includes the understanding that the person-in-charge (manager) is no longer the person-in-control (knowledge worker) of an operation (group of workers).

Part business person, part computer scientist, and part engineer, *knowledge workers* need an up-to-date understanding of the spectrum of telecommunications so as to be able to implement electronic highways that allow information to be created *anywhere*, and used *everywhere*, without delay. It is for these persons that this book has been written.

2

Basic Concepts

Signals are the manifestations of messages exchanged-at-a-distance. Their generation, transmission, and detection are the fundamental tasks performed by communication systems. The messages consist of:

- **Data**, as in file transfers between data processors
- **Text or graphics**, as in electronic mail facsimile, or information retrieval from a remote database
- **Spoken words** (voice), as in point-to-point telephone conversations or voice mail
- **Music**, as in broadcast programming
- **Television**, as in point-to-multipoint cable television, broadcast television, or video conferences.

For those who are surprised to find music and television in the list, we point out that advanced facilities now under development (such as broadband, integrated services digital networks, BISDNs, see §10.7) will deliver these signals in broadcast, or one-to-one modes. At the signal level, it is impossible to know for what purpose the messages are used. Accordingly, without reference to their operational context, we discuss them as voice, music, television, low-speed data, and higher-speed data signals. To them, we must add those signals that interfere with and distort the wanted (message) signals. Collectively known as *noise*, these signals degrade the performance of all communication systems. In this Chapter, we identify the parameters used to describe signals and signal-related topics, and discuss some of the properties of noise signals. In Chapter 3 we discuss voice, music, and television signals, and in Chapter 4, we discuss the properties of data signals.

2.1 SIGNAL CLASSIFICATION

Signals are real-time events that cannot be held, repeated, changed, or otherwise processed at the whim of an observer. To do this, it is necessary to build models in which mathematical functions represent real-world signals. These functions exhibit properties that permit us to classify them in various ways; by extension, the signals they represent may be classified in the same way.

2.1.1 Variation of Values with Time

Functions can be divided on the basis of the way in which their values vary over time:

- **Analog**: a continuous function that assumes positive, zero, or negative values. Changes in value occur smoothly and rates of change are finite. Analog signals are the natural forms of signals generated by many physical processes; they play a large part in the *transmission* of messages by electronic media. Indeed, over electrical transmission systems, data signals are coded in analog form
- **Discrete**: a non-continuous function whose values form a discrete set and occur at isolated points in the time continuum
- **Digital**: a function that assumes a limited set of positive, zero, or negative values. Changes of value are instantaneous, and the rate-of-change at that instant is infinite—at all other times it is zero.

A common class of digital functions is two-valued (binary), i.e., they exist in two states only. Binary signals are the natural form of signals generated by computers and similar devices. Digital techniques dominate the *processing* of messages. Binary signals are used exclusively in the transmission of messages by optical fibers. Digital functions may be converted to analog functions, and vice-versa.

From an operational point of view, the separation into analog or digital is probably the most important of all the divisions we shall describe. It implies a split in the way in which information is carried. Message information resides in the variations of amplitude or frequency or phase of an analog signal. In digital signals the message information is carried in the sequence of bits that it represents.

2.1.2 Degree of Certainty

Functions can be divided on the basis of the certainty with which their behavior is known:

- **Deterministic**: at every instant, a deterministic function exhibits a value (including zero) that is related to values at neighboring times in a way that can be expressed exactly. Thus we can determine the time at which the function will achieve a particular value, and the value the function will have at a particular time. As a consequence, deterministic signals can be described by classical mathematical functions
- **Probabilistic**: a function for which future values can be described only in statistical terms. While we may have a set of historical values, neither the value of the function at a specific future time, nor the future time at which the function will have a specific value, can be known for sure. Future values are estimated on the basis of the statistics associated with past values and the assumption that behavior will remain the same in the future. An important sub-class of probabilistic functions is random functions
- **Random**: a probabilistic function whose values are limited to a given range. Over a long period of time, each value within the range will occur as frequently as any other value.

2.1.3 Effect of Duration

Functions can be divided on the basis of their duration:

- **Transient**: a function that exists for a limited time only. It represents the class of real-world signals that are turned on, and then turned off. If the function is deterministic, it can be analyzed with the help of *Fourier transforms.* (Determinism is necessary because evaluating these transforms requires integrating $\pm\infty$ with respect to time, i.e., from way in the past to way in the future)
- **Eternal**: a function that exists for all time. It is used to describe the performance of communication systems that are operating in the steady state (i.e., any behavior solely due to turn on has long since disappeared so that it is as if they have been operating forever). An important sub-class of eternal functions is periodic functions
- **Periodic**: an eternal function whose values repeat at regular intervals. To know everything about it, only a single period need be studied. Periodic functions can be decomposed into a related set (fundamental plus harmonics) of sinusoidal functions through the use of *Fourier series.*

2.1.4 Real-World Signals

With few exceptions, the signals present in real-world, communication systems are probabilistic signals. They cannot be represented accurately by deterministic transient, or periodic functions, and we are left with nothing to say about their future behavior, except it is not predictable. To overcome this impasse, future values are estimated on the basis of the statistics associated with past measurements and the *assumption* that future behavior will remain the same. In mathematical terms, we assume that the process giving rise to the signals is *ergodic* and *wide sense stationary*. These terms are defined as:

- **Ergodic**: time-averages are equal to corresponding ensemble averages.
- **Wide sense stationary**: the expectation of the random process is constant, and the autocorrelation and covariance functions depend only on the *time interval* over which they are evaluated (not the *time* at which they are evaluated).

They are powerful concepts whose application requires substantial mathematical skills. Fortunately, a great many useful results can be obtained from deterministic transient and periodic approximations. In the remainder of this chapter, we use periodic functions to represent the signals with which we are concerned.

2.2 BANDWIDTH AND PASSBAND

The only acceptable goal for a communication system is to deliver the message signal exactly as it was sent. For this to happen, the system facilities must have sufficient passband to accommodate the bandwidth of the signal. To explain this point, we introduce the ideas of amplitude, frequency, and phase using a sinusoidal function.

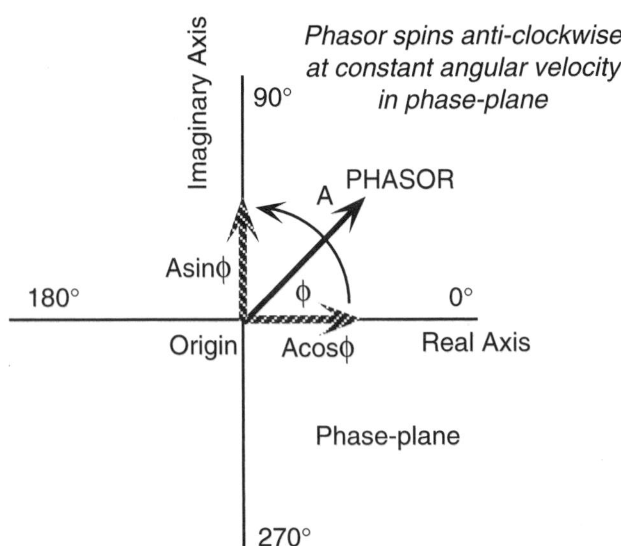

Figure 2.1 Argand Diagram Representation of Sinusoidal Function

2.2.1 Sinusoidal Functions

Sinusoidal functions occur in the modeling of many simple, natural processes such as the motion of a pendulum, and the vibrations of a structure. In the electric age, sinusoidal signals are present as single-tone audio signals, the 60 Hz alternating current present in public, power supply systems and carriers employed in radio and television broadcasting. Fundamental to an understanding of sinusoidal functions are the notions of amplitude, frequency, and phase. To illustrate the meanings of these terms, we start with the model shown in **Figure 2.1**. Called an *Argand* diagram (for its inventor, Jean-Robert Argand, an 18th century Swiss mathematician), it consists of a phasor of amplitude A signal units (su) that rotates anti-clockwise about the origin at a constant angular velocity and completes f rotations a second. For reasons that we need not go into, the horizontal axis is known as the *real* axis, and the vertical axis is known as the *imaginary* axis. Angles are measured in relation to the horizontal, right-pointing, real axis (marked 0°). As the phasor rotates, its projection on the real axis is $A\cos\phi$, and on the imaginary axis it is $A\sin\phi$. In **Figure 2.2**, we introduce time as the third dimension. In this diagram, the tip of the

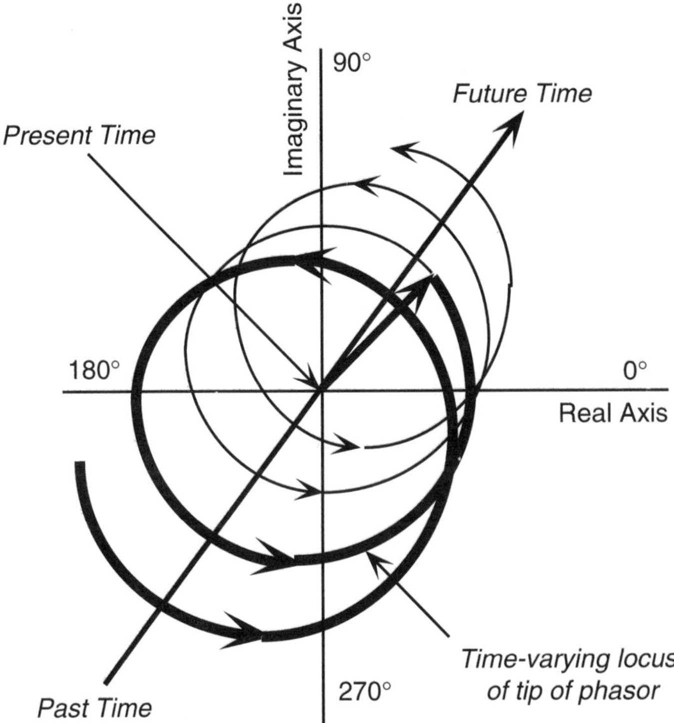

Figure 2.2 Three-Dimensional Model of Sinusoidal Function Including Phase and Time

phasor traces out the spiral path shown. When projected on to the plane defined by the real axis and the time axis, the spiral of Figure 2.2 forms the cosine function shown in the time-plane diagram at the upper right-hand side of **Figure 2.3.** If one rotation of the phasor takes τ seconds, the repetition frequency f is $1/\tau$ *times per second (i.e.,* $1/\tau$ Hertz). During each period, the phasor rotates through 360°, and the phase-angle can be associated with position in the period as shown. These angles are repeated each period. In the time plane, $A\cos\phi$ becomes $A\cos 2\pi ft$—a periodic function that repeats in synchrony with the rotating phasor. At the bottom of Figure 2.3, we show the frequency-plane representation of the cosine function. It consists of a single line of amplitude A su situated at a frequency fHz.

2.2.2 Complex Periodic Functions

When designing and testing communications equipment, the use of sinusoidal functions and their signal equivalents, single-frequency tones, simplifies calculations

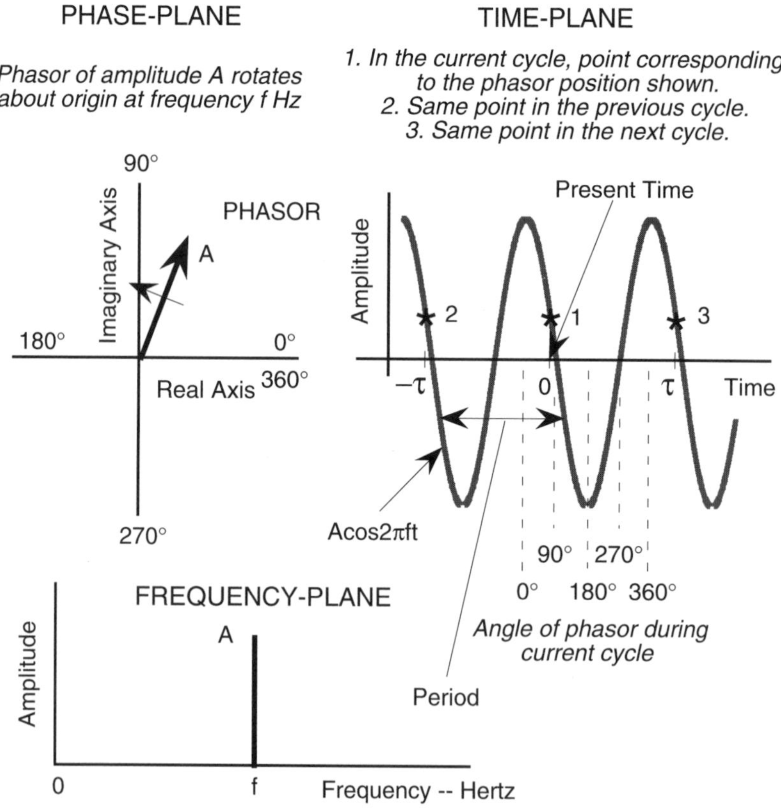

Figure 2.3 Phase-, Time, and Frequency-Plane Representations of Sinusoidal Functions

and measurements. However, real signals are more complicated than this. Fortunately, if they can be assumed to be periodic, they can be modeled by complex periodic functions formed from sinusoidal functions that comprise a Fourier Series. The function with the lowest frequency (but not zero) is called the fundamental, and the other functions are called harmonics. They occur at integer multiples of the fundamental frequency.

Figure 2.4 shows an example of a complex sinusoidal function. It approximates a signal generated by a musical instrument, or the output signal from equipment that is excited by the fundamental and operates improperly to produce the third harmonic (an action known as third harmonic distortion). The function is made up of a fundamental component at frequency f Hz of amplitude 1.25A su, and a third harmonic component at frequency $3f$ Hz of amplitude 0.56A su. The fundamental component establishes the basic periodicity of the function; the third harmonic component provides periodic variations about the fundamental component.

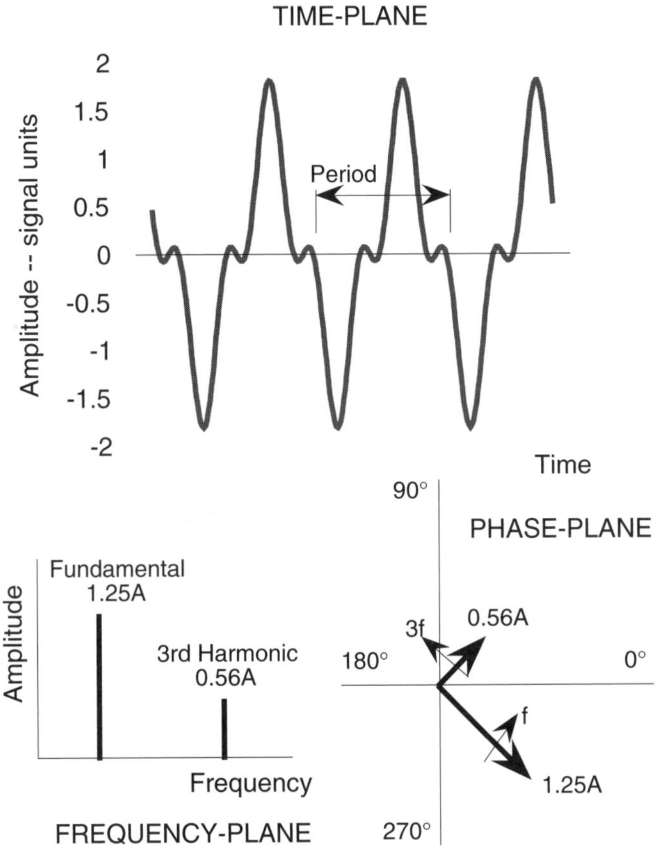

Figure 2.4 Complex Sinusoidal Function

In the Argand diagram, a phasor of magnitude 1.25*A* su rotates at the fundamental frequency (*f* Hz) and a phasor of magnitude 0.56*A* su rotates at the third harmonic frequency (3*f* Hz).

Figure 2.5 shows another complex periodic function—a symmetrical pulse train. A binary digital function that changes from *A* to 0 and 0 to *A* at regular intervals, it contains an infinite number of harmonic components. In Figure 2.5, we show the magnitudes of these components up to the ninth harmonic. That only odd-

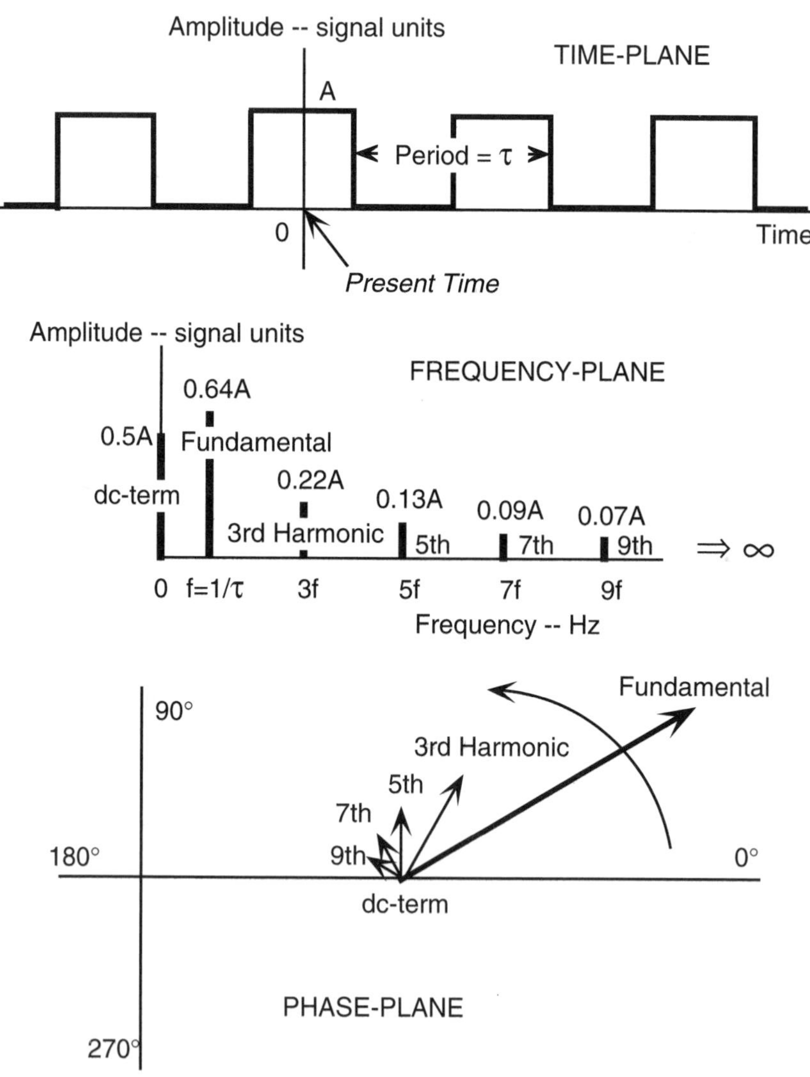

Figure 2.5 Symmetrical Pulse Train

harmonics are present results from defining zero time (i.e., $t = 0$) in the middle of a pulse so that the train is an even function (i.e., values at $\pm\, t$ are identical). In the general case, this class of functions contains both odd and even harmonics. Further, because the function assumes only positive ($+A$) and zero values, it contains a dc-term (0 Hz, $0.5A$ su) that accounts for the positive bias of the train. It is the average value of the function. In the phase-plane, Figure 2.5 shows an Argand diagram in which phasors representing the fundamental and harmonic components rotate about the point $0.5A$. This is the offset created by the dc-term. Because each phasor rotates at a different multiple of the fundamental frequency, the diagram is a snapshot of the action; it is repeated f times a second.

2.2.3 Bandwidth

Some of the power of the complex sinusoidal function shown in Figure 2.4 is carried by the fundamental component (0.78 signal watts, sw), and some by the third harmonic component (0.16 sw). To reproduce the signal exactly, these components must be transmitted to the receiver as they exist in the original signal. To this end, the bearer and associated equipment between the sender and receiver must carry frequencies of f and $3f$ Hz. We define the bandwidth of a signal as:

- **Bandwidth:** the range of frequencies that just encompasses all of the energy present in a given signal.

We may state that, for the signal to be received in the same condition as it was transmitted, the bearer (and associated equipment) must carry the signal bandwidth without distortion. In the case of the complex sinusoidal function, the range of frequencies that just encompasses all the power present in the signal is from f to $3f$ Hz, i.e., a bandwidth of $2f$ Hz.

(1) Passband

- **Passband**: the range of frequencies transmitted without distortion by a bearer and associated equipment.

Practical communication systems are endowed with finite properties, and this includes the frequencies transmitted by the bearers and associated equipment. Examples of their behavior are shown in **Figure 2.6**. Some will transmit energy from 0 Hz (dc) to gf Hz, but not at frequencies above gf Hz; they are said to exhibit *low-pass* characteristics and to have a passband of gf Hz. Others will transmit energy between mf Hz and nf Hz, but not at frequencies below mf Hz nor above nf Hz; they are said to exhibit a *band-pass* from mf Hz to nf Hz, and to have a passband of $(n - m)f$ Hz. Yet others will transmit energy at frequencies greater than hf Hz, but not at frequencies below hf Hz. They are said to exhibit *high-pass* characteristics. In reality, because there will always be a limit to the frequencies they transmit, the concept of a high-pass system is artificial. In practice, communication systems are

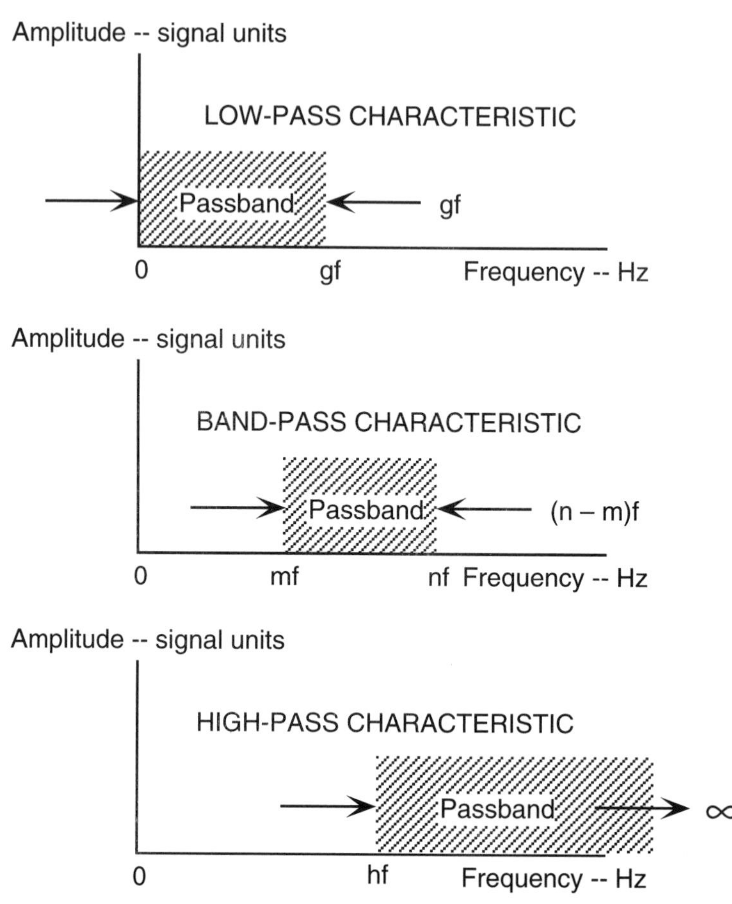

Figure 2.6 Passband Characteristics of Bearers and Associated Equipment

of two types—low-pass and band-pass. If it is to be transmitted without distortion, what is important is that the passbands of all the component parts of a transmission system must be greater than, or equal to, the bandwidth of the signal.

(2) Bandwidth > Passband = Distortion

When a signal with a bandwidth larger than the passband is presented to a communications system, the signal that is transmitted is a distorted version of the original. For the case of the symmetrical pulse train drawn in Figure 2.5, parts A, B and C of **Figure 2.7** show the result of progressively diminishing the passband (as measured by the number of harmonic frequencies that are transmitted) of a low-pass bearer that carries the pulse train. Figure 2.7A shows the original pulse train. The second signal (B) is the signal that will be received from a system that passes frequencies no higher than nine times the fundamental frequency. This system has a passband

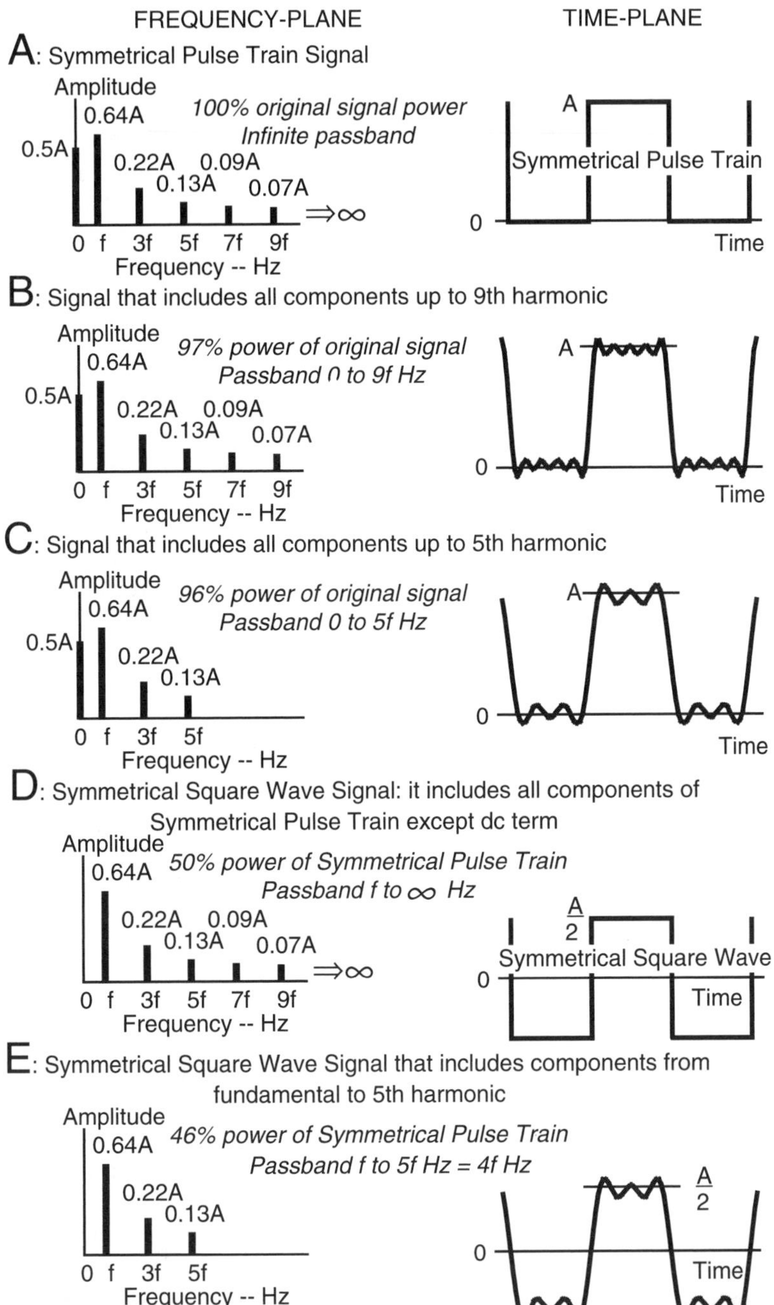

Figure 2.7 Effect of Limiting the Passband Available for the Transmission of Wideband Signals

of $9f$ Hz, and transmits approximately 97% of the power in the original signal. Figure 2.7B shows the signal that will be received from a system that passes frequencies no higher than five times the fundamental frequency. This system has a passband of $5f$ Hz, and transmits approximately 96% of the power in the original signal. The fourth signal (D) is that which is received from a high-pass system that passes frequencies of f Hz, and above. Without the dc-component, the pulse train becomes a symmetrical square wave with an average value of zero. Finally, Figure 2.7E shows the signal that is received from a bandpass system that passes frequencies from f Hz to $5f$ Hz. It has a passband of $4f$ Hz, and the signal is identical to (C) except for the elimination of the positive offset. With the same periodic features as the symmetrical pulse train, it requires only one-half of the power.

An important first point in our exploration of telecommunication signals is that the passband of the bearer must be large enough to accommodate the bandwidth of the signal. Thus, because coaxial cables have greater passbands than twisted pairs, they can carry signals of wider bandwidth, e.g., a coaxial cable will deliver an undistorted television signal (nominal bandwidth 6 MHz) to a point several miles away; this will not be the case if we try to send it over a twisted pair. Further, because optical fibers have a higher passband than electric bearers, optical fibers carry much higher speed data streams than coaxial cables, microwave radios, or twisted pairs.

2.3 MODULATION

- **Modulation**: a signal processing technique in which one signal (the *modulating* signal) modifies a property of another signal (the *carrier* signal) so that a composite wave (the *modulated* wave) is formed; later in the transmission chain the modulating signal is recovered (this action is known as *demodulation*). The bandwidth of the modulated wave is equal to, or greater than, the bandwidth of the modulating signal (known as the *baseband*).

Modulation is used to facilitate:

- **Long-distance transmission**, by impressing message signals on higher frequency carrier signals to create a modulated wave. Because of the higher frequency (and shorter wavelength) it enjoys, the modulated wave can be launched from, and received by, practical sizes of antennas, or conducted by moderate sizes of cables or waveguides. With efficient coupling to the communication medium, the high-frequency, modulated waves travel over greater distances than the original message signal can travel unaided.
- **Sending digital messages**, by creating sets of symbols that can be transmitted over a given, bandwidth-limited channel. Each symbol represents a specific sequence of bits and the symbol set covers all possible combinations of bits. The maximum symbol rate is determined by the passband of the bearer

and associated equipment. Use of this technique permits the transmission of low-speed bitstreams over telephone lines and radio channels.

Before discussing modulated waves and their applications, we shall describe the procedures used to allocate carrier frequencies.

2.3.1 Electromagnetic Spectrum

Unless guided, or contained by shielding, electromagnetic energy radiates freely in all directions. When used to effect telecommunication between two parties, the energy can spill over and interfere with others who are trying to communicate. For this reason, specific frequencies are assigned to specific stations, and part of the act of establishing telecommunication involves adjusting (tuning) the equipment at the receiving station to the frequency used by the transmitting station. In this way, transmitter and receiver operate on the same frequency. Others, within receiving distance of the first pair, will employ different frequencies so as not to interfere with them, or to be interfered with by them. In principle, selecting suitable frequencies should be a simple task. In practice, because of the large number of potential users, and that certain frequencies may propagate around the world, frequency assignments must be made carefully, and in full compliance with international agreements. The electromagnetic spectrum is a limited resource that is administered by governments to ensure that as many persons as possible are able to benefit from its use.

(1) Allocations and Assignments

Specialized terms are used to distinguish between all the frequencies allocated to a particular service, and the particular frequency assigned to a specific operator who provides the service:

- **Frequency allocation**: the set of frequencies allocated to all stations offering a specific service. Frequency allocations are administered by the International Telecommunications Union (ITU) with the help of information gathered at periodic, regional, and worldwide Administrative Radio Conferences (RARCs and WARCs).
- **Frequency assignment**: the specific frequency on which a particular station must operate to provide the service to which the frequency is allocated. In the United States, frequency assignments are made by the Federal Communications Commission (FCC) for non-government operations, and by the National Telecommunication and Information Agency (NTIA) for government operations.

(2) Operating License

As well as a specific frequency assignment and authorization to provide service(s), operating licenses usually include restrictions as to permissible radiated power, and type of modulation, and may limit times of operation and other matters. In **Table 2.1**, we indicate some of the important allocations in the United States.

2.3.2 Continuous Wave Modulation

Continuous-wave modulation brings together a higher-frequency sinusoidal carrier signal and a lower-frequency message signal to create a modulated wave in which

Table 2.1 Some Frequency Allocations in the United States

ONE-WAY RADIO SERVICE

- **AM broadcasting**: 107 channels, 10 kHz wide, 535 - 1605 kHz
- **FM broadcasting**: 100 channels, 200 kHz wide, 88 - 108 MHz
- **Television broadcasting**: 83 channels, 6 MHZ wide, in four frequency bands, 54 - 72, 74 - 88, 174 - 216, 470 - 806 MHz. Also, broadcast satellite services at 12.2-12.7 GHz (downlink) and 17.3-17.8 (uplink).

TWO-WAY RADIO SERVICE

Fixed Service: two-way radio service between fixed points

- **Common carrier point-to-point microwave radio service** (CCMRS): two-way, point-to-point, radio service in 40 bands between 928.0125 and 19620 MHz
- **Multipoint distribution service** (MDS): one-way, point-to-multipoint, radio service (up to 6 MHz channels) and limited return capabilities at frequencies around 4, 6, 11, 12, 13 and 23 GHz
- **Fixed satellite service** (FSS): voice, data, and video common-carrier services in C-band (3.700-4.200 GHz downlink, 5.925-6.425 GHz uplink) and Ku-band (11.7-12.2 GHz downlink, 14.0-14.5 GHz uplink)
- **Digital electronic message service** (DEMS) and Digital termination service (DTS): two-way, fixed service for exchange of digital data. 2.5 and 5 MHz channels in frequency bands 10.55 - 10.68, and 17.7 - 19.7 GHz

Mobile Service: two-way radio service between mobile stations, or between a mobile station and a fixed station

- **Air traffic control** (ATC): frequencies allocated in four bands, 118 - 121.4, 121.6 - 123.075, 123.6 - 128.8, and 132.025 - 135.975 MHz
- **Cellular radio service** (CRS): 832 frequency pairs spaced 45 MHz apart with 30 kHz channel spacing allocated in one band (base to mobile), 869.040 - 893.970 MHz, and one band (mobile to base), 824.040 - 848.970 MHz. Divided between wire-line and non-wire-line carriers
- **Wireless network service**: WARC'92/ITU–R allocated 1880 to 2025 MHz and 2110 to 2200 MHz for use by wireless networks. Two subbands -- 1980 to 2010 MHz and 2170 to 2200 MHz -- are designated for mobile satellite links. FCC has allocated frequencies in the 1850 to 1970 MHz and 2130 to 2200 MHz bands for personal communication services.

message signal energy is distributed at frequencies around the carrier signal frequency. A sinusoidal carrier signal can be modulated by a message signal in two ways. Consider the signal $s(t)$ that is modeled as a cosine function, i.e., $s(t) = A\cos(2\pi ft + \phi)$. When serving as a carrier, it may be modulated by varying the amplitude A in sympathy with the message signal; in this case we have *amplitude* modulation (AM). Also, it may be modulated by varying the frequency f or the phase angle ϕ; in these cases we have *frequency* or *phase* modulation (FM or ΦM). Collectively, frequency modulation and phase modulation are known as *angle* modulation.

(1) Amplitude Modulation

Amplitude modulated waves can take several forms—large-carrier AM (LCAM), double-sideband, suppressed-carrier AM (DSBSC), single-sideband AM (SSB), and vestigial-sideband AM (VSB). In this Section, we discuss them briefly. To illustrate time-domain behavior, we employ a single-tone message signal $m(t) = \cos\pi t$, and a sinusoidal carrier signal $c(t) = 1.5\cos5.3\pi t$. They are shown in **Figure 2.8**. In real life, the carrier will have a much higher frequency. The ratio of 5.3 to 1 for the frequencies of the carrier and message signals is chosen to make the diagrams easy to understand.

- **Large-carrier AM** (LCAM) is the modulation technique employed by all radio stations broadcasting information and entertainment in the frequency range 535 to 1605 kHz (the AM-band). It has been specially crafted to make it possible to receive the message signal with inexpensive circuits—a prereq-

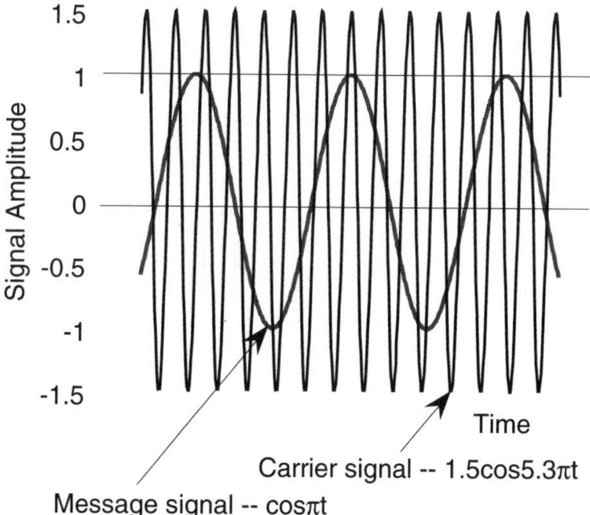

Figure 2.8 Single-Tone Message Signal and Carrier Signal

uisite for the development of large-scale, consumer-oriented, public, broadcast, radio services. Because the envelope of the modulated wave mimics the message signal, demodulation can be accomplished with a relatively simple envelope detector. LCAM waves are produced by multiplying the carrier signal by the factor $[1 + \mu m(t)]$. μ is known as the *modulation index*. It controls the *depth* of modulation. As shown in **Figure 2.9**, if $\mu = 0.7$, the amplitude of the modulated wave will swing from 0.45 to 2.55. To aid visualization, we have fixed the outline of the message signal on the modulated wave to emphasize that the envelope mimics it. The frequency-domain representation is shown in **Figure 2.10**. In the upper diagram, we show the single-tone message signal case. In the lower diagram, we show the general case of a message signal $m(t)$ with a spectral density $M(f)$. The modulated signal is *double-sideband* (i.e., the energy associated with the message signal is distributed on both sides of the carrier frequency), and contains a large-carrier signal (i.e., there is a separate signal at the carrier frequency). If the message signal (baseband) is band-limited to the frequency interval $0 \leq f \leq W$, the modulated wave is limited to the band $f_c - W \leq f \leq f_c + W$, and the transmission bandwidth is $2W$ Hz. Because less than 50% of the total power in the modulated wave is associated with the message-bearing sidebands, LCAM is considered to be an inefficient process.

- **Double-sideband, suppressed-carrier AM** (DSBSC) is used in FM stereo broadcasting systems (the channel that carries the difference signal uses DSBSC, not FM), and as the first stage in the production of frequency multiplexed signals (see § 8.1). DSBSC waves are produced by multiplying the carrier signal by the message signal. The time-domain representation is shown in **Figure 2.11**, and the frequency-domain representation is shown in

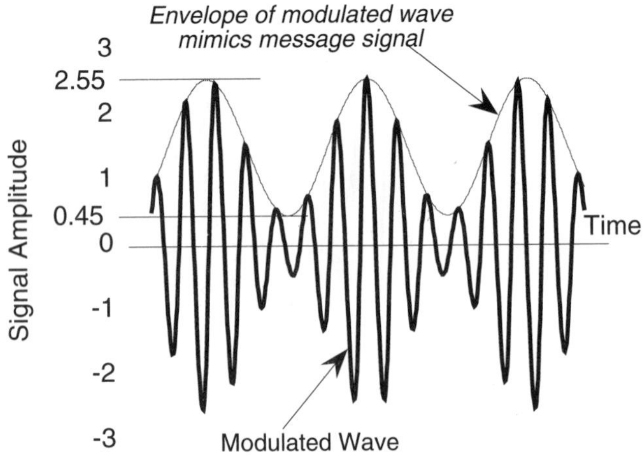

Figure 2.9 Large-Carrier Amplitude Modulation

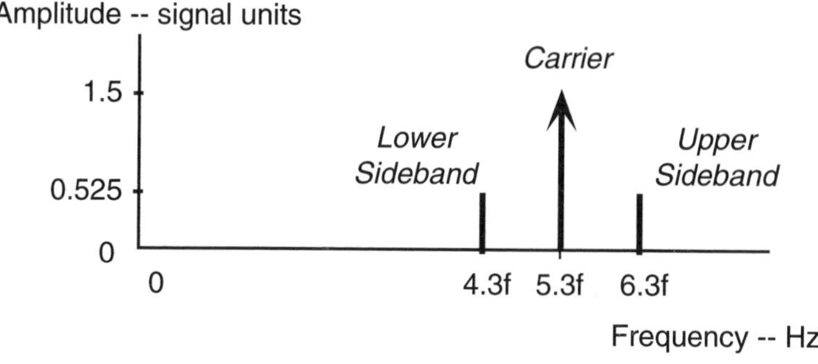

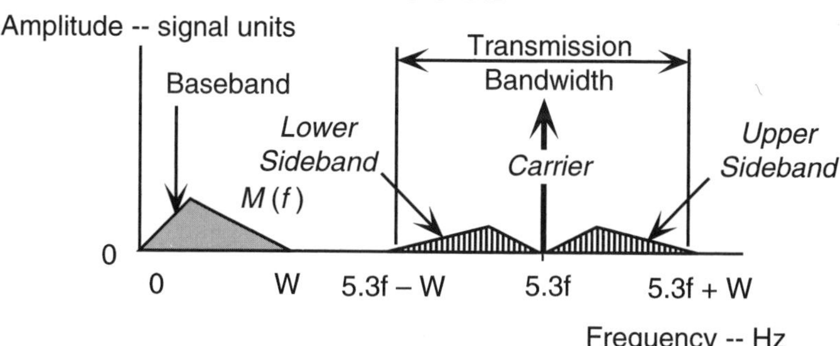

Figure 2.10 Frequency-Plane Representations of Large-Carrier, Amplitude-Modulated Wave

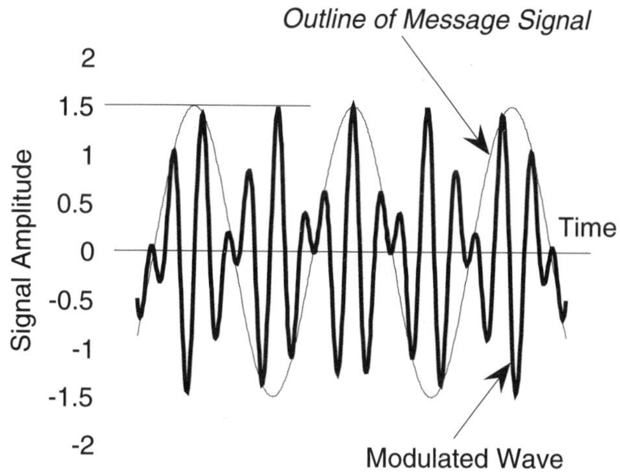

Figure 2.11 Double-Sideband, Suppressed-Carrier Modulation

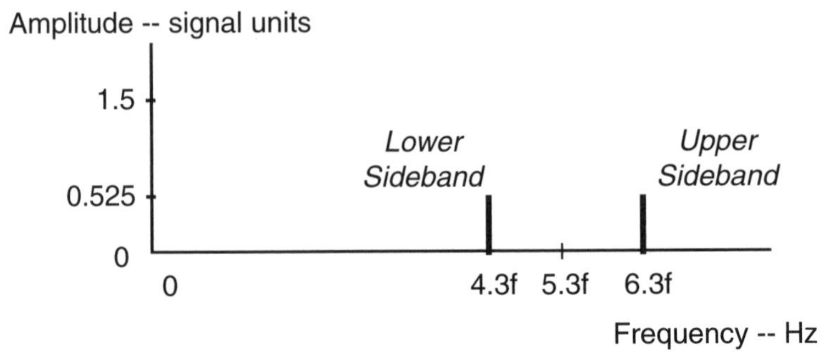

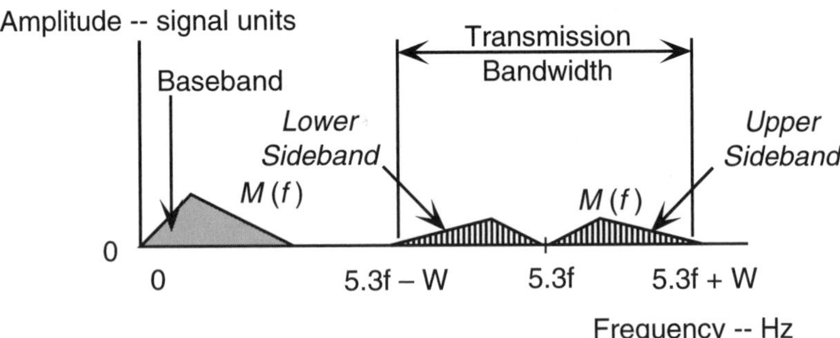

Figure 2.12 Frequency-Plane Representations of Double-Sideband, Suppressed-Carrier, Amplitude-Modulated Wave

Figure 2.12. The spectral density is *double-sideband*, i.e., energy associated with the message signal is distributed on both sides of the carrier frequency according to the frequency distribution defined by $M(f)$, but there is no separate carrier component. If the message signal is band-limited to the frequency interval $0 \leq f \leq W$, the modulated wave is limited to the band $f_c - W \leq f \leq f_c + W$, and the transmission bandwidth is $2W$ Hertz, the same as LCAM. Without the powerful carrier signal component of LCAM, detection is much more difficult. Usually, DSBSC (and other modulation schemes that eliminate the carrier component) employs pilot tones or other signals, from which the carrier signal can be derived at the receiver. All of the power in the wave is associated with the message-bearing sidebands.

- **Quadrature-carrier amplitude modulation** (QAM) is a special case of DSBSC modulation. Because sine and cosine functions are orthogonal, two DSBSC waves can occupy the same transmission bandwidth. The procedure employs two, independent message signals that modulate two carrier signals that differ in phase by 90° but have identical frequencies. The transmission bandwidth is $2W$ Hz, the same as LCAM and DSBSC—but two message signals share it. QAM is a bandwidth conservation procedure that is used in radio systems designed to carry digital signals (see § 4.7.1(1)).

- **Single-sideband AM** (SSB) is an efficient, bandwidth-minimizing, modulation technique that is used extensively in voice multiplexing systems (see § 8.1.1). All transmitted power is devoted to sending one copy of the message. When the higher-frequency sideband is employed, the modulated wave is described as upper-sideband (USB) SSB; when the lower-frequency sideband is employed, the modulated wave is described as lower-sideband (LSB) SSB. The frequency-domain representation of these waves is shown in **Figure 2.13**. If the message signal is band-limited to the frequency interval $0 \leq f \leq W$, the modulated wave occupies W Hertz, and the transmission bandwidth is W Hertz, i.e., one-half of that required by LCAM and DSBSC. Because the modulating technique produces sharp peak values when the message signal makes abrupt transitions, the use of SSB is restricted to well behaved, zero-mean, message signals such as are generated in voice applications. SSB is not recommended for digital signals or signals (such as television) that contain significant energy at frequencies close to 0 Hz, i.e., $m(t)$ contains important low frequencies.

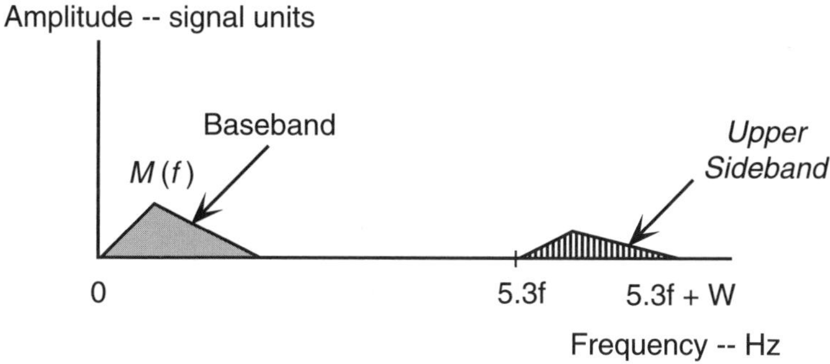

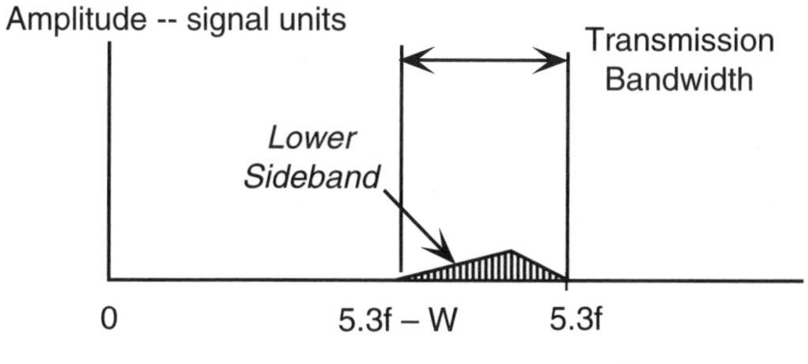

Figure 2.13 Frequency-Plane Representations of Single-Sideband, Amplitude-Modulated Wave

- **Vestigial-sideband AM** (VSB) is the modulation technique used for domestic television service broadcasting in the United States. Approximating the bandwidth efficiency of SSB, in VSB modulation one sideband is passed almost intact, together with a *vestige* of the other sideband. This is achieved by a bandpass filter designed to have odd-symmetry about the carrier frequency and a relative response of 50% at the carrier frequency. **Figure 2.14** shows how vestigial-sideband operation is applied to an LCAM signal. For equal total power, VSB provides less message power than SSB. However, the power at the carrier frequency makes signal detection less expensive and has made the development of consumer-oriented television services possible.

(2) Angle Modulation

Angle modulated waves exist in two forms—phase modulated and frequency modulated:

- **Phase modulation** (ΦM) is employed to convey color (chrominance) information in NTSC television signals, and for other specialized functions. A ΦM wave is produced by varying the phase angle in sympathy with the amplitude of the message signal. It is a constant power wave; as modulation occurs, power is shared among the carrier and the side frequencies. The total power remains constant at the value of the power in the unmodulated carrier signal. The time–domain representation of a phase modulated wave formed from the single-tone message signal and the sinusoidal carrier signal of Figure 2.8 is shown in the lower diagram of **Figure 2.15**. Phase modulation is closely related to frequency modulation.

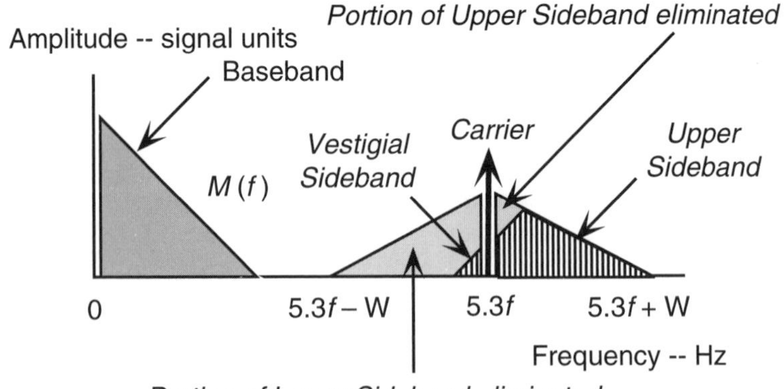

Figure 2.14 Frequency-Plane Representations of Vestigial-Sideband, Amplitude-Modulated Wave

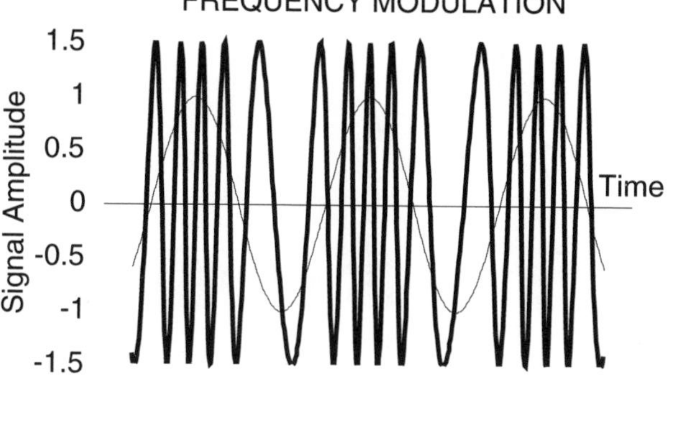

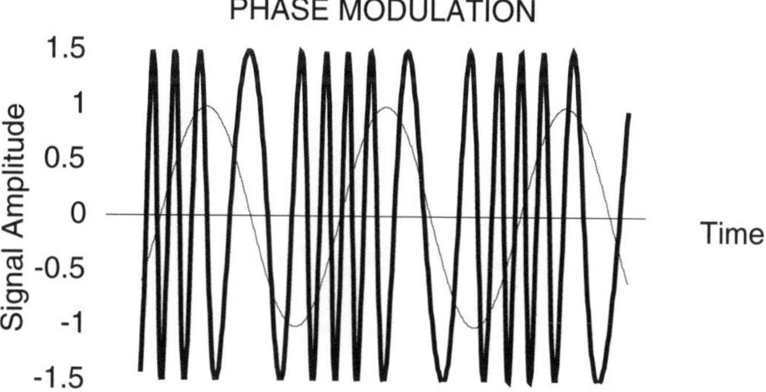

Figure 2.15 Time-Domain Representations of Angle-Modulated Waves

- **Frequency modulation** (FM) is used to implement point-to-point, microwave-radio links, and cellular, mobile, radio telephone service. In addition, it is employed by all radio stations broadcasting information and entertainment in the frequency range 88 to 108 MHz (FM-band). Further, FM is employed in the sound segment of NTSC television signals. An FM wave is produced by varying the frequency in sympathy with the amplitude of the message signal. As we shall see, it is a non-linear process. The time-domain representation of a frequency modulated wave formed from the single-tone message signal and the sinusoidal carrier signal of Figure 2.8 is shown in the upper diagram of Figure 2.15. It differs from the phase modulation diagram (under it) in the location of the maximum frequency. To aid in visualization we have included the outline of the message signal. In phase modulation, the maximum frequency of the wave occurs at the positive-going zero crossings of the message signal; in frequency modulation, the maximum frequency occurs at the positive peaks of the message signal. In the frequency-domain, even the case of single-tone modulation is complicated. **Figure 2.16** shows

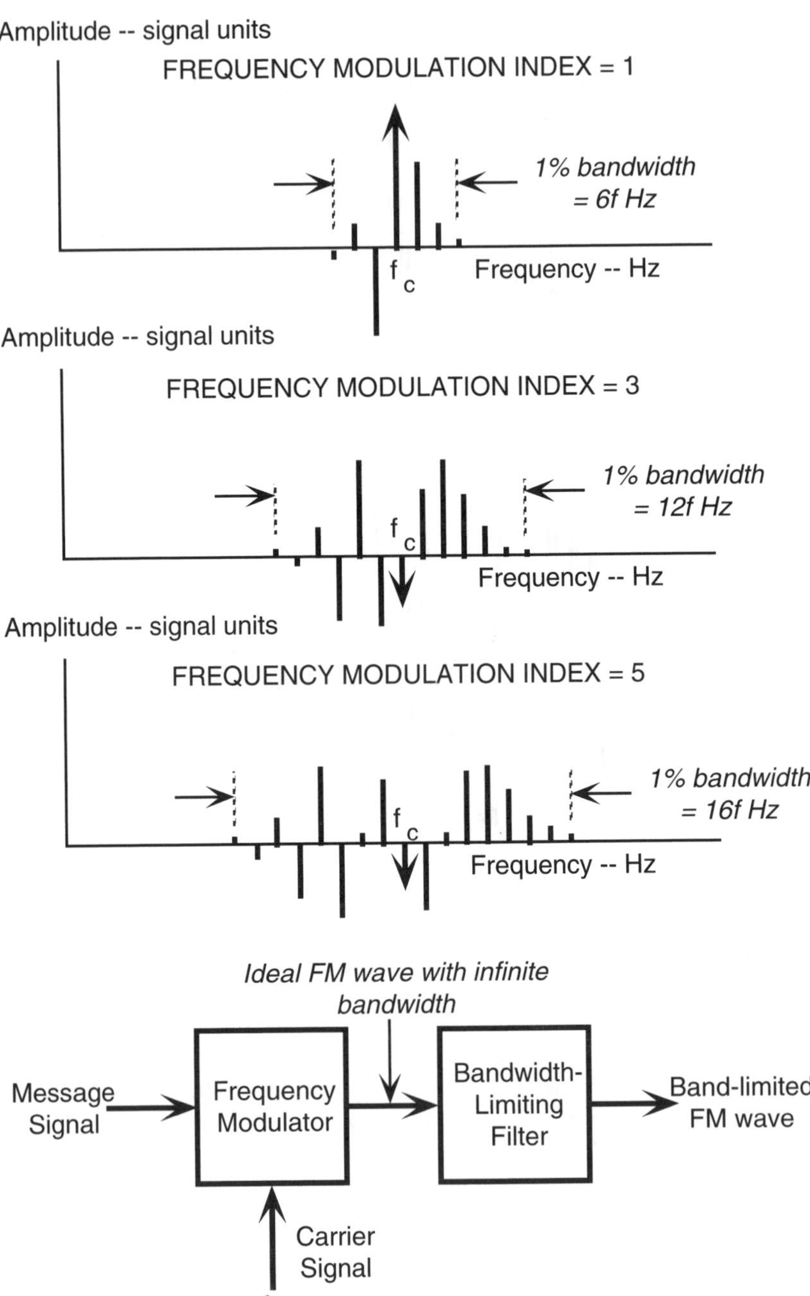

Figure 2.16 Frequency-Plane Representations of Frequency Modulated Waves of Increasing Modulation Index

spectral diagrams of single-tone-modulated FM waves for which the frequency modulation index is increasing. In theory, the side components extend to infinity. We limit them to the interval that just contains side components whose amplitudes are ≥ 1% of the amplitude of the unmodulated carrier. This is known as the *1% bandwidth.* An FM wave is a constant power wave; as modulation occurs, power is shared among the carrier and the side frequencies. The total power remains constant at the value of power in the unmodulated carrier signal.

(3) FM Radio

With a maximum permitted frequency deviation of 75 kHz, and a maximum audio signal frequency of 15 kHz, commercial, FM-band radio stations operate with a modulated wave whose bandwidth lies between 180 kHz and 240 kHz. As shown at the bottom of Figure 2.16, for operation in a spectrum-limited world, all FM systems contain a bandwidth-limiting filter to keep the radiation within authorized operating limits (200 kHz for FM-band radio stations). In normal operation, the transmitted signal is a bandwidth-limited version of the signal generated by the modulator. An important property of FM waves is their significant immunity to interference from noise. This capability is derived from the fact that noise signals add to the wanted signal to change the amplitude of the wave; they do not change the frequency. In practice, the signal-to-noise ratio of FM can be 1000 times (30 db; see **Table 2.2**), or more, better than LCAM. This is achieved by adding emphasis to the higher frequencies present in the message signal before modulation (known as pre-emphasis), and restoring the level during demodulation (known as de-empha-

Table 2.2 The Decibel

$$\text{Power Ratio} = \frac{P_2}{P_1} = 10\log_2 \frac{P_2}{P_1} \text{ decibels}$$

Thus, a power ratio of...		
10,000	is	40 db
1,000	is	30 db
100	is	20 db
10	is	10 db
2	is	3db
1	is	1 db
0.5	is	–3 db
0.1	is	–10 db
0.01	is	–20 db
0.001	is	–30 db
0.0001	is	–40 db

sis). Whenever sufficient bandwidth is available, frequency modulation is the technique of choice for high-quality radio communication links.

2.4 NOISE

2.4.1 Sources of Noise

Noise is the sum of all the unwanted signals that are added to the message signal in the generation, transmission, reception, and detection processes employed in effecting telecommunication. Noise consists of various signals:

- Signals picked up at the receiver
  - another modulated wave that interferes with the wanted signal
  - signals generated by natural, impulsive phenomena such as lightning, or by man-made impulsive sources, such as automobile ignition systems
  - galactic radiation
- Signals generated in the processes and equipment used to consummate telecommunication
  - the random motion of electrons, ions, or holes in the materials that compose the (receiving) equipment
- Signals that represent errors
  - approximations made in processing the signal for transmission that result in sending a signal that differs from the signal that the sender intended to send
  - errors made in detecting the signal at the receiver.

They add together to produce a signal at the receiver output that is a distorted version of the intended signal. The difference between the received (corrupted) signal and the intended (ideal) signal is called the *noise* signal.

2.4.2 White Noise

When it is not possible to take account of the sources of noise individually, we assume that they produce a single, random signal in which power is distributed uniformly at all frequencies. By analogy with white light (that contains all visible frequencies), this signal is called *white* noise. It is represented by the power spectral density (PSD) diagram shown in **Figure 2.17**. At all frequencies, the spectral density is N_0 signal watts/Hz. Thus, in a passband of $(m - n)f$ Hz, the equivalent noise power is $N_0(m - n)f$ signal watts. White noise has the property that different samples are uncorrelated, and that, if the probability density function of the amplitude spectrum is Gaussian, they are statistically independent. In short, white, Gaussian noise is truly random.

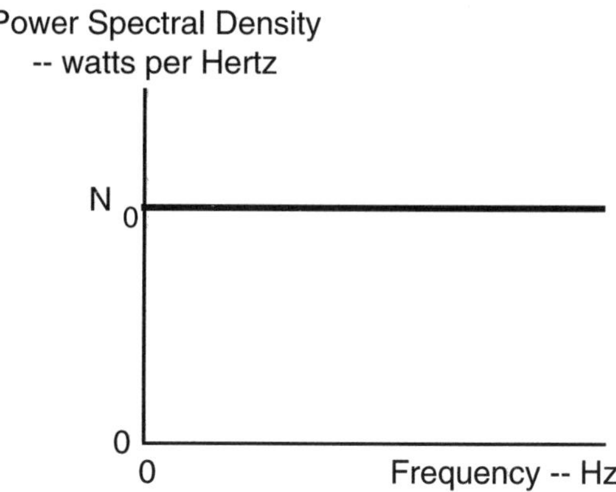

Figure 2.17 Power Spectral Density of White Noise

2.4.3 Signal-to-Noise Ratio

In a communication system, the ratio of the power in the message signal to the power in the noise signal is known as the signal-to-noise ratio (SNR). Because it makes calculations of overall performance easier, it is usual to express this ratio as a logarithm as stipulated in Table 2.2. The unit ratio is 1 decibel (db). Good quality communication systems achieve SNRs of 35 to 45 db, or better. Many people have difficulty recognizing words when the SNR in a telephone system drops below 15 db.

3

Voice and Video

We continue the examination of signals with discussions of the properties of voice, music, and television signals. Naturally analog, signals are converted to digital form so as to be transported over modern communication systems. We describe the process of digitizing them, and discuss ways of reducing bit rates.

3.1 SPEECH AND MUSIC

3.1.1 Speech Components

(1) Phonemes

Speech consists of a sequence of elementary sounds that are uttered at rates up to ten per second. Called *phonemes*, these sounds are produced by bursts of air that originate in the voice box, acquire pitch by passing through the vocal chords, and excite resonances in the vocal tract. The result is the string of sounds we call speech, spoken in a voice unique to the speaker.

(2) Frequency, Phase, and Amplitude

By means of a microphone, natural speech is transformed from sound pressure waves to electrical signals. At this point, the speech signal acquires technical descriptors: *frequency*, *phase*, and *amplitude*. Within reasonable limits, they correspond to the pitch, intelligibility, and intensity of the original sound. Of the three electrical parameters, phase appears to be the least critical. Significant phase shifts can be produced in the message signal without affecting our ability to recognize the speaker and understand the words spoken.

(3) Talkspurts

In a conversation, the participants share the task of talking. First one speaks, then the other. Occasionally, the talking activities will overlap as one interrupts the other. On average, each talks about 40% of the time. The remaining 20% represents the time neither party is talking. It is made up of pauses between words and sentences when the speaker is clearly going to continue, and pauses when one speaker has finished speaking and the other is preparing to respond. Considered from this perspective, speech signals consist of bursts of electrical energy interspersed with periods of no activity at all. More or less connected, periods of vocal activity are called *talkspurts*.

3.1.2 Electrical Parameters of Speech

Measurements on the speech signals produced by a large population of talkers indicate facts concerning natural speech:

- Contains energy between approximately 100 Hz and 7000 Hz
- Has peak energy around 250 Hz to 500 Hz
- Has a dynamic range of some 35 or 40 db, i.e., the ratio of the power produced by the loudest shout and the least whisper may be as high as 10,000
- Contains an amplitude distribution that can be approximated by a Laplace-type probability density function.

For transmission on telephone facilities, the signal energy is usually restricted to a range from 200 Hz to 3200 Hz in the United States and Japan, and 300 Hz to 3400 Hz in Europe. On toll facilities, the value of the ratio of the power of the voice signal to the power of all interfering signals (signal-to-noise ratio) is generally expected to be 30 db, or better. For transmission on AM radio, voice energy is restricted to a frequency range from about 100 Hz to something less than 5000 Hz. Only on FM radio is it possible to transmit the full range of voice frequencies.

3.1.3 Speech Quality

The quality of the speech signal delivered to the human ear is a matter for the exercise of individual judgment. While not measurable in any absolute way, three levels can been defined that are recognizable by most persons:

- **Network quality:** contains the full range of frequencies without perceptible noise or distortion. It is suited to the speech requirements of radio broadcasting (particularly FM broadcasting) and television

- **Communications quality**: contains a limited range of frequencies and may have discernible noise and distortion. The speech quality is adequate for easy telephone communication including speaker recognition. This level includes toll quality voice
- **Synthetic quality**: contains sounds generated by an electronic analog of the human speech production mechanism. May be more intelligible than communication quality voice, but speaker recognition values are missing.

3.1.4 Electrical Parameters of Music

Music is produced by instruments that emit essentially continuous sounds. The electrical signals they generate have the following properties:

- Contain energy between approximately 20 Hz and 20,000 Hz
- Have maximum energy between 200 Hz and 600 Hz
- Have a dynamic range of some 75 or 80 db.

AM radio programs contain speech and music in the frequency range from about 100 Hz to something less than 5000 Hz. FM radio programs contain speech and music in the frequency range from about 100 Hz to 15,000 Hz. Neither facility is able to reproduce a dynamic range of 80 db. **Figure 3.1** shows some of the properties of speech and music signals.

3.2 DIGITIZING VOICE SIGNALS

As digitalization is applied to all telecommunication facilities, the digital equivalents of analog signals become an important topic, particularly digital voice signals. Analog voice signals are changed to digital voice signals by sampling, and then expressing the sample values as binary numbers. Digital voice signals are changed back to analog voice signals by converting the binary numbers to amplitude-modulated pulses (a procedure known as pulse amplitude modulation, PAM) that are applied to a reconstruction filter to give an analog signal. Because of the dynamic range encountered in voice signals, a technique known as companding is required so that very loud signals do not overpower the system, and very soft signals are not lost in noise. Within limits, the reconstructed, analog signal mimics the original analog voice signal.

3.2.1 Sampling an Analog Signal

(1) Band-Limited Signal

If we define a signal simply as an analog signal, it may contain energy at any frequency. As such, it is not a practical signal—signals in the telecommunication

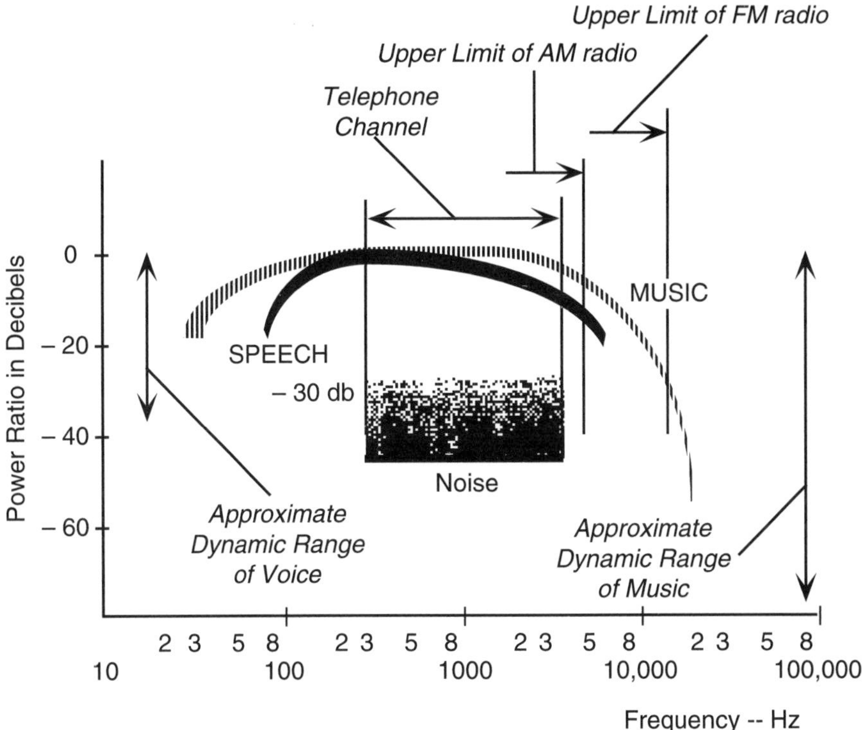

Figure 3.1 Some Properties of Speech and Music Signals

environment are limited to specific bandwidths by the equipment that processes them. Accordingly, we define:

- **Band-limited signal**: a signal that contains no energy at frequencies greater than W Hz, i.e., $0 < f \leq W$ Hz.

(2) Sampling

- **Sampling**: the process of determining the value of the amplitude of an analog signal at an instant in time.

Usually, sampling is done in a regular repetitive manner at equal time intervals. When a band-limited signal is sampled at a rate that is exactly twice the highest frequency present, sampling is said to occur at the Nyquist rate:

- **Nyquist rate**: i.e., sampling rate = $2W$ samples per second.

If the sampling rate (r) is equal to, or greater than, the Nyquist rate (i.e. $r \geq 2W$ samples/s), sampling yields a set of values that contains all of the information necessary to reconstruct the original, band-limited, signal. Using a process known as quantizing (see 3.2.3), sample values are expressed in octets to produce digital signals.

(3) Aliasing

If the sampling rate is less than $2W$ samples per second (i.e. $r < 2W$ samples/s), the reconstructed signal will be a distorted version of the original, with the energy that falls in the frequency range $r/2 < f \leq W$ Hz masquerading at lower frequencies. This effect is known as aliasing:

- **Aliasing**: condition that exists when signal energy that is present over one range of frequencies in the original signal appears in another range of frequencies after sampling and reconstruction.

Perhaps the most widely observed example of aliasing occurs in movie films and on television. Often, wheels on wagons and cars appear to be turning in the wrong direction. This is due to the 30 frames/s sampling rate being too slow to capture the high-speed motion of the wheels. To prevent aliasing, it is usual to *oversample* the band-limited signal (perhaps 110 to 120% of the Nyquist rate), and include a low-pass filter of bandwidth W Hz in the signal chain before the sampling operation.

3.2.2 Reconstructing an Analog Signal

To reconstruct the band-limited signal from the string of octets that represents the sample amplitudes, we form a train of narrow pulses whose amplitudes are modulated (PAM) so that they equal the values of the octets. The pulse repetition rate is equal to the sampling rate. Only with an exact match between these rates will it be possible to regain the frequencies present in the original signal. This pulse train is applied to a low-pass, reconstruction filter, that passes frequencies $\leq W$ Hz. Provided sampling was conducted at, or above, the Nyquist rate, the signal at the output of the filter will be a close mimic to the original, band-limited signal. **Figure 3.2** shows the steps in creating a digital signal from an analog signal, and the reconstruction of an analog signal from the digital signal. After passing through the anti-aliasing filter, the signal is bandwidth-limited to W Hz and sampled at a rate $r \geq 2W$ samples/s. The sampled signal consists of a series of impulses at the sampling rate whose envelope imitates the amplitude of the band-limited signal. At this point, the samples are converted to octets through a process known as quantizing (see 3.2.3). At the receiver, the octets become the amplitudes of narrow pulses that are generated at the sampling rate. The result is an amplitude-modulated pulse stream whose envelope mimics the envelope of the sampled signal. After passing through a low-pass reconstruction filter with a pass-band of W Hz, the signal emerges as a close likeness of the band-limited input signal.

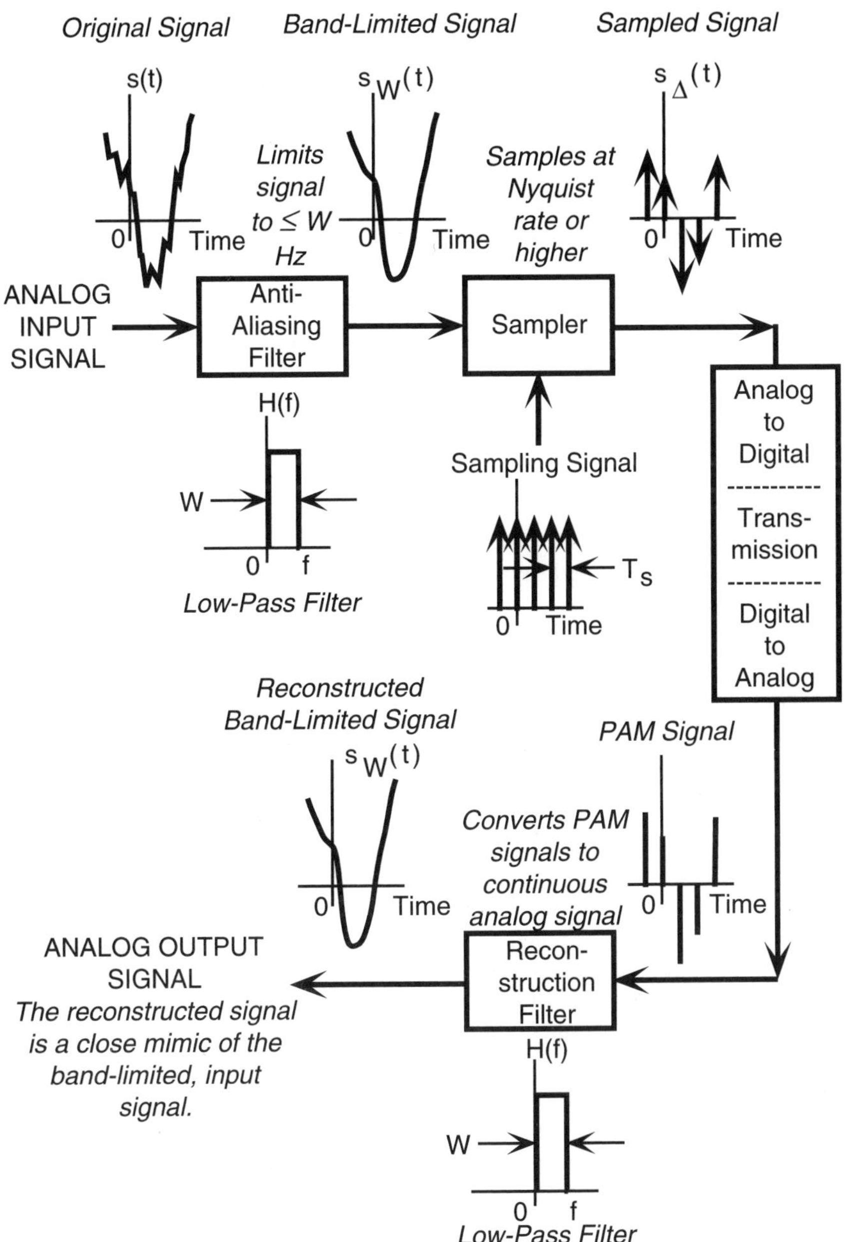

Figure 3.2 Sampling, Digitizing, and Reconstructing an Analog Signal

3.2.3 Quantizing

An octet is able to represent $2^8 = 256$ states. If we use them to describe the values of the train of impulses that is the sampled signal, we cannot describe each impulse value precisely, but must round it to the nearest binary value—in other words, the octets will only approximate the values of the real signal. This is known as quantizing:

- **Quantizing**: the process that segregates sample values into ranges and assigns a unique, discrete identifier to each range. Whenever a sample value falls within a range, the output is the discrete identifier assigned to the range.

In addition, so as to handle different levels of signals, we declare the greatest sample value to be ±1 and equal to the limit of the range to which the greatest discrete value is assigned. This process is known as *normalization*.

(1) Linear Quantizing

The process of linear quantizing is illustrated in **Figure 3.3**. The samples are represented by impulses whose areas (represented as amplitudes) correspond to the amplitude of the band-limited signal at the sample times. A set of uniformly-spaced ranges is shown. Each of the q ranges corresponds to an octet selected from all possible octets that can be made from an n-bit binary code. When the levels span ±1, the distance between each level is $2/q$. As the pulses arrive, they are assigned the octet that identifies the range in which they fall. The result is a series of octets that represents bands of sample values derived from the original signal. In transmitting the octet, the most significant bit is sent first. It is the bit that defines whether the sample is positive (1) or negative (0).

(2) Quantizing Noise Signal

At the output of the reconstruction filter, the signal can be considered to consist of two components, the band-limited version of the original signal, and a signal that is the difference between the actual output signal and the band-limited signal. The former is the ideal outcome of the digital conversion and analog reconstruction processes. The latter is the amount by which the processes have degraded the output—it is the quantizing *noise* signal. For a sinusoidal signal, given the number of quantizing levels, it is possible to calculate the magnitude of the signal-to-quantizing noise power ratio. In **Figure 3.4** we plot the SNR experienced by a sinusoidal signal quantized in 256, equally-spaced levels, against the ratio of the power of the signal and the power of a sinusoidal signal that just sweeps through all 256 levels. The latter is known as a *full-load* signal. From the graph, we see that the quantizing SNR experienced by a full-load signal is approximately 50 db (i.e. a ratio of 100,000); that an input signal with approximately one-hundredth (i.e. – 20 db) of the power of a full-load signal will experience a quantizing SNR of 30 db (i.e. a ratio of 1000); and that an input signal with approximately one-ten thousandth (i.e.

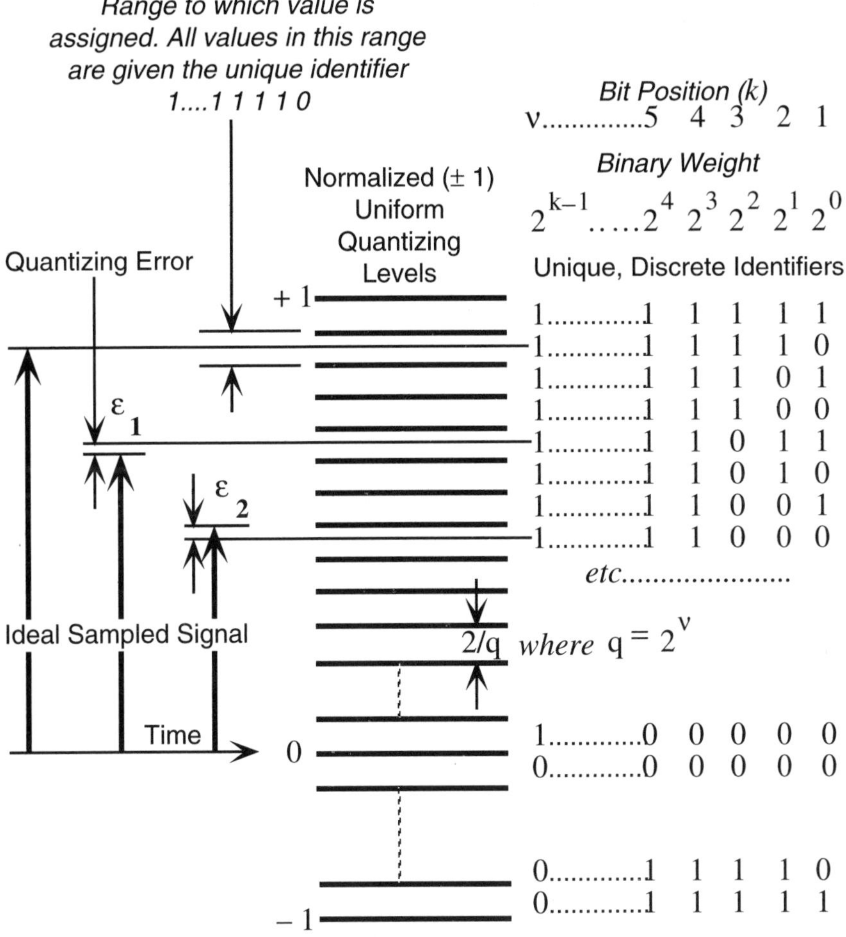

Figure 3.3 Quantizing

– 40 db) of the power of a full-load signal will experience a quantizing SNR of 10 db (i.e. a ratio of 10). Recalling that the dynamic range of voice signals can be as much as 40 db, and that toll-quality voice requires a SNR of 30 db, the performance of a linear quantizer with 256 levels is such that it can handle only part of the range with better than 30 db SNR.

(3) Changing the Number of Quantizing Levels

The quantizing SNR is improved by 6 db for each unit change in *n*, i.e., for 2^9 quantizing levels, a full-load, sinusoidal tone signal has a quantizing SNR of approximately 56 db, and an input signal with approximately one-hundredth of the

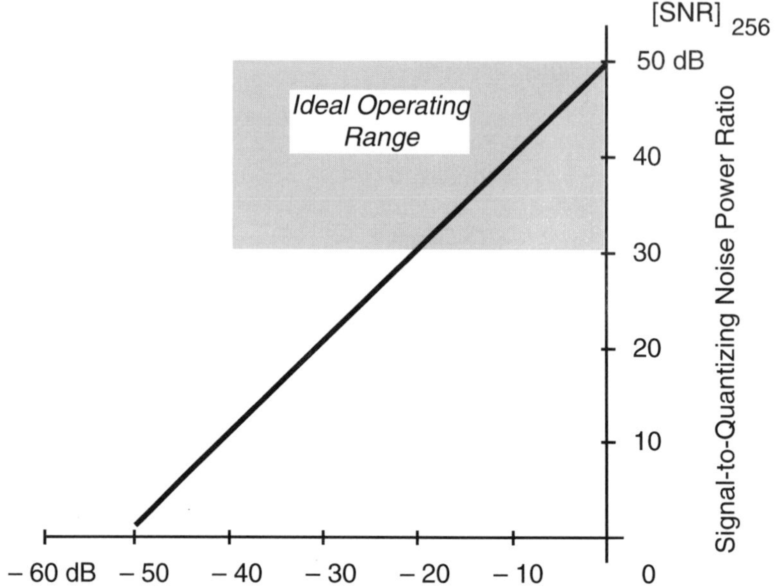

Figure 3.4 Signal Power and Signal-To-Quantizing Noise Power Ratio for Single-Tone Signals With 256 Equally Spaced Quantizing Levels

power of a full-load signal will experience a quantizing SNR of 36 db. If we increase n to 12, an input signal with approximately one-ten thousandth (i.e. – 40 db) of the power of a full-load signal will experience a quantizing SNR of approximately 34 db. This is the sort of performance required to achieve toll-quality with linear quantizing. To do it, the input signal samples must be matched against 4096 levels, and each discrete identifier will be 12 bits long. Until recently, such precision was impossible to attain in practical circuits, and a technique called companding was invented to improve the performance of 8-bit quantizing.

3.2.4 Companding

In digital voice systems, the dynamic range required within the digital portion is reduced through the use of a technique called companding (a contraction of the two words compressing and expanding). At the transmitting end, a compressor decreases the amplitude of the stronger signal components with respect to the weaker signal components. At the receiving end, an expander reverses the distortion introduced by the compressor and restores the voice components to their original relationship. Devices of this sort also act to normalize the signals—no matter how large the input signal becomes, the compressor produces a signal of limited amplitude.

(1) Producing the Same Quantizing SNR for All Signals

So that small power signals are not penalized with respect to large power signals, an ideal quantizing process should produce approximately the same signal-to-quantizing noise power ratio for all signals. This may be achieved in two ways:

- **Employ companding**: distort the amplitude relationship among the sample values so that the quantizing error associated with higher values is greater than the quantizing error associated with lower sample values. In **Figure 3.5**

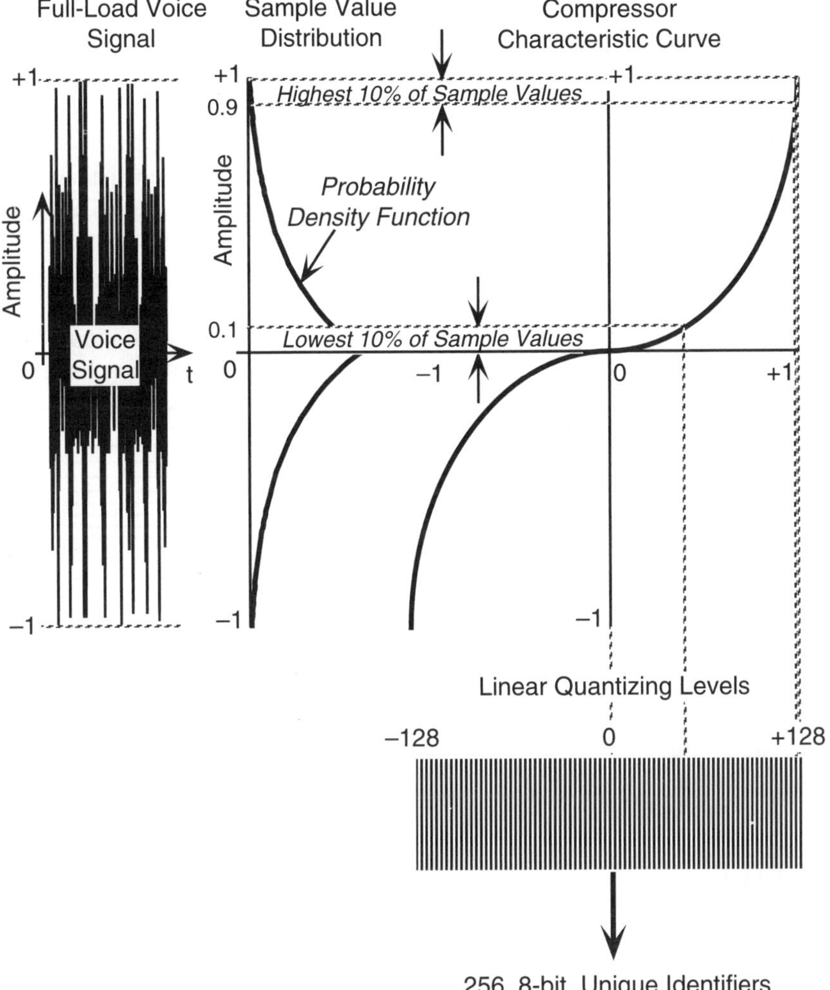

Figure 3.5 Use Of Compressor and Linear Quantizing to Produce Same Signal-To-Quantizing Noise Power Ratio for all Sample Values

we illustrate the use of this technique with a voice signal in which the sample amplitudes follow a Laplace-type, probability density function. By compressing the samples, the number of levels of a linear quantizer that are used by the highest 10% of the sample values is significantly less than the number of levels that are used by the lowest 10% of the sample values. Thus, on average, the round-off errors, and therefore the quantizing noise powers vary inversely with the sample values.

- **Grade the quantizing steps**: divide the normalized sample value space into unequal steps whose distribution provides more levels for smaller sample values and fewer levels for larger sample values. This technique is illustrated in **Figure 3.6**. Through the use of read-only memory (ROM), integrated circuit technology has made using graded quantizing steps the technique of choice. In a typical implementation, the voice signal is sampled and quantized in 4096 (2^{12}) levels, or more, in a straightforward, linear fashion; then, using a table stored in a ROM, 12-bit words are divided in a biased fashion among the 256 octets so as to taper the step size and implement a companding function that produces approximately equal SNR values at all levels. The quantiz-

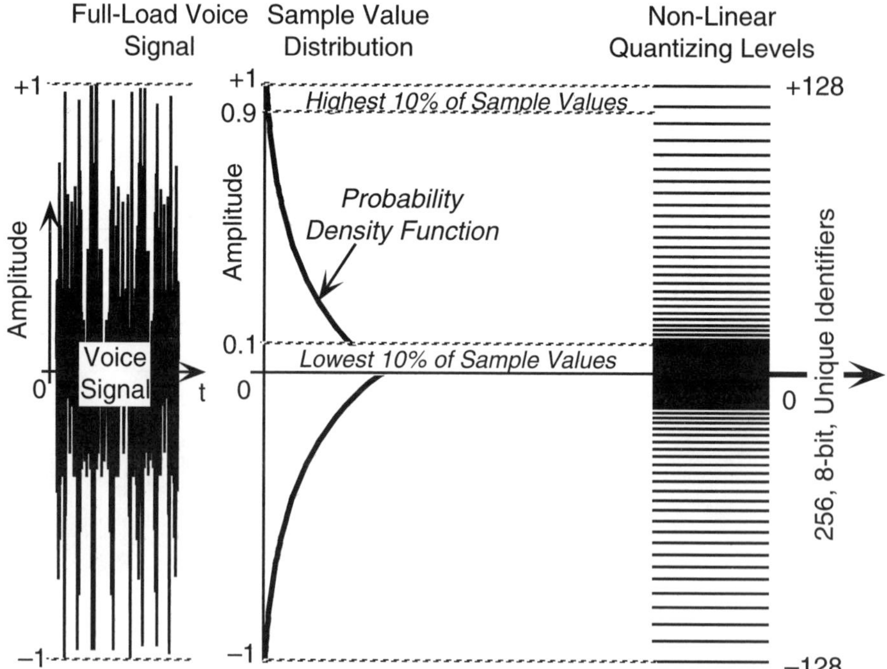

Figure 3.6 Use of Non-Linear Quantizing to Produce Same Signal-To-Quantizing Noise Power Ratio for all Sample Values

ing levels are closer together for small amplitude signals, and farther apart for large amplitude signals.

Irrespective of how it is done, the result is the same. In a non-linear way, the amplitudes of the signal samples are converted to quantizing levels and the process is reversed to reconstruct the signal samples. In **Figure 3.7**, the signal-to-quantizing noise ratio of a companded signal is plotted against the power of the input signal expressed in db relative to a full-load signal. Comparing this curve to that obtained earlier for linear quantizing (Figure 3.4), it is apparent that companding improves the SNR at lower signal levels and degrades it at higher levels and that it is a solution to achieving better than 30 db SNR over a 40 db dynamic range.

(2) μ-Law and A-Law Companding

Among the telephone administrations of the world, two companding functions are employed. One, called μ-law companding, is used in North America and regions of the world that follow North American practices. The other, called A-law companding, is used in Europe and regions of the world that follow European practices. **Figure 3.8** compares these two functions. Using compression according to one law, and expansion according to the other, will produce a distorted signal.

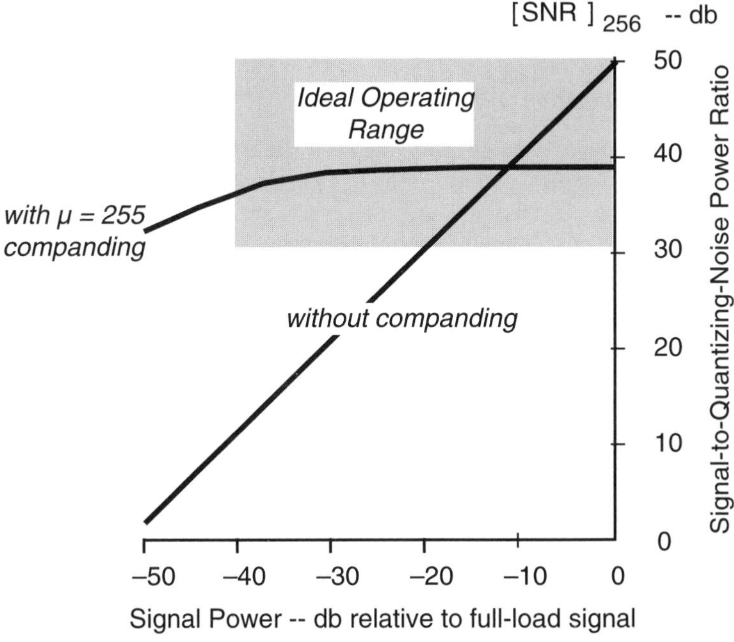

Figure 3.7 Effect of μ-Law Companding on Signal-To-Quantizing Noise Power Ratio

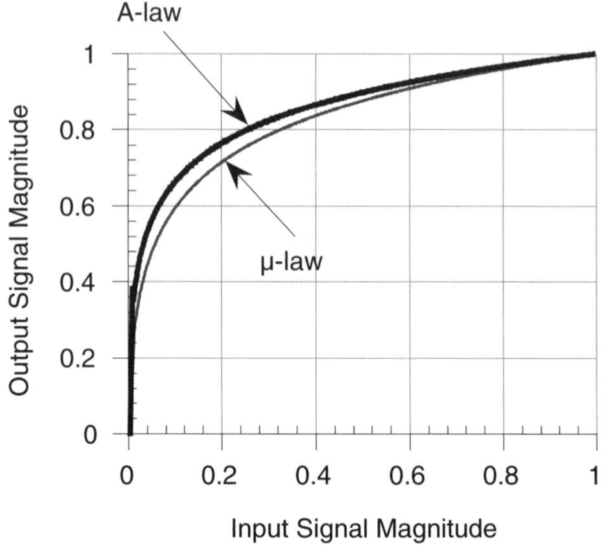

Figure 3.8 Comparison of μ-Law and A-Law Quantizing Characteristics

3.2.5 PCM Voice

(1) Pulse Code Modulation (PCM)

The energy in telephone voice is limited to between 200 and 3400 Hz by the equipment that creates, switches, and transmits it. Because it must withstand the exigencies of worldwide service that may include several analog-to-digital, and digital-to-analog, conversions, the telephone administrations of the world have adopted a robust coding scheme that employs a sampling rate of 8000 samples/s, codes each sample in 8-bits to produce a signal with a bit-rate of 64,000 bits/s, and includes μ-law or A-law companding. Called *pulse code modulation* (PCM), it provides communications quality voice signals that are easily understood and readily attributed to specific speakers. Because it is based on sampling at 8000 times per second, the analog signal reconstructed from the PCM stream cannot contain frequencies above 4000 Hz. The sequence of operations involved in creating, transmitting, and reconstructing 64 kbits/s, PCM, digital voice is shown in **Figure 3.9**.

(2) Decoding Noise

Signals that interfere with the digital signal stream can affect regeneration and produce erroneous values in the receiver. Inspection of Figure 3.3 shows the change in quantizing levels due to an error in bit position n is 2^{n-1} levels; this results in an absolute change in magnitude of $(2/q).2^{n-1}$, i.e., $2^n/q$. If the errors are uniformly

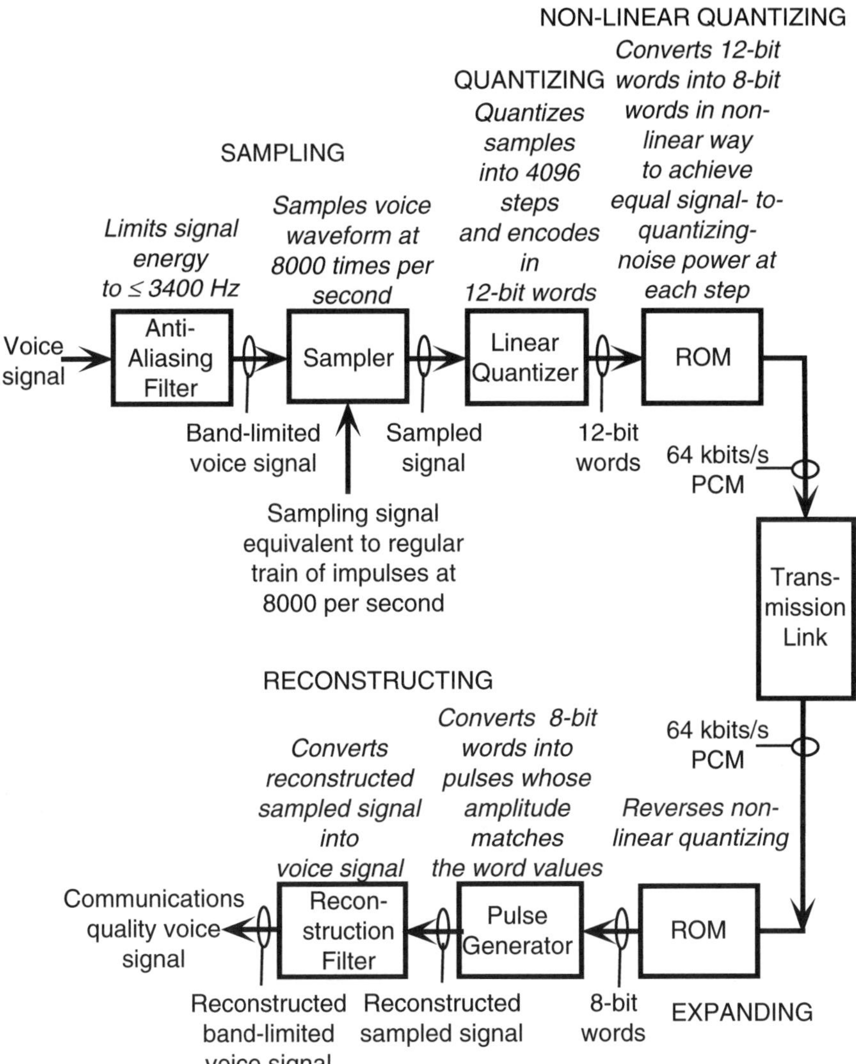

Figure 3.9 Sequence of Actions Involved in Processing 64 kbits/s, PCM, Digital Voice

distributed over the bit positions, and the probability of an error occurring in any bit is P_e, decoding introduces an additional error signal whose noise power can be shown to be approximately 1.33 P_e signal watts.

(3) Signal-to-Total Noise Power Ratio

The analog voice signal delivered by a PCM channel is corrupted by the sum of quantizing noise and decoding noise. In **Figure 3.10**, we plot values for the signal-

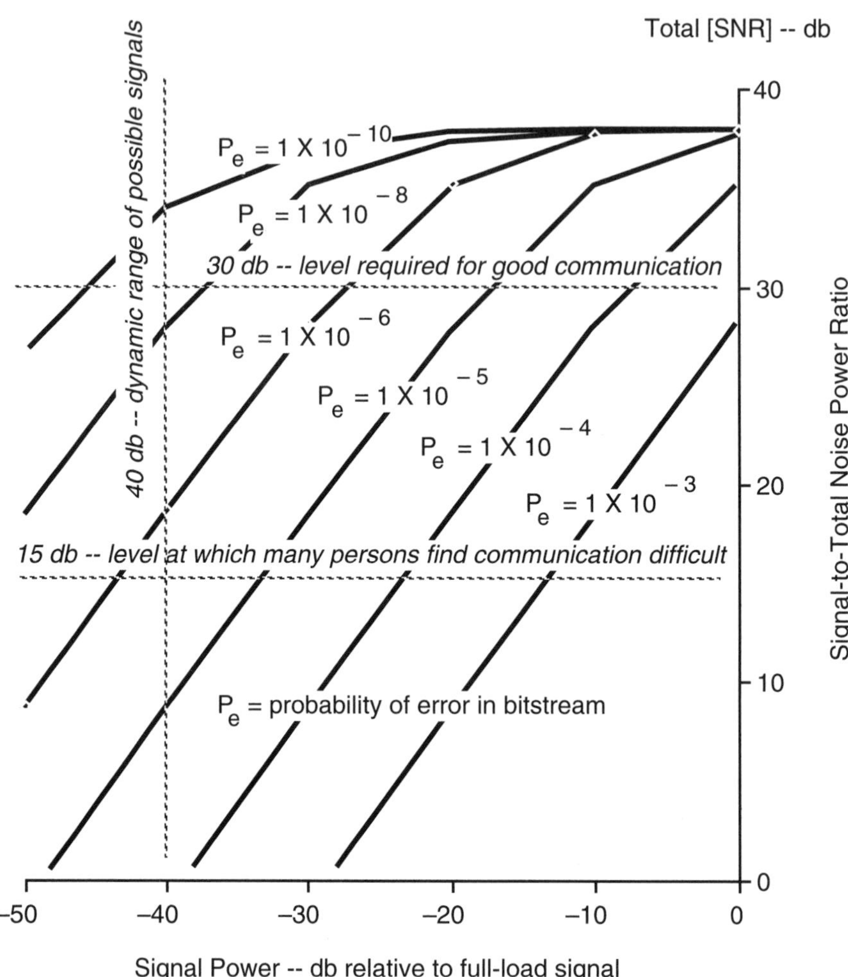

Figure 3.10 Ratio Of Signal-To-Total Noise Power as a Function of Voice Signal Power Relative to Full-Load for Range of Values of Probability of Bit-Error

to-total noise power ratio for a range of probabilities of bit errors. To achieve a channel with $[SNR]_T \geq 30$ db, and dynamic range ≥ 40 db, the probability of bit error must be greater than 1 bit in error in 100 million bits (1×10^{-8}). This is a value that is hard to achieve on wire lines, although it is readily achievable over optical fibers. If we reduce $[SNR]_T$ to 15 db—a value that many persons would describe as difficult to use—then a wire line with a probability of 1 bit in error in

100 thousand bits (1×10^{-5}) can accommodate a dynamic range of approximately 35 db. If we limit dynamic range to 20 db—a value that is practical in many situations—wire line error rates as low as 1×10^{-4} will produce a $[SNR]_T$ of approximately 20 db. Thus, operating conditions have a significant effect on the noise level that can be tolerated on a PCM channel.

3.2.6 Lower Bit Rate Schemes

Because it allows more users to make simultaneous use of the limited channels that are available, lower bit-rate coding is of increasing importance in long-distance transmission systems, mobile radio, and personal telephone systems. By increasing the complexity of signal processing, communication quality voice signals can be produced at rates significantly less than 64 kbits/s. The techniques employed are divided into *waveform coding* and *source coding*.

(1) Waveform Coding

In waveform coding, speech is reduced to a string of bits by operating on the signal in one of two ways:

- **Differential coding**: exploits the correlation between adjacent signal samples
- **Frequency domain coding**: divides the signal into a number of frequency bands, and codes each separately.

(2) Differential Coding

If the voice signal is sampled at the Nyquist rate, the sample values do not vary rapidly from one sample to the next so that sending the differences between consecutive samples will result in a lower bit rate than sending the signal values. At the receiver, the differences are summed (integrated) to reconstruct the voice signal. The process that is employed leads to several differential coding schemes:

- **Delta modulation**: a coding technique in which the sampling rate is many times the Nyquist rate for the input signal. (The ratio of the sampling frequency to the Nyquist rate is known as the oversampling index. Values of the index may range from 2 to 32.) As a result, the correlation between consecutive samples is very high, and the difference between them is very small. A one-bit quantizer is employed and the transmitted signal is a series of + 1s, 0s and – 1s that represent a positive, zero, or negative unit difference between adjacent samples. Integrating these signals produces a good approximation to the input signal.
- **Adaptive delta modulation**: as an alternative to increasing the sampling rate so as to match the step size to the region of maximum change, step size adaptation can be employed in regions that contain the faster signal changes.

- **Differential PCM**: a technique in which the difference quantizer has two or more levels so that larger differences between the codes representing successive samples can be used.
- **Adaptive DPCM**: an adaptation algorithm is employed to vary the quantizer levels in accordance with the rate of change of the input signal, producing adaptive DPCM.

These changes are used to improve both dynamic range and signal-to-noise ratio.

The most important of these techniques are differential PCM (DPCM), and adaptive DPCM (ADPCM). In recent years, ITU-T has developed ADPCM standards for 4 kHz voice at bit rates of 40, 32, 24, and 16 kbits/s. With minor modifications to suit the North American network, ANSI-T1 has adopted these standards, placing particular emphasis on 32 and 16 kbits/s. In addition, ITU-T has developed an ADPCM standard for 7 kHz voice at 64 kbits/s, i.e., a coding that carries the full range of voice frequencies in 64 kbits/s.

(3) Frequency Domain Coding

Because speech is not usually energetic across the entire frequency spectrum simultaneously, it is possible to dynamically allocate more bits (number of quantizing levels) to those bands in which the activity is significant at the expense of those bands that are quiescent. Two techniques of this sort are subband coding (SBC) and adaptive transform coding (ATC):

- **Subband coding**: the input signal frequency band is divided into several subbands (typically 4 or 8). Each subband is encoded using an adaptive step size PCM (APCM) technique. SBC produces communications quality voice at 9.6 kbits/s.
- **Adaptive transform coding**: the input signal is divided into short time blocks (frames) that are transformed into the frequency-domain using techniques based on Fourier series. The result is a set of transform coefficients that describe each frame. These coefficients are quantized into numbers of levels based on the expected spectral levels of the input signal. The number of bits for any quantizer, and the sum of all the bits per block are limited to a predetermined maximum. ATC produces communications quality voice at 7 kbits/s.

(4) Source Coding

In source coding, speech sounds are identified and encoded. By transmitting a description of the speech and emulating the mechanism of the vocal tract in the receiver, it is possible to produce artificial speech at greatly reduced bit rates:

- **Vocoding**: in spectrum channel vocoders, the speech signal is divided into several separate frequency bands (e.g., 16 bands of 200 Hz each), and the magnitude of the energy in each band is transmitted to the receiver together with information on the presence or absence of pitch. At the receiver, voiced or unvoiced energy is applied to a set of receivers that produce recognizable speech.

- **Linear prediction models**: speech analysis and synthesis are performed using models in which successive samples of speech are compared with previous samples to obtain a set of linear prediction coefficients (LPCs) that are used to reproduce the sound. In effect, the transmitter listens to the input speech, calculates LPC strings, and transmits them to the receiver. It regenerates the sound using an artificial vocal tract. Reducing voice to a string of codes representing these sounds can produce a signal of the order of a few hundred bits per second. To achieve this rate, however, significant analysis of the signal is required, and it results in considerable delay.

In recent years, ITU-T has developed standards for 16 kbits/s low-delay, linear prediction voice coding; ANSI has developed standards for 7.9 kbits/s, vector-sum excited, linear prediction, voice coding for digital mobile telephone; and the European digital mobile radio community has standardized a residual excited, linear prediction, voice coding technique at 7 kbits/s. In addition, the Federal Telecommunications Standards Program has developed linear prediction, voice coding standards at 2.4 and 4.8 kbits/s that serve the secure communications needs of the U.S. government.

(5) Processing Delay

In order to process the signal, it must remain within the encoding/decoding equipment for a finite time. The result is that the output train is displaced from the input train by a time delay. For 64 kbits/s PCM, this delay is the same as the sampling interval, i.e., 1/8000 s, or 125 μs. For ADM and ADPCM, the delay may be 250 μs, for SBC it may be 20 ms, for ATC over 100 ms, and for vocoders up to one-half a second (500 ms). Interactive, duplex communication becomes progressively more difficult as the magnitude of the delay extends into the millisecond region. For very long delays, the users must employ a formal procedure to notify the other party that they may speak.

3.2.7 Comparison of Techniques

In **Figure 3.11**, we summarize the performance of some of the techniques discussed. The figure is divided into horizontal fields that embrace transparent (i.e., indistinguishable from the original), network, communication, and synthetic quality. Curves are given for telephone voice (4 kHz voice), high-quality speech (7 kHz speech) for AM and FM radio commentary, and high-fidelity sound (20 kHz audio) for FM broadcasting. It is important to note that the telephone voice, ADPCM technique at 32 kbits/s is judged by many to be more robust than the original 64 kbits/s PCM technique. A similar level of confidence is developing in the 16 kbits/s technique.

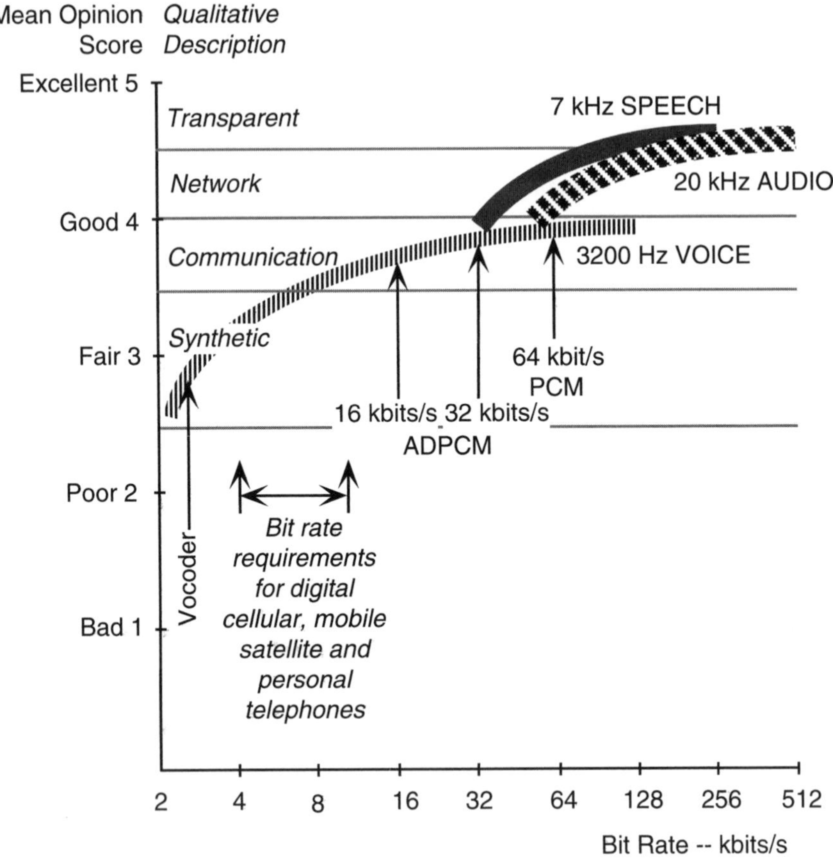

Figure 3.11 Comparison of Techniques Used to Encode Audio Signals

3.3 TELEVISION

3.3.1 Production of Video Signals

(1) Monochrome

While an electron beam scans it horizontally from top to bottom, scenes are projected on the optically active target of a camera tube. Each scene is converted to a signal, composed of segments (that correspond to the horizontal lines) whose intensity is modulated in accordance with the details of the image. When the beam reaches the bottom of the target it returns to the top to start again. In the time (known as the vertical interval) required for this action (known as vertical retrace) a small number of unmodulated lines are generated. They are used to transmit reference information to the receiver circuits, and to provide information to the viewer (closed captioning, for instance). In order to reduce the bandwidth required by the signal, yet provide a relatively flicker-free picture, the target scene is scanned in halves; all of the even lines are scanned and then all of the odd lines are scanned. The two frames (odd and even) make up a complete picture. In those countries with a power line frequency of 60 Hz, the picture usually consists of 525 lines (495 lines picture; 30 lines vertical retrace) and a complete picture is produced 30 times a second. In countries with a power line frequency of 50 Hz, the picture usually consists of 625 lines and a complete picture is produced 25 times a second.

(2) Color

Color information is produced by separating the natural colors of the scene into red, green, and blue components, imaging them on separate targets, and scanning them with synchronized electron beams to produce three color signals. The red, green, and blue signals are encoded to form a luminance signal and a chrominance signal that has in-phase and quadrature components (QAM). Because of the repetitious scanning used to generate the color signals, the energy is concentrated at the line frequency (15,734.264 Hz) and its harmonics, with levels generally decreasing as the frequencies rise. The chrominance signals are modulated on to a carrier signal at 3.579545 MHz. The choice of this frequency (usually referred to as 3.58 MHz) causes the spectral lines representing the amplitude spectrum of the chrominance signal to fall between the lines representing the luminance signal.

3.3.2 Television Signal

In standard television, the signal consists of a color video signal and a sound signal. Television sound takes the form of a 50 Hz to 15,000 Hz baseband that frequency modulates a carrier signal of 4.5 MHz to produce a frequency deviation of 25 kHz. The upper diagram in **Figure 3.12** shows the frequency domain representation of a baseband, television signal. To produce a modulated signal that will fit into the 6 MHz channels assigned to broadcast television service, a picture carrier whose frequency is 1.25 MHz above the frequency of the lower boundary of the assigned channel is modulated by the composite signal to produce a vestigial sideband, amplitude-modulated signal. This is shown in the lower diagram in Figure 3.12.

3.4 DIGITIZING VIDEO SIGNALS

Video pictures can be encoded in three ways:

- **Waveform encoding**: the video signal is reduced to a string of bits by sampling the original signal and encoding the values of the samples in some way (8-bit PCM, for example). Further processing may be used to reduce the bit rate (differential PCM, for example)

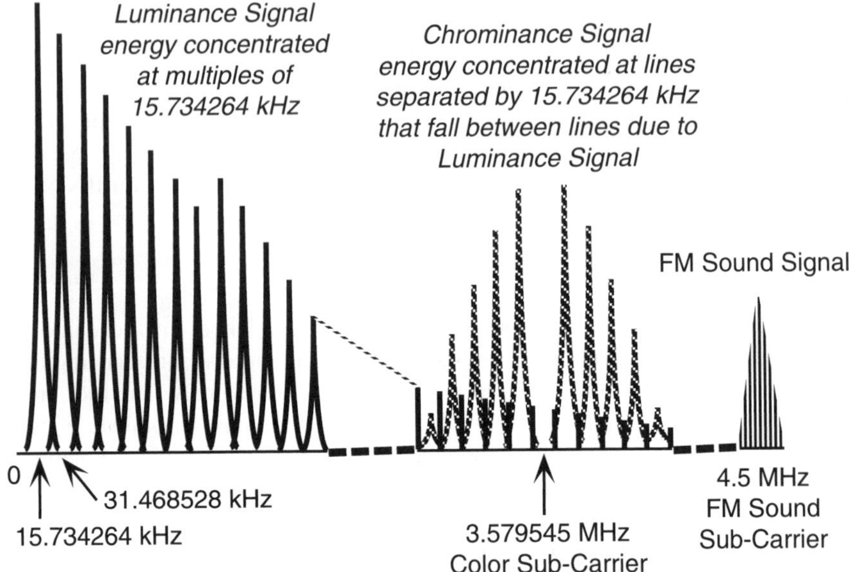

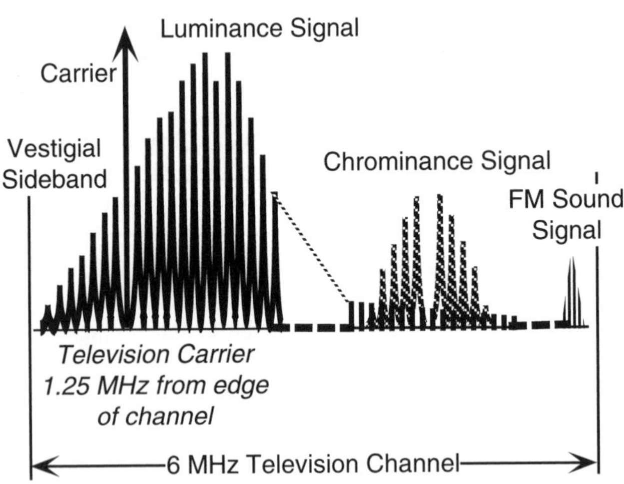

Figure 3.12 Some Properties of Television Signals

- **Source encoding**: operates on the picture material in some manner so as to take advantage of correlations within frames, or between frames.
- **Entropy encoding**: uses the statistical properties of the signal to implement variable–length coding and reduce the entropy of the coding source.

3.4.1 Waveform Encoding

A color television signal can be digitized using the same waveform sampling procedures as we have described for voice. Consider an NTSC system:

- Taking 6 MHz as the maximum frequency to be reconstructed (analogous to 4 kHz for voice), and 256 quantizing levels, the signal can be converted to a 96 Mbits/s, PCM bit stream
- Without jeopardizing picture quality, this rate may be reduced to 72 Mbits/s, PCM by recognizing that the video information is contained in 4.5 MHz and the sound can be transmitted on a separate channel
- With only a slight degradation of picture quality, the rate can be decreased further to approximately 57 Mbits/s, PCM. This is achieved by sampling at twice the color carrier frequency (3.58 MHz) so that only the lower sideband of the chrominance signal is preserved.

Further reductions in these bit rates can be achieved by using differential PCM coding.

3.4.2 Source Encoding

Source encoding makes a tradeoff between picture quality and motion-handling capability. Because a typical videoconference consists of a few persons seated at a table, and a videotelephone call consists of a talking head, such tradeoffs are possible without degrading performance to a level that is unacceptable to most users. Source encoding of video signals begins by separating the luminance and chrominance signals, sampling and encoding each component, and storing the digital information in a frame memory. Next, the digital image is divided into 8 × 8 or 16 × 16 pixel (i.e., picture element) blocks that are processed individually using a discrete-cosine transform (DCT) that is applied in the vertical and horizontal dimensions. The result is a series of values that represent the spatial frequencies contained in the image. When the entire frame has been processed in this way, a reconstructed image is stored in the frame memory. The succeeding frame is matched against this stored information and the difference between the blocks is sent to the receiver where it is integrated with the previous block to form an estimate of the current block. In order to achieve a specific bit rate, odd or even fields may be discarded, and full frames may be eliminated. Source encoding is used to produce video conference connections of varying fidelity at 1.536 Mbits/s, 768 kbits/s, and 384 kbits/s.

3.4.3 Entropy Encoding

To reduce the number of bits of coded information, a variable-length code can be employed. Known as entropy coding, the technique is based on two concepts:

- **Self-information**: the self-information of a symbol is inversely proportional to the probability of the symbol being transmitted
- **Entropy**: the entropy of a source is the average self-information of the symbols it emits.

To create an efficient code, we need to encode the symbols in a fashion that takes account of their self-information. Encoding all symbols with binary words of equal length (as in ASCII and EBCDIC, for instance) is only efficient if all symbols have equal probability of being transmitted. Such would be the case with a source that emits randomly. However, in pictures, and with other practical source materials, there is a large component of correlated information. To achieve efficient coding, it is necessary to use a variable-length code whose members are assigned to the symbols in some way in proportion to their use. Symbols transmitted frequently (i.e., self-information $\Rightarrow 0$), should be represented by shorter code words, and symbols transmitted infrequently (i.e., self-information $\Rightarrow 1$), should be represented by longer code words. In addition, in order to guarantee unique decodability, the code should be constructed so that no member is the prefix of any other member. If this condition cannot be achieved, a unique identifier must be used to separate the members—diluting the advantage gained. Huffman codes (see § 4.3.2) meet this condition. For pictures, compressions of up to 4:1 over equal-length coding can be achieved with one-dimensional codes, and factors greater than this are achieved with two-dimensional coding.

3.4.4 Comparison of Techniques

For video signals that are likely to be used in video telephony and video conferencing, **Figure 3.13** compares the bit rates achieved through the use of waveform coding and the combination of source and entropy coding. They are divided into those that produce transparent quality, studio quality, communications quality, and synthetic quality, pictures.

3.4.5 Other Methods of Video Coding

(1) Digital Modulation

Converting a standard color television signal to an 8 Mbits/s digital stream that takes advantage of coding and compression efficiencies, and using QAM (see § 2.3.2(1)) to create a modulated wave, produces signals whose bandwidths are approximately 2.5 MHz (for 16–QAM), 1.9 MHz (for 32–QAM), and 1.6 MHz (for 64–QAM). Thus, using digital modulation, three high-quality TV signals can be accommodated in the 6 MHz channel allocated to standard television. Furthermore, using 32–QAM digital modulation, the same three TV signals can be delivered to a residence over a drop wire (twisted pair) up to 800 feet in length.

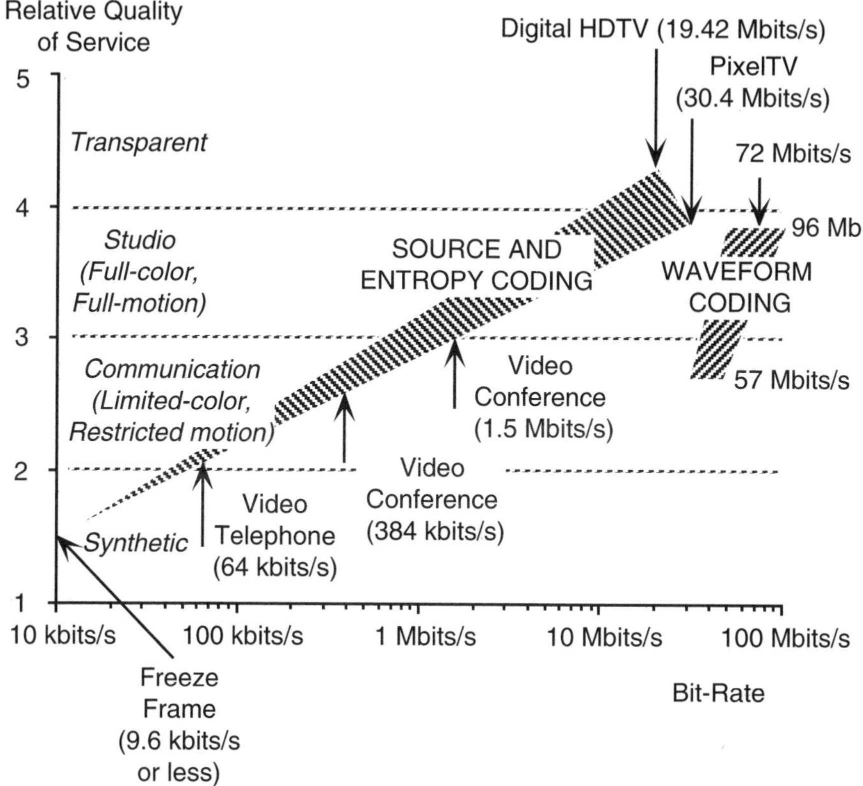

Figure 3.13 Comparison of Techniques Used to Encode Video Signals

(2) Compression Standards

The subject of video compression is engaging the attention of several professional groups:

- **SMPTE** (Society of Motion Picture and Television Engineers):
  - *Task Force on Headers/Descriptors.* Concerned with the development of a universal header that can be attached to any video stream. The purpose is to provide information required by the receiver to decompress the stream and reconstruct the image(s).
  - *Task Force on Digital Image Architecture.* Concerned with the development of digital standards to facilitate interoperation of high resolution display systems.
- **ISO JPEG** (Joint Photographic Experts Group): concerned with developing coding algorithms for still pictures.

- **ISO MPEG** (Moving Picture Experts Group): concerned with compression of full-motion video to about 1.5 Mbits/s and audio to 64, 128, and 192 kbits/s. A 1.5 Mbits/s standard (MPEG–I) has been adopted in which the luminance signal is derived from an input picture of 360 × 240 lines; the chrominance signal is derived from an input picture of 180 × 120 lines. The frame rate is 29.97 frames/s for NTSC compatible systems.
- **ITU-T Working Group H.261** (also called p×64): concerned with compression for videotelephone (p = 1 or 2) and videoconference ($6 \geq p$) services over ISDN (see § 10.5) circuits. H261 describes a real-time system operating at 1.5 Mbits/s with a delay of less than 150 milliseconds. For videoconferences, the luminance signal is derived from an image consisting of 360 × 288 lines and the chrominance signal is derived from an image containing 180 × 144 lines. Frames are non-interlaced and the rate is 29.97 frames/s for NTSC compatible systems. For videotelephone, the luminance signal is derived from an image consisting of 180 × 144 lines and the chrominance signal is derived from an image consisting of 90 × 72 lines.

4

Data

The final Chapter on Signals is concerned with signals that represent data. We describe some well-known codes, methods of error control, data compression, data encryption, the use of analog symbols to represent streams of low-speed data, and digital modulation methods.

4.1 BINARY CODES

4.1.1 Binary Digits

Between machines, information is exchanged by binary digits (bits). Since each bit is two-valued—either 0 or 1—*per se*, it can describe two situations (states). To describe more states, it is necessary to use the bits in sequences—to create unique codes—so that a set of sequences can represent a range of information. Further, in order for machines to communicate, they must employ sequences whose meanings are known to both of them. Two sets are in common use today:

- **ASCII**: the *American Standard Code for Information Interchange* employs a sequence of seven bits. Since each bit may be 0 or 1, ASCII contains 2^7, or 128 unique patterns. Some are assigned to represent the alphanumeric characters and symbols found in messages, and others represent the special characters required to control communication.
- **EBCDIC**: the *Extended Binary Coded Decimal Interchange Code* employs a sequence of eight bits. Since each bit may be 0 or 1, EBCDIC contains 2^8, or

256 unique patterns. Some are assigned to represent the alphanumeric characters and symbols found in messages, others represent the special characters required to control communication, and the remainder may be used for purposes defined by the users.

Table 4.1 lists representative members of ASCII and EBCDIC. Examples are given of the decimal digits, upper and lower case letters, and some of the special characters used to control communication. The bit patterns are arranged so that Bit #1 (the least significant bit), is on the right-hand side of the member. It is transmitted first—a procedure that is opposite to digital voice octets where the most significant bit is transmitted first. So that there shall be no ambiguity in the meaning of ASCII and EBCDIC bit patterns, we use an arrow to show the direction of transmission, and thus identify the position of the least-significant bit.

4.1.2 Asynchronous and Synchronous Operation

In some terminals with keyboard input, characters are generated and transmitted singly, one after the other. Given that no two persons type at the same speed, and that individual keystrokes may be separated by different time delays, the stream will be irregularly paced with spaces between each character. In other terminals, the characters are collected until a complete line of text is created, or the return key is pressed, causing the line to be sent as a burst of continuous characters. Whether sent one-by-one as they are generated, or sent line-by-line as each line is completed, each character is framed by a start bit (0) and one or two stop bits (1 or 11). The start bit provides a reference from which the receiver measures time in order to sample the succeeding bits approximately in their center. The stop bit(s) provide(s) a positive indication of the end of the character. Called *character-framed data*, timing begins anew with each character received. Thus, the receiver (see § 6.1.1(1)) does not have to maintain precise timing over more than ten bits (7 or 8-bit character + stop bit, or 7 or 8-bit character + two stop bits). Because the data are sent as soon as they are ready, this style of operation is known as *asynchronous*. **Figure 4.1** shows the datastream formats employed in asynchronous operation.

In some terminals, particularly those capable of sophisticated processing, characters are run together without start and stop bits to form a block of data. Then, the data block is framed by a start sequence and a stop sequence. The start sequence is called the *header*—it contains synchronizing, address, and control information. The stop sequence is called the *trailer*—it contains error checking and terminating information. The entire data entity is called a *frame*, and the arrangement is described as *message-framed data*, i.e., the message data are framed by data devoted to the transmission process. In response to a command from *external* equipment, the frame is transmitted as a continuous stream. At the receiver, the header is used to signal the beginning of a frame and to synchronize the receiving equipment to

TABLE 4.1 Representative Members of ASCII and EBCDIC

Alphas	ASCII	EBCDIC	Alphas	ASCII	EBCDIC
a	1100001⇒	10000001⇒	A	1000001 ⇒	11000001⇒
b	1100010	10000010	B	1000010	11000010
c	1100011	10000011	C	1000011	11000011
d	1100100	10000100	D	1000100	11000100
e	1100101	10000101	E	1000101	11000101
f	1100110	10000110	F	1000110	11000110
g	1100111	10000111	G	1000111	11000111
h	1101000	10001000	H	1001000	11001000
i	1101001	10001001	I	1001001	11001001
j	1101010	10001010	J	1001010	11001010
k	1101011	10001011	K	1001011	11001011
l	1101100	10001100	L	1001100	11001100
m	1101101	10001101	M	1001101	11001101
n	1101110	10001110	N	1001110	11001110
o	1101111	10001111	O	1001111	11001111
p	1110000	10010000	P	1010000	11010000
q	1110001	10010001	Q	1010001	11010001
r	1110010	10010010	R	1010010	11010010
s	1110011	10010011	S	1010011	11010011
t	1110100	10010100	T	1010100	11010100
u	1110101	10010101	U	1010101	11010101
v	1110110	10010110	V	1010110	11010110
w	1110111	10010111	W	1010111	11010111
x	1111000	10011000	X	1011000	11011000
y	1111001	10011001	Y	1011001	11011001
z	1111010	10011010	Z	1011010	11011010

Control	ASCII	EBCDIC	Numerics	ASCII	EBCDIC
SYN	0010110⇒	00110010⇒	0	0110000 ⇒	11110000⇒
SOH	0000001	00000001	1	0110001	11110001
STX	0000010	00000010	2	0110010	11110010
ETX	0000011	00000011	3	0110011	11110011
EOT	0000100	00110111	4	0110100	11110100
ENQ	0000101	00101101	5	0110101	11110101
ACK	0000110	00101110	6	0110110	11110110
NAK	0010101	00111101	7	0110111	11110111
DLE	0010000	00010000	8	0111000	11111000
ETB	0010111	00100110	9	0111001	11111001

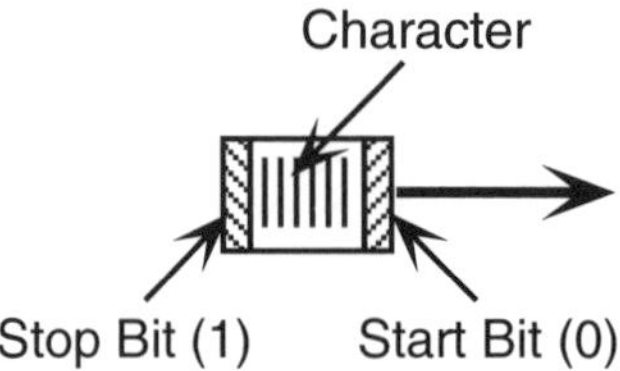

In asynchronous transmission, each character is framed by one start bit and one or two stop bits.

Framed characters sent as they are created -- a datastream typical of keyboard input to a terminal or communications controller.

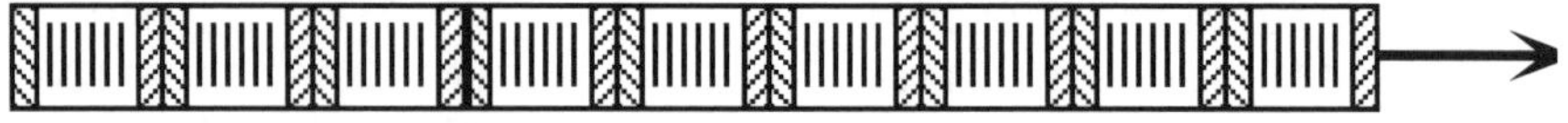

Framed characters that are concatenated and sent when a string is completed -- a datastream typical of a terminal sending keyboard input line-by-line to a communications controller.

FIGURE 4.1 Asynchronous Transmission Format

cause all the bits in the frame to be sampled at exactly the same instant in the bit interval. For one frame in common use, precise timing must be maintained over 1088 bits (5 characters in the header + 128 characters in the text + 3 characters in the trailer = 136 characters × 8-bits = 1088 bits). Because transmission is paced by an external source, this style of operation is known as *synchronous*. **Figure 4.2** shows the datastream format used in synchronous operation.

Characters are assembled into a datablock that is framed by a header and a trailer to produce a frame. The header and trailer contain information needed to send the frame. The frame is sent when a command is received from the controlling unit in the communication system.

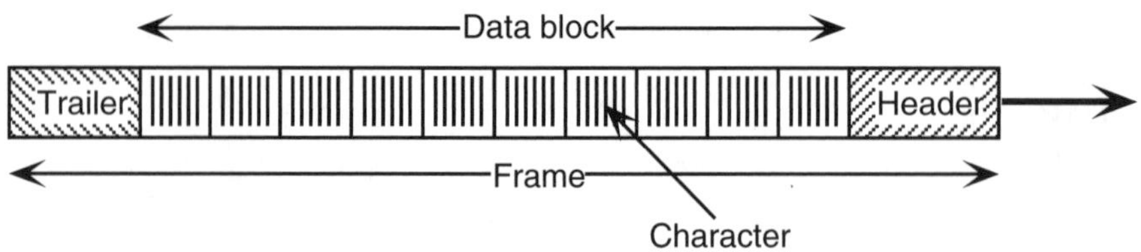

FIGURE 4.2 Synchronous Transmission Format

There is a significant difference in the coding efficiency of these two transmission methods:

- **Asynchronous operation**: assuming an 8-bit character sequence, one start bit and one stop bit, the ratio of information bits to total bits is 8 ÷ 10 = 0.8, i.e., the coding efficiency is 80%. With two stop bits it is 73% efficient.
- **Synchronous operation**: assuming a 5 character header, a 128 character text (information characters), and a 3 character trailer, the ratio of information bits to total bits is 1024 ÷ 1088 = 0.94, i.e., the coding efficiency is 94%.

Thus, when working asynchronously under the conditions stated, the data stream is made up of 73 or 80% information bits and 27 or 20% overhead bits. When working synchronously under the conditions stated, the data stream is made up of 94% information bits and 6% overhead bits.

4.2 ERROR CONTROL

4.2.1 Importance of Accurate Reception

Because each member of a code is assigned a unique identity, it is extremely important that it be sent and received without error. For instance:

- The EBCDIC code for *c* is 10000011 ⇒
- An error in bit #1 produces $1000001\underline{0}$ ⇒ which is the code for *b*
- An error in bit #2 produces $100000\underline{0}1$ ⇒ which is the code for *a*
- An error in bit #8 produces $\underline{0}0000011$ ⇒ which is the code for *ETX*, the character reserved to signify that the text is completed (*ETX* = end of text).

If communication is being pursued under the procedure known as binary synchronous protocol (BSC or Bisync, discussed in § 6.1.2), the occurrence of the last error will cause the receiver to regard all succeeding characters as part of the trailer. Clearly, it is important to know whether the message has been corrupted in transmission.

4.2.2 Error Detection

Error detection is a cooperative activity between the sender and the receiver in which the sender adds information to the character or frame to assist the receiver in determining whether an error has occurred in transmission or reception. The essentials of these activities are shown in **Figure 4.3**. Before sending, the sender and receiver must agree on the information that will be added, and how it will be used.

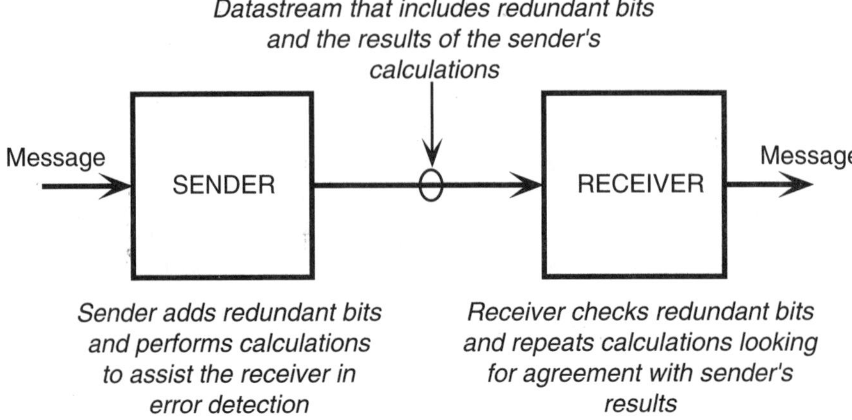

FIGURE 4.3 Error Detection Process

(1) Parity Checking

One method adds parity bits. Parity checking is a process that adds a bit (0 or 1) to a sequence of bits to make the total number of 1s in the sequence either even or odd:

- **Even parity check**: the parity bit makes (or leaves) the number of 1s even
- **Odd parity check**: the parity bit makes (or leaves) the number of 1s odd.

The addition of a parity bit to a 7-bit code produces an 8-bit code. Sometimes the action is described as 7B/8B, i.e., a 7-bit code is replaced by an 8-bit code.

Parity bits may be added to individual characters, in which case the process is known as vertical redundancy checking (VRC), or to bit sequences formed from bits assigned to specific positions in a block of characters, in which case the process is known as longitudinal redundancy checking (LRC). The use of the directions, vertical and longitudinal, presupposes a model. It has its origin in the days when punched paper tape was popular. To check the code, the tape was laid across the desk (longitudinally). This put the characters vertical with the parity bits along the bottom, and placed block checks at the right-hand side of the blocks they were intended to check.

(2) VRC Checking

VRC adds a parity bit to each character. Most often, it is used with ASCII; the parity bit occupies bit position number eight. Thus:

- ASCII for *a* is ⇐ 1000011
- ASCII for *a* with even parity is ⇐ $1000011 1_p$; the parity bit (1_p) is a 1 so as to make a total of four 1s (an even number of 1s)

- ASCII for *a* with odd parity is $\Leftarrow 10000110_p$; the parity bit (0_p) is a 0 so as to leave a total of three 1s (an odd number of 1s).

When operating asynchronously, the start (0_s) and stop (1_s) bits frame the entire eight bits:

- ASCII for *a* sent asynchronously is $\Leftarrow 0_s10000111_s$
- ASCII for *a* with even parity sent asynchronously is $\Leftarrow 0_s10000111_p1_s$
- ASCII for *a* with odd parity sent asynchronously is $\Leftarrow 0_s10000110_p1_s$.

The state of the parity bit is determined only by the number of 1s in the sequence of 7 bits that represents *a*, and the decision whether to use odd or even parity. The start and stop bits have no role in the VRC process; they simply frame the data for the convenience of the receiver.

(3) Coding Efficiency

Adding a parity bit to ASCII characters sent asynchronously reduces the transmitted coding efficiency to 70%. Thus, to send 7 information bits requires 1 start bit + 7 character bits + 1 parity bit + 1 stop bit, i.e., a total of 10 bits. If two stop bits are employed, the efficiency drops to 64%. Employing parity bits with ASCII in a standard, 136 octet frame (5 header, 128 text, and 3 trailer) sent synchronously produces an efficiency of 82%.

(4) Limitations of VRC Checking

The use of VRC will not detect all errors. Consider what happens if one, two, three, and four bit errors occur in the ASCII character *a* (with even parity):

- If an error occurs in the #2 bit, the character *a* (with even parity) becomes $\Leftarrow 1\underline{1}000111_p$; there is an odd number of ones. But even parity requires an even number of 1s. Something is wrong! The use of VRC detects that the character is in error.
- If an error occurs in the #2 and #3 bits, the character *a* (with even parity) becomes $\Leftarrow 1\underline{11}00111_p$; there is an even number of ones—and that is expected with even parity. However, the character decodes to *g* not *a*. The use of VRC has not detected that the character is in error.
- If an error occurs in the #2, #3 and #4 bits, the character *a* (with even parity) becomes $\Leftarrow 1\underline{111}0111_p$; there is an odd number of ones. Something is wrong. The use of VRC detects that the character is in error.
- If an error occurs in the #2, #3, #4 and #5 bits, the character *a* (with even parity) becomes $\Leftarrow 1\underline{1111}111_p$; there is an even number of ones. The character decodes to *DEL*, a control character for delete, not *a*. The use of VRC has not detected that the character is in error.

The use of VRC only detects the presence of an odd number of errors.

(5) Longitudinal Redundancy Checking

LRC adds parity bits to bit sequences formed from bits assigned to specific bit positions in a block of characters. The process produces an additional character, the block check character (BCC), that is transmitted in the trailer of the frame. As with VRC, the use of LRC will only detect the presence of odd numbers of errors.

(6) VRC and LRC Combined

Table 4.2 shows vertical and longitudinal redundancy checks on a block of five ASCII characters. The parity bits added to the characters, provide even parity checking of each character. The parity bits added to the rows, provide even parity checking of each row of bits; they form the block check character. At the receiver, independent checks are made to confirm the veracity of the VRC bits and the BCC. If agreement is found, the block of data is assumed to have been received without transmission error(s). In certain circumstances, the combination of VRC and LRC can locate the exact position of an error. The bottom half of Table 4.2 shows the determination of the location of a specific bit error. However, the technique fails to uncover many multiple errors. For instance, the second example at the bottom of Table 4.2 shows a combination of four errors that will not be detected.

(7) Checksum

Another technique that the transmitter can use to assist the receiver in detecting the presence of errors employs addition. Called a checksum, the process adds the characters together and expresses their sum as a 16-bit binary number to produce a 2-octet checksum. **Table 4.3** shows the principle of a checksum calculation on the block of five ASCII characters. However, because few persons can work directly with binary numbers, the five octets in Table 4.3 are reduced to their decimal equivalents, added together, and then converted back to a 16-bit binary number.

EXAMPLE 4.1 Decimal to Binary and Binary to Decimal Transformations

Decimal Number. The decimal number 345678 is shorthand for

$$3\times10^5+4\times10^4+5\times10^3+6\times10^2+7\times10^1+8\times10^0.$$

Binary Number. The binary number 101100 is shorthand for

$$1\times2^5+0\times2^4+1\times2^3+1\times2^2+0\times2^1+0\times2^0.$$

Binary Sequence to Decimal Number. Suppose the binary sequence we wish to express as a decimal number is

$$\Leftarrow 1\;0\;1\;1\;0\;0\;1\;0$$

The arrow shows the direction of transmission (from right-to-left).

TABLE 4.2 Vertical and Longitudinal Redundancy Checking

Bit#	A	b	C	d	E	BCC
1	1	0	1	0	1	1
2	0	1	1	0	0	0
3	0	0	0	1	1	0
4	0	0	0	0	0	0
5	0	0	0	0	0	0
6	0	1	0	1	0	0
7	1	1	1	1	1	1
P+	0	1	1	1	1	0

Longitudinal Redundancy Checking adds parity bit to each row to produce block check character (BCC)

Vertical Redundancy Checking adds Parity Bit to 7 bit ASCII Code

ASCII characters A b C d E with even parity and block check character

⇐ 10000010 |01000111 |11000011 | 00100111 |10100011 |10000010

A b C d E BCC

	A	b	C	d	E	BCC	
	1	0	1	0	1	1	
	0	1	<u>0</u>	0	0	0	← Error
	0	0	0	1	1	0	
	0	0	0	0	0	0	
	0	0	0	0	0	0	
	0	1	0	1	0	0	
	1	1	1	1	1	1	
P+	0	1	1	1	1	0	
			↑ Error				

Use of VRC and LRC to detect and locate single bit error

	A	b	C	d	E	BCC
	1	0	1	0	1	1
	0	1	<u>0</u>	0	<u>1</u>	0
	0	0	0	1	1	0
	0	0	0	0	0	0
	0	0	<u>1</u>	0	<u>1</u>	0
	0	1	0	1	0	0
	1	1	1	1	1	1
P+	0	1	1	1	1	0

Set of four errors that are not detected by combination of VRC and LRC

If the sequence is a *byte*, the least significant bit is transmitted first. In this scenario, the sequence must be rewritten as a binary number by reversing the sequence so as to place the least significant bit at the right-hand end, i.e.,

0 1 0 0 1 1 0 1

TABLE 4.3 Principle of Checksum Calculation

Bit#	A	b	C	d	E
1	1	0	1	0	1
2	0	1	1	0	0
3	0	0	0	1	1
4	0	0	0	0	0
5	0	0	0	0	0
6	0	1	0	1	0
7	1	1	1	1	1
P+	0	1	1	1	1
Decimal	65	226	195	228	197 = 911

Checksum is formed by summing decimal values of characters (with parity bit) and expressing result as 16 bit binary equivalent

Checksum = 0 0 0 0 0 0 1 1 1 0 0 0 1 1 1 1 ⇒

ASCII characters A b C d E with even parity and checksum

⇐ 10000010 | 01000111 | 11000011 | 00100111 |
A b C d

10100011 | 11110001 | 11000000
E Checksum (2 bytes)

Expressing each bit as the appropriate power of 2:

$$0\,1\,0\,0\,1\,1\,0\,1 = 0+1\times2^6+0+0+1\times2^3+1\times2^2+0+1\times2^0 = 64+8+4+1 = 77.$$

If the sequence is an *octet*, the most-significant bit is transmitted first so that the equivalent binary number will be the same as the sequence, i.e.,

$$1\,0\,1\,1\,0\,0\,1\,0.$$

Expressing it in powers of 2:

$$1\,0\,1\,1\,0\,0\,1\,0 = 1\times2^7+0+1\times2^5+1\times2^4+0+0+1\times2^1+0 = 128+32+16+2 = 178.$$

Decimal Number to Binary Sequence. The reverse procedure breaks up a decimal number into a binary sequence. Suppose the number we wish to express as a binary sequence is 163.

We find the largest value of n so that $2^n \leq 163$. It is 7, i.e., $2^7 = 128$. Thus, in the eighth bit position there will be a 1. Subtracting 128 from 163 leaves 35.

We find the largest value of n so that $2^n \leq 35$. It is 5, i.e., $2^5 = 32$. Thus, in the sixth bit position there will be a 1. Subtracting 32 from 35 leaves 3.

We find the largest value of n so that $2^n \leq 3$. It is 1, i.e., $2^1 = 2$. Thus, in the second bit position there will be a 1. Subtracting 2 from 3 leaves 1.

We find the largest value of n so that $2^n \le 1$. It is 0, i.e., $2^0 = 1$. Thus, in the first bit position there will be a 1.

Putting the bits in sequence and entering 0s in those bit positions not mentioned above, the binary number equivalent to 163 is

$$1\,0\,1\,0\,0\,0\,1\,1.$$

(8) Cyclic Redundancy Checking

Another mathematical technique treats the datastream as one continuous binary number. Called cyclic redundancy check (CRC), the process is illustrated in **Figure 4.4**. Given a k-bit message M_k, the sender divides it by a prime number G_{n+1} to produce an n-bit remainder, F_n, known as the *frame check sequence* (FCS). G_{n+1} is called the *generator function*; it is n + 1 bits long, and is known to both the sender and receiver. The value of the FCS calculated by the sender is placed in the trailer of the frame. In the incoming frame, the receiver divides the k-bit message by G_{n+1} to produce the frame check sequence F_n' If $F_n' = F_n$, it is almost certain that M_k has

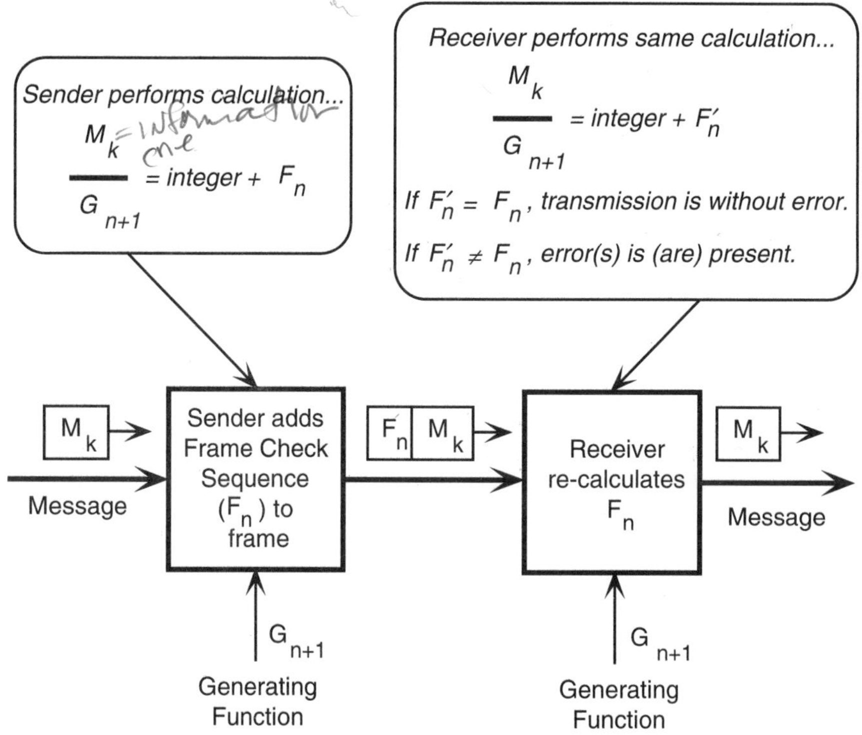

FIGURE 4.4 Cyclic Redundancy Check

been received without error. For common use, two generating functions have been standardized:

- CRC-16: 11000000000000101
- CCITT V.41: 10001000000100001.

Both functions are 17-bit binary prime numbers. They produce 16-bit FCSs that are unique for all practical datastreams. They detect all single-bit errors, all double-bit errors, any odd number of errors, any burst errors for which the number of bits in error is less than 16, and most of the errors caused by larger bursts. Cyclic redundancy checking is by far the strongest of the common error checking procedures. Some estimates place the likelihood of an undetected error at 1 in 10^{14} bits. For a system transmitting continuously at 9600 bits/s, this amounts to approximately one undetected error every 300 years.

4.2.3 Error Correction

Once detected, an error must be corrected. Two basic approaches to error correction are available:

- **Automatic-repeat-request** (ARQ): ARQ procedures require the transmitter to re-send the portions of the exchange in which errors have been detected
- **Forward error correction** (FEC): FEC techniques employ special codes that allow the receiver to detect and correct a limited number of errors without referring to the transmitter.

(1) ARQ Techniques

Generally, ARQ procedures include the following actions by the receiver or the sender:

- **Receiver**:
  - discards those frames in which errors are detected, or discards all frames in the incoming stream beginning with the first one in which an error was detected
  - for frames in which no error was detected, the receiver returns a positive acknowledgment to the sender
  - for the first frame in which errors have been detected, or all frames in which errors have been detected, the receiver returns negative acknowledgment(s) to the sender.
- **Sender**:
  - retransmits the individual frames in which the receiver has identified errors, or re-sends all frames from the first frame in which errors were detected

– after a pre-established time, the sender retransmits a frame that has not been acknowledged.

Three common ARQ techniques include:

- **Stop-and-wait**: the sender sends a frame and waits for acknowledgment from the receiver. The technique is slow and is suited to half-duplex connections.
- **Go-back-n**: the sender sends frames in a sequence, and receives acknowledgments from the receiver. On detecting an error, the receiver discards the corrupted frame, and ignores any further frames. The receiver notifies the sender of the number of the frame it expects to receive (i.e., the number of the first frame in error). On receipt of this information, the sender begins re-sending the data sequence starting from that frame. For reasonable levels of errors, this technique is faster than stop-and-wait. When used on half-duplex connections the sender must stop sending after an agreed upon number of frames, turn the channel around, and wait for the receiver to respond with the status of the received frames. On duplex connections, the sender must not send any more frames than there are numbers to identify them (1 through 7 and 1 through 127 are common ranges), before receiving an acknowledgment from the receiver.
- **Selective-repeat**: used on duplex connections, the sender only repeats those frames for which negative acknowledgments are received from the receiver, or no acknowledgment is received. However, the appearance of a repeated frame out of sequence may provide the receiver with additional complications. Also, the sender must not send any more frames than there are numbers to identify them, before receiving an acknowledgment from the receiver. In cases where the rate of errors is low, the extra operational complexity it introduces, makes this algorithm only marginally better than go-back-n.

The principles involved in these procedures are diagrammed in **Figures 4.5** and **4.6**. In Figure 4.5, we show examples of two exchanges over half-duplex connections that employ stop-and-wait and go-back-n ARQ. In Figure 4.6, we show examples of two exchanges over duplex connections that employ go-back-n and selective repeat ARQ. With all of them, a fundamental assumption is that repeating the frame is likely to result in the reception of an error-free frame. If it does not, the re-send procedure is repeated. Obviously, there comes a point at which intervention is required to prevent the sender from continuing to re-send, and the receiver from continuing to reject, the frame.

(2) Forward Error Correction

Provided the number of errors is less than a value determined by the coding, it is possible for the receiver to detect and correct errors without reference to the sender. This convenience is bought at the expense of adding more bits. For instance, if we only have two messages to send, we can represent one (A) by the octet 10101010,

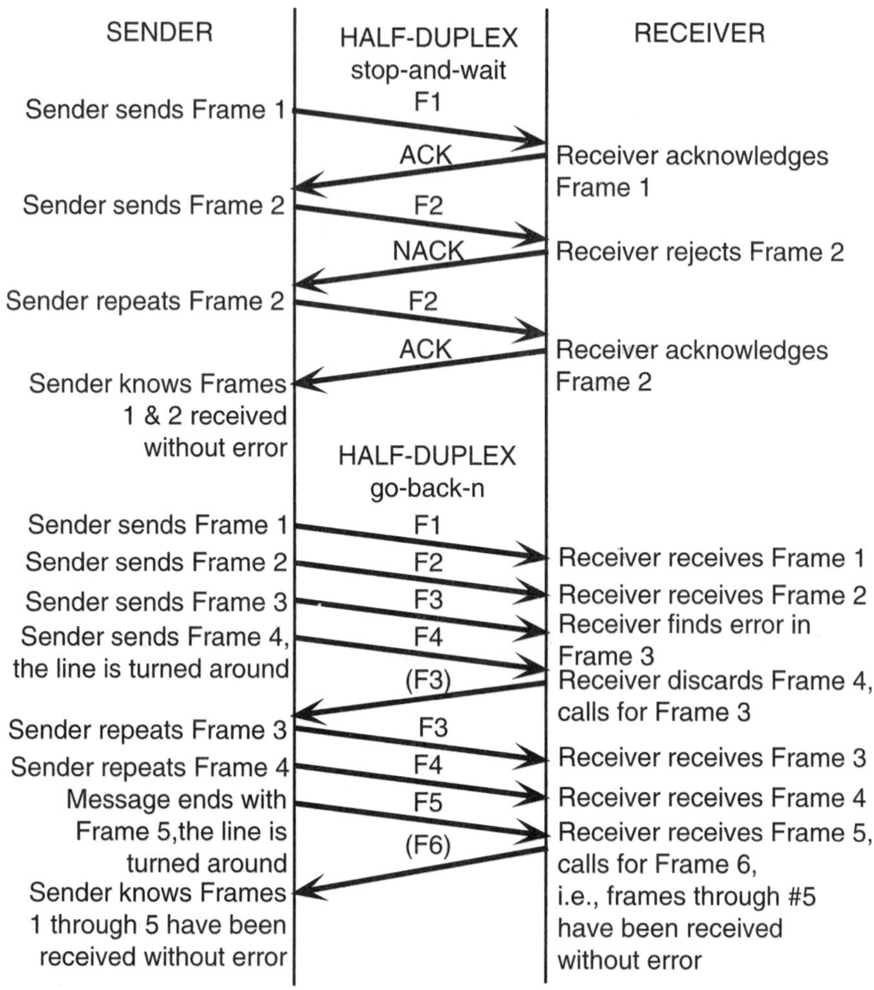

FIGURE 4.5 Half-Duplex, ARQ Procedures

and the other (B) by the octet 01010101. If the receiver knows that the message is A or B and no other, and it is provided with the ability to determine the logical distance between each incoming message and the two known messages, this strategy will allow the receiver to correct for up to three bits in error. (The logical distance between two binary sequences is determined by finding the number of bits that must be changed to make one sequence into the other.) The proof is as follows:

- Suppose A is in error by 1 bit, so that A′ = 00101010, the logical distance between the received pattern and A is 1 and the logical distance between the received pattern and B is 7; thus, A′ is likely to be A.

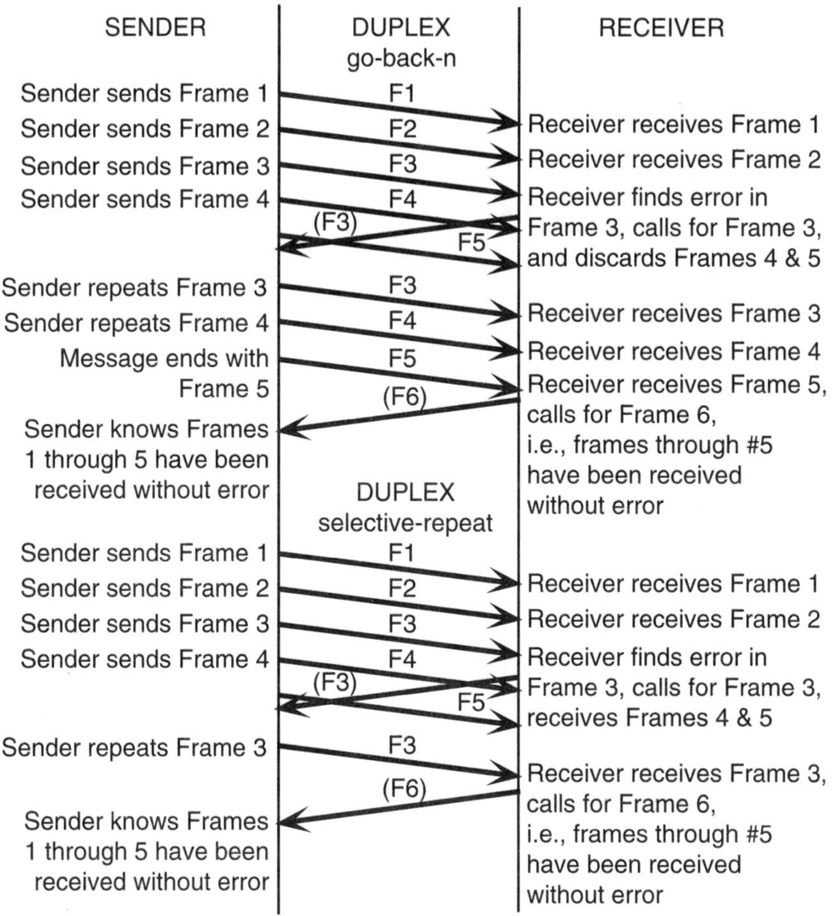

FIGURE 4.6 Duplex, ARQ Procedures

- Suppose A is in error by 2 bits, so that A′ = 01101010, the logical distance from A is 2 and the logical distance from B is 6; thus A′ is likely to be A.
- Suppose A is in error by 3 bits, so that A′ = 01001010, the logical distance from A is 3 and the logical distance from B is 5; thus A′ is likely to be A.
- Suppose A is in error by 4 bits, so that A′ = 01011010, the logical distance from A is 4 and the logical distance from B is 4; thus A′ is equally likely to be A or B.

Continuing the sequence to higher levels of error makes A′ more likely to be B than A. For this particular case, then, the limit of correction is 3-bits in error. To do this we have used a code with 8-bits to represent one of two states (A or B)—a situation handled by one bit in an error-free environment.

Codes used to provide forward error correction (FEC) are more sophisticated than our example. They can be divided into two types:

- **Linear block codes**: in linear block coding, a block of k information bits is presented to the encoder; the encoder responds with a unique code block (codeword) of m bits ($m > k$). The set of codewords is selected so that the logical distance between each codeword and its neighbors is approximately the same. If (d_{min} - 1), represents the minimum distance between codewords, the code detects any combination of 1/2 d_{min} - 1, or fewer, errors, and can correct any combination of three, or fewer, errors. The example given above is a block code that corrects any combination of three or fewer, errors. Furthermore, parity checking is a special case of a block code; it is a single error detecting code—as such, it cannot correct errors.
- **Convolutional codes**: convolutional codes are generated by continuously performing logic operations on a moving, limited sequence of bits (n bits, say) contained in the message stream. For each bit in the message stream, the encoder produces a fixed number (two or more) of bits; their values depend on the value of the present message bit and the values of the preceding bits (i.e., the previous n - 1 bits). The output of the encoder depends upon the length of the sequence, the number of logic operations performed (this sets the number of output bits produced by each input bit), and on the way in which the logic operations are performed. Optimal encoding and decoding strategies are determined by exhaustive computer searching.

Which coding scheme to use is a complex decision that depends on the characteristics of the channel to which they are applied. As with other techniques discussed in this Section, forward error correction depends upon cooperation between sender and receiver, and on the receiver having exact knowledge of the processing performed by the sender. Also, if the number of errors is larger than the correction span, the receiver must resort to an ARQ technique to achieve retransmission of the corrupted message.

(3) To Code or Not to Code?

Developed during the early years of data communication when the performance of terminal and transmission equipment was marginal, and a great deal of time was consumed in repeating frames to correct errors, forward error correcting codes are no longer a feature automatically considered for today's transmission systems. The expense of the coding and decoding equipment, and the inefficiency created by an increased number of bits, cannot be justified when transmission systems are implemented over optical fibers. It is cheaper and more efficient to rely on the receiving equipment to detect errors in the frame (or total message) and to request an occasional retransmission. In § 6.4, we show that ARQ techniques achieve very close to maximum throughput in error rate environments of 1 in 10^8—the sort of performance achieved by optical fibers. However, on radio circuits such as those employed with communication satellites and the signaling channel for mobile communications, FEC is still employed (see § 14.2.2). Further, FEC is used in high-performance modems (see § 5.1.3).

4.3 DATA COMPRESSION

Data compression reduces the number of characters transmitted during a communication session so that more information can be transmitted over a given facility. Data compression techniques are divided into those that employ character-oriented compression, and those that employ statistical encoding.

4.3.1 Character-Oriented Compression

By substituting special characters, or small groups of characters, for larger strings of characters that appear in a message, the length of the datastream can be made shorter than the original. Some of the techniques employed include:

- **Null compression**: in datastreams that describe formatted documents, there are likely to be blank spaces (null characters) denoted by all-zeros characters (i.e., NUL = 0000000 in ASCII, and 00000000 in EBCDIC). To reduce the number of zeros transmitted, a null-compression-indicating character is used followed by the number of all-zeros characters in the string. Thus, if the original character string is

⇐NAMEzzzzzzzzzzzzzzzzzzzzzzzzzzzzzzCRSS#zzzzzzzzzzzzzzzzzzzzCR

  where z represents the all-zeros or null character, and CR is carriage return, the compressed character string might be

⇐ NAME©30CRSS#©20CR

  where © is the null-compression-indicating character. The numbers 30 and 20 are the numbers of all-zeros characters in the two segments of the string. If they are expressed as 8-bit binary numbers, i.e., 00011110 and 00010100, the number of characters in the compressed string is 15. Since the number of characters in the original string is 61, the compression ratio is approximately 4:1.
- **Run-length compression**: a superset of null compression that is used to compress any string of repeating characters (not just zeros).
- **Pattern substitution**: common character pairs, character triplets, and entire words in a message are replaced by special characters. Thus, in the last statement the word *character* appears three times. Represent it by ©. The statement will then appear as…*common © pairs, © triplets and entire words in a message are replaced by special ©s*. In this case, the compression ratio is approximately 4:3.

- **Binary number substitution**: for decimal numbers of two digits and more, significant character saving can be achieved by converting the numbers to binary. An 8-bit sequence taken from the code table can represent one decimal digit, or up to 255 (11111111) if full use is made of the binary representation. For 8-bit coding, this represents a compression of 2:1 for 10 ≤ number ≤ 99, and 3:1 for 100 ≤ number ≤ 255.

4.3.2 Huffman Coding

The idea of entropy coding was introduced in § 3.4.3. Based on self-information and entropy, it employs a variable-length code whose members are assigned to the symbols in some way, in proportion to their use. Symbols transmitted frequently (i.e., self-information ⇒ 0), are represented by shorter code words, and symbols transmitted infrequently (i.e., self-information ⇒ 1), are represented by longer code words. In addition, to guarantee unique decodability, no member is the prefix of any other member. A process that constructs such a code is shown in **Figure 4.7**. Called Huffman coding, the variable-length codewords approach the theoretical

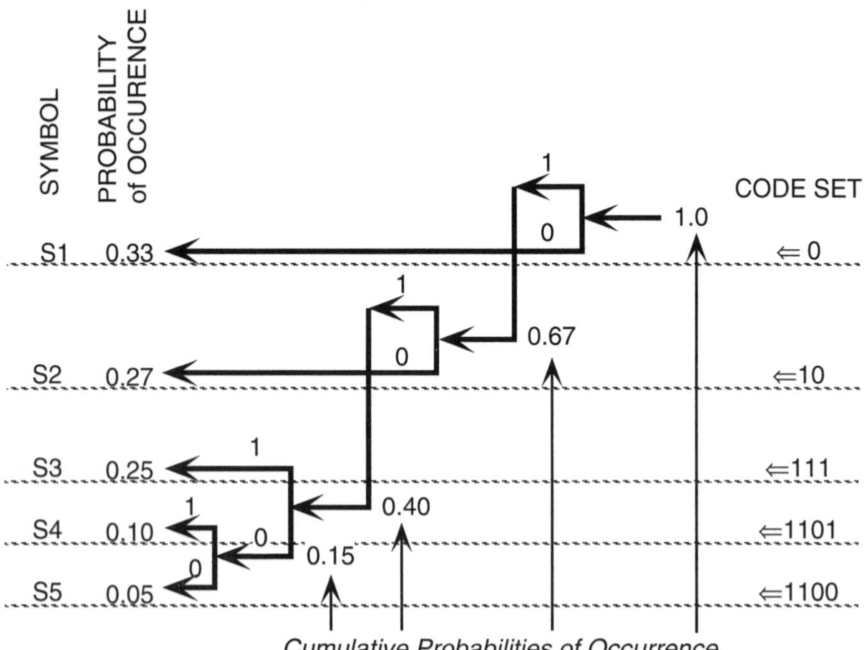

FIGURE 4.7 Building a Huffman Code Set for Five Symbols
Short codes are assigned to frequently used symbols, and longer codes are assigned to less frequently used symbols.

limit of coding efficiency. The symbols are arrayed in order of decreasing probability of occurrence. To construct a Huffman code, connect the two lowest symbols in a node whose probability is the sum of their probabilities of occurrence. Assign 0 and 1 to the two branches. Connect this node to the next highest symbol to form another node that takes on the sum of the probabilities again. Be certain to draw the connection so that the branch with the higher probability component is uppermost. Assign 0 and 1 to the new branches in the same way that 0 and 1 were assigned to the first pair of branches. Continue to form nodes in this way until all the symbols have been connected. To find the codeword that represents a symbol, start at the root node (probability 1.0) and trace out the path to the selected symbol. Its codeword consists of the sequence of 0s and 1s along the route. In the simple example shown in Figure 4.7, the five symbols S1, S2, S3, S4, and S5 have probabilities of occurrence of 0.33, 0.27, 0.25, 0.10, and 0.05, respectively. The Huffman code set is listed in the right-hand column in the Figure. For long messages in which the five symbols achieve average occurrences, the ratios of the numbers of bits per message are

$$\text{EBCDIC} : \text{ASCII} : \text{Huffman} = 100 : 87.5 : 27.7,$$

i.e., a compression of approximately 3:1 is achieved with Huffman coding for this set of symbols. No codeword is the prefix of any other codeword so any sequence of symbols can be sent without additional markers.

4.3.3 Techniques for Facsimile

A letter-size document scanned with a horizontal and vertical resolution of 200 dots per inch (dpi) could generate 3.2 million bits. Transmitting this amount of data over a 64 kbits/s channel, will take 50 seconds. To achieve faster service, data compression techniques must be employed. Two techniques are in general use:

- **Modified Huffman code**: employs a combination of run-length coding and variable-length coding. In any document scanned horizontally, there will be many sequences (runs) of all white picture elements (pels) and all black pels. In modified Huffman (MH) coding a run of length R pels is expressed as

  $$R = 64\,m + n,$$

  where, m = 0, 1, 2,...27, and n = 0, 1, 2,...63. m and n are encoded according to tables of Huffman codes developed to suit a variety of documents. Average compression ratios of 8 are obtained with this technique.
- **Modified READ code**: in the Modified Relative Element Address Designate (READ) Code, run lengths are encoded by noting the position of change elements (i.e., where white changes to black, or black changes to white) in relation to a previous change element. The use of this coding procedure produces average compression ratios from 12 to 15.

4.4 DATA ENCRYPTION

Successful businesses must communicate all sorts of data among their many locations. Because a communication network can be compromised, encryption should be considered as a means of keeping sensitive data private (see § 15.3.1(4)). Even when network access is gained, encrypted messages are unintelligible, and may be so well encrypted that potential thieves may reconsider their schemes.

4.4.1 Encryption Defined

- **Encryption**: the action of disguising information so that it can be recovered relatively easily by persons who have the *key*, but is highly resistant to recovery by persons who do not have the key.

A message in cleartext (also called plaintext) is disguised (encrypted) through the use of an encryption key to create a cryptogram (i.e., create a facsimile of the plaintext message in ciphertext). **Figure 4.8** shows some of these basic relationships. The encryption key may be changed from time to time to make an intruder's task more difficult. Restoration of a cryptogram to cleartext is achieved by the action of decryption using a decryption key. In a symmetric, or single-key cryptosystem, the encryption and decryption keys are the same. In an asymmetric, or two-key cryptosystem, the encryption and decryption keys are different, and one cannot be derived readily from the other.

4.4.2 Encryption Methods

Encryption is accomplished by scrambling the bits, characters, words, or phrases in the original message. Scrambling involves two activities:

- **Transposition** in which the order of the bit patterns, characters, words, or phrases is rearranged
- **Substitution** in which new bit patterns, characters, words, or phrases are substituted for the originals without changing their order.

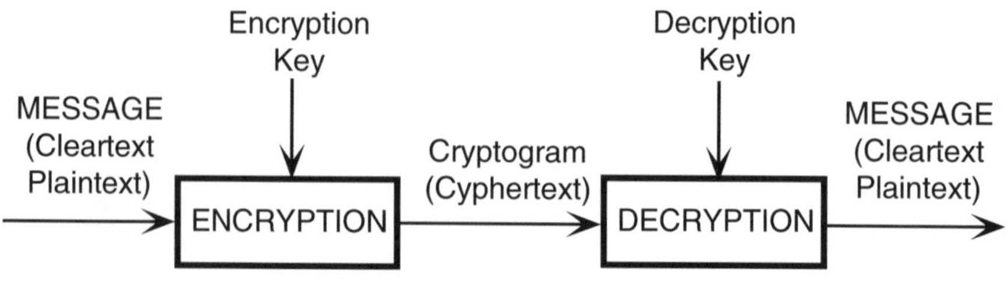

FIGURE 4.8 Encryption and Decryption

They are said to employ:

- **Ciphers**, when the bit patterns or characters in the original text are transposed or substituted
- **Codes**, when the words or phrases in the original text are transposed or substituted.

Because the use of a code requires a large element of human judgment, computer-based encryption and decryption algorithms employ ciphers.

(1) Data Encryption Standard Cryptosystem

Probably the most widely used encryption algorithm is the Data Encryption Standard (DES). Pioneered by IBM, it is employed in symmetric cryptosystems. Originally developed to protect sensitive, unclassified, US government, computer data, DES is used to provide authentication of electronic funds transfer (EFT) messages, and is available in both hardware and software implementations for the protection of other private data. Based on a 56-bit cipher key, a single, randomly selected, specific key (selected from the 2^{56}, 56-bit patterns afforded by the cipher key) is employed in both encryption and decryption. DES uses a complicated algorithm that thoroughly scrambles the plaintext.

In one version of DES, 64-bit blocks of data are scrambled using a sequence of bit substitutions and bit transpositions. Encryption begins by transposing the 64-bit block of plaintext using the specific cipher key selected. The resulting block is split into two halves. One half is transformed by a non-linear function defined by the specific key in use. It is then added to the other half of the transposed text. Next, the halves are interchanged, and the process is repeated using an intermediate key derived from the original key. After 16 operations of this kind, a final transposition is applied to create a 64-bit block of ciphertext. Decryption is the inverse of encryption; the intermediate keys are used in reverse order. The principle of encryption and decryption using DES is shown in **Figure 4.9**. The success of DES depends on random key selection, and frequent key changes. There is no known way to break DES except to test possible keys one-by-one. If all keys (2^{56}) are equally likely to have been used, on average, 2^{55} tests will be required. If greater security is desired, a 112-bit cipher key may be employed to create 2^{112} possible keys.

(2) Public Key Cryptosystem

A public key cryptosystem is an asymmetric cryptosystem. First announced in 1976, public key algorithms offer a radically different approach to encryption. Based on the difficulty of factoring very large integers, the idea depends on the use of a pair of keys that differ in a complementary way. However, of the several algorithms originally proposed, only one, the RSA algorithm (named for its inventors Rivest, Shamir, and Adelman), is considered to be highly secure. The input is

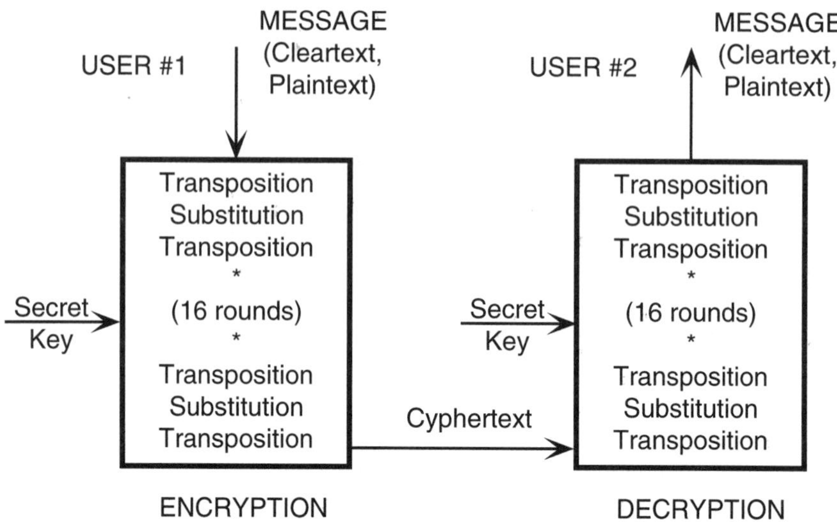

FIGURE 4.9 Encryption and Decryption Using DES

recovered by successive transformations based on the two keys. A substantial level of protection is achieved by holding one key secret, and significant operating flexibility is achieved by allowing the other key to be public knowledge. Because of the computational complexity involved, even though the public key (PK) is known, and messages encrypted through its use may be available to those who would steal the information, it is impossible to decipher the encrypted message in any reasonable time without the use of the secret key (SK). **Figure 4.10** shows the use of PKS to achieve the following:

- **Privacy**: when User #1 wishes to send a private message (M_{12}) to User #2, it is encrypted as $PK_2{}^*M_{12}$ (i.e., M_{12} is encrypted by User #1 using the public key belonging to user #2). When User #2 wishes to read the message, it is decrypted as $SK_2{}^*PK_2{}^*M_{12}$ to yield M_{12} (i.e., $PK_2{}^*M_{12}$ is decrypted by User #2 using the secret key known only to User #2). Without SK_2, others cannot read M_{12}.
- **Authentication**: if User #1 wishes to send message M_{12} to User #2, and wishes User #2 to know without a doubt that it comes from User #1, the message is encrypted as $SK_1{}^*M_{12}$. When User #2 receives the message it is decrypted as $PK_1{}^*SK_1{}^*M_{12}$ to yield M_{12}. In this case, all users have access to PK_1 and can read M_{12}. However, authentication is the issue for User #1 and User #2, not privacy.

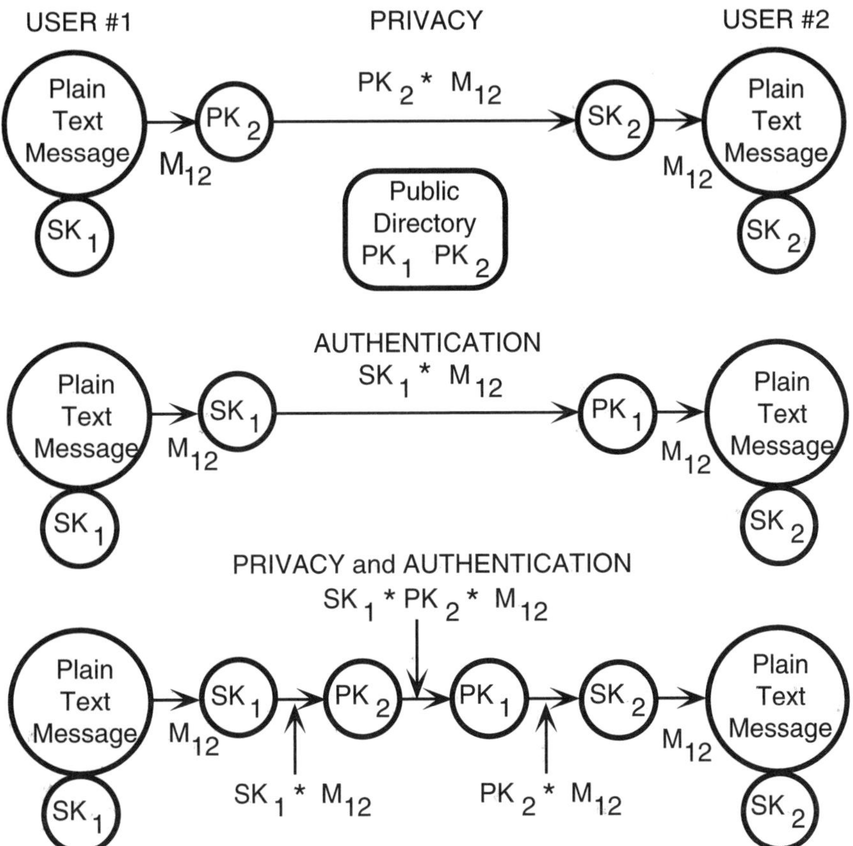

FIGURE 4.10 Using PKS to Achieve Privacy, Authentication, and Privacy and Authentication

Messages encrypted using a Public Key can only be decrypted by the person who has the corresponding Secret Key. Combinations of Public Keys and Secret Keys are used to achieve Privacy, Authentication and Privacy and Authentication.

- **Privacy and authentication**: if User #1 wishes to obtain both privacy and authentication, the message is encrypted as $PK_2{}^*SK_1{}^*M_{12}$. On receiving the message, User #2 forms $SK_2{}^*PK_2{}^*SK_1{}^*M_{12}$ to yield $SK_1{}^*M_{12}$, and then forms $PK_1{}^*SK_1{}^*M1_2$ to yield M_{12}. Only User #2 has access to both PK_1 and SK_2.

In § 15.3 we pursue the use and management of encryption in commercial networks.

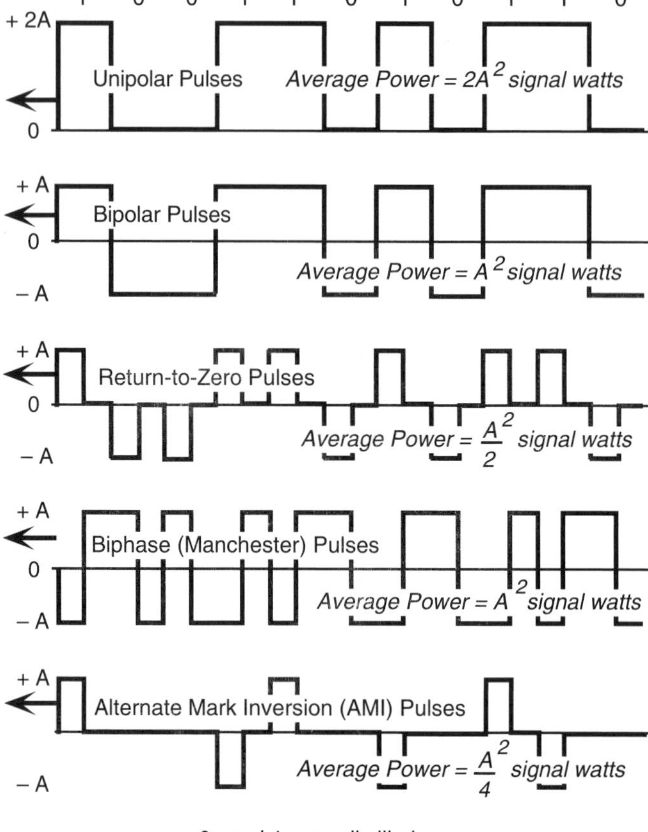

FIGURE 4.11 Digital Signal Waveforms
The five pulse trains convey the same information. They vary considerably in the power they require.

4.5 BINARY SIGNALS

4.5.1 Representation of Binary Sequences

Figure 4.11 shows five examples of binary signaling applied to the sequence ⇐ 10011010110. They include the following:

- **Unipolar**: for 1s, positive current flows, and for 0s, no current flows. It is the technique that requires most power. In optical fibers, because pulses are generated by turning infrared sources (lasers, or light-emitting diodes) on and off, unipolar is the only way of signaling.
- **Bipolar**: for 1s, positive current flows, and for 0s, negative current flows. In an electronic circuit, it requires one-half the power of the unipolar sequence.

- **Return-to-zero**: bits are represented by a positive or a negative current pulse that is one-half the width of bipolar signals. The current level returns to zero before the next bit in the sequence is handled. By halving the duration of each bit, the power required is reduced by one-half over bipolar signaling, i.e., to one-quarter the power required by the unipolar sequence.
- **Biphase or Manchester**: a 1 is a positive current that changes to a negative current, and a 0 is a negative current that changes to a positive current. Since the changeover occurs exactly at the middle of the bit period, the receiver is better able to sample the signal—this reduces the chance for error. Manchester signaling is employed between collocated equipment that must operate at high speeds and maintain close synchrony; it is a popular signaling technique for local area networks. The power required is equal to that required by bipolar signaling, i.e., it requires one-half the power of the unipolar sequence.
- **Alternate mark inversion** (AMI): 1s are represented by return-to-zero pulses that alternate between positive and negative levels and 0s are represented by the absence of pulses. AMI signaling requires one-eighth the power consumed by the unipolar sequence. Furthermore, the average repetition rate is one-half of the average repetition rate of the other sequences so that the peak power density is centered at one-half the bit rate (expressed in Hz); virtually the entire power is contained within a bandwidth in Hertz equal to the bit rate, and the dc-power is zero. AMI is used extensively within public and private communication networks.

In **Figure 4.12**, we show power spectral density functions for streams of random data using bipolar, Manchester, and AMI signaling. They represent the way in

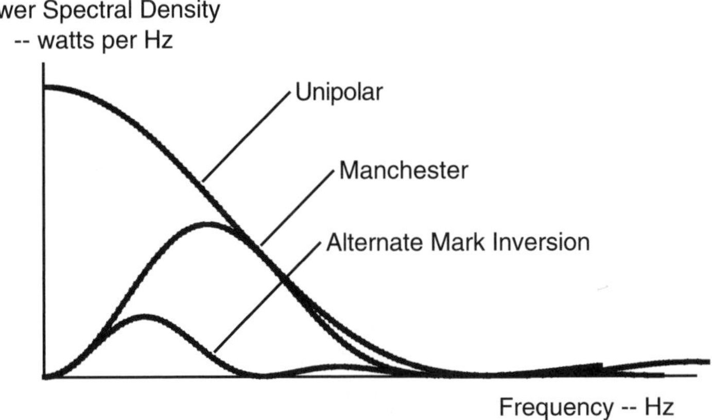

FIGURE 4.12 Power Spectral Density Diagrams for Unipolar, Manchester, and AMI Signals

which power density varies with frequency. The total power required by each signaling technique is measured by the area under the appropriate curve. From Figure 4.12 we conclude that unipolar and Manchester signaling occupy approximately twice the bandwidth that AMI requires, and that the power relationships between the techniques are those enumerated above.

4.5.2 Scrambling

Electronic equipment works best when the datastream contains an equal number of 1s and 0s and the composite signal is zero-mean (i.e., the dc-term is zero). Because real datastreams contain long strings of 1s or 0s that do not produce these conditions, many transmission systems *scramble* them; i.e., the input stream is multiplied by a semi-random sequence to produce an acceptable ratio of 1s and 0s. After the scrambled stream is transmitted from sender to receiver, the receiver multiplies it by the inverse of the semi-random sequence to produce the original datastream. The procedure is transparent to the users.

4.5.3 Circuit Speeds and Bit Sequences

The number of bits transmitted over a circuit in one second is the speed of the circuit. Circuit speeds vary from a few bits per second (bits/s) to several billion bits per second (gigabits/s or Gbits/s). The values determine the times required to send specific quantities of data from sender to receiver. In **Figure 4.13** we assume a circuit speed of 300 bits/s and calculate the time to transmit:

- One bit is 0.0033 seconds (i.e., 3.3 milliseconds)
- One octet (8-bits) is 0.0267 seconds (i.e., 26.7 milliseconds)
- One message of 136 octets (1088-bits) is 3.6267 seconds.

If the circuit speed increases to 9600-bits/s, the times are 0.1042 ms (i.e., 104.2 μs), 0.8333 ms (i.e., 833.3 μs), and 0.1133 ms (i.e., 113.3 μs) respectively.

Digital signaling is concerned with the exchange or distribution of well-defined streams of 1s and 0s. Each digit is carefully selected and properly positioned in the stream so that, upon the application of rules agreed to before the process began, it will yield precise information. A given grouping of bits will convey the same information whether delivered over a telephone line at a circuit speed of 300-bits per second, a circuit speed of 9600-bits/s, or over an optical fiber at a circuit speed of 3 billion bits per second. What matters is that the binary digits succeed each other in order—not how quickly one bit succeeds another—although speed determines how rapidly a given datastream can be generated and received. Thus, the bit sequence determines the information a datastream contains, and the circuit speed determines the time required for its delivery.

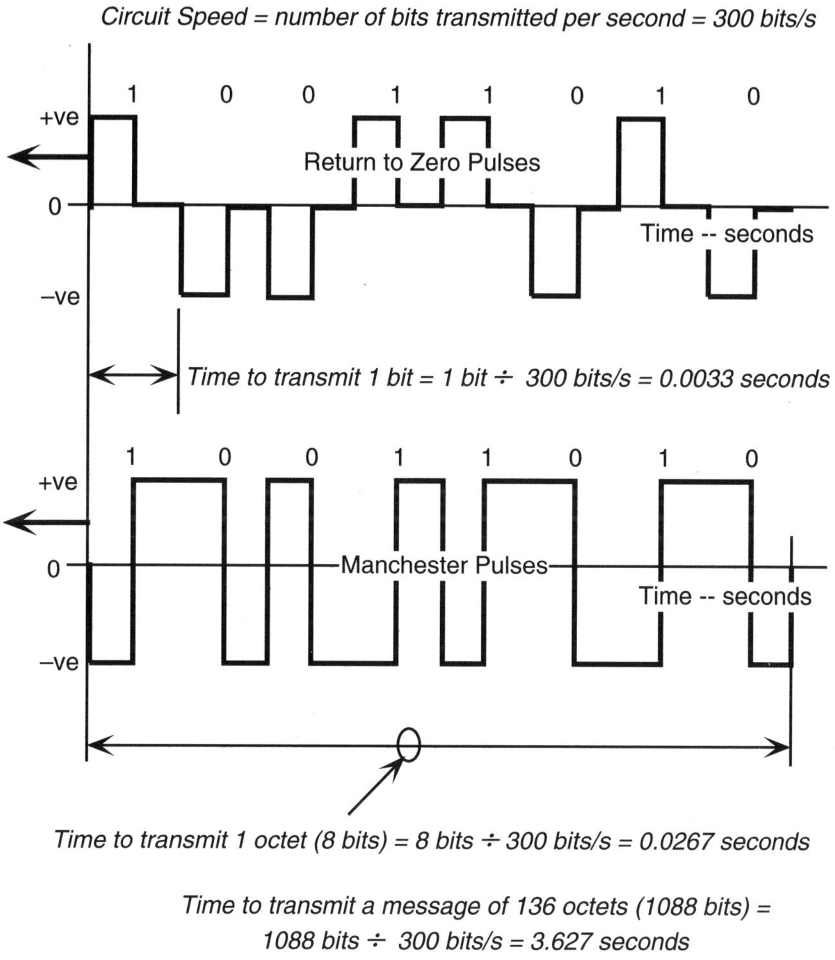

Circuit Speed = number of bits transmitted per second = 300 bits/s

1 0 0 1 1 0 1 0

+ve

0

–ve

Return to Zero Pulses

Time -- seconds

Time to transmit 1 bit = 1 bit ÷ 300 bits/s = 0.0033 seconds

1 0 0 1 1 0 1 0

+ve

0

–ve

Manchester Pulses

Time -- seconds

Time to transmit 1 octet (8 bits) = 8 bits ÷ 300 bits/s = 0.0267 seconds

Time to transmit a message of 136 octets (1088 bits) =
1088 bits ÷ 300 bits/s = 3.627 seconds

FIGURE 4.13 Circuit Speed and Time to Send Data

4.5.4 Intersymbol Interference

In the early days of telegraphy, it became obvious that the dots and dashes of the Morse code could be sent only so fast. Speeding them up beyond a certain point made the symbols indistinguishable. What telegraphers did not find out until later was that their pulses became rounded, diminished, and stretched out, and individual pulses ran into one another before they reached the receiver. To reduce the level of *intersymbol interference* (ISI), the further the distance, the longer the sender must wait between pulses. Expressed another way, for practical bearers, passband diminishes with length; as a consequence, the maximum signaling rate a bearer can support decreases as the transmission distance increases.

(1) Pulse Shaping

To improve the receiver's ability to discriminate between pulses, one way of reducing intersymbol interference is to substitute a less demanding pulse for the rectangular pulses shown in Figure 4.11. Such a pulse is shown in **Figure 4.14**. Called a raised cosine pulse, it consists of one period of a cosine function offset by its amplitude so that the values are always ≥ 0. For the same bit-rate, a stream of raised cosine pulses will retain its identity over a greater distance than a stream of rectangular pulses. Alternatively, for the same levels of ISI, a stream of raised cosine pulses can deliver a higher bit-rate than a stream of rectangular pulses.

(2) Adaptive Equalizer

Another technique employed to reduce intersymbol interference uses a device that corrects channel induced distortion. Called an adaptive equalizer, it shapes the received pulses according to information it has derived from the distortion induced in a standard, training pulse. Located at the receive end of a channel, it consists of a transversal filter trained to produce a distortion-free signal from the distorted signal it receives. Training is accomplished by adjusting the characteristic of the filter to produce a sequence of standard pulses from a test stream introduced at the transmitting end. When training is completed, the filter is able to correct the distortion suffered by pulses sent over the channel.

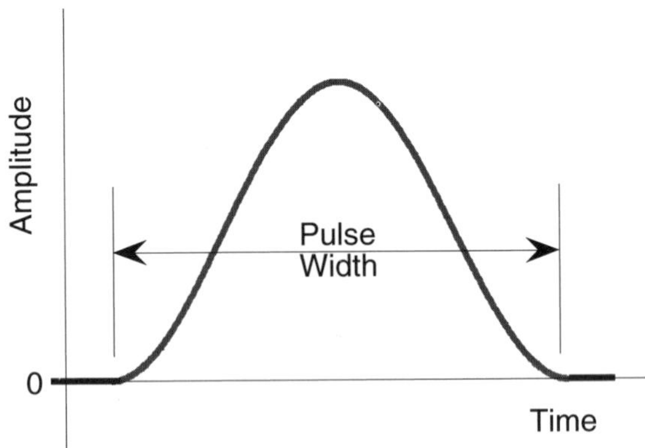

FIGURE 4.14 Raised Cosine Pulse
A pulse shape that reduces bandwidth requirements and is less susceptible to intersymbol interference.

(3) Regenerator

A different strategy that achieves higher signaling rates is to reduce the distance over which the symbol stream is transmitted before processing. This can be done by introducing regenerators (also known as repeaters) at regular intervals along the transmission path. Their purpose is to read the pulse stream before it degrades substantially, regenerate a new stream, and pass it on to the next unit. In this way, the pulse stream is restored to (almost) original condition at points along the transmission route so that it never becomes unreadable. As a result, while the average, voice-grade, twisted pair, telephone line (with equalizers) may support a signaling rate of 2400 baud (see below), a twisted pair line with regenerators every 6000 feet will support a T-1 channel with a signaling rate of 1.544 Mbaud (see § 8.2.1).

(4) Echo Cancellation

In an electrical circuit, it is important to ensure that each segment appears as an infinite line to the signal entering it. If this is not so, signal energy will be reflected back to the source, in proportion to the impedance mismatch that exists. The reflected energy creates an echo that may garble or destroy the signal being sent. Known as *echo suppressers*, electronic devices are available that create inverted echos to cancel the echo signals generated at the near-end and far-end of a data link. Using these devices, it is possible to send data both ways over two-wire circuits (see § 5.1.3 and 10.3.1).

4.6 LOW-SPEED DATA

The pervasive use of electronic mail, facsimile, and similar text- or graphics-based, person-to-person services, requires signaling methods that permit low-speed data signals to be sent over the voice bandwidth of the public switched telephone network.

4.6.1 Signaling Rate

The signaling rate of a transmission system is the number of independent symbols sent per second. Its unit is the *baud*, i.e., one symbol per second is a signaling rate of 1 baud. For binary symbols (1s and 0s) the circuit speed in bits per second is equal to the signaling rate in bauds.

4.6.2 Nyquist's Theorem

In the early 20th century, Nyquist (a 20th century Dutch mathematician) established that, given a bearer that provides a low-pass channel of passband B Hz, it is possible to transmit independent symbols at a rate no greater than 2B baud:

- **Nyquist's theorem**: the maximum signaling rate (r) in bits/s over a channel having a passband (B) in Hertz is given by $r = 2B$.

Nyquist based his conclusion on conditions impossible to duplicate on real-world bearers. As a result, the practical limits to signaling rates are quite a bit less than the Nyquist relationship predicts. For instance, for voice-grade telephone lines, the practical maximum signaling rate is around 2400 baud; Nyquist's formula predicts a maximum of 8000 bauds for a channel with 4 kHz passband.

4.6.3 Overcoming the Signaling Rate Limitation

Does Nyquist's result, and evidence that the maximum signaling rate on voice-grade circuits is 2400 bauds, mean that circuit speed can be no higher than 2400 bits/s? Certainly, this will be the case if each symbol represents a 1 or a 0 (i.e., we employ a scheme in which each symbol represents 1-bit). To achieve data rates higher than 2400 bits/s, we must employ schemes that make each symbol represent several bits. This is not as difficult as it may sound. If we have four unique symbols (S1, S2, S3, and S4), they can represent all possible combinations of 2 bits. Thus

S1 = 00; S2 = 01; S3 = 10; and S4 = 11.

If we have 16 unique symbols, they can represent all possible combinations of 4 bits:

S1 = 0000; S2 = 0001; S3 = 0010; S4 = 0011; S5 = 0100; S6 = 0101;

S7 = 0110; S8 = 0111; S9 = 1000; S10 = 1001; S11 = 1010; S12 = 1011;

S13 = 1100; S14 = 1101; S15 = 1110; and S16 = 1111;

and, if we have 2^n unique symbols, they can represent all possible combinations of n bits. As an example, consider the datastream

⇐ 011111100101101000000000000000000100010000111110.

- Using 2 symbols, S1 = 1 and S2 = 0 (a signaling efficiency of 1 bit per baud), and a signaling rate of 2400 bauds, we can transmit 2400 bits/s.
- Using 4 symbols (a signaling efficiency of 2 bits per baud) we can represent the data stream as

  ⇐ S2S4S4S3S2S2S3S3S1S1S1S1S1S1S1S1S3S1S3S1S2S4S4S3

  and, if the signaling rate is 2400 bauds, transmit 4800 bits/s.

- Using 16 symbols (a signaling efficiency of 4 bits per baud) we can represent the data stream as

⇐ S8S15S6S11S1S1S1S1S9S9S8S15

and, if the signaling rate is 2400 bauds, transmit 9600 bits/s.

In all three examples, the signaling rate is the same—2400 bauds (i.e., 2400 symbols/s). The increase in circuit speed is obtained solely at the expense of complicating the symbol structure.

4.6.4 Creating Unique Symbols

Suppose 9600 bits/s are sent over a 2400 baud line. The number of signal states required to implement a full symbol set is $2^4 = 16$. They are created from 16 sinusoidal signals of the same frequency that are characterized by a particular amplitude and phase angle. For instance, we may choose to create 16 unique signals from two amplitude values (1 and 2) and eight phase angle values (0°, 45°, 90°, 135°, 180°, 225°, 270°, and 315°) applied to an 1800 Hz signal. Known as 16-QAM (quadrature-amplitude modulation, see § 2.3.2), the 16 signal states are plotted in polar form in **Figure 4.15**. Such a diagram is known as a signal *constellation*. Four bit sequences can be assigned to the signal states in any order. Of course, once assigned, they must be made known to both sender and receiver so that they will associate the correct four bits with the same symbol. Sixteen unique signal states can be created in other ways. For instance, we can use these combinations:

- Four amplitude values (1, 2, 3 and 4) and four phase angle values (0°, 90°, 180°, and 270°)
- Sixteen amplitude values and one phase angle
- One amplitude and sixteen phase angle values
- etc.

Constellations for each of these combinations of amplitudes and phase angles are included in Figure 4.15.

A unique symbol is formed from each of the QAM waves by selecting a segment of the signal. With an 1800 Hz sinusoidal wave as carrier, and a 2400 baud requirement, each symbol will consist of a 270° segment of the signal. Using them, we can produce a stream of analog symbols that changes 2400 times per second (2400 baud), fits within the bandwidth of a voice grade line, and transmits 9600 bits/s. For the 48-bit datastream introduced above, **Figure 4.16** shows the physical signal that carries the message on the voice-grade line. Created from the 16-QAM signals formed from the combinations of four amplitudes and four phase angles drawn in Figure 4.15, the symbols are concatenated according to the 4-bit symbol stream derived from the 48-bit datastream. We pursue applications of this technique including trellis-code modulation (TCM) for forward error correction in § 5.1.3. Not limited to voice grade circuits, the principle is applied to microwave radio channels. They are described in the next Section.

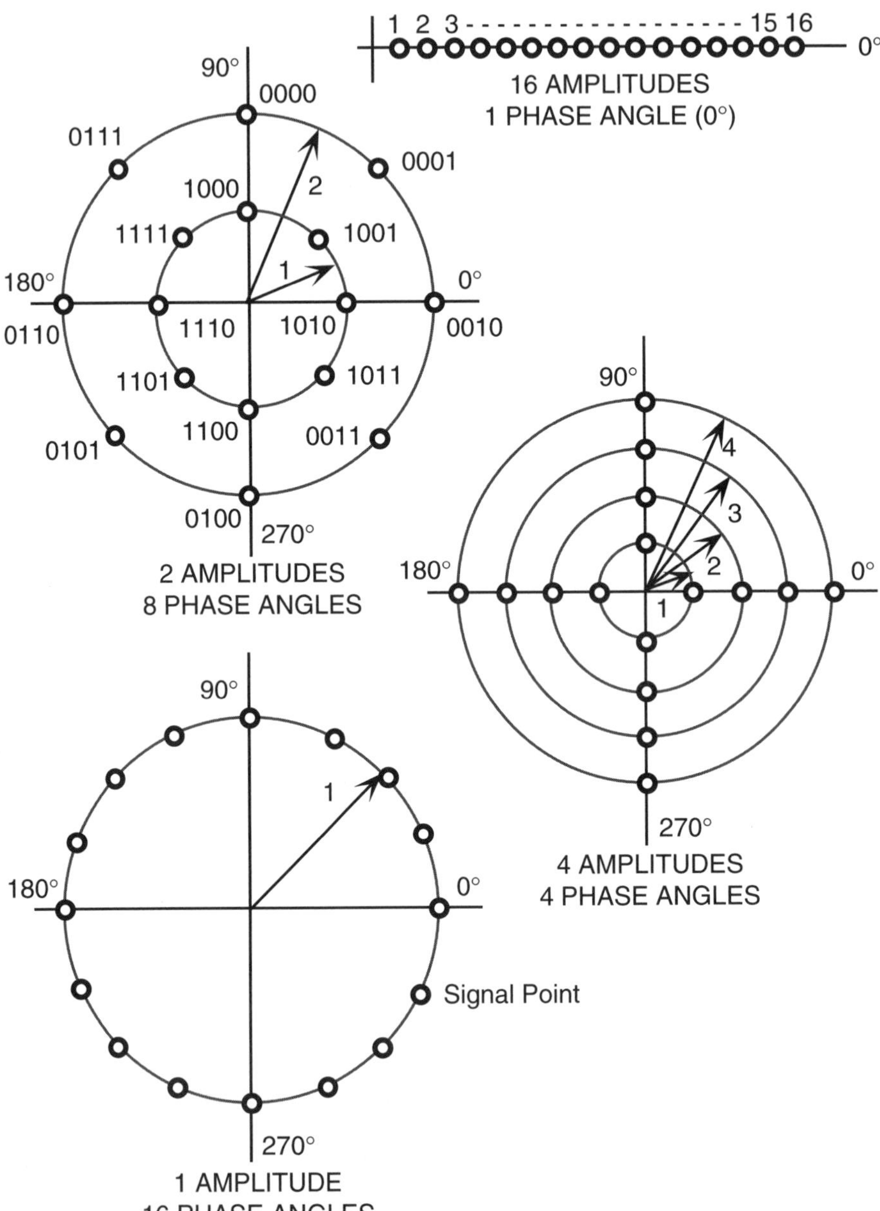

FIGURE 4.15 Constellations That Create 4-Bit Symbol Sets
Each constellation consists of 16 signal points that can be used to represent the 16 combinations of 4-bits.

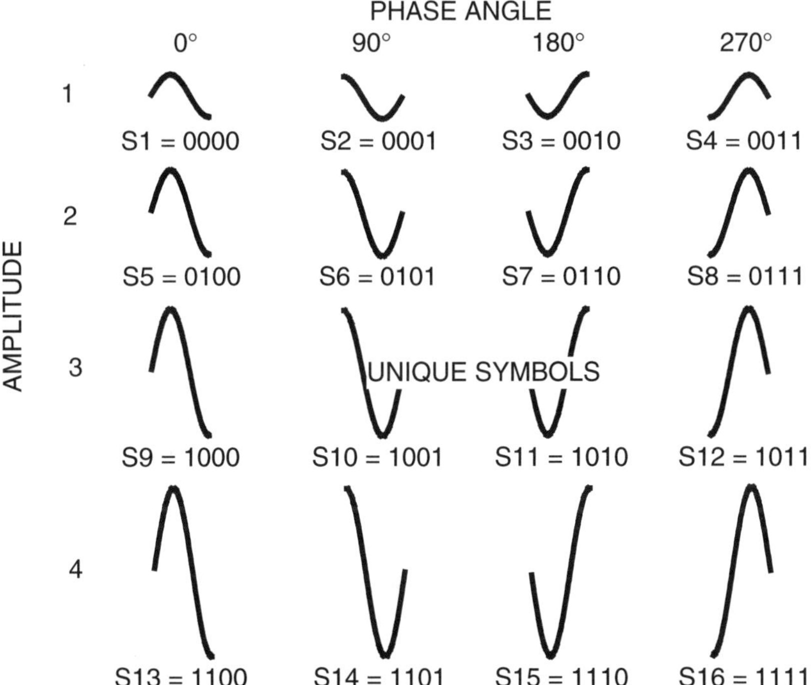

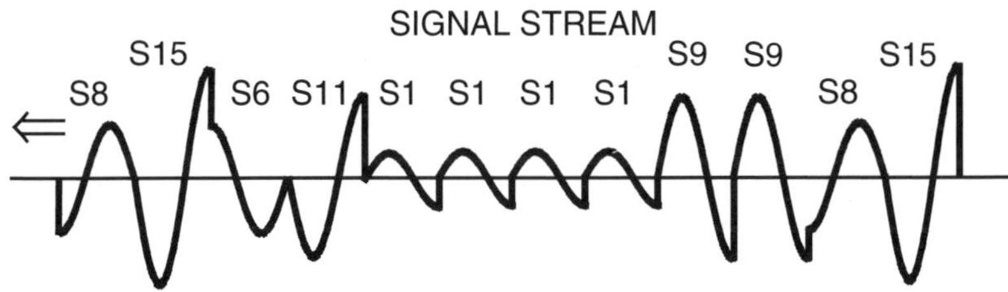

FIGURE 4.16 Creation of a 16-QAM Signal Stream

4.7 DIGITAL MODULATION

In Section 4.6.3, some techniques used to impress data signals on carriers compatible with the characteristics of telephone circuits were described. They serve as a practical introduction to *digital modulation*. By modulating continuous-wave (CW) carriers with the datastream, data signals are transported over radio channels. The fixed transmission bandwidth assigned to a particular channel sets an upper limit to the symbol rate, and noise causes errors in the received message. The amplitude, frequency, and phase of the carrier are changed by the data signal to produce a modulated wave. If the modulating (message) signal consists of NRZ rectangular pulses (see Figure 4.11), the modulated parameters will be switched from one state to another in sympathy with the values of the bits in the message stream. This process is known as *keying*; it can be used to change the amplitude, phase, and frequency of the carrier. Respectively, the actions are known as ASK (amplitude-shift keying), PSK (phase-shift keying) and FSK (frequency-shift keying).

4.7.1 Amplitude-Shift Keying

In amplitude-shift keying, the transmitted signal is a sinusoid whose amplitude is modulated in discrete steps. As shown in **Figure 4.17**, ASK can be generated by *on-off* keying (OOK) so that the presence of the carrier signal corresponds to a 1-state, and no carrier corresponds to a 0-state. When only two states are employed, the ASK wave is known as a *binary* ASK wave. To produce more complex symbols that represent groups of bits, several amplitude levels are generated. For *M*-levels (*M* - 1 discrete *on* amplitudes and one *off* state), the signal is described as *M*-ary ASK.

(1) Power Spectral Density of Binary-ASK

For binary-ASK created by a random datastream, the power spectral density (PSD), i.e., the distribution of power density (in signal watts/Hz) about the carrier frequency, is shown in **Figure 4.18**. The impulse function at the carrier frequency repre-

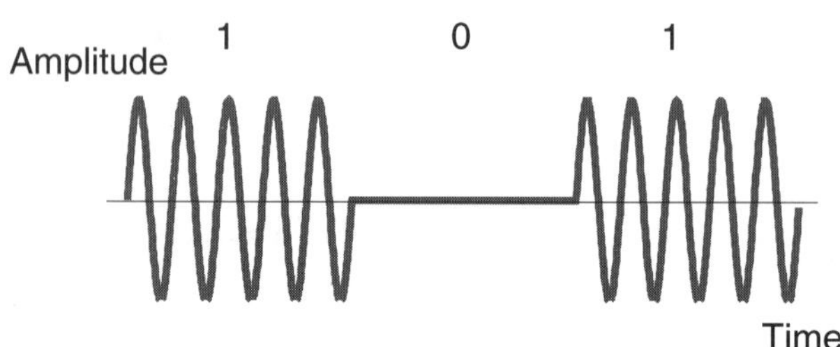

FIGURE 4.17 Binary-ASK

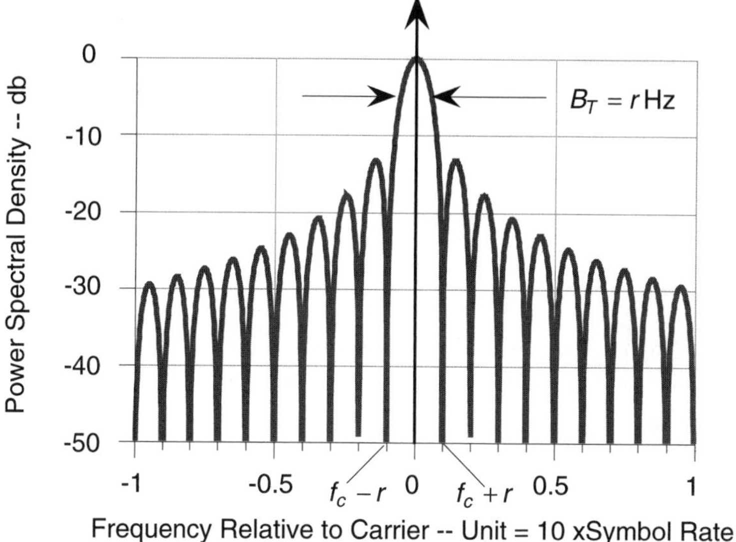

FIGURE 4.18 PSD of ASK Signal Modulated by Random Datastream
The impulse at the carrier frequency is produced by power associated with the keying process.

sents power associated with the keying process. With a symbol rate of r (symbols/s), the transmission bandwidth B_T is equal to r Hz and most of the signal power is contained within the range $f_c - r/2$ to $f_c + r/2$ Hz. The main lobe of the signal spans the frequency range $f_c - r$ to $f_c + r$Hz, and the power in the sidelobes drops off relatively slowly. Because the spectrum extends well beyond the transmission bandwidth, *spillover* is an important concern. To reduce interference with other signals (interchannel interference) bandpass filters are used. At the risk of degrading the signal by increasing intersymbol interference, a bandpass filter may be applied to limit the energy of the NRZ pulses that form the datastream. By restricting the baseband signal to a frequency range that just encompasses the primary lobe of the spectral density diagram, approximately 90% of the datastream energy is available to the modulation process. Another bandpass filter is placed at the output of the transmitter.

(2) Quadrature-Carrier AM

In quadrature-carrier amplitude modulation (QAM, see § 2.3.2(1)), we transmit symbols at rate $2r$ in the bandwidth required by binary-ASK (r Hz). By dividing the datastream in two streams (even-numbered bits and odd-numbered bits) that are used to amplitude-modulate orthogonal carriers ($\sin\omega_c t$ and $\cos\omega_c t$, for instance) derived from the same source, we produce two modulated waves containing r symbols/s that occupy the same bandwidth of r Hz. **Figure 4.19** compares signal space diagrams for binary-ASK and QAM. For binary-ASK, signal points (1 and 0) are arranged on the horizontal axis. For QAM, signal points (1 and 0) occur on the horizontal and vertical axes. The set on the horizontal axis represents the signal states assumed by $\cos\omega_c t$ when amplitude-modulated by one of the datastreams, and the set on the vertical axis represents the signal states assumed by $\sin\omega_c t$ when amplitude-modulated by the other datastream. The four points 00, 01, 10, and 11 are the states assumed by the composite signal.

4.7.2 Spectral Efficiency

For an M-ary signal, the number of bits represented by a symbol is $\log_2 M$, so that the binary data rate $r_b = r\log_2 M$, where r is the symbol rate (symbols/s), and $r_b/B_T = \log_2 M$. Expressed in bits/s/Hz the ratio of the binary bit rate (in bits/s) to the transmission bandwidth (in Hz) is known as the *spectral efficiency* of the modulation. For binary signals (i.e., $M = 2$), $r_b/B_T = 1$ bits/s/Hz; for quaternary signals (i.e., $M = 4$), $r_b/B_T = 2$ bits/s/Hz; etc. Thus, binary-ASK has a spectral efficiency of

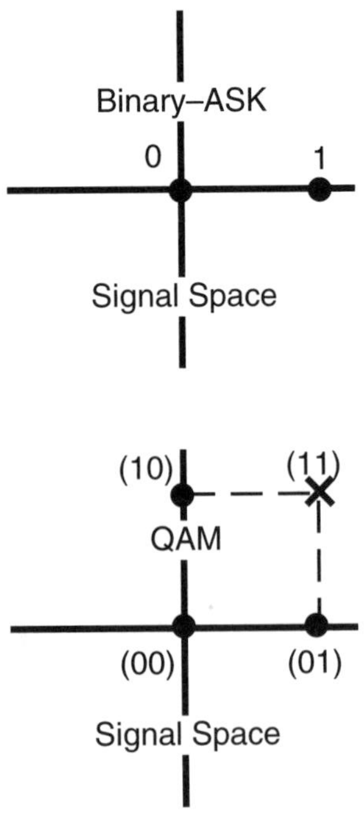

FIGURE 4.19 Signal Space Diagrams for Binary-ASK and QAM

1 bit/s/Hz and for QAM it is 2 bits/s/Hz. Spectral efficiency is a readily understood measure of digital system performance. In practical situations, care must be taken to ensure that comparisons are made under the same conditions.

4.7.3 Phase-Shift Keying

In phase-shift keying, the transmitted signal is a sinusoid of constant amplitude. As shown in **Figure 4.20**, PSK can be generated by OOK so that the presence of the carrier signal in one phase condition corresponds to a 1-state, and the presence of the carrier signal 180° out of phase with it, corresponds to a 0-state. This is known as *binary*-PSK (BPSK) and sometimes as *phase-reversal* keying (PRK). For a random datastream and binary-PSK, the distribution of power density is the same as that shown in Figure 4.18 without the impulse at the carrier frequency. Thus, the percentage of the power devoted to signaling is greater in PSK than ASK, and the spectral efficiency is 1 bit/s/Hz.

(1) Differential PSK

Called *synchronous* demodulation, recovery of the datastream from a PSK-modulated wave requires a relatively complicated receiver that must reconstruct the carrier so as to be able to detect phase changes in the received signal. By modifying the transmitted signal, *differential* phase-shift keying (DPSK) eliminates the need for the synchronous carrier in the demodulation process. At the transmitter, the datastream is processed to produce a modulated wave in which the phase changes by 180° when one symbol (0 or 1) is transmitted, and does not change when the other symbol (1 or 0) is transmitted. This behavior can be detected by a relatively simple receiver.

(2) Quadrature PSK

QPSK (known as quadrature phase-shift keying, quaternary-PSK, or quadriphase PSK) employs orthogonal sinusoidal carriers derived from a common source. Using

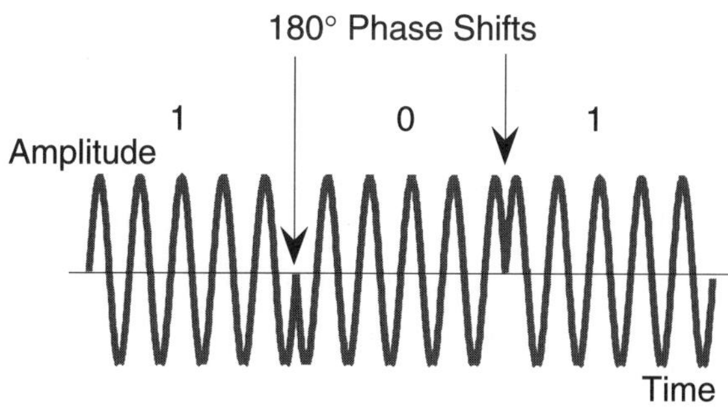

FIGURE 4.20 Binary-PSK

QPSK, we achieve a spectral efficiency of 2 bits/s/Hz by transmitting symbols at rate $2r$ symbols/s in a bandwidth of r Hz. To do this, the datastream is divided in two streams (of rate r symbols/s) and applied to the orthogonal carriers ($\cos\omega_c t$ and $\sin\omega_c t$). Improved operation is achieved by delaying one of the datastreams by one *bit* period. Known as *offset*-QPSK (OQPSK), phase changes are limited to 90°; this produces a beneficial effect which will be discussed later. **Figure 4.21** shows signal space diagrams for binary-PSK and QPSK. For binary-PSK, signal points (1 and 0) are arranged on the horizontal axis equidistant from the origin. The point representing 1 is used as the reference phase (0°) and the point representing 0 is 180° out of phase. For QPSK, signal points (1 and 0) are arranged on the horizontal and vertical axes. The set on the horizontal axis represents the signal states assumed by $\cos\omega_c t$ when phase-modulated by one of the datastreams, and the set on the vertical axis represents the signal states assumed by $\sin\omega_c t$ when phase-modulated by the other datastream. The four points 00, 01, 10, and 11 are the states assumed by the composite signal.

Phase-shift keying produces abrupt changes in phase that translate into substantial changes in the amplitude of the modulated wave on the output side of the bandwidth-limiting transmit-filter. In a communication system, when passed through equipment with non-linear characteristics (the power amplifier of a

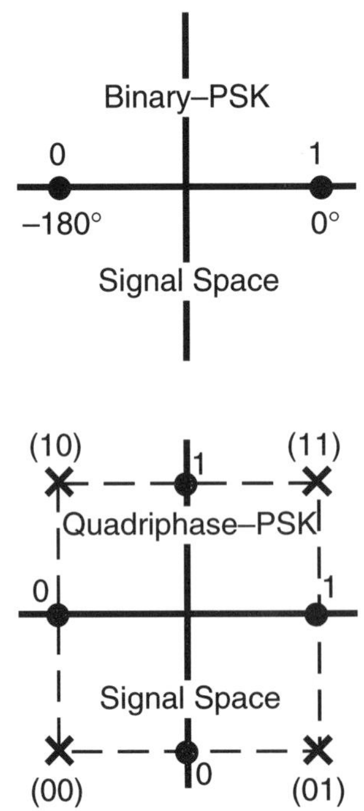

FIGURE 4.21 Signal Space Diagrams for Binary-PSK and QPSK

transponder in a communication satellite, for instance), these amplitude variations are suppressed—giving rise to additional spectral components outside the authorized bandwidth, and producing interchannel interference. Because the phase changes in OQPSK are limited to 90°, its performance in this sort of environment is better than QPSK which relies on phase changes of 180°.

4.7.4 Frequency-Shift Keying

In frequency-shift keying, the modulated wave is a sinusoid of constant amplitude. As shown in **Figure 4.22** FSK can be generated by OOK so that the presence of a signal at one frequency corresponds to a 1-state, and the presence of a signal at another frequency corresponds to a 0-state. When only two frequencies are employed ($M = 2$), the FSK wave is known as a *binary*-FSK (BFSK) wave. To produce more complex symbols, several frequencies are employed. Usually, they are spaced at equal intervals from one another and distributed symmetrically about the carrier frequency. Thus, a set of frequencies can be described as $f_n = f_c + a_n f_d$, where f_n is the frequency corresponding to symbol n, f_c is the carrier frequency, $a_n = \pm 1, \pm 3, \ldots \pm (M - 1)$ with M even, and f_d is the spacing between consecutive fiequencies.

For binary-PSK created by a random datastream and $f_d = {}^{r}/_{2}$, the distribution of power density about the carrier frequency is shown in **Figure 4.23**. Impulse functions associated with discontinuities caused by the keying process occur at the frequencies employed to represent the 1 and 0 states ($fc \pm {}^{r}/_{2}$). The main lobe of the signal spans the frequency range $fc - {}^{3r}/_{2}$ to $fc + {}^{3r}/_{2}$ Hz, a range that is 50% greater than with ASK; however, the power in the sidelobes drops off more quickly. This particular version of FSK is known as Sunde's FSK, after its inventor E. D. Sunde, a 20th century American engineer.

(1) Continuous-Phase FSK

How the symbol frequencies are generated has an important influence on the modulated signal. If the symbol frequencies are derived from a bank of M independent sources, discontinuities will occur when one symbol changes to another. Under this

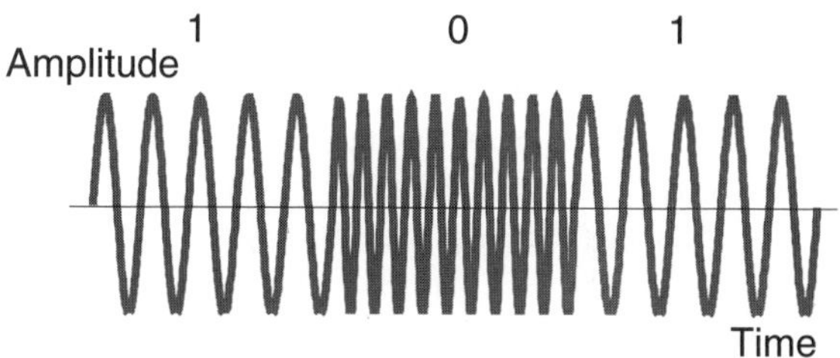

FIGURE 4.22 Binary-FSK

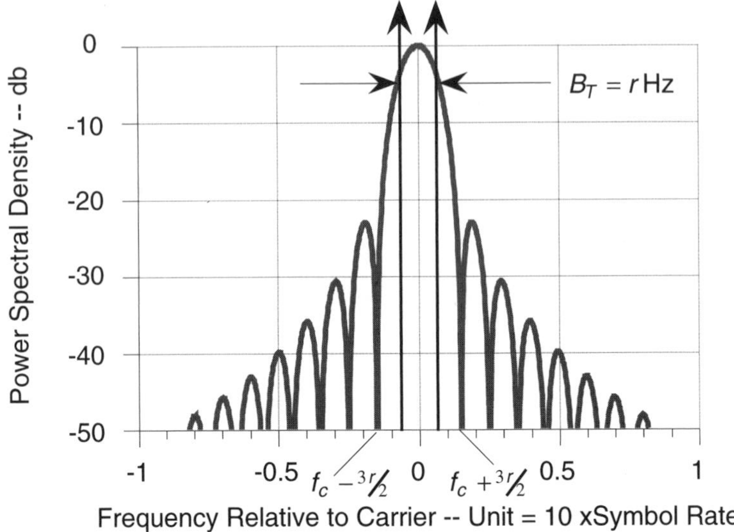

FIGURE 4.23 PSD of FSK Signal Modulated by Random Datastream
The impulses at the frequencies representing the two symbols are produced by power associated with the keying process.

circumstance, the modulation is described as *discontinuous* FSK. If the symbol frequencies are derived from a single source by frequency modulation, discontinuities can be avoided and the modulation is described as *continuous-phase* FSK (CPFSK). The amplitude of the modulated wave remains constant and interchannel interference effects are avoided. In order to achieve continuity of phase across the boundary between symbols, the phase of the frequency representing the next symbol must be adjusted to compensate for a phase shift introduced by the frequency keying process. Thus, CPFSK embodies both frequency shifts and phase shifts. In binary-CPFSK, the phase shift between symbols is 90°.

(2) Minimum-Shift Keying

In minimum-shift keying (MSK), advantage is taken of phase continuity. To represent the datastream, the waveforms impressed on the quadrature carriers are sinusoidal (rather than NRZ pulses). Employing $f_d = r/4$, i.e., one-half of the spacing used in Sunde's FSK, the distribution of power density about the carrier frequency is shown in **Figure 4.24**. Because of the continuous-phase property, the impulses present in FSK are eliminated and the sinusoidal datastream waveforms produce smaller sidelobes. The main lobe spans the frequency range $f_c - 3r/4$ to $f_c + 3r/4$, a range that is 25% less than ASK and 50% less than FSK. The power in the sidelobes drops off very quickly; more quickly than FSK and much more quickly than ASK.

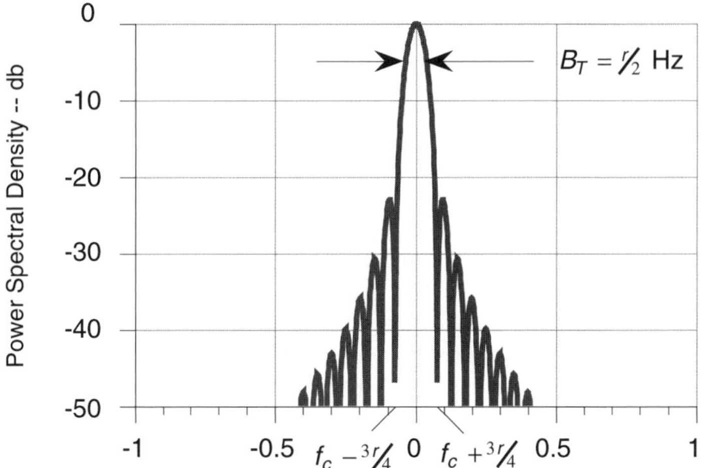

FIGURE 4.24 PSD of MSK Signal Modulated by Random Datastream
MSK achieves improved performance by ensuring phase continuity as the symbols change, and employing sinusoidal signals to represent the datastream.

4.7.5 Duobinary Encoding

From the discussion of CPFSK and MSK, we can imply that the PSD (power spectral density) of a digitally-modulated wave depends on the style of waveform employed to represent the datastream at modulation. The greater the maximum frequency of the baseband signal, the wider the PSD of the ensuing modulated wave. *Duobinary encoding* is a technique that, applied to the baseband signal, effects a reduction in the maximum frequency it contains, and doubles the transmission capacity of the system. Acting as a specific embodiment of *correlative coding* (also known as *partial response signaling*), it adds intersymbol interference to the transmitted signal in a known way so that its effect can be interpreted in the receiver. In practical terms, duobinary signaling achieves a signaling rate of $2B_T$ symbols/s in a bandwidth of B_T Hz. This is the maximum rate predicted by Nyquist (see § 4.6.2).

4.7.6 *M*-ary Systems

When using M-ary signaling, M distinct symbols are employed; each represents $\log_2 M$ bits. As noted (§ 4.7.2), the binary data rate $r_b = r\log_2 M$, where r is the symbol rate (symbols/s), $r_b/B_T = \log_2 M$ and . The symbols are generated by changing the amplitude, phase, or frequency (alone, or in combination) of a carrier (either a single carrier, or a coherent quadrature-carrier arrangement) in M discrete ways. M-ary signaling schemes are used to transmit datastreams over bandpass channels when the desired data rate is more than can be achieved using binary-ASK, -PSK, or -FSK. With increasing amounts of power in a fixed bandwidth channel, it is possible to increase the binary bit-rate several fold. Thus: when $B_T = 1$ MHz and $M = 2$, $r = 1$ Msymbols/s and $r_b = 1$ Mbits/s; when $B_T = 1$ MHz and $M = 4$, $r = 1$ Msymbols/s and $r_b = 2$ Mbits/s; and, when $B_T = 1$ MHz and $M = 256$, $r = 1$ Msymbols/s and $r_b = 8$ Mbits/s. The spectral efficiencies for these examples are 1, 2, and 8 bits/s/Hz respectively. Because noise affects the amplitude of the received signal, ASK techniques are inefficient in noisy environments. Because they are constant amplitude signals in which the information resides in parameters not greatly affected by amplitude perturbations, PSK and FSK perform better in noisy environments. As M increases, however, the transmitted power must increase to maintain the signal-to-noise ratio above an operating limit. Thus, the selection of a digital modulation scheme is a complex tradeoff among bandwidth (symbol rate), bit-rate, and permissible radiated power. In practice, QPSK is widely used because it offers attractive power/bandwidth performance for acceptable levels of error.

4.7.7 Spread Spectrum Signals

First used by the military because it is hard to detect and almost impossible to jam, spread spectrum modulation (SSM) is now employed in global positioning systems (GPSs), mobile telephones, personal communication systems (PCSs), and very-small-aperture terminal satellite communication systems (VSATs).

- **Spread spectrum modulation** (SSM): modulation technique in which the modulated wave occupies a much larger bandwidth than the minimum bandwidth required to transmit the information it carries. Commonly, this is accomplished through the use of a pseudo-noise sequence as *spreading* function. Because spread spectrum signals use a wide frequency band, the *average* signal power per unit bandwidth (in watts/Hz) can be very low.

(1) Pseudo-Noise Sequences

- **Pseudo-noise** (*pn*) **sequence**: a periodic binary sequence with a noiselike distribution of 1s and 0s. Generated by a feedback shift-register, the parameters of the sequence are determined by the following:
  - *length*: depends on the number of stages in the shift register. Commonly expressed in *chips* (see below).
  - *composition*: depends on the initial state of the register, the number of feedback connections, and the logic in the feedback loop(s).
  - *period*: depends on the length (in chips) and the rate at which the register is driven (in chips per second).

- **Chip**: the output of a code generator during one clock period. The name given to each element of a *pn*-sequence.
- **Maximum-length** (*m*) **sequence**: a *pn*-sequence produced by an *m*-stage feedback shift-register whose period is 2^m - 1 chips.

To illustrate the generation of *pn*-sequences, **Figure 4.25** shows a 3-stage shift-register in which the state of the second stage is combined with the output state to provide feedback to the first stage in accordance with the truth table shown. Suppose the stages are set to contain 1, 0, 0. On the first shift command, the sequence 1, 0, 0 is shifted to the right so that the state of stage 2 becomes 1, stage 3 becomes 0, and 0 is present on the output line. With inputs of 0 and 1, the feedback logic produces a 0 that is entered in stage 1 to make the register read 0, 1, 0. On the second shift command, this sequence is shifted to the right so that stage 2 becomes 0, stage 3 becomes 1, and 0 is present on the output line. With inputs of 0 and 0, the

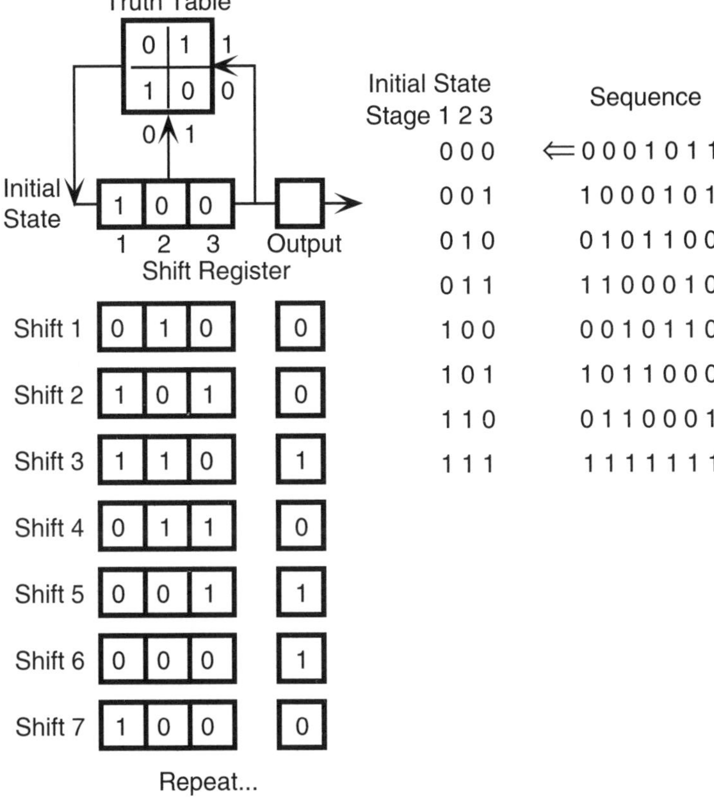

FIGURE 4.25 PN-Sequences Created by 3-Stage Shift Register

feedback logic produces a 1 that is entered in stage 1 so that register reads 1, 1, 0,...and so on. On the seventh shift command, the register stages return to 1, 0, 0 and the process repeats. The simple arrangement of Figure 4.25 produces a repetitive code made up of the seven chips 0 0 1 0 1 1 0.

Different initial conditions produce different results. In Figure 4.25, we list the eight possible sequences. Ignoring the all 1s case, we have seven sequences that contain three 1s and four 0s. It is important to emphasize that they are unique to the feedback arrangement shown. Changing the feedback logic to an exclusive-or produces a set of seven, 7-chip sequences in which there are four 1s and three 0s, and an all 0s sequence (which is discarded). These simple examples mimic the properties of the general *m*-sequence produced by a shift-register of *n*-stages:

- **Length**: 2^n - 1 chips per sequence
- **Number**: 2^n - 1 different sequences
- **Composition**: (number of 1s) = (number of 0s) ± 1.

An immense range of register sizes and feedback connections have been explored to find useful codes of almost any length.

(2) Spreading Techniques

To create a spread spectrum signal, a *pn*-sequence is employed to convert the (narrowband) information signal to a very wideband modulated signal. Two basic techniques are used:

- **Direct-sequence** (DS) **spread spectrum**: the information signal modulates a *pn*-sequence to produce a signal that extends over the wideband spectrum of the digital sequence. (Figure 2.5 shows the wideband spectrum of a symmetrical pulse train). In turn, this signal is used to modulate a sinusoidal carrier (PSK, for instance). The speed of the digital sequence is known as the *chip* rate. It may be as high as 100 Mchips/s. The DS signal occupies the entire spectrum space, the power density is low everywhere, the appearance is noise-like, and the signal level can be significantly below ambient noise. At the receiver, the spread spectrum signal is modulated a second time by the *pn*-sequence to recapture the original information signal. The use of this spreading code during demodulation, spreads interfering signals and reduces their effect on the wanted signal. The number of simultaneous users is limited by the cross-correlations of their spreading sequences. By restricting knowledge of the *pn*-sequence to the transmitter and receiver, the communicators can be assured of a significant level of privacy.
- **Frequency hopping** (FH) **spread spectrum**: the information signal modulates a sinusoidal carrier to produce a modulated wave. Subsequently, the frequency of the modulated wave is changed by mixing with individual frequencies that are selected by a *pn*-sequence (the spreading signal). As a result, the

transmitter *hops* from one frequency to another over a wide frequency band; the FH signal occupies a narrow bandwidth for a very short time before shifting to another position in the spectrum. While the average power density across the spectrum space is low, it is high in the bandwidth that happens to be occupied and must be above the noise level for the signal to be detected. To reclaim the information signal, the receiver must know the frequency-hopping sequence of the transmitter. When several cooperating transmitters are operating, multiple signals may appear at the same time in the same bandwidth. This creates interference and errors; they set a limit to the number of channels that may be operated simultaneously. As before, by restricting knowledge of the hopping sequence to the transmitter and receiver, the communicators can be assured of a significant level of privacy.

Example 4.2: Power Densities in Spread Spectrum Signals

Given. A 1 watt, information-bearing signal of bandwidth 10 kHz (average power per unit bandwidth = 100 μwatts/Hz).

Problem. Determine the average power density when the signal is spread by a 10 Mchip/s *pn*-sequence, 1) in a bandwidth of 10 MHz using a direct sequence technique, and 2) in a bandwidth of 100 MHz using a frequency hopping technique (assume the frequency makes 10,000 hops/s).

Solution. 1) Employing a 10 Mchip/s *pn*-sequence to spread the signal over 10 MHz creates a signal with an average power per unit bandwidth of $\frac{1\text{ watt}}{10\text{ MHz}} = 0.01\mu\text{watts/Hz}$. The average power in 10 kHz will be 0.1% of the transmitted power (a reduction of 30 db).

2) Without overlap, the signal can be occupy any one of $\frac{100\text{ MHz}}{10\text{ kHz}} = 10{,}000$ slots. If the hopping rate is 10,000 times per second, on average, a slot will be selected once per second. Each time it is selected, the signal will dwell on it for 100 μs, i.e., the signal is present for 0.01% of the time. The average power density is $\frac{100\text{ MHz}}{10\text{ kHz}} = 0.01\mu\text{watts}$.

However, the power is concentrated at one frequency or another, and is a total of 1 watt in the occupied slot.

(3) DS and FH Systems

Figure 4.26 shows the principles incorporated in direct sequence and frequency hopping systems. In DS spreading, the *pn*-sequence is employed at the first modulator in the transmitter to produce a wideband, spread signal, and at the last demodulator in the receiver to *despread* the wideband signal. At the second modulator in the transmitter, the spread signal modulates the carrier to produce the direct sequence spread spectrum signal. In the wideband channel, this signal is distributed

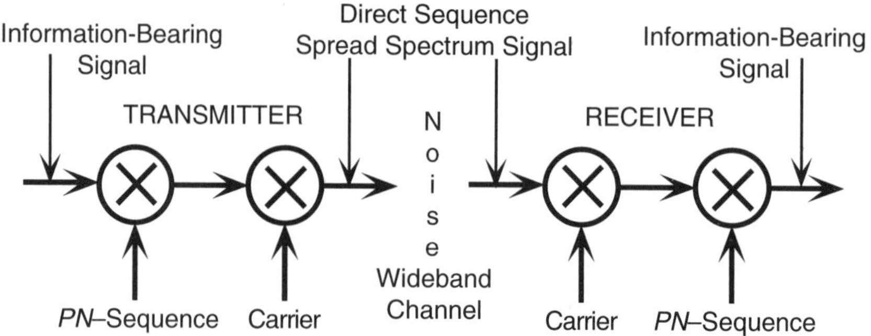

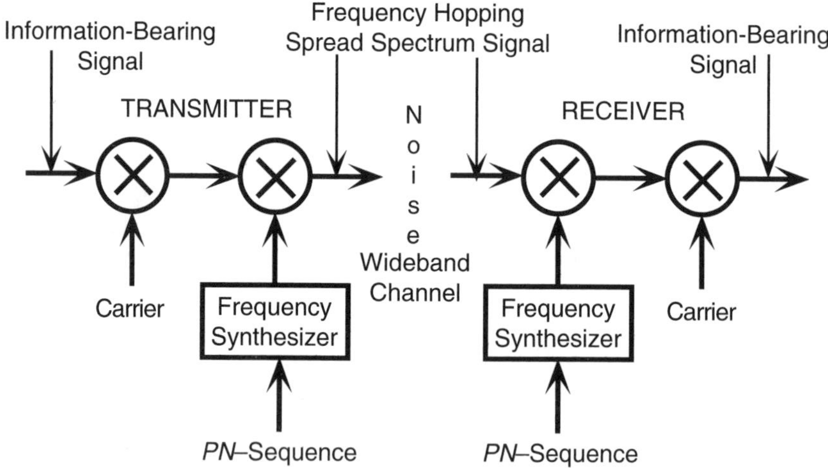

FIGURE 4.26 Principle of Direct Sequence and Frequency Hopping Spread Spectrum Systems

across the channel and picks up noise and interference from other transmitters. At the first demodulator in the receiver, the carrier is removed to yield the spread signal and noise. The final demodulator in the receiver despreads the wanted signal and spreads any interfering signals. In FH spreading, the *pn*-sequence is used to create a pseudo-random set of frequencies that are mixed with the modulated carrier to achieve frequency hopping. In the wideband channel, the hopping signal is narrowband. It picks up noise, and interference from other transmitters that may occasionally be on the same frequency.

In **Figure 4.27**, we provide waveform information to illustrate the use of a *pn*-sequence in the direct spreading application of Figure 4.26. At the transmitter a

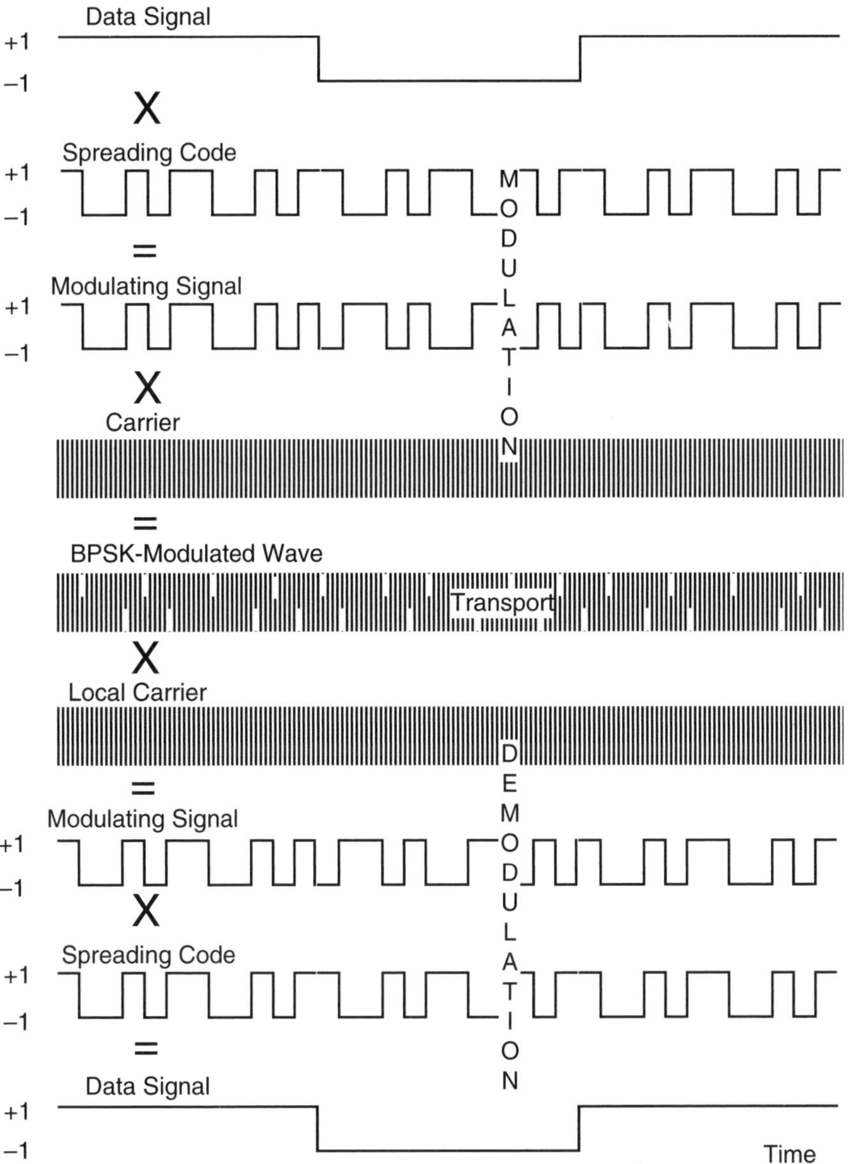

FIGURE 4.27 Principle of Direct Sequence Modulation and Demodulation

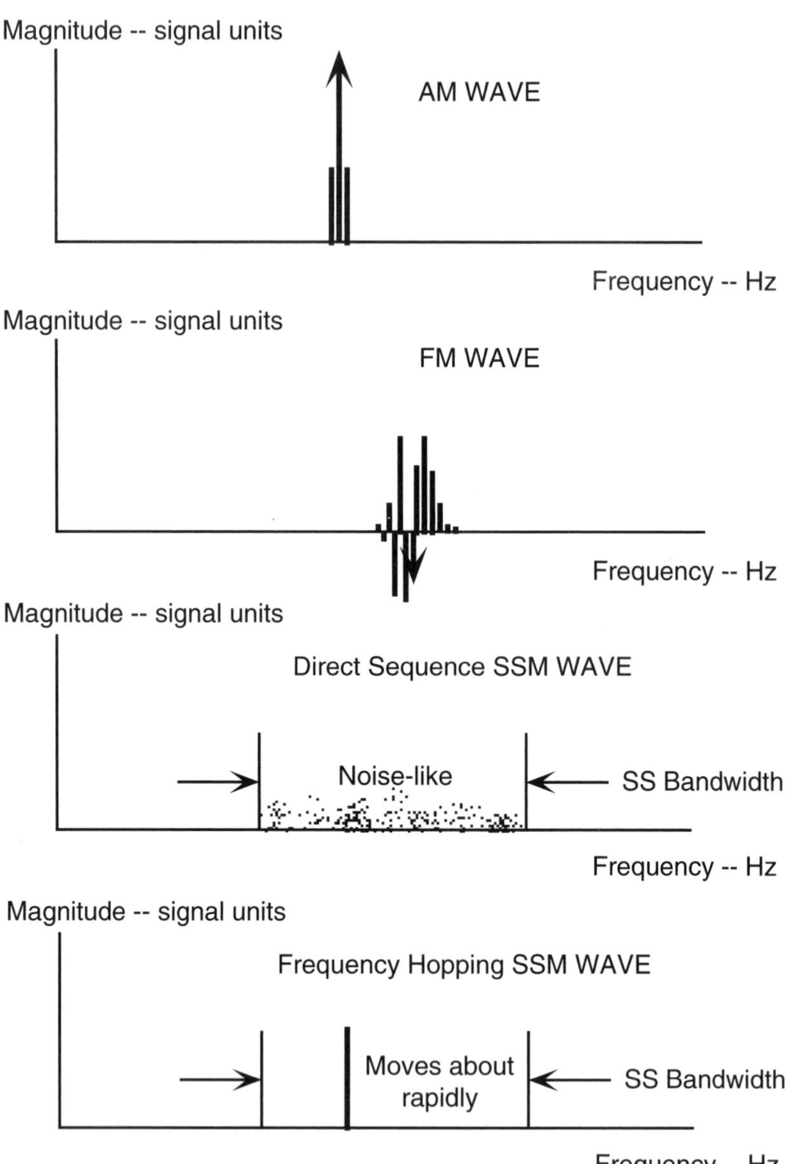

FIGURE 4.28 Frequency Plane Representations of Amplitude, Frequency, and Spread Spectrum Modulated Waves

data signal modulates a *pn*-sequence. In turn, the composite digital signal is used to create BPSK of a sinusoidal carrier. At the receiver, the composite digital signal is reclaimed. On multiplication by the original *pn*-sequence, it yields the original data signal. Should interfering signals be present, multiplying them by the *pn*-sequence spreads their spectrums and reduces their negative effect. A prerequisite for successful operation is that the spreading signal and the carrier signal at the receiver are synchronized with their counterparts in the transmitter.

By way of summary, in **Figure 4.28**, we compare spectral density representations of AM, FM, and SSM signals.

(4) Code Division Multiple Access

Using unique spreading sequences that are orthogonal to each other, more than one user at a time can transmit in the same bandwidth. Called *code division multiple access* (CDMA), cooperating stations transmit at the same carrier frequency with approximately equal power and the same chip rate using DS or FH techniques. Each transmitting station employs a code that is uncorrelated with the codes used by the other stations. In practice, the spreading sequences in the set will have a cross-correlation with the rest of around -20 db. The receivers see the sum of the individual signals as uncorrelated noise. Each station can demodulate one, or more, signals if it has been provided with the associated spreading code(s) and the carrier frequency. For reliable performance in a multi-user environment, it is important to control the power level of competing signals to prevent higher power signals from capturing the receiver.

5

Data Communication

Data communication involves the distribution or exchange of data messages between machines that process data. In this activity the sequences of bits must be preserved while the signals that represent them are converted to forms compatible with the different phases of generation, transport, and reception. Specialized equipment makes these transitions and the processors follow common procedures. In this Chapter, we describe the hardware and software functions required to implement communication between data processing machines.

5.1 CONNECTIONS

5.1.1 DTEs and DCEs

- **Data terminal equipment** (DTE): devices that send or receive data messages.
- **Data circuit terminating equipment** (DCE): signal conversion devices.

DTEs create, send, receive, and interpret data messages. They are connected to DCEs that condition signals received from the DTE for transmission over communication connections, and restore signals received from the network so as to be compatible with the receiving DTEs. **Figure 5.1** shows a range of DTE-to-DCE-to-network connections. We explain the configurations below.

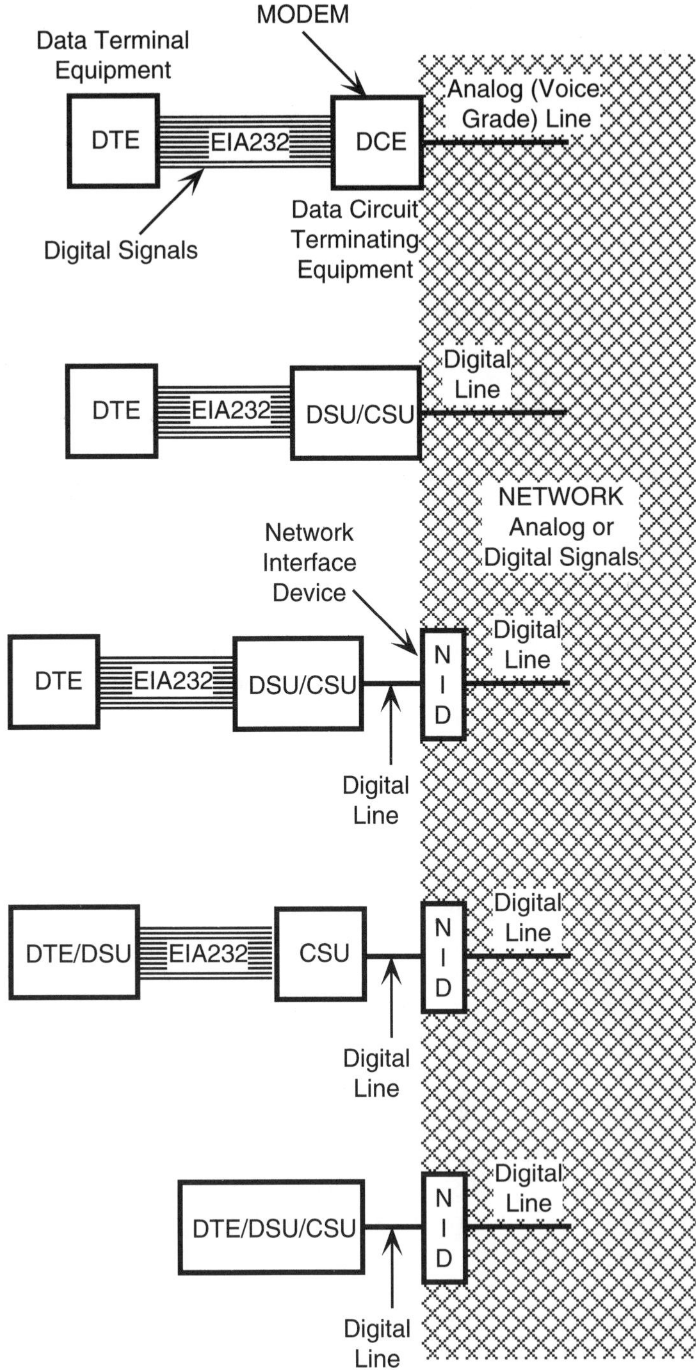

Figure 5.1 DTE–DCE–Network Connections

(1) Data Terminal Equipment (DTE)

DTEs are digital devices. Internally, their signals are likely to be simple, unipolar pulses; externally they may use one of the more sophisticated digital signaling schemes described in § 4.5.1. Most DTEs are called *terminals*. They can include:

- **Teletypewriters** (TTYs): devices that accept keyboard input, print hard copy output, and store and forward messages
- **Video display terminals** (VDTs): devices that accept keyboard input and display output on a video screen
- **Transaction terminals**: devices that capture and process on-line transactions; for example, automatic teller machines and supermarket checkout terminals with bar-code readers
- **Intelligent terminals**: devices programmed to perform some of the tasks usually assigned to a host machine—such as screen formatting and data editing, or more complex assignments—so as to reduce dependence on the central processor
- **Smart terminals**: devices that contain memory and logic to perform specific tasks—such as facsimile machines or order entry devices
- **Dumb terminals**: devices that depend on the host for all but the most routine functions.

However, the designation DTE includes a broader category of equipment including:

- **Workstations** or **personal computers**: devices that contain significant processing power in their own right, but have need to access other processors
- **Front-end processors**: equipment that performs all of the communication-related functions at a host site. The front-end processor (FEP) relieves the much faster host of activities associated with the relatively slow processes required for data communication.

(2) Data Circuit Terminating Equipment (DCE)

A DCE is a signal conversion device that assists DTEs to send or receive data messages over data circuits. When sending, DCEs modify the digital signals supplied by the DTEs to match the transmission characteristics of the circuits that effect communication-at-a-distance:

- If the connections are analog (voice-grade) connections, the DCE is called a *modem*. When sending, it converts the digital signals received from the DTE to analog signals to match the bandwidth of the line. When receiving, the DCE converts the analog signals to digital signals and passes them on to the DTE.

- If the connections are digital connections, the DCE consists of two parts:
  - *digital service unit* (DSU). Over EIA 232 and similar cables (see § 5.1.2), a DSU receives unipolar digital signals from the DTE at 2.4, 4.8, 9.6, 19.2, 56, or 64 kbits/s, and 1.544 Mbits/s. The DSU converts them to bipolar signals. When sending, a DSU performs special coding to make the signals compatible with the operation of the transport system—such as zeros substitution codes for subrate data and 56 kbits/s applications (see § 8.2.2). When receiving, a DSU removes any special codes inserted by the sending DSU and converts the bipolar signals to unipolar signals compatible with the DTE.
  - *channel service unit* (CSU). A CSU provides line conditioning, loopback (for testing), limited diagnostic capabilities, and, if needed, span powering for repeaters. When sending, it converts bipolar signals to AMI; when receiving, it converts AMI signals to bipolar. When operating at 1.544 Mbits/s, a CSU performs eight zeros substitution coding, and supports special framing requirements—such as superframe and extended superframe (see § 8.2.1).

Figure 5.1 shows DSU/CSU installations. DSUs and CSUs may be combined in the same piece of equipment, DSUs may be incorporated in DTEs, and DSUs and CSUs may be incorporated in DTEs. Furthermore, because the FCC defines CSUs as customer provided equipment (CPE), carriers may install a network interface device (NID) that performs carrier-controlled testing, loopback, and powering functions.

5.1.2 EIA232 Interface

Commonly, a DTE is connected to a DCE by a cable that conforms to EIA standard EIA232. Figure 5.1 shows the arrangement. EIA232 describes a multi-wire cable that terminates in 25-pin connectors. The cable supports asynchronous or synchronous operation at speeds up to 19.2 kbits/s. At 19.2 kbits/s, cable length is limited to 50 feet. The EIA232 circuits linking DTE and DCE carry signals that initiate, maintain, and terminate communication between the two. **Table 5.1** provides information on the signals that flow over the cable. In the parallel transmission of data between the terminating devices:

- Data from the sending DTE (DTE1) appear on pin #2 and are transmitted over circuit AB to the sending DCE
- The data stream is conditioned to suit the circuit being used and passes through the network to the receiving DCE (DCE2)
- At DCE 2, the signal is converted to a digital datastream, placed on pin #3 and transmitted over circuit BA to the receiving DTE (DTE2).

Before transmission can start, the system must be *ready*. Marks (1s) present on circuits CA (request to send), CB (clear to send), CC (DCE ready), and CD

Table 5.1 EIA232 Interface Circuit Definitions

PIN#	CIRCUIT	NAME	DIRECTION	COMMENTS
		Ground Connections		
1	AA	Protective Ground		Machine ground
7	AB	Signal Ground		Ground reference for all circuits
		Data Signals		
2	BA	Transmit Data	DTE to DCE	Data generated by DTE
Circuit BA may only be used if a mark condition exists on CA (Request to send), CB (Clear to send), CC (DCE ready), and CD (DTE ready).				
3	BB	Receive Data	DCE to DTE	Data received by DTE
		Control Signals		
4	CA	Request to Send	DTE to DCE	DTE has message to transmit
5	CB	Clear to Send	DCE to DTE	DCE is ready to accept and transmit DTE data
6	CC	DCE Ready	DCE to DTE	DCE is ready to operate
20	CD	DTE Ready	DTE to DCE	DTE is ready to operate
22	CE	Ring Indicator	DCE to DTE	DCE is receiving a ringing signal
8	CF	Signal Detect	DCE to DTE	DCE is receiving a carrier signal
21	CG	Signal Quality Detector	DCE to DTE	OFF indicates may be errors in received data
23	CH/CI	Data Signaling Rate Selector/Indicator	CH to DCE; CI to DTE	Selects higher of two signaling rates
		Timing Signals		
24	DA	DTE Transmit Clock	DTE to DCE	To help detect center
15	DB	DCE Transmit Clock	DCE to DTE	of signaling element on BA
17	DD	DCE Receive Clock	DCE to DTE	To help detect center of signaling element on BB
		Special Signals		
14	SBA	Secondary Transmit Data	DTE to DCE	)
16	SBB	Secondary Receive Data	DCE to DTE	)
19	SCA	Secondary Request to Send	DTE to DCE	) Special Operations
13	SCB	Secondary Clear to Send	DCE to DTE	)
12	SCF	Secondary Signal Detect	DCE to DTE	)

(DTE ready), fulfill this condition. Four versions of the EIA232 standard have been issued. The latest, EIA232D is slowly displacing EIA232C. Similar standard interfaces have been adopted by ITU-T. Known as V.24 and V.28, they are functionally similar to EIA232. In addition, ITU-T's V.35 supports speeds to 4.8 kbits/s and X.21 provides connections to public data networks.

(1) Synchronizing DTE and DCE

It is imperative that the DTE/DCE combinations at each end of the communication circuit be synchronized. To ensure the integrity of the bit stream:

- DCE1 generates a timing signal (clock signal) and sends it on pin 15 of an EIA232 cable to DTE1;
- As each timing pulse is received, DTE1 responds on pin 2 with the next bit in the message sequence so that the timing of the data bits corresponds to the timing of the clock bits.

Thus, DTE1 and DCE1 keep in lockstep and DCE1 is certain to send the intended bit stream on to DCE2. At the other end:

- DCE2 receives the datastream
- Derives a timing signal from it
- Adjusts the timing signal relative to the datastream
- Sends the datastream to DTE2 on pin 3
- Sends the timing signal (clock) to DTE2 on pin 17.

Thus, DTE2 keeps in lockstep with DCE2. The upper half of **Figure 5.2** shows the process of synchronizing DTE1 to DCE1, and then DCE2 to DTE2. In the lower half of the diagram, the process is reversed. In both cases, DTE1 is synchronized to DCE1, and DTE2 is synchronized to DCE2; DCE1 is not synchronized to DCE2.

(2) EIA449

To serve longer distances and higher speeds, EIA has adopted EIA449. It permits operation at 2 Mbits/s at distances up to 4000 feet. Considerably more complicated than EIA232, it contains two electrical specifications:

- EIA423, an unbalanced circuit version
- EIA422, a balanced circuit version

A more recent standard, EIA530, achieves the same performance as EIA449 with a smaller connector.

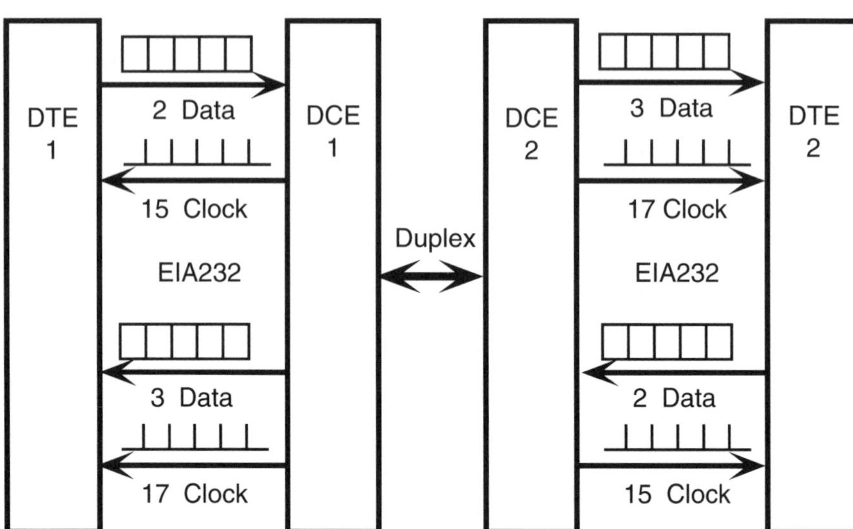

Figure 5.2 Synchronizing the Ends of a Data Circuit
In the upper half of the diagram, DTE1 transmits data to DTE2. DTE1 is synchronized with DCE1 and DTE2 is synchronized with DCE2. In the lower half, DTE2 transmits to DTE1.

(3) Direct Interconnection of DTEs

For short distances, direct interconnection of DTEs fitted with EIA232 ports is possible. However, the cable must be modified by routing the transmit data circuit of one DTE to the receive data circuit of the other DTE, and *vice versa*. Such a cable is known as a *null modem*.

(4) Higher Speed Interconnections

EIA232 and its derivatives cannot handle all of today's requirements. As internal data rates increase, there is a need for higher speed connections. Some systems that fill the gap include:

- **Enterprise systems connection** (ESCON): an optical fiber connection operating up to 40 kilometers at 17 Mbits/s. Introduced by IBM, ESCON extends the mainframe, high speed, I/O channel (see § 12.2.1) to remote peripherals.
- **High-performance parallel interface** (HIPPI): introduced by Los Alamos National Laboratory, and standardized by ANSI, HIPPI provides a connection operating up to 120 feet at 800 Mbits/s on a single cable, and 1600

Mbits/s on two cables. HIPPI includes flow control and error correction capabilities. Over optical fiber, HIPPI operates up to 10 kilometers.

- **High-speed serial interface** (HSSI): supports data transfer at speeds up to 52 Mbits/s. Standardized by ANSI, HSSI describes the physical and electrical interface at the DTE. It can be used with T-3 multiplexers (see § 8.2.3).
- **Fiber channel standard** (FCS): designed to operate up to 10 kilometers at speeds up to 800 Mbits/s on optical fiber. FCS includes error control and switching.

As high-speed interconnection needs develop, more specialized equipment will be introduced. Eventually, a family of FCS units may be standardized.

5.1.3 Modems

(1) Practical Modems

Data are sent over voice-grade lines with the help of a *modem* that creates a stream of unique analog symbols that fits within the bandwidth of the analog line. Section 4.6.4 describes the generation of one such stream. In addition, in Figure 4.15, examples are given of constellations that create 4-bit symbol sets to carry 9600 bits/s over a 2400 baud connection (16-QAM). However, none of them is likely to be used in real-world systems. Instead, so that the margin for error can be about the same for all symbols, the signal points are distributed in a regular fashion over the signal space. **Figure 5.**, shows a regular constellation formed from 16 combinations of 3 amplitudes and 12 phase angles.

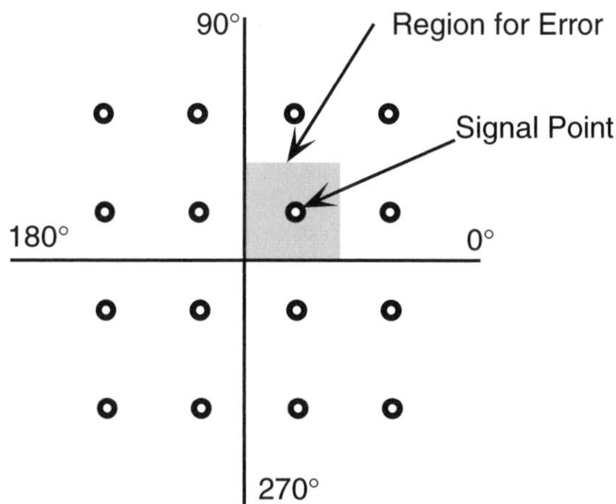

Figure 5.3 Constellation for Practical 16-QAM Modem

Not all modems generate symbols using QAM. Other techniques are:

- **Phase-shift keying** (PSK): a technique in which marks and spaces are represented by a difference in the phase of the carrier. PSK is described in § 4.7.3.
- **Frequency-shift keying** (FSK): a technique in which a carrier at one frequency represents a mark (1), and a carrier at another frequency represents a space (0). It is described in § 4.7.4.

Some modems use an extension of QAM known as trellis code modulation:

- **Trellis code modulation** (TCM): a special form of QAM that adds an extra bit to the n-bit symbol sequence to create twice as many signal points in the constellation as are needed to represent the data. Thus 16-QAM with TCM requires $2^5 = 32$ signal states, not $2^4 = 16$. The last two bits of the symbol set and the TCM-bit, define 8 subsets of the constellation with 4 symbols each. The first two bits of the symbol set identify which member of the subset (i.e., which signal state) is used. Doubling the number of signal points in this manner ensures that contiguous symbols fall far enough apart to be distinguished readily. In the demodulation process, the trellis decoder compares the received symbol with all possible valid symbols, and selects the best fit. As a result, errors are reduced, and TCM modems are able to operate at higher speeds than their non-TCM counterparts (for the same error rate).

Yet other modems use adaptive techniques:

- **Adaptive-bandwidth, adaptive-rate modulation**: on the basis of measurements of the channel characteristics made by the receiver,
  - the sender selects an optimum transmission bandwidth (W Hz, say)
  - the sender and receiver cooperate to equalize the channel signal-to-noise ratio over this bandwidth
  - the sender selects a QAM coding such that the number of bits per symbol meets the error probability objective established for the communication
  - the sender sends W symbols/s.

Information theory estimates that the theoretical maximum speed for a telephone channel is between 20,000 and 30,000 bits/s. Using adaptive-bandwidth, adaptive-rate modulation techniques, speeds in this range have been achieved.

(2) Application of Techniques

Divided roughly according to speed, the use of these techniques is as follows:

- **300 bits/s modems**: most employ FSK. Two frequencies are assigned to one direction, and two different frequencies are assigned to the other. In all, four carrier frequencies are used over a 2-wire, duplex connection. In North America, AT&T 103/113 modems, and in Europe ITU-T V.21 modems, employ this procedure.

- **Multiple-speed modems**: may employ FSK or PSK. AT&T 212A modems employ BFSK at 300 bits/s and BPSK at 1200 bits/s. AT&T 202 modems operate at 1200 and 1800 bits/s with FSK. ITU-T V.22 modems use BPSK at 600 bits/s, and QPSK at 1200 bits/s. ITU-T V.23 modems operate at 600 or 1200 bits/s with FSK.
- **2400 bits/s modems**: most employ PSK. AT&T 201 modems and ITU-T V.26 modems operate at 2400 bits/s with BPSK.
- **4800 bits/s modems**: most employ 8-PSK. AT&T 208 modems and ITU-T V.27 modems operate at 4800 bits/s with 8-PSK.
- **9600 bits/s modems**: AT&T 209 modems use 16-QAM to operate at 9600 bits/s. ITU-T V.29 modems operate at 9600 bits/s using a technique in which the data stream is divided into segments of 4-bits. The first bit determines the amplitude of the symbol signal and the next three bits determine the phase of the symbol signal (i.e., 2 amplitudes and 8 phase angles). ITU-T V.32 modems employ TCM (32-QAM) and echo cancellation to operate at 9600 bits/s over a 2-wire circuit (i.e., 2-wire, duplex).
- **TCM modems**: these modems operate at 2400 bauds, 2-wire, duplex, at speeds of 9600 bits/s (V.32), and 14,400 bits/s (V.32bis). They incorporate trellis coding and echo-cancellation. Other modems operate over private lines to achieve 14,400 bits/s (V.33), 16,800 bits/s, and 19,200 bits/s. If data compression is also used, effective throughput rates of 40,000 to 50, 000 bits/s are reported. The data rate achieved depends on the susceptibility of the datastream to the compression technique employed.
- **TCM modems with precoding and equalization**: using adaptive-bandwidth, adaptive-rate modulation, operation at 24,000 bits/s has been achieved over 4-wire, private lines, and 19,200 bits/s over 2-wire, dialup circuits.
- **packetized-ensemble-protocol multicarrier modems**: perhaps the ultimate in adaptive-bandwidth, adaptive-rate modulation modems, one embodiment employs a comb of 512 carrier tones within the voice band to send QAM symbols in parallel. To start, the originating modem transmits all 512 tones; the receiver evaluates the signals and notifies the transmitter of tones that cannot be used because of noise, or for other reasons. With this information, the originating modem selects a signaling rate, breaks the datastream into packets that contain a fixed number of frame check sequence symbols and as many message symbols as there are acceptable carrier tones, and transmits a packet-at-a-time, in parallel, on the available carriers. If the receiver detects an error, it generates a request for retransmission of the packet. Rates up to 19.2 kbits/s can be achieved with these modems.

(3) Modem Compatibility

In listing some popular types of modems, we have purposely kept away from the details. It is important that the reader not assume that modems are compatible just because the speeds are the same. There is a wide range of operating parameters among them. Besides operating speeds, and exact operating frequencies and signal constellations, the performance depends on operating mode (i.e., half-duplex or duplex), facility type (i.e., dial-up or private line), and asynchronous or synchronous operation. Although many modern modems adhere to ITU-T standards, the wide variety of in-place modems makes compatibility an issue to be considered whenever voiceband data circuits are being established. For users with international requirements, the compatibility of in-place modems can be particularly troubling. Except for 2400 bits/s, V.26 modems that are functionally compatible with AT&T 201B modems, reliable matches do not exist. The only prudent strategy is to use ITU-T modems exclusively on international circuits.

(4) Microcom Networking Protocol

When passing data through a modem and over dial-up links, error detection and correction is not available (except for parity checking). It can be obtained through the use of a proprietary product known as MNP (Microcom Networking Protocol). By assembling the data in blocks and performing cyclic redundancy checks, MNP permits the receiving modem to verify the integrity of the received stream. If an error is detected, the receiving modem requests re-transmission of the block. Later versions of MNP include data compression, automatic negotiation to achieve the highest data speed compatible with link conditions, and management of half-duplex connections so as to optimize throughput in both directions. MNP-configured products exist in several versions (known as Classes) that focus on different operating environments. Some of them have been incorporated in ITU-T V.42, a recommendation covering error correcting protocols for modems.

(5) Modem Training

For error-free performance, training signals are sent from transmitting modem to receiving modem to ensure that, at the receiver:

- The equalizing and detecting circuitry are adjusted to differentiate among the amplitude and phase components of the signal points in the same manner as the transmitter
- The recognition circuitry assigns the same sequence of bits to each symbol as are assigned at the transmitter.

Modem training is an important part of the set up of a data circuit. As the complexity of modem processes increases, synchronizing, stabilizing, and calibrating the receiving modem so that it can identify all the received symbols correctly becomes increasingly arduous. A 2400 baud, 9600 bits/s modem employs 16 unique symbols and the receiving modem must be able to recognize all 16 without errors. For higher speeds, the recognition problem expands rapidly. **Figure 5.4** shows typical training times for 2400 baud modems. As might be expected, they increase exponentially with speed.

5.2 COMMUNICATION PROCEDURES

In person-to-person communication, the basic rules were worked out thousands of years ago. For machine-to-machine communication, there is little history, and the rules must be all-encompassing so that nothing is left to chance. *If it can happen, it will happen* is the maxim that must guide those who design protocols for the machine-centered environment. In this section we discuss the rules and procedures needed to send data over a data link.

5.2.1 Range of Protocol Functions

Communications protocols must include rules and procedures to cope with all normal and abnormal operating situations. Some of them are:

- **Startup**: how does communication begin? Does the sender gain access to the line by being polled, or by competing for it? Can any station initiate commu-

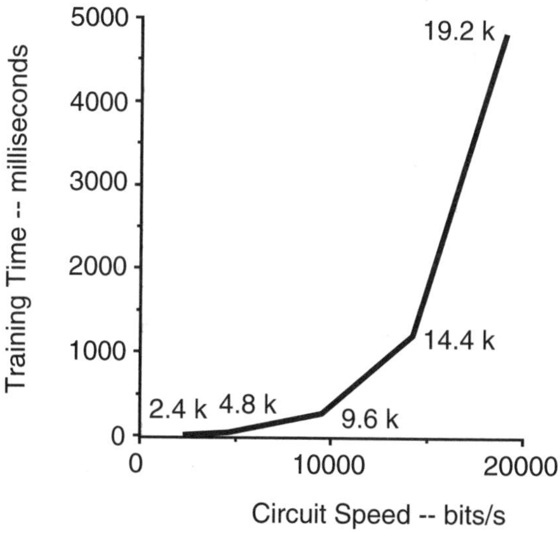

Figure 5.4 Training Times for 2400 Baud Modems

nication, or is one station in control? Once interrupted, how is communication re-established?

- **Message format**: how are data organized and in what sequence are they sent? How are long messages handled? How are short messages handled?
- **Identification and framing**: how does the receiver separate the bit stream into characters? How does the receiver distinguish between control information and data characters? How does the receiver separate header information from text information, and text information from trailer information? How are frames identified?
- **Line control**: in half-duplex operation, under what circumstances is the line turned around?
- **Error control**: what process will be used for the detection and correction of errors? How, and when, does the receiver inform the sender that no errors have been detected in the frame(s) received, or, conversely, that errors have been detected?
- **Termination**: under normal conditions, how is communication terminated? Under abnormal conditions, how is communication terminated?

In addition, to be user friendly, communication procedures should have the following attributes:

- **Transparency**: they must function in the presence of all bit sequences that occur in the text portion of the frame. Specifically, if data bytes and control octets share the same bit patterns, there must be a way to distinguish between a bit sequence that is data, and the same bit sequence that is used for a control purpose.
- **Independence**: their performance should not be influenced by the code employed.
- **Efficiency**: the number of bits added to the frame for control purposes must be small so as to devote most of the transmission time to sending information.

(1) Half-Duplex Line Discipline

In half-duplex transmission, data flow in one direction at a time:

- Usually, one station is the primary station; it determines when secondary stations shall send, and when they shall receive;
- Sometimes, information is sent frame by frame with the sending station pausing between frames for the receiver to acknowledge receipt of an error-free frame, or to request re-transmission of the frame. This is stop-and-wait ARQ error correction (see § 4.2.3(1)).
- Often, the sending station sends a predetermined number of frames before pausing for acknowledgment. If a frame contains an error, the receiver dis-

cards it, ignores any subsequent frames sent, and, when able to do so, requests resending of all frames from the one found in error. This is go-back-n ARQ error correction (see § 4.2.3(1)).

- Because a limited number of bits are allocated to frame numbers, the receiver must acknowledge error-free receipt of frames at least once within each numbering cycle. If acknowledgment is not made, the sender must stop sending until it is.

(2) Duplex Line Discipline

In duplex transmission, data flow in both directions at the same time, and the stations may be receiving and transmitting simultaneously:

- For equal status stations, either station may initiate communication. To begin communication the station with something to send must inquire whether the other station is ready to receive the message.
- For unequal stations, one station will be designated primary and the other(s) will be designated secondary station(s). The primary station manages message flow (see polling, below).
- Information is transmitted as a sequence of frames. Sending is interrupted to repeat frames in which the receiver has detected an error. Go-back-n ARQ and selective repeat ARQ error correction are used (see § 4.2.3(1)).
- Because a limited number of bits is allocated to frame numbers, the receiver must acknowledge error-free receipt of frames at least once within each numbering cycle. If acknowledgment is not made, the sender must stop sending until it is.

(3) Polling

For stations of equal status connected on a point-to-point, or multipoint-to-multipoint basis, the stations poll each other:

- **Peer polling**: the stations poll one another, in sequence, continually. If the polled station has something to send, it sends it before polling the next station; if it has nothing ready, it polls the next station in the sequence. When all stations have been polled by another station, the process is repeated. To ensure equal access for all stations, and to provide continuing operation even when some stations have failed, it is essential that one station monitors the total process and is supplied with procedures to restart it if, for any reason, it fails.

For unequal stations connected on a point-to-point, or point-to-multipoint basis, the primary station manages communications by polling the secondary station(s):

- **Primary polling**: in turn, the primary station asks each of the secondary stations if they have anything to send. If a secondary station has a frame (frames) to send, when polled, it transmits it (them) to the primary; if the secondary station has nothing ready to send, it indicates nothing to send and the primary station polls the next station. Under normal operation, the primary station interrogates the secondary stations continually. When the primary station has a frame, or frames, to send, it ceases to poll the secondary stations, and sends its frames. Often, the primary station is a front-end processor that manages the communication process for a host.

5.2.2 Making a Data Call

Suppose an application process running on System 1 is compiling the bill-of-materials for a new product designed in several locations of a corporation on a variety of computers. At a point in the compilation, when the application calls for data on a certain subassembly, the data are found to be absent from the product source file. The activity to locate and to obtain this data might proceed as follows:

- The application process running on System 1 generates a request for the missing data (D).

To emphasize that it is expressed in terms that are derived from the organization of System 1 (i.e., internal naming conventions, storage location numbers, etc.), we shall call the request R(D)S1.

- R(D)S1 is passed to the operating system in System 1 (OSS1).

By means of the subassembly number code, a data directory, or other means, OSS1 determines that the source of D is System 2. (System 2 is a computer remote from System 1.)

- OSS1 converts R(D)S1 to R(D)S2.

Using information describing System 2 (which must be available in System 1) OSS1 converts the request from System 1 understandable terms [R(D)S1] to System 2 understandable terms [R(D)S2] for presentation to System 2. This conversion is essential if System 2 is to know what it is that System 1 wants.

- OSS1 proceeds to establish communication with System 2.

A session procedure is initiated that, with its counterpart in System 2, will supervise communication between System 1 and System 2 for the time required to request, find, and transfer D.

- From among the available resources, a network is selected that connects System 1 and System 2.

The need to send data between System 1 and System 2 causes System 1 to consult a routing table, or other information, to select a communication network (e.g., public data network, public switched telephone network, or private network) over which to transport R(D)S2.

- For the chosen network, the address of System 2 is determined.

This action requires access to a directory that describes the chosen network and identifies the specific connection to be used. In turn, this information defines the transmission mode to be employed (packet or circuit, for instance) and any message requirements that are special to the network (protocols, block length, etc.).

- The front-end processor selects a data link that connects to System 2 through the chosen network.

Using the network information provided, the FEP selects a suitable data link from those it has access to, and provides the resources to support message flow.

- The FEP and System 1 allocate and prepare to manage buffer memory.

Because the speed of the communication circuit is significantly less than the internal speed of System 1, the combination of FEP and System 1 allocates buffer memory to be able to make the necessary circuit speed conversions in the FEP and in the host, and to exercise flow control in both directions.

- The FEP establishes a communication connection to System 2.

Finally, the front-end processor attached to System 1 establishes a physical connection to the front-end processor attached to System 2.

- FEPS1 requests permission to send.

FEPS1 sends a frame to System 2 that requests permission to communicate with System 2.

- FEPS2 refers the request to OSS2.

FEPS2 receives the frame, checks for errors, and refers the request to OSS2. If OSS2 determines that communication with System 1 is permitted, it initiates a session process that will link with that started in System 1 to manage the communication session. Also, OSS2 authorizes FEPS2 to reply affirmatively to System 1.

- FEPS2 and OSS2 allocate and prepare to manage buffer memory.
- FEPS2 replies to FEPS1 with *ready to receive*.
- FEPS1 sends R(D)S2 to FEPS2.
- FEPS2 checks for transmission errors, strips the text from the frame, and passes R(D)S2 to OSS2.
- OSS2 validates R(D)S2.

If OSS2 determines that System 1 can receive D, OSS2 instructs the application on System 2 to prepare it for presentation to System 1.

- The application creates A(D)S2 and sends it to OSS2.

The file server copies the data D to memory, adds any additional data required for the application program on System 1 to use it, and releases control to OSS2. We call this answer A(D)S1.

- OSS2 conditions A(D)S1 for the chosen network.

OSS2 causes the telecommunication software in System 2 to divide A(D)S1 into convenient datablocks and authorizes FEPS2 to send them to System 1.

- FEPS2 encapsulates the datablocks in frames, and sends them in sequence to System 1.
- FEPS1 receives the sequence of frames, checks for errors, strips off protocol bits, transfers control of error-free datablocks to OSS1, and replies to FEPS2 with frames acknowledging error-free receipt, or requesting re-transmission of corrupted blocks.
- FEPS2 completes transmission of datablocks to FEPS1.
- FEPS1 receives the final frame, confirms receipt of all frames to FEPS2, and sends the final datablock to OSS1.
- OSS1 reassembles datablocks into A(D)S1, confirms all datablocks are present, and sends A(D)S1 to the application.
- The combination of OSS1 and OSS2 terminate the communication session.

The hardware and software associated with System 1 and System 2 terminate the communication session, and the application running on System 1 proceeds with compiling the bill-of-materials for the new product.

What we have described is only the bare skeleton of the procedure—to say the least, the steps for effecting communication between cooperating systems to achieve file-transfer, are complicated. Moreover, it is likely that not one but many communications will be going on at the same time; there must be means to prevent one message being confused with another. Too, each mainframe will be executing many jobs simultaneously, some that require communication support, and some that do not; they must be carefully managed to ensure their integrity. Communication among machines is a daunting task.

5.3 OPEN SYSTEMS ARCHITECTURE

5.3.1 OSI Model

Over many years, the world's telecommunication organizations have studied scenarios such as we have presented to develop an architecture for implementing data communication between cooperating systems. One is called *open systems architecture* (OSA). Standardized by ISO, it is embodied in the Open Systems Interconnection Reference Model (OSI model, also referred to as OSIRM). The OSI model shows how to provide a wide range of communication services to a global community of computer systems of different origins and to achieve a state of hardware and software transparency so that data communication can proceed without reference to its implementation. Under these circumstances, an *open system* exists.

The OSI reference model provides a framework for communication that relies on standard interconnection protocols. Because it does not assume a specific system architecture, nor require a particular implementation, the model is able to accommodate evolving technology and changing user requirements. It consists of seven layers that serve to decompose the complexity of information flow between cooperating machines into consecutive steps that are substantially independent of each other. This independence is achieved by allowing the superior level to depend on the functions (services) performed by the next lower level without regard to the way in which the level is implemented. Thus, equipment from different manufacturers can work together, and new equipment can be introduced as technology changes. To ensure success, it is important that each of the layers performs a well-defined set of functions, and that similar functions are assigned to the same layer. This process is illustrated in **Figure 5.5**, and the activities of each layer are summarized in **Table 5.2**.

5.3.2 Description of Layers

Communication between cooperating end systems is accomplished by layer entities that perform the functions defined by the layer standard. The layer standard consists of two parts—a *service definition* (i.e., what functions does the layer perform) and a *protocol specification* (i.e., what procedures are employed to execute the functions). Shown in **Figure 5.6**, a layer consists of the layer entities in the cooperating systems. Each layer entity is implemented to send, receive, and act on messages exchanged with a peer layer in the other end-system. Passing a message between user's applications is achieved by a sequence of activities that descends through the seven layers implemented in one system, and ascends through the seven layers implemented in the other system. It is important to realize:

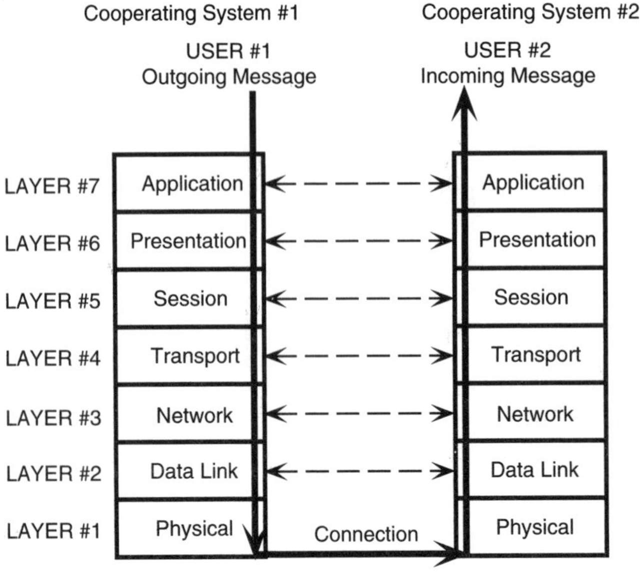

Figure 5.5 OSI Model of Communication Between Cooperating Systems

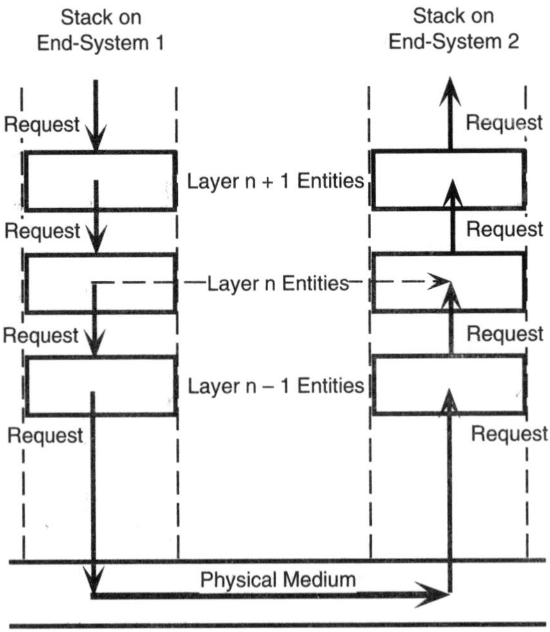

Figure 5.6 Relationships Between Layer Entities*Each layer entity performs services for its neighbors and passes messages through the stacks to its peers.*

Table 5.2 Summary of Actions Taken by OSI Model Layers to Implement Machine-and-Machine Communication—Application, Presentation, Session, Transport, and Network Layers

	ACTION WHEN SENDING	ACTION WHEN RECEIVING	PEER-TO-PEER COMMUNICATION
APPLICATION	*Organizes information from application agent into block of user data. Adds AH with identifier unique to this communication and these participants. Passes user data to Presentation.*	*Removes AH, converts user data block to information. Passes information through application agent to addressee.*	*Agreement on identifier unique to this communication and these participants.*
PRESENTATION	*Translates user data into recipient's code, or an agreed on intermediate code. Performs compression and/or encryption, if used. Adds PH with encoding, compression and encryption information. Passes data to Session.*	*Removes PH, decyphers and/or decompresses if appropriate, converts to recipient's code if intermediate code used. Passes user data to Application.*	*Agreement on code to be used for transmission; also schemes for compression and encryption, if used.*
SESSION	*Manages line discipline, defines beginning and ending of message, provides checkpoints for failure recovery. Adds SH with markers.*	*Manages line discipline. Removes SH, notes any markers, holds until all data is received. Passes PPDU to Presentation.*	*Agreement on line discipline and markers.*
TRANSPORT	*May break user data into segments. Adds TH with sequence numbers, and FCS to form TPDU. Copies TPDU, retains until acknowledged.*	*Removes TH, uses FCS to verify TPDU. Acknowledges error free TPDU, discards and requests resend of corrupted TPDU. If message was broken up, waits for all TPDU, reassembles message. Passes SPDU to Session.*	*Agreement on communication parameters such as size of TPDU, etc.*
NETWORK	*May break TPDUs into smaller segments to match size limits of chosen network. Adds NH with sequence numbers and destination address to form NPDUs, or packets.*	*Removes NH, verifies destination address and sequence number. If TPDU was broken up, waits for all packets, reassembles TPDU. Passes TPDU to Transport.*	*Agreement to initiate, maintain and terminate network level connection.*

AH = Application Header; PH = Presentation Header; SH = Session Header; TH = Transport Header; NH = Network Header; TPDU = Transport PDU.

Table 5.2 (continued): Summary of Actions Taken by OSI Model Layers to Implement Machine-and-Machine Communication—Data Link and Physical Layers

	ACTION WHEN SENDING	ACTION WHEN RECEIVING	PEER-TO-PEER COMMUNICATION
DATA LINK	*Adds DH and DT to each packet to form frame. DH includes sequence number for DLC protocol. DT includes FCS. Copies frame, retains until acknowledged.*	*Removes DH and DT, uses FCS to determine if error has occurred in frame. Passes packet to Network.*	*Agreement on DLC protocol parameters, including error control procedures.*
PHYSICAL	*Transmits each frame as a sequence of bits.*	*Receives each frame and reconstructs datastream.*	

DH = Data Link Header; DT = Data Link Trailer; FCS = frame check sequence

- All the information exchanged between the end-systems enters or leaves the model through the application layers
- Even though we refer to messages passing between peer entities, the actual messages flow down one stack of layers and up the other
- Communication signals are exchanged between the systems only over the connection between the physical layers.

(1) Layer 7: Application (or User) Layer

The application layer makes system independent application functions available to users of the end-system on which it resides. These services are managed by an entity embedded in the local operating system called the *application agent*. The application agent interfaces to the user program and to the operating system of the end-system on which it is resident. It provides access to specific end-system resources that may be required to complete the transaction between the end-systems.

The application layer differs from the other layers of the OSIRM. Its users are not entities of the next higher layer but programs that provide services to the users of the end-system. No other layer is accessed directly by a user. Because the application layer interfaces with a wide variety of user programs, it contains many user-specific services that are hard to standardize. Some application service elements that have been standardized by ISO include:

- File transfer, access, and management (FTAM)
- Electronic messaging (X.400)
- Job transfer and manipulation (JTM)

- Virtual terminal (VT).

They cover many of the reasons for engaging in machine-and-machine communications. In addition, to assist the interworking of open systems by providing the information they need to communicate, Recommendation X.500 describes a Directory. Intended to provide *location transparency*, it holds information concerning many things (people, distribution lists, applications entities, etc.), and may be consulted by a user to obtain addressing information (telephone numbers, electronic mail address, postal address, facsimile number, service access point information, etc.).

When sending, the application layer organizes information received from the application agent into a user data block, adds a header (application header, AH) that identifies this communication between specific participants, and passes the application protocol data unit (APDU) on to the presentation layer. When receiving, the application layer removes the application header from the application PDU, converts the data block to user information, and passes it to the application agent for delivery to the user application identified by the header. Peer-to-peer communication is required to agree on the unique identifier for the communication.

(2) Layer 6: Presentation Layer

The presentation layer translates the information to be exchanged into terms that are understood by the end-systems. If the users are supported by the same (or compatible) equipment, this conversion may be trivial. However, if the systems are not compatible, conversion may be much more difficult. To make the translation capability required by each system less onerous, a common, intermediate form may be used for the information. Thus, in the previous scenario, R(D)S1 may be converted to R(D)X, transmitted, and converted from R(D)X to R(D)S2 in System 2. In this way each end-system need only contain the capability to convert its requests to one other form [R(D)X].

When sending, the presentation layer performs translation services (and may perform data compression and encryption), adds a header (presentation header, PH) that identifies the specific encoding, compression, and encryption employed, and passes the presentation PDU on to the session layer. When receiving, the presentation layer removes the presentation header from the presentation PDU, performs decoding, decompressing, and decrypting, and passes the application PDU to the application layer. Peer-to-peer communication is required to agree on encoding, compression, and encryption. The ISO connection-oriented, presentation layer is defined in ISO 8822 (Service Definition) and ISO 8823 (Protocol Specification).

(3) Layer 5: Session Layer

The session layer directs the establishment, maintenance, and termination of the connection. It controls data transfer and manages the insertion of synchronization points into the information flow. These points structure the data exchange into a series of dialog units so that the exchange can restart if service is interrupted.

When sending, the session layer manages line discipline, defines the beginning and ending of the exchange, adds a header (session header, SH) that identifies any specific markers employed, and passes the session PDU on to the transport layer. When receiving, the session layer removes the session header from the session PDU, notes any specific markers, and passes the presentation PDU to the presentation layer. Peer-to-peer communication is required to agree on line discipline and the use of markers. The ISO connection-oriented, session layer is defined in ISO 8326 (Service Definition) and ISO 8327 (Protocol Specification).

(4) Layer 4: Transport Layer

The transport layer ensures the integrity of end-to-end communication, independent of the performance and number of networks involved in the connection between end-systems. It serves as interface between the three lower layers that are concerned with providing transparent connections between users, and the three upper layers that ensure that the data are delivered in correct and understandable form. The transport layer selects the networks that will be employed, provides network addresses for the parties, and may multiplex messages from several sources for transmission on a common, physical connection. The transport layer is responsible for detecting the loss or duplication of message segments, and for requesting resends as needed. In conjunction with the data link layer (that supervises message flow at the frame level), it controls the flow of message segments so that overloads (in the host, or in the communication network) are prevented.

When sending, the transport layer may break the user data block into segments, adds a header (transport header, TH) that contains the sequence number, calculates a frame check sequence (FCS) that checks the entire segment, copies the transport PDU and passes it on to the network layer. When receiving, the transport layer removes the transport header from the transport PDU, verifies the FCS, acknowledges an error-free transport PDU or discards and requests resend, and passes the session PDU to the session layer. Peer-to-peer communication is required to agree on the maximum size of the transport PDU, and the network(s) that will be used for this communication. Typical transport layer protocols are Internet's TCP (see § 6.3.1(1)) and OSI's TP0, TP1, TP2, TP3, and TP4 (see § 6.3.2(2)).

(5) Layer 3: Network Layer

The network layer provides communications services to the transport layer. To do this, it makes use of the underlying data link layer services to establish, maintain, and terminate transmission across the chosen network(s). It determines how to route the segments within the network chosen by the transport layer, selects the circuit on which they shall be sent, and breaks segments into data blocks of a size suited to the chosen circuit.

When sending, the network layer adds a header (network header, NH) to each block that provides destination addresses (in a format understood by the chosen network) and sequence number, and passes the network PDU to the data link layer. When receiving, the network layer removes the network header (NH), verifies destination address and sequence number, reassembles the transport PDU, and passes it on to the transport layer. Peer-to-peer communication is required to initiate, maintain, and terminate the network level connection. Typical network level protocols are ITU-Ts X.25 Packet Layer Protocol (see § 6.2.1); Internet's IP (see § 6.3.1(3)); and ISO's 8348 and 8648 (see § 6.3.2(1)).

(6) Layer 2: Data Link Layer

The data link layer transfers data over a single communication link without intermediate nodes. The link can be point-to-point, point-to-multipoint, broadcast, switched, or non-switched. The data link layer protocol prepares the data in a specified format (SDLC, Bisync, etc., see § 6.1.2(1)) for transmission over the selected circuit that may use a variety of media including twisted pairs, coaxial cables, optical fibers, and communications satellites or other radio links.

When sending, the data link layer adds a header (DH) and a trailer (DT) to the network PDU to form the data link PDU. The header includes the flag, class of frame identifier, sequence number for use in the control byte, and other information. The trailer includes a FCS to check the frame, and a flag. The data link layer copies the frame and passes it on to the physical layer. When receiving, the data link layer reconstructs the data link PDU from the bitstream received from the physical layer, removes both header and trailer from the data link PDU, verifies FCS and other layer information, requests resend, if necessary, and passes the network PDU on to the network layer. Peer-to-peer communication is required to agree on data link protocol parameters and error correction procedures. In conjunction with the transport layer (which supervises message flow at the segment level), the data link layer controls the flow of frames so that overloads (in the host, or in the communication network) are prevented.

(7) Layer 1: Physical Layer

The physical layer provides the connection over which bitstreams flow between the users. It defines the physical (electrical and mechanical) standards and signaling required to make and break the data communication path and to send and receive the data signals over twisted pairs, coaxial cables, optical fibers, and communications satellites or other radio links. The meaning of the bitstream is of no significance to the physical layer. When sending, the physical layer transmits a sequence of bits that represents each frame received from the data link layer. When receiving, the physical layer receives a sequence of bits, and passes them to the data link layer. Typical physical layer protocols are EIA 232/449 (see § 5.1.2); ITU-T X.21—that uses balanced signaling; ISDN I.430 that provides basic rate access, and ISDN I.431 that provides primary rate access (see § 10.5); and a range of procedures for different media components used to implement IEEE 802 local-area networks (see § 11.1.2).

5.3.3 Functional Structure of OSI Model

Figure 5.7 shows more details of the functional structure of the OSI model. The application process, i.e., that part of an end-system that processes information for an application task that requires communication in an OSI environment, consists of two parts:

- **Application agent**: exists between the user and the operating system of the processing resource to provide access to specific processor capabilities
- **Application entity**: exists within the application layer. As required, provides services to the application agent by exchanging PDUs with peer entities (application entities in other machines).

They cooperate to transfer the data required to initiate communication activity with another end-system. As these data descend the stack associated with end-system #1, headers are added by the application, presentation, session, transport, and network entities, and both header and trailer are added by the data link layer entity to produce the data link PDU. Converted to a datastream, it is transferred to end-system #2 over the physical connection between them. As the information ascends the stack associated with end-system #2, the data link entity strips off the data link header and trailer—they constitute a message from the peer entity in end-system #1. The data link entity acts on the information they contain. In turn, the network, transport, session, presentation, and application entities strip off headers, and act on the information they contain. Finally, the presentation entity delivers the information from the application running in end-system #1 to the operating system running in end-system #2.

5.3.4 Reality of OSIRM

The OSI model is the result of many years of work by standards committees representing the world's major computer and telecommunication organizations. It is an exhaustive solution to the problem of establishing data communication between any two machines, located anywhere, and contains many more options than any individual communication situation requires. Data communication is largely a matter of exchanging messages between many terminals and a few hosts, or among a few hosts, for purposes defined by a single enterprise. Usually, the community of interest exists within an organization, and equipment is bought that communicates effectively. The result is that there are many more quasi-homogeneous (more or less compatible) systems than heterogeneous systems. Nevertheless, in addressing the general heterogeneous case, the OSI reference model makes it possible to construct better communications equipment and networks, and to improve the average overall performance of data communication systems.

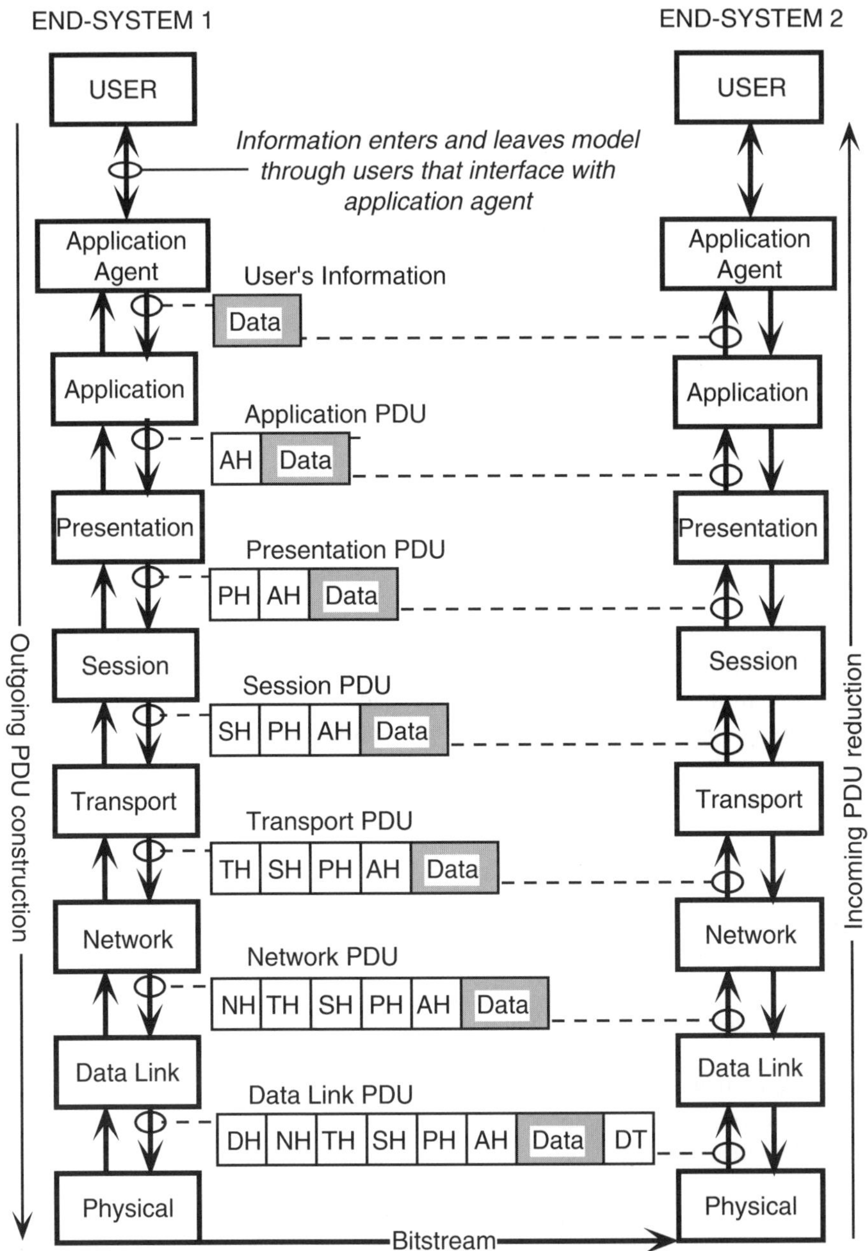

Figure 5.7 Construction and Reduction of PDU in OSI Model

6

Protocols

In Figure 5.7, we traced the development of the data link protocol data unit (DPDU) that becomes the signal exchanged by the physical layers of cooperating machines. In this Chapter we describe the headers and trailer associated with the protocols that serve the data link, network, and transport layers—the layers that have to do with communication.

6.1 DATA LINK PROTOCOLS

- **Data link control** (DLC) **protocol**: a set of rules that governs the exchange of messages over a data link.

  DLC protocols are divided into two classes, as follows:

- **Asynchronous** operation:
  - start-stop DLC protocol
- **Synchronous** operation:
  - character-oriented DLC protocol (e.g., BISYNC)
  - byte-count-oriented DLC protocol (e.g., DDCMP)
  - bit-oriented DLC protocol (e.g., SDLC).

6.1.1 Start-Stop DLC Protocols

Start-stop, DLC protocols are used by terminals, and other input-output devices, over relatively slow-speed links that employ asynchronous transmission. They manage a stream of characters framed by start and stop bits. The framed characters are transmitted one at a time and are independent of each other.

(1) USART

In contemporary systems, an integrated circuit chip called a *universal synchronous/asynchronous receiver transmitter* (USART), is employed to implement start-stop protocol. As a transmitter, USART:

- Adds a parity bit (if so instructed)
- Frames the character sequence with start and stop bits
- Generates unipolar signals.

As a receiver, USART:

- Extracts timing information from the start bit
- Converts the incoming signal to a bit sequence
- Checks parity
- Strips off the start, stop, and parity bits.

In the idle state, the data link is held at a voltage level that corresponds to 1. Transmission begins by dropping the line level to zero to send a start bit (0). The start bit is followed by bits that represent the character being sent. At the end of the character sequence, a stop bit returns the line to the level representing 1. It remains at that level until the next character is transmitted. **Figure 6.1** shows this sequence.

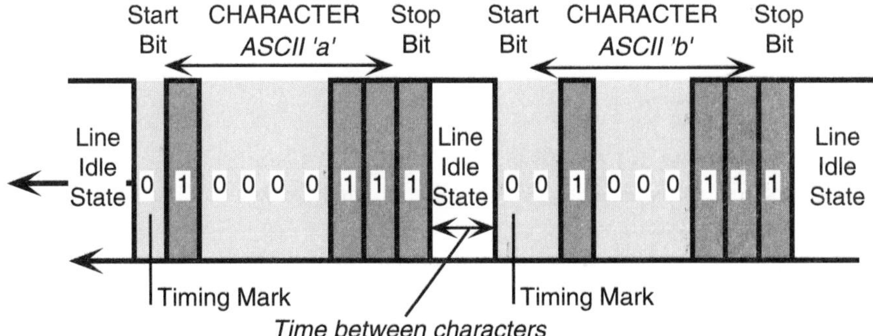

Figure 6.1 Transmission Format for Start-Stop (Asynchronous) Signaling
In idle state, the line is maintained at the 1 level. The start bit (0) reduces the level to zero signaling the commencement of activity.

(2) Synchronization

At the receiver, bit synchronization is achieved by developing a timing mark at the center of each start pulse. Because the characters consist of only a few bits, sampling the ensuing voltage levels at nominal pulse-width intervals from this mark permits the receiver to test the incoming signal at approximately the center of each bit. Because operation is asynchronous, synchronization at byte (octet) and frame levels is not possible.

(3) Error Control

Error control relies on parity checking. If parity is not confirmed, the USART is likely to flag the condition. In some cases, the received bit sequence is echoed back to the sending terminal and displayed. If it is the character the user intended to send, it confirms that the correct character was received. If it is not the intended character, the user corrects it. In many cases, the user is not conscious of echoing and assumes that the appearance of a character, different from the one typed, was caused by an improper keyboard input. Because each character is only framed by a start bit and a stop bit, the receiver cannot determine when all the characters have been transmitted. Also, unless echoing is employed, the sender has no indication of whether the characters have been received. Start-stop, DLC protocols are the simplest of communications procedures.

6.1.2 Character-Oriented DLC Protocols

Character-oriented, DLC protocols employ special characters that are reserved in the code set to facilitate communication. A common data link control protocol is known as BISYNC (binary synchronous data link control). Developed by IBM, it is the oldest data link control protocol in common use; as a consequence there are many user-specific variations in its implementation.

(1) Binary Synchronous Data Link Control Protocol

BISYNC supports ASCII and EBCDIC codes; it uses specific control characters identified by mnemonics:

- **ACK0** and **ACK1** (positive acknowledgment): no error detected in datablock; ACK0 is used for even-numbered datablocks; ACK1 is used for odd-numbered blocks
- **DLE** (data link escape): changes the meaning of following character(s)
- **ENQ** (enquiry): requests a response from another station
- **EOT** (end of transmission): concludes a transmission that may have contained several datablocks
- **ETB** (end of text datablock): end of a text datablock—requires receiver to acknowledge receipt

- **ETX** (end of text): terminates text portion of frame
- **ITB**(end of intermediate text datablock): does not require receiver to acknowledge receipt; BCC follows ITB
- **NAK** (negative acknowledgment): error detected in datablock
- **SOH** (start of header)
- **STX** (start of text)
- **SYN** (synchronizing character): begins frame—usually sent in pairs.

(2) BISYNC Frame

A BISYNC frame has the form:

⇐ *SYN SYN SOH Header STX Text ETX BCC.*

- It employs synchronous transmission, i.e., no start or stop bits and transmission is in response to an external event
- The frame begins with a repeated *SYN* character
- Control characters are used as needed to identify portions of the frame
- A block check character is used to check the frame from *SOH* to *ETX*. *BCC*s are used only with information frames (i.e., those that contain text)
- On point-to-point connections:
  - no addressing is required
  - if stations have equal status, the station with a frame to send asks the other station for permission to send
  - the header is optional and is specified by the users.
- On multipoint connections:
  - the header includes the address of the secondary station taking part in the communication
  - if the secondary station is sending, its address tells the primary station from which secondary station the frame comes
  - if the primary station is sending, it is the address of the secondary station for which the frame being sent is intended
  - if several frames are included in the exchange, the header is likely to include the frame number.

An exchange that employs stop-and-wait, error correction on a half-duplex, multipoint connection between primary station A and secondary stations B, C, and D is portrayed in **Table 6.1**. When polled by A, station B has nothing to send, and responds with *EOT*. Station C has a two frame message to send. When polled by A, C sends the first frame. A finds it to contain error(s), destroys it, and requests C to repeat the frame. On being sent the second time, it is good, and A sends a positive acknowledgment to C. C then sends the second frame and signifies there are no more frames to come. A confirms receipt of frame 2 and polls D. If D has nothing to send, A polls B again…and so on. Even though there are no messages to send,

Table 6.1 Bisync on Half-Duplex Multipoint Circuit With Stop-And-Wait ARQ

An exchange that employs stop-and-wait error correction on a half-duplex, multipoint connection between primary station A and secondary stations B, C, and D might proceed as follows:

A polls B.

ENQ 'B' SYN SYN ⇒

The control character *ENQ* asks if B has any information to send. After sending, the line is turned around, A becomes the receiver, and B becomes the sender. B responds to A.

⇐ *SYN SYN 'B' EOT*

'B' tells A the reply is from station B. The control character *EOT* (end of transmission) indicates no information to send. After sending, the line is turned around, B reverts to receiver, and A becomes the sender again.

A polls C.

ENQ 'C' SYN SYN ⇒

After sending, the line is turned around, A becomes the receiver, and C becomes the sender. C responds to A.

⇐ *SYN SYN SOH 'C' '1' STX Text1 ETB BCC*

C sends frame 1 (*'1'* in the header indicates the frames are numbered). It includes text block 1 (*Text1*); the use of *ETB* (end of text block) indicates more blocks to come. After sending, the line is turned around, C becomes the receiver, and A becomes the sender. A finds the VRCs and LRCs it calculates are not in agreement with the values sent. An error exists. A destroys the frame and sends *NAK*, a negative acknowledgment to C.

NAK 'C' SYN SYN ⇒

After sending, the line is turned around, A becomes the receiver, and C becomes the sender. C repeats frame 1.

⇐ *SYN SYN SOH 'C' '1' STX Text1 ETB BCC*

After sending, the line is turned around, C becomes the receiver, and A becomes the sender. Having received the frame and detected no error this time, A sends a positive acknowledgment to C.

ACK1 'C' SYN SYN ⇒

After sending, the line is turned around, A becomes the receiver, and C becomes sender. C sends frame 2 which includes text block 2.

⇐ *SYN SYN SOH 'C' '2' STX Text2 EOT BCC*

EOT indicates there are no more blocks to follow and sending is complete. After sending, the line is turned around, C becomes the receiver, and A becomes the sender. A acknowledges error-free receipt of frame 2.

ACK0 'C' SYN SYN ⇒

At this point, A retains the role of sender. A polls the next station (*'D'*).

ENQ 'D' SYN SYN ⇒

After sending, the line is turned around, A becomes the receiver, and D becomes the sender.... and so on.

---

the circuit is in constant activity as A polls each station in turn, repeatedly—until there is a message. When sending is completed, A returns to polling.

(3) BISYNC Transparent Operation

Defining certain characters as control characters means that sequences in the text portion of the frame may provoke a control action. For instance, in EBCDIC, *ETX* is represented by 00000011 ⇒ —this is also the binary code for 3. Thus, if the text contains *b3* (binary number 3), the receiver will take it to mean *ETX*, expect that the next octet(s) are error checking information, and that this is the last frame of the series.

(4) Use of DLE

In the text, to preserve the integrity of bytes whose bit patterns duplicate control characters, the escape character (*DLE*) is employed. It is inserted before all control characters in the text portion of the message; thus, *DLE STX*, *DLE ITB*, *DLE ETB*, etc. In this way a bit pattern in the text portion of the message that happens to duplicate any control character except *DLE*, will be treated as a character intended to be part of the text. In EBCDIC, *DLE* is 00010000 ⇒, i.e., the bit pattern for the binary number 16. If nothing else is done, the occurrence of *b16* in the text portion of a message will cause the receiver to expect the number following to be a control character. If the number following is *b3* = *ETX*, the receiver will understand that the text portion of the message has ended. If the character that follows *b16* is not the same as a control character, the receiver will understand it is dealing with an unknown, and assume an error has been made. In either event, the text will not be processed properly.

(5) Use of DLE DLE

A technique employed to prevent this confusion and to obtain completely transparent operation, is as follows. After inserting *DLE* in front of the *STX* character, the transmitter scans the text for occurrence of the *DLE* bit pattern. Upon detection of *b16* (the *DLE* bit pattern), the transmitter inserts *DLE* ahead of it. At the receiver, receipt of *DLE STX* signifies the beginning of the text of a message that is being sent in transparent mode. Hereafter, until *DLE ETB*, *DLE ETX*, or *DLE EOT* is received, the receiver will interpret the sequence *DLE DLE* as the binary number 16. In this way, control characters are interpreted correctly, and the text is protected. As an example, if a frame contains a portion of text that includes the binary numbers 3 and 16, the transmission will be:

⇐ *SYN SYN SOH Header DLE STX b29 b36 DLE b16 b3 b29 DLE ETX BCC.*

DLE is used ahead of control characters that define the text portion of the frame, and ahead of *b16*—the binary number that mimics *DLE*—to ensure that it is interpreted as the binary number 16. Because it is not preceded by *DLE*, the next number *b3* is treated as *b3* not *ETX*. *DLE ETX* closes the text segment.

(6) BISYNC Synchronization

Usually, two synchronizing characters are sent at the beginning of the frame. Two are needed to ensure that the receiver can establish character synchronization reliably. Some systems employ more *SYN* characters. Also, *SYN* characters may be transmitted when the line is idle to maintain character synchronization between the end equipment. In addition, if the text block is long, they may be sent in the middle of the text to ensure continued synchronization throughout reception of the frame, e.g.:

⇐ *SYN SYN SOH Header STX Text1 ITB BCC SYN SYN STX Text2 ETB BCC.*

The first *BCC* checks the frame up to its position in the frame; the second *BCC* checks the remainder of the frame.

6.1.3 Byte-Count-Oriented DLC Protocols

In byte-count-oriented DLC protocols, the number of bytes (or octets) in the text field is carried in a byte-count field in the header. Set numbers of bits are present in all fields except the text field. The leading example of such a protocol is DDCMP, a protocol developed by Digital Equipment Corporation.

(1) DDCMP Frame

DDCMP (Digital Data Communications Message Protocol) employs a header that contains exactly 10 octets, a text portion that contains a stated number of octets, and a trailer that consists of a 2-octet error detection sequence (FCS). The fields in a DDCMP frame are shown in **Figure 6.2**:

- Two *SYN*s are used to establish character synchronization.
- The **Class field** contains one of three control characters:
  - *SOH* if the frame is an information frame
  - *ENQ* if the frame is a supervisory frame
  - *DLE* if the frame is a maintenance frame.
- Supervisory frames do not contain a text field.
- The **Text Byte Count field** contains the number of bytes that are present in the text field. With 14 bits in the field there can be 2^{14}, i.e., a maximum of 16,383 bytes in the text.
- The **Flag field** contains 2 bits:
  - *Q*-bit (Quick Sync) tells the receiver whether this frame will be followed by *SYN* characters
  - *S*-bit (Select), tells the receiver whether the line can be turned around (in half-duplex operation).
- The **Receive Count field** notes the number of the last frame received without error.

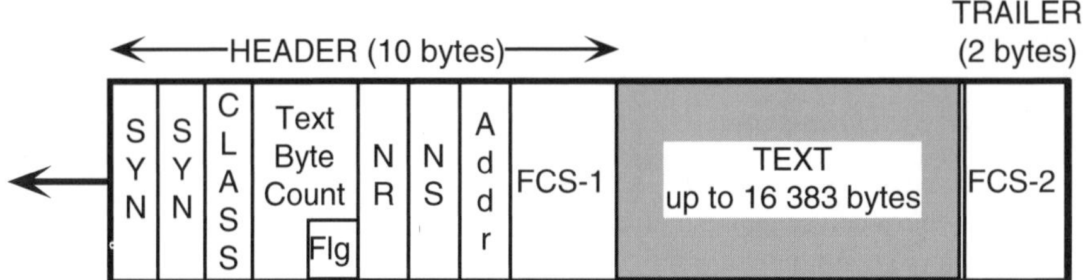

SYN (8 bits) = synchronizing character
CLASS (8 bits) = SOH for information frames; = ENQ for supervisory frames (no text); and = DLE for maintenance frames
Text Byte Count (14 bits) = number of bytes in text portion of frame -- maximum = 16 383
Flag (2 bits) = Quick Sync (this frame will be followed by SYN characters or Select (to control transmission on half-duplex circuits)
NR (8 bits) = receive count i.e., frame NR and all previous frames have been received correctly
NS (8 bits) = send count i.e., this is frame NS
Addr (8 bits) = address
FCS-1 (16 bits) = frame check sequence on header
FCS-2 (16 bits) = frame check sequence on text

Figure 6.2 DDCMP Frame Format
The number of bytes in the header and trailer are defined. The number of bytes of text are recorded in the header.

A count of *x* implies that frame *x*, and all lower-numbered frames have been received without detecting errors. Thus, one receive count can acknowledge a number of frames.

- The **Send Count** is the number of the frame being transmitted. Each station maintains its own send count.
- *FCS-1* is the frame check sequence for the header.
- *FCS-2* is the frame check sequence for the text field.

Supervisory frames are used to start communication and to initialize frame counts. Also, if only one station is sending information frames, supervisory frames are used for acknowledgments.

(2) DDCMP Transparency

Because the exact number of bits or bytes is known for each field, there can be no confusion between control information and text information so long as the receiver keeps accurate count. Thus, special arrangements to ensure transparency are not needed in DDCMP.

(3) DDCMP Synchronization

Two *SYN* characters are sent at the beginning of the frame. Two are needed to ensure that the receiver can establish character synchronization reliably. No other *SYN* characters are employed.

6.1.4 Bit-Oriented DLC Protocols

Bit-oriented DLC protocols make use of a special character, the flag character, to mark the beginning and ending of the frame. Between these markers, the header and the trailer fields are of predetermined lengths. Whatever data lie between them is the text field. There is no need for other special characters to separate the frame into its major divisions. The original bit-oriented protocol was IBM's:

- **SDLC** (Synchronous Data Link Control Protocol).

Introduced in 1972, SDLC was modified and standardized by ITU-T and ISO as:

- **HDLC** (High Level Data Link Control Protocol).

  Since then, other bit-oriented protocols have been standardized:
- **ADDCP** (Advanced Data Communications Control Procedure)—an ANSI standard
- **LAP-B** (Link Access Procedure—Balanced)—a component of ITU-T's X.25 standard,
- **LAP-D** (Link Access Procedure—Channel)—for ISDN D-channel data transport
- **LAP-E** or **LAP-F** (Link Access Procedure—Extended or Frame Relay)—a version of LAP-D used in frame relay applications.

Different in the detailed meaning of specific control field bits, all of these protocols share a common structure. In the order that they are transmitted, they consist of the following fields: flag, address, control, text, frame check sequence, and flag. We shall use SDLC to illustrate the use of bit-oriented protocols.

(1) Synchronous Data Link Control Frame

An SDLC frame is shown in **Figure 6.3**:

- It employs synchronous transmission.
- The flag is the sequence 01111110. It defines the beginning and ending of the frame.

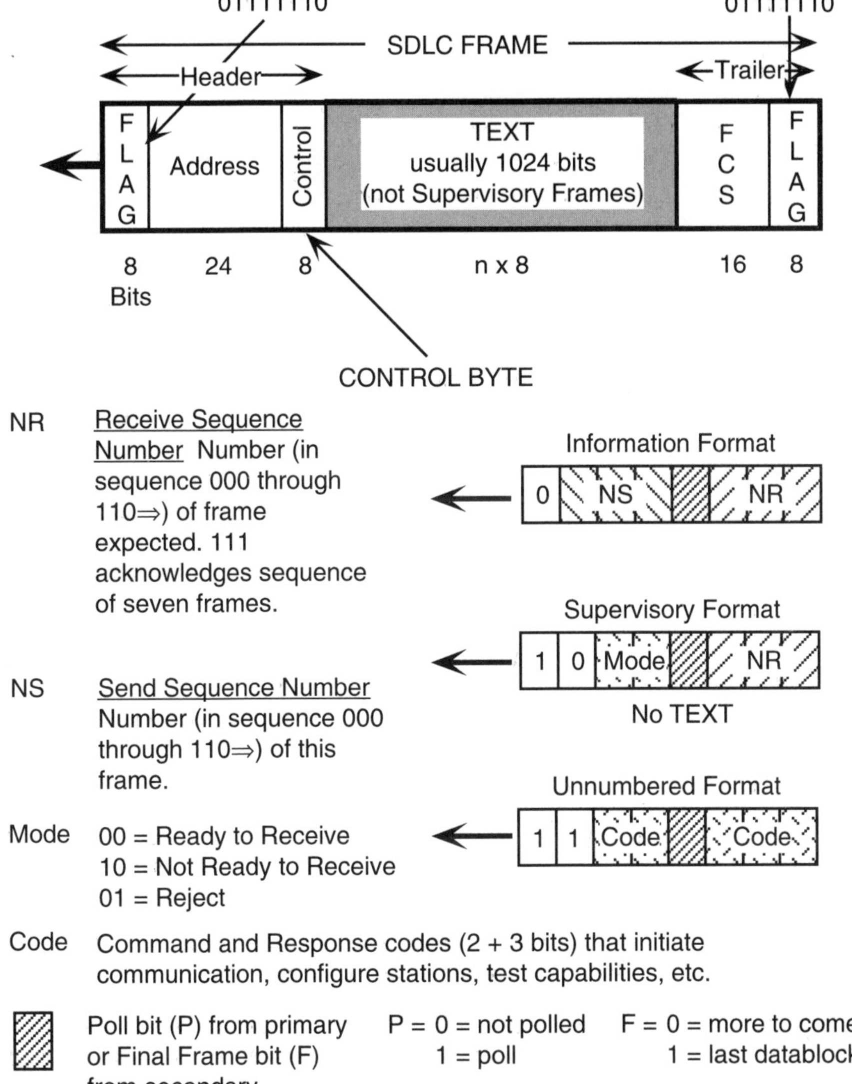

Figure 6.3 SDLC Frame Format

The flag character marks the beginning and ending of the frame. Its use requires a zero-stuffing strategy to prevent data in the frame appearing to be a flag.

- Usually, the address field is three octets long.
- Usually, the text portion of the frame is limited to 1024 bits (128 characters).
- The frame check sequence (FCS, the Cyclic Redundancy Check remainder) follows immediately on the text.

- 0 as the first bit in the control byte identifies an information frame, i.e., one that contains text. The state of the #5 bit in the control byte indicates whether information transfer is complete with the present frame (F = 1), or whether there is more to come (F = 0).
- ⇐ 10 as the first two bits in the control byte identifies a supervisory frame. It initiates and manages communication. It does not contain text. The state of the #5 bit in the control byte indicates whether a reply is required (P = 1) or not required (P = 0).
- 11 as the first two bits in the control byte identifies an unnumbered frame (also called a nonsequenced frame). An unnumbered frame sets the operating mode and initializes transmission. The state of the #5 bit in the control byte indicates whether the access method employs polling (P = 1), or contention (P = 0). It may contain data in the text field.
- NS (send sequence number) and NR (receive sequence number) are 3-bit numbers used to track individual frames. NS is the number of this frame in the sending sequence (the sequence is unique to the sender). NR is the number of the frame next expected in the receiving sequence. If it is one more than the last frame sent (i.e., NR = NS + 1), it indicates that all frames received so far are without detectable errors. If it is the number of the last frame transmitted, or the number of an earlier frame (i.e., NR ≤ NS), it indicates that error has been detected in frame NR, and that the receiver is still expecting a good NR. Using go-back-n, it will be the frame from which the sequence must be restarted. Using selective repeat, it will be the number of the frame that must be repeated. NR = 111 acknowledges the error-free receipt of all frames within a sequence of seven. In this context, the frames are numbered:

  frame 1 = 000 ⇒; frame 2 = 001 ⇒; frame 3 = 010 ⇒;

  frame 4 = 011 ⇒; frame 5 = 100 ⇒; frame 6 = 101 ⇒;

  frame 7 = 110 ⇒; all seven frames = 111. ⇒.

An exchange between primary station A and secondary station B that employs SDLC and go-back-n error correction on a half-duplex, point-to-point connection, over which a maximum of four frames is sent before pausing for acknowledgment, is given in **Table 6.2**. A polls B. B has a 5 frame sequence to send. In frame 3, A detects an error and destroys frames 3 and 4. After 4 frames are sent, B queries A. A responds that it is still waiting for frame 3. B repeats frames 3 and 4 and sends frame 5. A sends an information frame to B that acknowledges correct receipt of all five frames and sends the first block of a two block message. A sends the second block and B responds with a positive acknowledgment for the two frames. With go-back-n ARQ, B's 5 text block message to A, and A's two block message to B, were received in sequence.

Table 6.2 SDLC on Half-Duplex Point-To-Point Circuit With Go-Back-N ARQ

An exchange between primary station A and secondary station B that employs SDLC and go-back-n error correction on a half-duplex, point-to-point connection, over which a maximum of four frames are sent before pausing for acknowledgement, might proceed as follows:

A sends a supervisory frame (*01* ⇒) to B.

Flag | FCS | 000 1 00 01 | 'B' | Flag ⇒

The mode bits (*00*) indicate A is ready to receive. The poll bit is set — a response is required. After sending, the line is turned around, A ceases to be sender and becomes receiver, and B becomes sender. B responds with information frame 1 (⇐ *0 000*) that contains text 1.

⇐ *Flag | 'B' | 0 000 0 000 | Text1 | FCS | Flag*

The final bit is not set — there is more to come. NR = *000* indicates that B is expecting the first frame from A i.e., A has sent no information frames to B. B continues with information frame 2 (⇐ *0 100*).

⇐ *Flag | 'B' | 0 100 0 000 | Text2 | FCS | Flag*

The final bit is not set — there is more to come. NR continues to indicate no information frames have passed from A to B. B continues with information frame 3 (⇐ *0 010*).

⇐ *Flag | 'B' | 0 010 0 000 | Text3 | FCS | Flag*

The final bit is not set — there is more to come. NR continues to indicate no information frames have passed from A to B. However, A detects error in frame 3 and destroys it. B continues with information frame 4 (⇐ *0 110*).

⇐ *Flag | 'B' | 0 110 0 000 | Text4 | FCS | Flag*

The final bit is not set — there is more to come. NR continues to indicate no information frames have passed from A to B. However, because of error in Frame 3, A destroys frame 4. B sends a supervisory frame (⇐ *10*) that indicates B is ready to receive A's acknowledgment of the first four frames.

⇐ *Flag | 'B' | 10 00 0 000 | FCS | Flag*

After sending, the line is turned around, B ceases to be the sender and becomes the receiver, and A becomes sender. A sends a supervisory frame to B (*01* ⇒).

Flag | FCS | 010 1 00 01 | 'B' | Flag ⇒

The mode bits indicate A is ready to receive. The poll bit is set — a response is required. NR is set to 3 (*010*) indicating that A is expecting frame 3, i.e., an error was detected in frame 3 and A has discarded frames 3 and 4. After sending, the line is turned around, A ceases to be sender and becomes receiver, B becomes sender. B re-sends information frame 3.

Table 6.2 (continued) SDLC on Half-Duplex Point-To-Point Circuit With Go-Back-N ARQ

⇐ *Flag / 'B' / 0 010 0 000 / Text3 / FCS / Flag*

B resends information frame 4.

⇐ *Flag / 'B' / 0 110 0 000 / Text4 / FCS / Flag*

B sends information frame 5 (⇐ *001*).

⇐ *Flag / 'B' / 0 001 1 000 / Text5 / FCS / Flag*

The final bit is set — the message is complete. NR continues to indicate no information frames have passed from A to B. After sending, the line is turned around, B ceases to be sender and becomes receiver, A becomes sender. A sends an information frame to B.

Flag / FCS / Text1 / 101 0 000 0 / 'B' / Flag ⇒

NS = *000* — this is the first information frame A has sent to B. The final bit is not set — there is more to come. NR is set to *101* indicating that A is expecting frame 6 from B thereby confirming that frames 1 through 5 have been received without errors being detected. A sends a second information frame to B.

Flag / FCS / Text2 / 111 1 001 0 / 'B' / Flag ⇒

NS = *001* — this is the second information frame sent from A to B. The final bit is set — there is no more to come. NR is set to *111* indicating all frames from B have been received and no errors have been detected. After sending, the line is turned around, A becomes receiver and waits for B's acknowledgement. B responds with a supervisory frame.

⇐ *Flag / 'B' / 10 00 0 010 / FCS / Flag*

The mode bits indicate B is ready to receive. NR is set to *010* indicating that B is expecting frame 3 from A thereby confirming that frames 1 and 2 have been received without errors being detected. After sending, the line is turned around, B reverts to receiver status while A reverts to sender status.

An exchange that employs the same scenario as Table 6.2 but with selective repeat ARQ is given in **Table 6.3**. A polls B. B begins to send the 5 frame message. When A receives frame 3 and detects an error, it destroys frame 3 and immediately requests B to repeat frame 3. After sending frame 4, B repeats frame 3 and then sends frame 5. At the same time, A acknowledges frames through 4 and begins to send a message to B. B responds with an all frames acknowledgment to A. Through the use of selective repeat ARQ, B's 5 text block message to A was received in the sequence 1, 2, 4, 3, and 5. Station A must reorganize the frames to reconstruct the message.

(2) Expanding the Frame Size

The address, control, text, and frame check sequence fields of the frame can be expanded to provide greater operational flexibility:

Table 6.3 SDLC on Half-Duplex Point-To-Point Circuit With Selective Repeat ARQ

An exchange that employs SDLC and selective repeat error correction on a duplex connection between primary station A and secondary station B might proceed as follows:

A sends a supervisory frame to B.

Flag / FCS / 000 1 00 01 / 'B' / Flag ⇒

The mode bits indicate A is ready to receive. The poll bit is set — a response is required. NR is set to *000*, i.e., A is expecting frame 1. Since this is a duplex connection, A and B can send or receive at any time. B responds with information frame 1 (*000*) that contains text 1.

⇐ *Flag / 'B' / 0 000 0 000 / Text1 / FCS / Flag*

The final bit is not set — there is more to come. NR = *000*, i.e., B is expecting frame 1 from A. B sends information frame 2 (⇐ *100*).

⇐ *Flag / 'B' / 0 100 0 000 / Text2 / FCS / Flag*

The final bit is not set — there is more to come. NR continues set to *000* indicating B is still expecting A's first frame. B continues with information frame 3 (*010*).

⇐ *Flag / 'B' / 0 110 0 000 / Text3 / FCS / Flag*

The final bit is not set — there is more to come. NR continues to indicate no information frames have passed from A to B. B continues with information frame 4 (⇐ *110*).

⇐ *Flag / 'B' / 0 110 0 000 / Text4 / FCS / Flag*

The final bit is not set — there is more to come. NR continues to indicate no information frames have passed from A to B. While B is sending information frame 4 to A, A destroys frame 3 and sends a supervisory frame to B.

Flag / FCS / 010 1 00 01 / 'B' / Flag ⇒

The mode bits indicate A is ready to receive. The poll bit is set — a response is required. NR is set to 3 (*010*) indicating that A has found an error in frame 3. B re-sends information frame 3.

⇐ *Flag / 'B' / 0 010 0 000 / Text3 / FCS / Flag*

A sends an information frame to B.

Flag / FCS / Text1 / 100 0 000 0 / 'B' / Flag ⇒

NS = *000* — this is the first information frame A has sent to B. The final bit is not set — there is more to come. NR is set to *100* ⇒, i.e., A is expecting information frame 5. This indicates that information frames through 4 were received without detecting an error. B sends information frame 5 (⇐ *001*).

⇐ *Flag / 'B' / 0 001 1 100 / Text5 / FCS / Flag*

The final bit is set — the message is complete. NR = ⇐ *100*, i.e., B is expecting information frame 2; also acknowledging error-free receipt of information frame 1 from A. A sends a second information frame to B.

Flag / FCS / Text2 / 111 1 001 0 / 'B' / Flag ⇒

NS = *001* ⇒, the final bit is set — there is no more to come. NR is set to *111* indicating all frames from B have been received and no errors have been detected. B sends a supervisory frame.

⇐ *Flag / 'B' / 10 00 0 111 / FCS / Flag*

The mode bits indicate B is ready to receive. NR is set to *111* indicating all frames from A were received without detecting an error.

- **Address**: the basic address module is an octet. It may be extended to any number of octets (SDLC uses three). To do this, bit #1 in the first and succeeding octets (except the last) is set to 0. In the last octet of the address, bit #1 is set to 1.
- **Control**: in the basic control octet, NS (send sequence number) and NR (receive sequence number) are limited to three bits. In an extended mode format, the control field is expanded to two octets and NS and NR are set at seven bits each. Thus, the number of frames that can be sent before acknowledgment must be made is increased from 7 to 127 (assuming 111 and 11111111 are reserved for acknowledgment of entire sequences of frames).
- **Information**: usually, in SDLC the information field is limited to 1024 bits (128 characters). However, because the header format is fixed, and the trailer consists of the FCS that occupies the last two octets before the terminating flag, the information frame can be as long as the users require. What is more, the text need not have a regular byte structure, nor need it contain an integral number of bytes. This is important when transmitting data that have been compressed using Huffman coding, and similar techniques.
- **Frame check sequence**: each frame contains an FCS in the last two octets before the terminating flag. For most purposes, a 16-bit sequence is adequate. However, if greater certainty of error detection is required, a 32-bit FCS may be employed.

(3) SDLC Communication Modes

Originally developed to function in an environment dominated by a single mainframe, SDLC designates one station (usually the mainframe, or the front-end processor attached to the host) as the primary station, and all other stations as secondary stations. Frames that pass from primary to secondary are called *commands*, and frames that pass from a secondary station to the primary are called *responses*. Only one primary station may be associated with a link at any one time. These conventions lead to the unbalanced point-to-point link and the unbalanced multipoint link shown in **Figure 6.4**. In the normal response mode, a primary station manages the exchange; secondary stations respond when polled. On a duplex connection, if there is only one secondary station, it may initiate transmission without explicit permission from the primary station to produce an asynchronous, unbalanced, response mode. With the expansion of HDLC to a communications environment that includes national and international networks, a balanced point-to-point link structure has been added. Shown in Figure 6.4, it consists of two stations that combine primary and secondary status. Over a duplex connection between stations with equal communication responsibilities, transmission may be initiated by either station. When sending a command, the station acts as a primary. When receiving a command, or sending a response, the station acts as a secondary. This is known as an asynchronous, balanced, response mode.

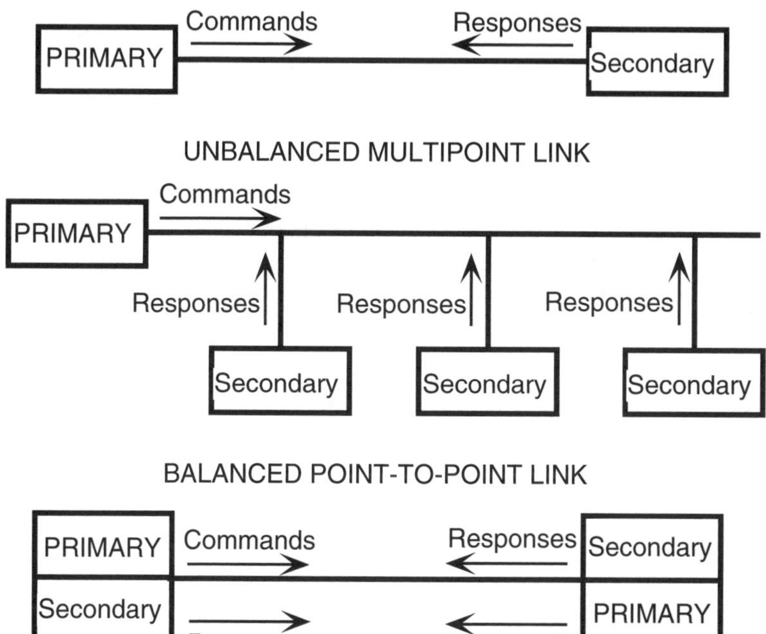

Figure 6.4 Link Structures and Communication Modes
Primary stations command; secondary stations respond.

(4) Classes of Communication Procedures

Bit-oriented, DLC protocols are subject to many options. Before reliable communication operations can be established in a specific environment, all stations must be set up to implement the same class of communication procedures. For instance, it will not do for some stations to be expecting 32-bit FCSs and others to be expecting 16-bit FCSs. For HDLC, international standards define three basic classes of procedure and 14 options that can be invoked.

(5) SDLC Transparency

In SDLC, the only special character is the flag, 01111110. Between the beginning flag and the ending flag, whenever a sequence of five 1s is detected, the transmitter inserts a 0. When a sequence of five 1s is detected by the receiver, the receiver deletes the following 0. In this way, bit patterns that mimic the flag do not provide false information to the receiver. Generally, this technique is described as *zero-stuffing*. As an example, consider the following datastream:

⇐ 01111110010111011111111111100000100111111011100101111110.

The first eight bits and the final eight bits are underlined—they are the beginning and ending flags (01111110). In between, there is a section of the datastream that mimics the flag (also underlined). Before transmission, between the beginning and ending flags, the transmitter inserts a 0 (denoted Ø for clarity) after sequences of five 1s. This makes the transmitted datastream:

⇐ 0111111001011101111 1Ø11111Ø11000001001111 1Ø1011100101111110.

The section of code in the datastream that mimicked a flag no longer does so. At the receiver, the inserted Øs are removed to produce the original signal. Note that zero-stuffing takes place without regard to the octet structure.

(6) SDLC Synchronization

The flag character is used to synchronize the receiver; no synchronizing characters are inserted in the data.

6.2 X.25 PROTOCOLS

ITU-T Recommendation X.25 describes the user-network interface employed with packet networks (see § 12.1.1). We describe it here to illustrate the interaction of network, data link, and physical layers.

6.2.1 X.25 Architecture

Between the originating DTE and the DCE that serves as entry to the packet network, X.25 defines three layers that are analogous to the network, data link, and physical layers of the OSI reference model. As shown in **Figure 6.5**, at the DTE we find:

- In the **packet layer**, or **X.25–3 layer** (network layer), packets are formed by adding control information to user information segments,
- In the **X.25–2 layer** (data link layer), the packet is encapsulated in an HDLC frame that implements LAPB (a bit-oriented protocol that operates in asynchronous balanced mode, i.e., either station may send without waiting for permission from the other, and all information frames are treated as commands)
- In the **X.25–1 layer** (physical layer), the frame is transmitted synchronously to the network DCE.

At the DCE we find the following:

- The **physical layer** reconstructs the datastream.

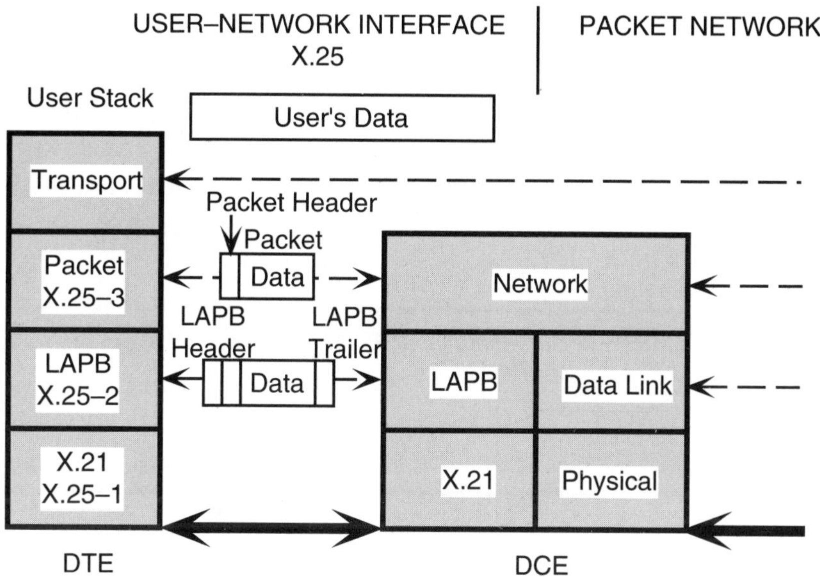

Figure 6.5 X.25 Architecture
X.25 specifies the user-network interface to a packet network.

- The **data link layer** interprets the header and trailer in accordance with LAPB rules, and passes the packet to the network layer.
- The **network layer** hands off the packet to the proprietary protocols that make whatever modifications are necessary for it to enter the packet network.

6.2.2 What is a Packet?

A packet is a sequence of bits that is divided in two—one part contains the user's information, and the other part contains control information. Usually, the packet is relatively short so that a user's message must be sub-divided before it is sent. The division of user's data into packets, and encapsulation in a LAPB frame, is shown in **Figure 6.6**. LAPB is the vehicle for communication between the originating DTE and the DCE at the packet network boundary. Each packet may be encapsulated in several different frames during its journey through the network and must carry along sufficient information to ensure it will traverse the network successfully. This information is contained in the three octets of control information that transit the route with the user's data segment.

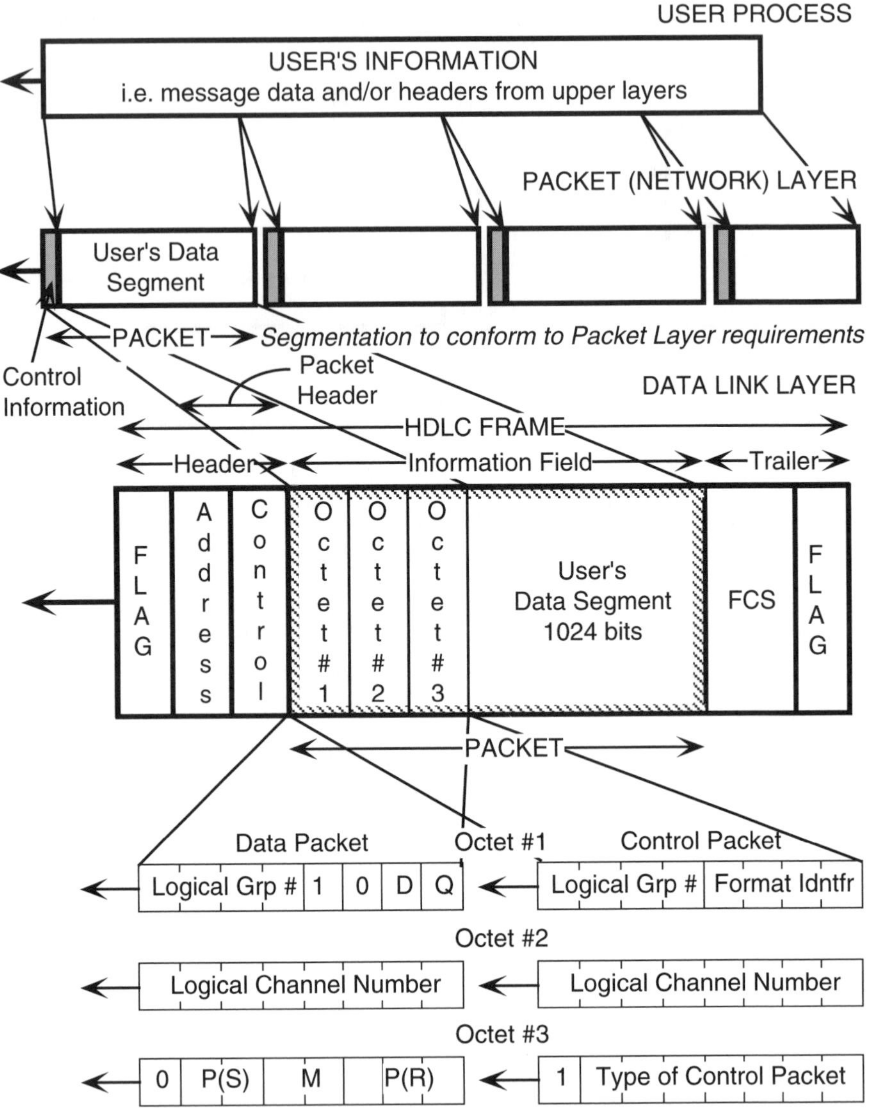

Figure 6.6 X.25 Packet and Frame Format

6.2.3 Frame and Packet Sequence Numbers

The send and receive sequence numbers—N(S) and N(R) in the control octet of the HDLC frame, and P(S) and P(R) in Octet #3 of the packet—are different. N(S) and N(R) counts are assigned consecutively to all frames crossing the link irrespective of the call to which the packets they contain are assigned. P(S) and P(R), on the other hand, refer to the sequence of packets that comprises a specific call.

6.2.4 Virtual Circuits

The network resources required to move a packet through the network are allocated on a link by link basis as each packet in a call transits the network. These resources are used by other packets at other times. Octets #1 and #2 of the packet header contain a 4-bit logical group number, and an 8-bit logical channel number; together, they provide a 12-bit identifier. If the all-zeros possibility is excluded, this identifier can discriminate among 4095 channels. Logical channel numbers are assigned at the start of each session by the originating DTE and terminating DCE to identify the stream of packets in the coming exchange of data. The originating and terminating logical channel numbers identify a virtual circuit in the same way that the telephone numbers of the calling and called parties identify a telephone circuit. It is up to the network administration resource to coordinate the network portion of the journey to satisfy the requirements of the communicating DTEs.

6.3 TRANSPORT AND NETWORK LAYER PROTOCOLS

Packet networks have expanded, proliferated, and interconnected to form *Internet*, the largest data network in the world. Users share a layered set of rules by which computer processes are able to communicate effectively across this network of networks (see § 12.1.2). Called TCP/IP, the name is shorthand for a suite of protocols whose major members are TCP (transmission control protocol), a transmission layer procedure, and IP (Internet protocol), a network layer procedure.

6.3.1 TCP/IP

(1) Transmission Control Protocol

The transport layer protocol used by the Internet community can operate with a variety of network layer protocols. Known as TCP, and most often used with IP datagram service, it was the first, successful, general-purpose, transport layer protocol. TCP is a connection-oriented protocol responsible for the reliable transfer of information between source and destination hosts. In spite of possible problems in the network layer protocol that may cause packet duplication, packet loss, or the delivery of packets out of sequence, TCP guarantees delivery of the exact datastream passed to it by the source. This is accomplished by sequence numbers, and timers that look for acknowledgments within specified time periods. At the destination host, TCP performs flow control by sending a window value to the source host. It authorizes the source to send just so many octets. When that number is reached, the source ceases to transmit until another authorization is received. **Figure 6.7**A shows the makeup of a TCP data unit (TCPDU). While most of the entries are read-

A: TCP Data Unit Format

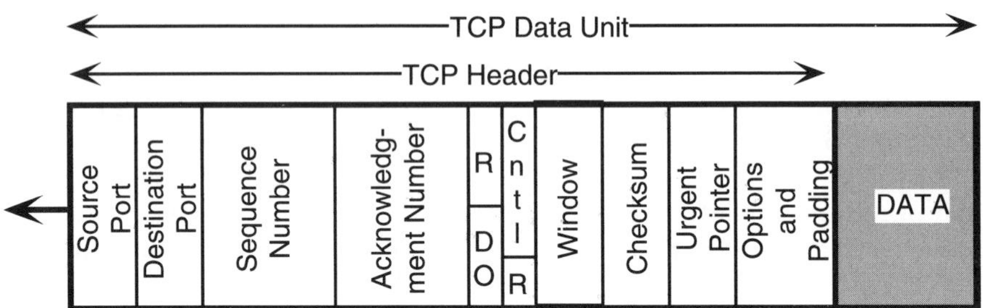

Source Port (16 bits) -- identifies source service access point (SAP), i.e., address of an application within the source node or host
Destination Port (16 bits) -- identifies source service access point (SAP)
Sequence Number (32 bits) -- number of first data byte in information PDU
Acknowledgment Number (32 bits) -- sequence number of next byte expected by TCP entity
Cntrl = Control Bits (6 bits) -- bits are set to indicate urgency, request immediate delivery, request reset, synchronize sequence numbers, and end of datastream
R = Reserved (6 bits) -- reserved for future uses
Window (16 bits) -- for flow control, number of bytes that receive port will accept before acknowledgment
Checksum (16 bits) -- see text for explanation
Urgent Pointer (16 bits) -- indicates sequence number of byte following urgent data
Options (variable) -- requests specific receive buffer size, and other services
Padding (variable) -- ensures that Header is a multiple of 4 octets long.

B: UDP Data Unit Format

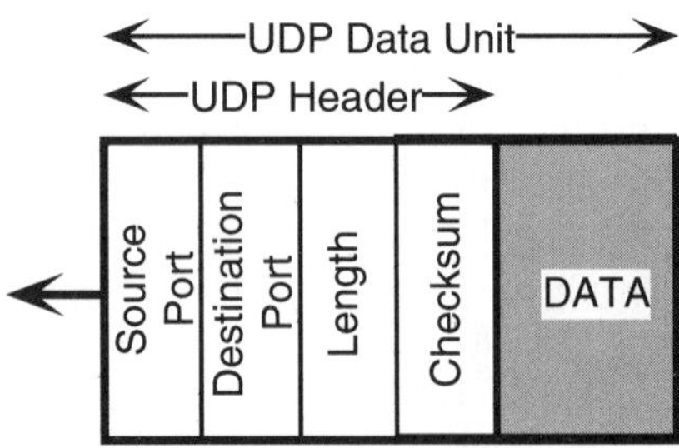

Figure 6.7 TCP and UDP Data Unit Formats
Part of TCP/IP suite, Transport Control Protocol and User Datagram Protocol are transport layer protocols.

ily understood, the checksum needs an explanation. It is computed by placing the 32–bit source address (contained in the IP Header), the 32–bit destination address (contained in the IP Header), and other information, ahead of the TCPDU and performing the calculation over these items and the entire TCPDU. The inclusion of the addresses in the checksum is a means of ensuring that the complete PDU reaches the correct host attached to the correct network. TCP employs virtual circuits (defined by the combination of source and destination identifiers) and ensures that all, in-process, data operations are completed before breaking a connection.

(2) User Datagram Protocol

For many applications, the capabilities of TCP are more than are required. This is particularly true for relatively short messages. For them, a connectionless procedure called the user datagram protocol (UDP) can be used. Without flow control and error recovery procedures, UDP employs the PDU shown in Figure 6.7B. It contains a destination port number that identifies the point of termination in the upper layer protocol, and may contain a source port number that identifies the port in the source host to which any reply is sent. The checksum, which is calculated in the same manner as the checksum for TCP, is used to check incoming datagrams. If the checksum is invalid, the datagram is destroyed.

(3) Internet Protocol

IP provides a robust, connectionless service between hosts in which data from the sending host is joined to an IP header to form an IP datagram. Shown in **Figure 6.8** the datagram is passed to a subnetwork in which it is encapsulated in the appropriate data link layer frame. For the particular arrangement shown, the datagram is passed across the subnetwork to an IP gateway that links to a second subnetwork. In turn, this network is connected to the destination host. At the gateway, the subnetwork information is stripped off the datagram, and it is encapsulated in a data link frame appropriate to the next leg of the journey. The gateway uses the destination address in the datagram header to route it to the intended host. Eventually, the datagram arrives at its destination and is delivered to the host. Practically, as long as the gateways know how to access them, the route can traverse any number of subnetworks and the subnetworks can employ different data link control protocols (DLCPs). To prevent datagrams from circulating indefinitely, the header contains a *time to live* number that is reduced by one every time the datagram traverses a gateway. If the number becomes zero, the datagram is destroyed. For lost, missing, or destroyed datagrams, it is up to the transport layer to provide corrective action.

6.3.2 ISO Protocols

(1) ISO Network Layer Protocol

For connection-oriented operation, ISO has not developed a complete network layer protocol. The principal difficulty is the dependence of routing and flow con-

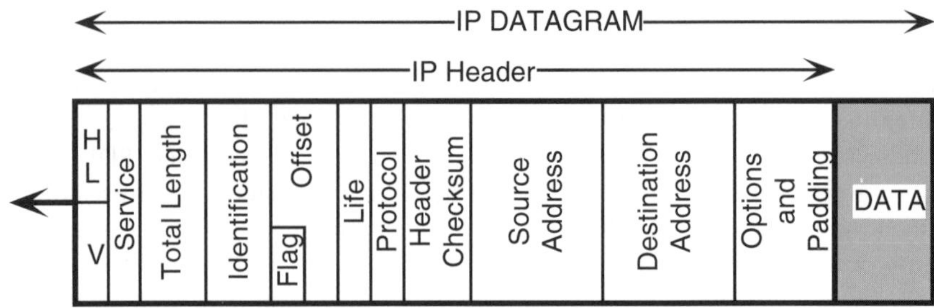

V = Version (4 bits) -- identifies version of IP in use
HL = Header Length (4 bits) -- specifies length of IP Header in multiples of 4 octets
Service = Type of Service (8 bits) -- specifies parameters such as desired reliability and throughput
Total Length (16 bits) -- identifies length of datagram (or current fragment); maximum length 65 535 bits
Identification (16 bits) -- sequence number
Flag (3 bits) -- permit, or prohibit, fragmentation of datagram
Offset (13 bits) -- i.e., fragment offset; indicates where in PDU this fragment belongs
Life = Time to Live (8 bits) -- measured in gateway hops; ensures fragment does not loop indefinitely
Protocol (8 bits) -- identifies next-level protocol to receive data at destination
Header Checksum (16 bits) -- performs error check on Header
Source Address (32 bits)
Destination address (32 bits)
Options (variable) -- requests specific routing, handling, and other services
Padding (variable) -- ensures that Header is a multiple of 4 octets long.

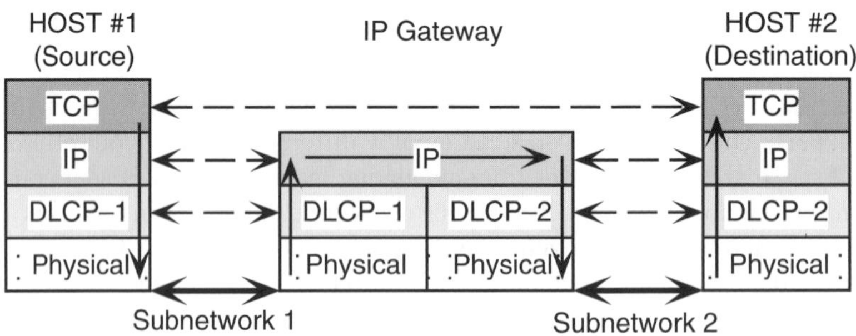

Figure 6.8 IP Datagram Format and Protocol Stacks for Data Transfer
Part of TCP/IP suite, Internet Protocol is a network layer protocol. The lower figure shows the use of TCP/IP to facilitate communication across different networks.

trol functions on implementation. However, standards do exist that describe the organization of the network layer (ISO 8648), the services the network layer provides to the transport layer (ISO 8348), and network layer addressing.

ISO 8648 divides the network layer into three sublayers:

- **Subnetwork independent convergence functions** (SNICF): concerned with internetwork routing and relay functions, and global network protocol. The SNICF sublayer contains functions that implement internetwork protocols; they do not depend on specific networks.
- **Subnetwork dependent convergence functions** (SNDCF): concerned with correcting network anomalies and deficiencies. The SNDCF sublayer is included to accommodate situations in which some networks do not provide all the services assumed by SNICF.
- **Subnetwork access functions** (SNAF): concerned with data transfer within individual networks. The SNAF sublayer defines how network layer entities make use of the functions networks provide.

ISO 8348 describes a connection-oriented process. Divided into three phases—connection setup, data transfer, and disconnection—the standard describes services that establish the connection, transfer data, recover from interruptions, confirm receipt, expedite transfer, and release the connection.

ISO has defined a connectionless network layer protocol (CLNP, ISO 8473). It is a *best efforts* datagram service in which data units are routed across the network by nodal decisions (distributed routing) without establishing a connection. Delivery is not guaranteed, and when the datagrams are delivered, they may not be in the sequence in which they were sent.

(2) ISO Transport Layer Protocol

ISO has defined five transport layer protocols for use with three different levels (A, B and C) of network performance:

- The high standard of **type A networks** eliminates the need for failure recovery services, and services to handle loss of data and other delivery problems.
- In contrast, the poor standard of **type C networks** makes it necessary to detect and recover from network failures, and to detect and correct for loss of data, out-of-sequence data, duplicated data, misdirected messages, and other delivery problems.
- **Type B networks** lie somewhere between these extremes.

The transport layer standards define the following connection-oriented headers:

- **TP0**: assumes a type A network with error control, sequencing and flow control. It provides minimal connection establishment, data transfer, and connection termination facilities. TP0 was developed for teletex applications (see § 7.3.2).

- **TP1**: assumes a type B network. It provides basic error recovery services including recovery from resets, clears, and restarts.
- **TP2**: assumes a type A network. It provides the ability to multiplex several transport layer connections over a network layer circuit, and transport layer flow control.
- **TP3**: assumes a type B network. It adds the error recovery services of TP1 to the multiplexing and flow control services of TP2.
- **TP4**: assumes a type C network. It provides comprehensive error detection and recovery services.

Five headers are defined for *transport-connection management* (connection request, connection confirm, disconnect request, disconnect confirm, and reject), and five for *data transfer* (data, acknowledgment, expedited data, expedited acknowledgment, and TPDU error). These functions are similar to those performed by TCP, but TP4 and TCP are not compatible. In **Figure 6.9**, we show the formats of connection request, connection confirm, data, and acknowledgment headers. Most of the field names are self-explanatory. Ones that require explanation are:

- *Class* (CLS): protocol class (i.e., 0, 1, 2, 3, or 4).
- *Credit* (CDT): credit allocation for flow control, i.e., the number of TPDUs that can be transmitted from the last acknowledgment.
- *Option* (OPT): specifies flow control fields. In normal operation, 7 bits are allocated to sequence numbers (i.e., possible sequence of 128 TPDUs) and 4 bits are reserved for credit (i.e., up to 16 TPDUs may be sent from last acknowledgment). In extended operation, 31 bits are allocated to sequence numbers (i.e., possible sequence of over 2 billion TPDUs) and 16 bits are reserved for credit (i.e., up to 65536 TPDUs may be sent from last acknowledgment).
- *EOT:* If the session layer data unit has been segmented, the bit is set to 1 in the TPDU containing the last segment.

ISO has defined a connectionless transport layer protocol (CLTP, ISO8602). It is a *best efforts* datagram service in which data units are transported between sender and receiver on the basis of destination address without establishing a connection. Delivery is not guaranteed, and when the datagrams are delivered, they may not be in the sequence they were sent.

6.3.3 A Bottleneck?

As demand increases for higher-speed data transfers over high-speed connections (such as optical fibers), it is becoming obvious that a significant delay is created by

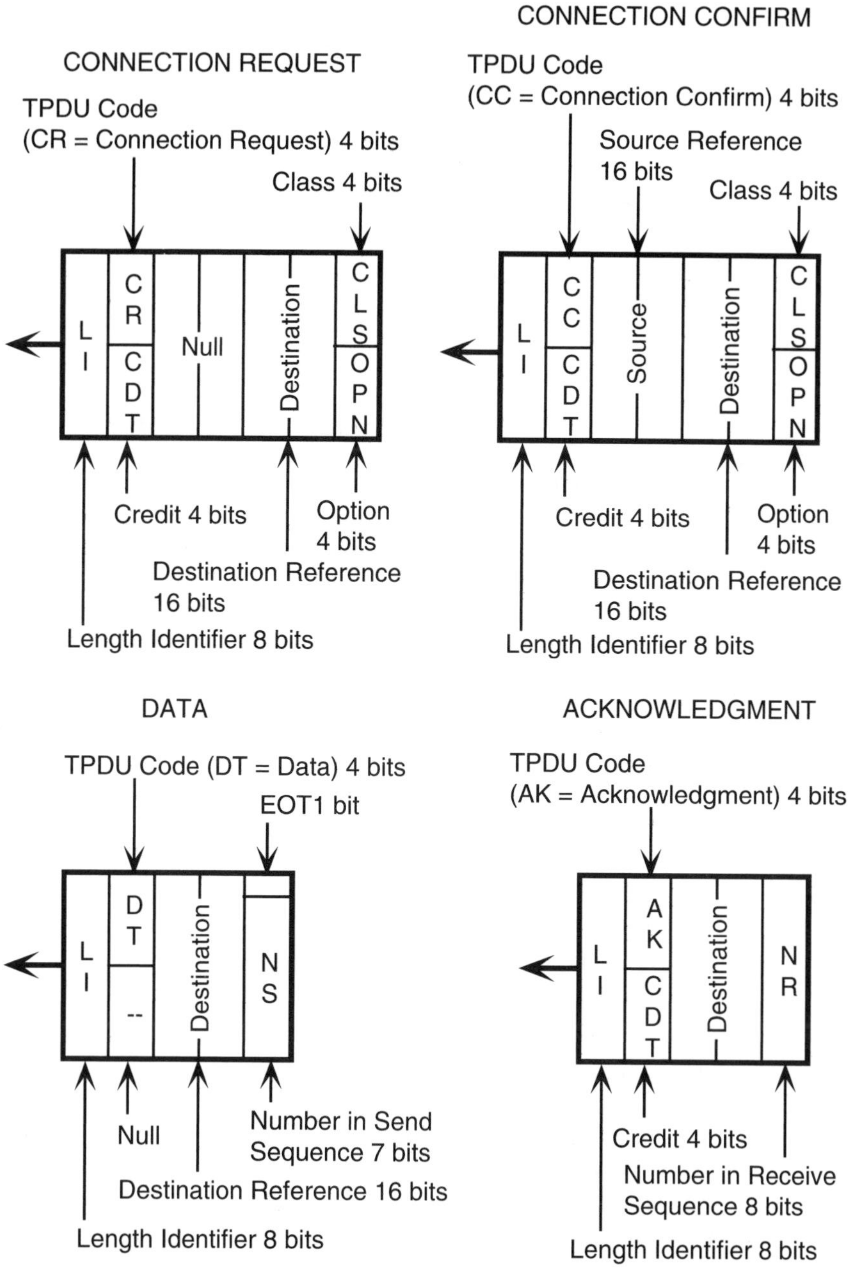

Figure 6.9 Examples of Headers For TP4
TP4 is an ISO Transport Layer Protocol.

the processing associated with the communication protocols resident in the terminals and the nodes. At the lower levels, speedup is accomplished by eliminating link-by-link error control and flow control, implementing as many functions as possible in hardware, and relying on the transport layer to reconstruct the message faithfully. With regard to the transport layer, critics are calling for a *new generation* of protocols while supporters call for better implementations of what exists, not new procedures. The result is increasing attention to the efficiencies of the transport and network layer protocols, and the emergence of new, proprietary, protocol suites.

6.4 THROUGHPUT

To the network designer, the rate at which each link transports data is of vital importance. However, to the sender, the internal network speeds are much less important than the time it takes to deliver a message. As this Chapter has shown, along with message information must go overhead, to alert the receiver that a message is coming, to carry the address, to manage the data link, and to check for and correct errors. In communication systems, the rates at which user's data flow are always less than the line rates. To evaluate this effect, we define throughout:

- **throughput**: the rate at which user's data bits are delivered:

$$\text{Throughput} = \frac{\textit{number of user' s data bits delivered}}{\textit{time to deliver user' s data bits}}.$$

Throughput depends on several factors. The following questions must be asked:

- What is the format of the frame employed?
- Does the data link PDU contain headers from other layers?
- Is the link half-duplex or duplex?
- Is ARQ used for error correction; if so, what type of ARQ?
- What is the bit error-rate of the connection?

To calculate throughput requires answers to these questions, and assumptions must be made as to the distribution of errors and the efficacy of error correction. We illustrate the situation by calculating throughputs over a range of situations. To keep things simple, we shall ignore the possibility of delays encountered in the upper layers of the protocol stacks of the sending and receiving equipment.

EXAMPLE 6.1 Simplex Link

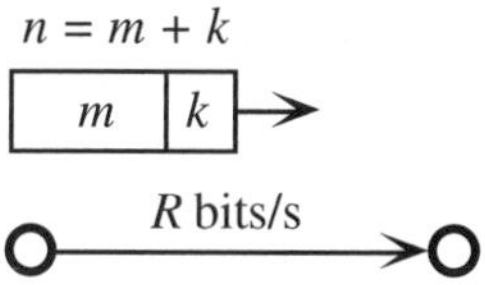

$n = m + k$

m	k	→

R bits/s

o———→o

Assumptions: the link is short enough that propagation delay is negligible and no processing is required at the receiver. Frame = n bits = m bits of user's data + k bits of overhead, so that frame coding efficiency, $\eta_f = \frac{m}{n}$.

Derivation: time to send frame = $\frac{n}{R}$ seconds. Hence,

$$\text{THROUGHPUT} = \frac{m}{\text{time to send}} = \frac{mR}{n} = \eta_f R \text{ bits / s} \ .$$

Numerical Values: For an SDLC frame sent at 56 kbits/s in which n = 1088 bits, m = 1024 bits and $\eta_f = 0.94$:

Throughput = 52.64 kbits/s.

For a DDCMP frame sent at 56 kbits/s in which n = 1120 bits and m = 1024 bits:

Throughput = 51.2 kbits/s.

EXAMPLE 6.2 Simplex Link with Delay

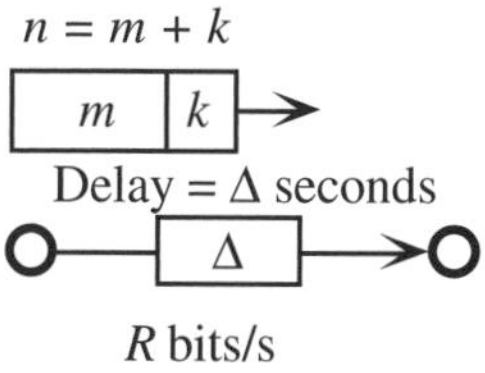

$n = m + k$

m	k	→

Delay = Δ seconds

o—[Δ]—→o

R bits/s

Assumptions: there is a delay of Δ seconds somewhere in the transmission process. It may be propagation delay, or processing delay at the receiver (verifying FCS, for instance), or both.

Derivation: time to send frame = $\frac{n}{R} + \Delta$ seconds. Hence,

$$\text{THROUGHPUT} = \frac{m}{\frac{n}{R} + \Delta} = \frac{mR}{n + \Delta R} = \eta_f \frac{nR}{n + \Delta R} \quad \text{bits/s.}$$

Numerical Value: for an SDLC frame sent at 56 kbits/s and a processing delay equal to 1 frame time:

$$\text{Throughput} = 26.32 \text{ kbits/s.}$$

Extension 1: suppose verification proceeds simultaneously with reception. There will be no delay, and throughput will be 52.64 kbits/s, the value calculated in Example 6.1.

Extension 2: suppose the sender sends f frames one after the other and that the receiver validates frame 1 while receiving frame 2, validates frame 2 while receiving frame 3, etc. There will be no delay for the evaluation of frames 1 through f - 1, and one frame delay while evaluating frame f. Hence,

$$\text{Time to send } f \text{ frames} = \frac{fn}{R} + \Delta \text{ seconds , and}$$

$$\text{THROUGHPUT} = \frac{fm}{\frac{fn}{R} + \Delta} = \eta_f \frac{fnR}{fn + \Delta R} \text{ bits/s.}$$

Thus, for 7 SDLC frames sent one after another at 56 kbits/s and a processing delay equal to 1 frame time for each frame.

$$\text{Throughput} = 46.1 \text{ kbits/s.}$$

EXAMPLE 6.3 Half-Duplex Link with Delay and Stop-and-Wait ARQ

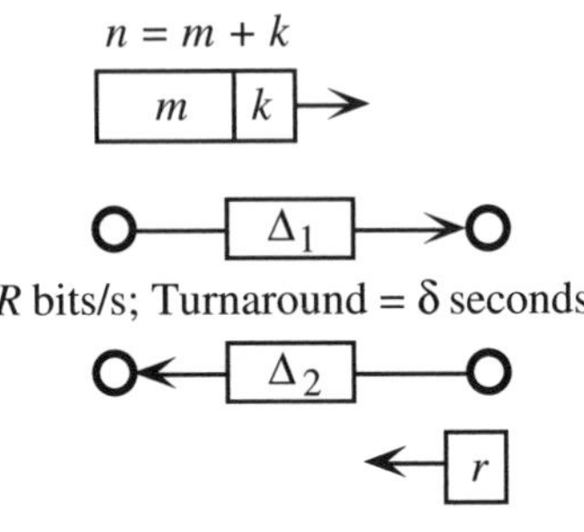

Assumptions: at the receiver there is a delay of one frame time for processing (checking FCS, for instance), and it is $\leq \delta$ seconds, so that $\Delta_1 = \Delta_2 = \delta$.

Derivation: the time to send the user's data is the time that elapses between sending one frame and the next frame.

$$\text{Time to send frame} = \frac{n}{R} + \Delta_1 \text{ seconds}.$$

$$\text{Time to send acknowledgment} = \frac{r}{R} + \Delta_2 \text{ seconds.}$$

Hence, total time that elapses between sending frames = $\frac{n}{R} + \delta + \frac{r}{R} + \delta$ seconds, and

$$\text{THROUGHPUT} = \frac{m}{\frac{n+r}{R}+2\delta} = \eta_f \frac{nR}{n+r+2R\delta} \text{ bits/s.}$$

Numerical Value: for an SDLC frame sent at 56 kbits/s, a response frame of 48 bits, and a turnaround time of 0.1 seconds,

$$\text{Throughput} = 4643 \text{ bits/s.}$$

Extension: suppose the frame is sent at 4.8 kbits/s, and there is a delay of one frame time for processing at the receiver, i.e., 0.227 seconds. This is larger than the turnaround time so that $\Delta_1 = 0.227$ and turnaround occurs during processing. For the response frame, there is no FCS, and no processing, just recognition of the frame as an acknowledgment, so that $\Delta_2 = \delta$. Under these circumstances,

$$\text{time between sending one frame and the next} = \frac{n+r}{R} + 0.327 \text{ seconds, and}$$

$$\text{Throughput} = 2996 \text{ bits/s.}$$

EXAMPLE 6.4 Half-Duplex Link with Delay, Errors, and Stop-and-Wait ARQ

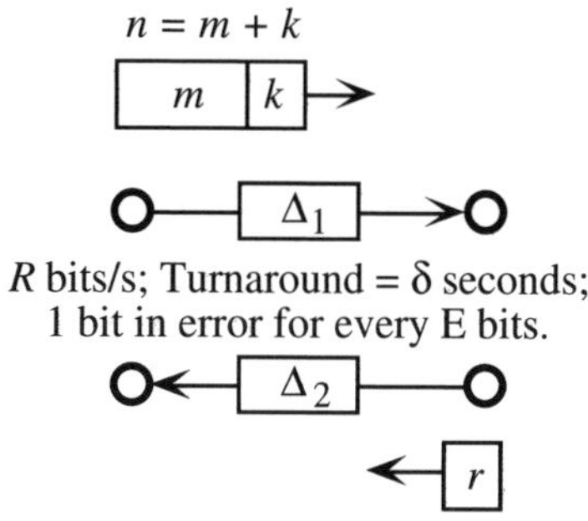

Assumptions: the errors are uniformly distributed, and the acknowledgment frame and the repeated frame are always good. Also, time to evaluate ≤ time to turnaround so that $\Delta_1 = \Delta_2 = \delta$.

Derivation: for every frame sent, we must send $\frac{n}{E}$ additional frames to correct for errors.

From Example 6.3, the total time that elapses between sending frames = $\frac{n+r}{R} + 2\delta$ seconds so that the total time that elapses between sending $1+\frac{n}{E}$ frames is $\left(1+\frac{n}{E}\right)\left(\frac{n+r}{R}+2\delta\right)$ seconds. Hence,

$$\text{THROUGHPUT} = \eta_f \frac{nR}{\left(1+\frac{n}{E}\right)(n+r+2R\delta)} \text{ bits/s.}$$

Numerical Value: for an SDLC frame sent at 56 kbits/s, a response frame of 48 bits, a turnaround time of 0.1 seconds, and an error rate equal to 1 bit in error in 10^4 bits,

$$\text{Throughput} = 4454 \text{ bits/s.}$$

EXAMPLE 6.5 Half-Duplex Link with Delay and Go-Back-N ARQ

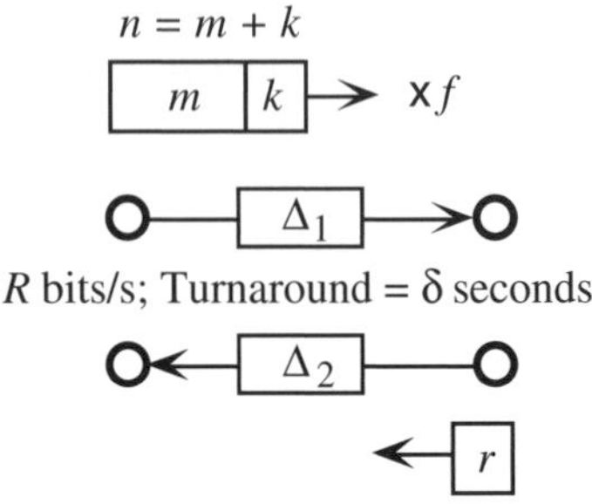

Assumptions: the sender sends *f* frames one after the other, then waits for acknowledgment. Because the process is error free, after receipt of the acknowledgment, the sender sends another *f* frames, etc. The receiver validates frames 1, 2,...(*f* - 1) while receiving frames 2, 3,...*f*. The time to evaluate ≤ time to turnaround so that $\Delta_1 = \Delta_2 = \delta$.
Derivation: there will be a frame delay while frame *f* is evaluated. However, this is done while the line is being turned around, so that the total time that elapses between beginning to send a burst of *f* frames, and beginning to send a second burst is $\frac{fn+r}{R} + 2\delta$. Hence,

$$\text{THROUGHPUT} = \eta_f \frac{fnR}{fn + r + 2\delta R} \text{ bits/s.}$$

Numerical Value: For bursts of 7 SDLC frames sent at 56 kbits/s, a response frame of 48 bits, and a turnaround time of 0.1 seconds,

$$\text{Throughput} = 21.2 \text{ kbits/s.}$$

EXAMPLE 6.6 Half-Duplex Link with Delay, Errors, and Go-Back-N ARQ

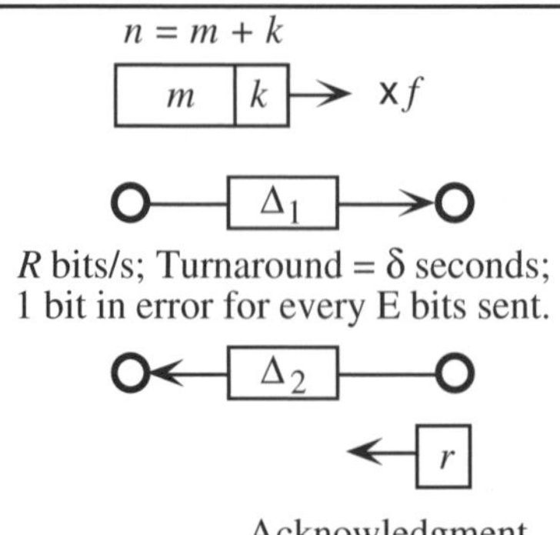

Assumptions: the errors are uniformly distributed and the acknowledgment frame and the repeated frames are always good. The sender sends f frames, then waits for acknowledgment. The receiver validates frames 1, 2,...(f - 1) while receiving frames 2, 3,...f. The time to evaluate ≤ time to turnaround so that $\Delta_1 = \Delta_2 = \delta$.
Derivation: on receipt of an acknowledgment that contains a frame number other than the one the sender is about to send, the sender must repeat between 1 and f frames. On average, when an error is detected, the sender must repeat $\frac{f}{2}$ frames, so that, for every user frame, the system must send $\left(1+\frac{fn}{2E}\right)$ frames. Using Example 6.5, the total time to send this number of frames is $\left(1+\frac{fn}{2E}\right)\left(\frac{fn+r}{R}+2\delta\right)$. Hence,

$$\text{THROUGHPUT} = \eta_f \frac{fnR}{\left(1+\frac{fn}{2E}\right)(fn+r+2\delta R)} \text{ bits/s.}$$

Numerical Value: for 7 SDLC frames sent at 56 kbits/s, a response frame of 48 bits, a turnaround time of 0.1 seconds, and an error rate equal to 1 bit in error in 10^4 bits,

Throughput = 15.3 kbits/s.

EXAMPLE 6.7 Duplex Link with Delay, Errors, and Go-Back-N ARQ

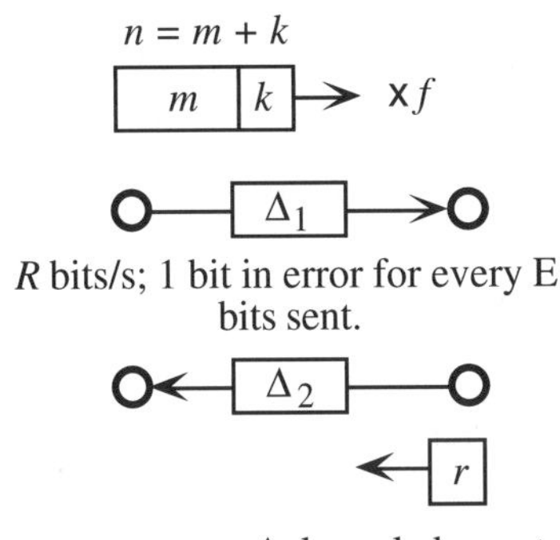

Assumption: the errors are uniformly distributed and the acknowledgment frame and the repeated frames are always good. Acknowledgments and requests to repeat frames are made without interrupting the flow of user's frames. This case is the same as Example 6.6 except that $\delta = 0$.
Derivation:

$$\text{THROUGHPUT} = \eta_f \frac{fnR}{\left(1+\frac{fn}{2E}\right)(fn+r)} \text{ bits/s.}$$

Numerical Value: for 7 SDLC frames sent at 56 kbits/s, a response frame of 48 bits, and an error rate equal to 1 bit in error in 10^4 bits,

$$\text{Throughput} = 38.1 \text{ kbits/s}.$$

Extension 1: if the error rate is 1 bit in error in 10^6 bits,

$$\text{Throughput} = 52.1 \text{ kbits/s}.$$

Extension 2: if the error rate is 1 bit in error in 10^8 bits,

$$\text{Throughput} = 52.6 \text{ kbits/s}.$$

EXAMPLE 6.8 Duplex Link with Delay, Errors, and Selective Repeat ARQ

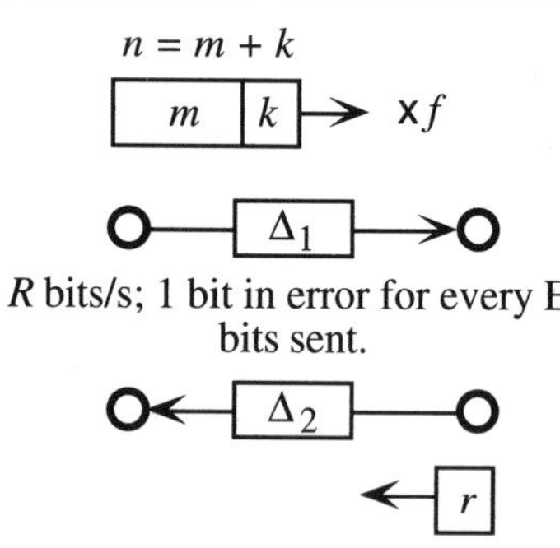

Assumptions: the errors are uniformly distributed and the acknowledgment frame and the repeated frames are always good. Acknowledgments and requests to repeat frames are made without interrupting the flow of user's frames.

Derivation: for every user's frame, we must send $1+\frac{n}{E}$ frames to correct for errors.
Hence, the time to send a frame is $\frac{n}{R}\left(1+\frac{n}{E}\right)$, and

$$\text{THROUGHPUT} = \eta_f \frac{R}{\left(1+\frac{n}{E}\right)} \text{ bits/s}.$$

Numerical Value: for SDLC frames sent at 56 kbits/s, and an error rate equal to 1 bit in error in 10^4 bits,

$$\text{Throughput} = 47.5 \text{ kbits/s}.$$

Extrapolating from these examples, and scenarios earlier in the Chapter, it is possible to conclude the following:

- The maximum possible throughput is $\eta_f R$ bits/s.

- It is important to know how the receiver will process incoming frames. When a string of frames is sent, the receiver should process one frame while receiving the next. If simultaneous reception and verification do not occur, the multiple processing delays depress throughput significantly.
- In half-duplex operation, critical items are turnaround and processing times. The longer they are, the less effective is higher speed in achieving greater throughput. This is particularly true for stop-and-wait ARQ.
- Error rates in the vicinity of 1 bit in error in 10^4 bits affect throughput substantially.
- With error rates in the vicinity of 1 bit in error in 10^4 bits, half-duplex, go-back-n ARQ, produces significantly better throughput than half-duplex, stop-and-wait ARQ.
- With error rates in the vicinity of 1 bit in error in 10^4 bits, duplex, selective repeat ARQ, produces significantly better throughput than duplex, go-back-n ARQ.
- Error rates in the vicinity of 1 bit in error in 10^6 bits affect throughput slightly.
- Error rates in the vicinity of 1 bit in error in 10^8 bits have an insignificant effect on throughput.
- With error rates in the vicinity of 1 bit in error in 10^8 bits on duplex circuits, throughputs with selective repeat ARQ and go-back-n ARQ are very close to the maximum.
- Go-back-n ARQ assures the receiver will receive multiple frames in order. This is not true for selective repeat ARQ.

Bearing in mind performance and processing complexity, for good circuits ($P_e \leq 1 \times 10^{-8}$), go-back-n ARQ may be the best error control technique to employ. For poor circuits ($P_e \geq 1 \times 10^{-4}$) selective repeat (duplex) and go-back-n (half duplex) ARQ are preferred. For such circuits, consideration should be given to forward error correction (see § 4.2.3(2)).

7

Personal Telecommunication

Personal telecommunication involves the distribution or exchange of messages among persons, or between persons and machines. Because very few of us can listen and talk simultaneously, person-and-person communication is usually one-way at a time. So too, is communication between person and machine. Mediated by a terminal, the equipment arrangement usually prevents simultaneous presentation of input and output data; and, even if this is not so, we can only give our attention to one *window* at a time.

7.1 PERSON-AND-PERSON COMMUNICATION

7.1.1 Modes of Communication

(1) Natural Communication Mode

The natural mode of person-and-person communication is caricatured in **Figure 7.1**; it can be described as two-way, interactive, *face-to-face*. For persons to engage in this style of communication, they must be close enough so that their speech may be heard and their body language may be seen. In practice, face-to-face communication occurs when the parties are within a few feet of each other. Pressure waves representing the words of the speakers are transmitted between the parties by gas molecules in the air; visible information is conveyed by light that is reflected from the parties and their visual aids. Thus, acoustic and optical waves are the physical media that make face-to-face communication possible. They accomplish action-at-a-*short*-distance so that the communication process can proceed. In this arrange-

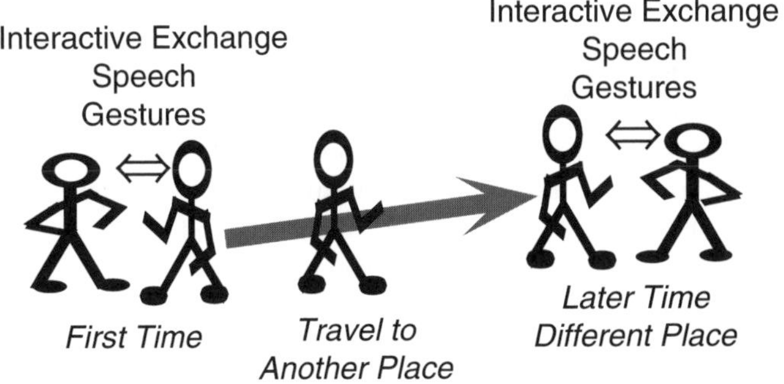

Figure 7.1 Communication Using Natural Face-To-Face Mode

ment, each party uses the full range of expressions of which they are capable—words and phrases emitted with appropriate emphasis, emotion, gestures, and body language. In addition, they may employ pencil and paper, or chalk and blackboard, to illustrate points; use physical models or simulations displayed on computer terminals to demonstrate sequences; play audio or video recordings to reinforce perceptions; etc. Once communication is completed, the participants may move to other places and use the information they have just gained in another encounter. In this way information and ideas circulate within a community. The distance between the first and second meeting is determined by the traveler, and the speed at which the information propagates is the speed of the traveler.

(2) Communication-at-a-Distance

The essential property of a *medium* is its ability to produce action-at-a-distance. Persons who are remote from one another must make use of media that can overcome interference from solid objects and that can span greater distances than the sound and reflected light waves of face-to-face encounters. This action requires conversion of the natural message to a form compatible with the medium employed. In this conversion, much of the *richness* of face-to-face communication is sacrificed to obtain convenience and low cost. Nevertheless, for most purposes, telecommunication facilities provide acceptable extensions of personal communication capabilities so that persons can maintain communities of interest in increasingly mobile situations, and can embrace an expanding array of information sources.

(3) Record and Real-Time Media

We can distinguish between record media and real-time media:

- **Record media**: by storing a limited amount of information for future reference, record media make it possible to communicate *at a later time*. Further, if the storage medium can be transported readily, communication can take place *at a distance, at a later time*; and, if the record medium can be duplicated in quantity, independent communication can occur with a large number of persons, at a distance, at their convenience. For example, after a book is written and printed, copies carry the author's ideas to a wide, contemporary audience that reads it at their convenience, and future readers can consult copies lodged in libraries for this purpose. By this mechanism, new generations of readers are able to build on the ideas, the ideals, and the discoveries of their forebears. In a similar fashion, audio and video tapes, and digital discs, capture episodes of sights and sounds that can be reproduced in other places at later times. The caricatures in **Figure 7.2** illustrate the process of communication using a record medium.

- **Real-time media**: make it possible to communicate-at-a-distance, *at the same time*. Telephone, television, and radio are prime examples of communication systems that employ real-time, communication media. Thus, so that persons who are separated from one another can discuss matters, the telephone can be used to connect them in pairs or conference groups. Moreover, as events happen, radio and television transfer eyewitness reports to audiences as big as the world. From remote receivers, they deliver a continuous stream of information that mimics the observations of persons in proximity to what is going on. However, there is a price to be paid for real-time performance—the information present at one instant is destroyed by the information that succeeds it. Although segments from the information stream may be independently recorded using an appropriate record medium, no storage capability exists in a real-time medium. The caricatures in **Figure 7.3**, show the process of telecommunication using a real-time medium. By way of illustration, the connection is divided into transducers (telephones), bearer, and medium, a division that is discussed below.

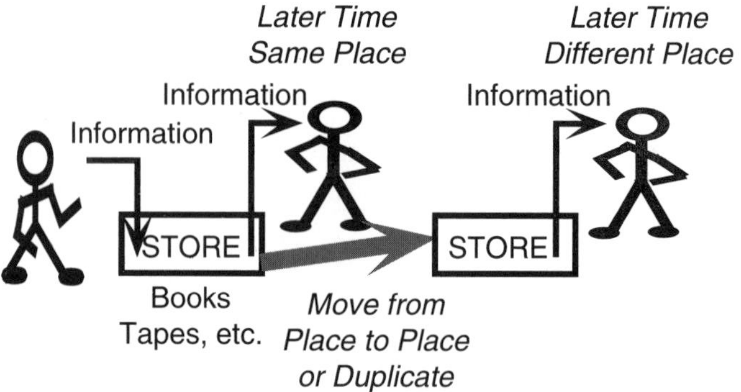

Figure 7.2 Communication Using Record Medium

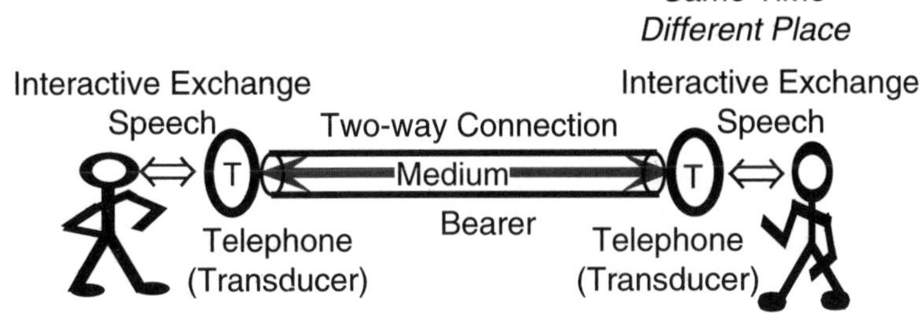

Figure 7.3 Communication Using Real-Time Medium

(4) Store-and-Forward Operation

Record and real-time communication media may be used in combination to achieve *store-and-forward* operation. Electronic mail and voice messaging are examples of store-and-forward services. In these systems, messages are generated, sent to an intermediate location, stored, and forwarded to the recipient on demand.

7.1.2 Electronic Communication

(1) Media

In electronic communication systems, *real-time* telecommunication is produced by electrical, electromagnetic, and optical energy; they are the media that make telecommunication possible:

- **Electric currents** flowing in telephone circuits enable signals representing the voices of talkers to pass to the telephones of listeners where they are reconstructed into sound waves that mimic those originally generated by the talkers.
- **Electromagnetic waves** that propagate through free-space, or are guided by coaxial cables, carry television signals to millions of receivers, and deliver steady streams of news, views, entertainment, and commercial messages.
- **Optical waves** that propagate along optical fibers carry high speed streams of digital data between computer systems, delivering data files or downloading programs, or they transport thousands of telephone calls between major telephone centers.

In electronic communication systems, *record* communication is facilitated by the ability to store information so that it can be retrieved at the location in which it was recorded, or at another location, at a later time. Consider the following:

- **Magnetostatic fields** are used to create records on tapes and disks.

- **Electrostatic fields** are used to create records in electronic arrays.
- **Mechanical** or **chemical imprints** are used to create holographic records.

(2) Bearers

Real-time, electronic media are carried by *bearers*:

- **Copper wires** (twisted pairs) are bearers—they transport electric currents for the delivery of voice and low-speed data messages between parties separated by some distance.
- **Coaxial cables** are bearers—within their sheaths they guide electromagnetic waves to complete intercommunication connections. They are well suited to very high-speed data communication between host computers, and they distribute television signals to groups of users with ease.
- **Free-space** is a bearer—when excited by *omni-directional* antennas, it distributes electromagnetic waves to provide mass communications. Such arrangements are well suited to broadcasting television or radio programs to mass audiences. Also, when excited by *directional* antennas, free-space can be made to transmit pencil beams of electromagnetic energy to complete intercommunication circuits. Such arrangements are well suited to carrying signals within telephone networks.
- **Optical fibers** are bearers—within themselves they transport optical energy to complete intercommunication connections. They are well suited to very high-speed data communication between host computers, and they carry voice and video signals between facilities with ease.

In summary, copper wires, coaxial cables, free-space, and optical fibers, do well as bearers of information signals in point-to-point applications; free-space does well as a bearer of information in mass communication applications; and coaxial cables do well as bearers in one-to-many applications.

(3) Transducers

Devices that convert messages from one medium to another are called *transducers*. Some are found at nodes of the network where they link dissimilar media so that signals may flow freely. Common conversions are electric current to microwave or optical energy, and *vice versa*. They are performed by solid-state devices (such as diodes, lasers, waveguides, and other devices) that are embedded in the network equipment. Other transducers are found at the edges of the network where they mediate user interaction with transmission facilities. Common examples include the following:

- **Telephone instruments**: required at each end of a telephone circuit, telephone instruments convert sound waves to electrical currents, and *vice versa*. They limit the message to the audio component of a face-to-face encounter.

- **Keyboards** and **video display units**: required at the user's end of a connection to a database facility, they convert keyboard inquiries to digital signals, and display the information retrieved.
- **Television receivers**: required at the user's end of a television channel in order to receive television programs. Television receivers convert complex signals to moving pictures and sounds.

Other examples are radios, facsimile machines, and personal computers. Combinations of media, bearers, and transducers, must be put together in a cost-effective way to provide communication channels whose characteristics are matched to the provision of specific services.

7.2 PERSONAL TELECOMMUNICATION SERVICES

7.2.1 The Services Plane

(1) Message Components

Personal communication services cater to the exchange or distribution of messages that are associated with an extensive range of human interests. The messages may employ symbols, sounds, or sights:

- **Symbols**: they are likely to specify facts, to define entities, or to seek to inform. The information they contain is in the arrangement of letters, numbers, and shapes. If the arrangement is sequential, we have *one-dimensional* communication exemplified by the output of printers and video display terminals that support electronic mail services and other services that employ *text*. If the arrangement covers an area, we have *two-dimensional* communication exemplified by the output of facsimile machines that support graphics mail and other services that employ *graphics*.
- **Sounds**: they are likely to discuss points, to describe situations, or to inspire feelings. The information is contained in the sequence and intensity of the sounds and in their pitch. Together with the sequence and intensity of the elementary sounds, communication depends on time—unless the sounds are executed at the correct rate, the message is distorted and may be lost. This is *three-dimensional* communication. Three–dimensional messages are the products of telephone, radio, and similar entities that employ *audio*.
- **Sights**: they are likely to present vistas, to witness events or to assist with the visualization of ideas. The information is contained in the intensity of the images and in static or dynamic relationships among objects arranged in space. Thus, the messages contain two- or three- dimensional space, intensity, and time. The result is *four-* and *five-dimensional* communication. Four-dimensional messages are the products of cameras and displays that employ

video. Five-dimensional messages may be the stuff of which *virtual reality* will be made.

(2) Richness

Using the concept of dimensions, it is possible to construct an empirical measure we call richness:

- **Richness**: a message is richer than another if it requires a more complex channel to deliver it. Channel complexity depends on the number of message dimensions. Thus, video messages are richer than audio messages, audio messages are richer than graphics messages, and graphics messages are richer than text messages.

To put form to these statements we structure a table by assigning bit positions to each dimension. In **Figure 7.4**, bit position 1 is associated with one-dimensional communication, i.e., services that employ text. Bit position 2 is associated with two-dimensional communication, i.e., services that employ graphics. Bit position 3

Message

CODING	Video	Audio	Graphics	Text	Richness
	1	0	0	0	8
		1	0	0	4
Designation			1	0	2
				1	1

EXAMPLES of SERVICES

Designation	Service	Richness
1 1 1 0	Augmented Videoconference	14
1 1 0 0	Videoconference	12
1 1 0 0	Television	12
1 0 0	Telephone	4
1 0	Facsimile	2
1	Electronic Mail	1

Figure 7.4 Text, Graphics, Audio and Video Messages, and Combinations, Arranged in Order of Richness

Richness is an arbitrary measure of the complexity of the channel required to deliver a person-oriented message.

is associated with services that employ audio, and bit position 4 is associated with services that employ video. The presence of a 1 in these positions defines the presence of the parameter. For instance, services that employ video are designated 1000; services that employ graphics are designated 10; etc. This notation permits us to construct designators for services such as television that employ both video and audio (1100), videoconferences that are augmented with facsimile (1110), telephone (100) and electronic mail (1). Also, the notation permits us to interpret the designator as an empirical richness number. Thus, telephone is of richness 4 and electronic mail is of richness 1. We shall make use of these values later in this section.

(3) Asymmetrical Connections

Connections among pairs or groups of persons for a common purpose need not have the same richness in both directions. Moreover, they are not constrained to assume the same topology in the forward and reverse directions. For instance, **Figure 7.5** shows a concept for a telecommunication system that provides education-at-a-distance (also known as distance-learning). The sound and sight of the instructor, and the images employed by the instructor are transmitted to all receiving sites, and, when authorized to do so, students at the sites may ask questions of the instructor. Under these circumstances, messages consisting of video, sound, and graphics (of richness 14) are distributed in simplex fashion from the instructor to the remote sites; when a student is given permission to ask a question, the site at which the student is located is connected to the instructor's site by a duplex voice connection (of richness 4). The discussion between student and instructor is re-transmitted from the instructor's site to all other sites on the existing simplex connection. Thus, in one direction, the system employs a bearer that connects one-to-many and supports a medium that can transport messages of richness 14. In addition, on demand, the facility employs a bearer that connects one-to-one, duplex between the instructor and a receiving site, and supports a medium that transports messages of richness 4. When the duplex connection is made, that part of the richness 14, simplex channel that carries sound to the site at which the question originates, is disconnected (to prevent echoes and oscillations). It is replaced by local sound.

(4) Directions and Information Space

The direction in which the more rich messages flow (i.e., instructor to receiving sites) is called the *downstream* direction and the direction in which the less rich messages flow (i.e. questioner to instructor) is called the *upstream* direction. Usually, one (perhaps more) communication mode is common to both directions. It is referred to as the common *information space*. In our tele-education example, the common information space is voice.

(5) Construction of Services Plane

Figure 7.6 shows a plane containing orthogonal axes that divide real-time media (right-hand side) from record media (left-hand side), and intercommunication

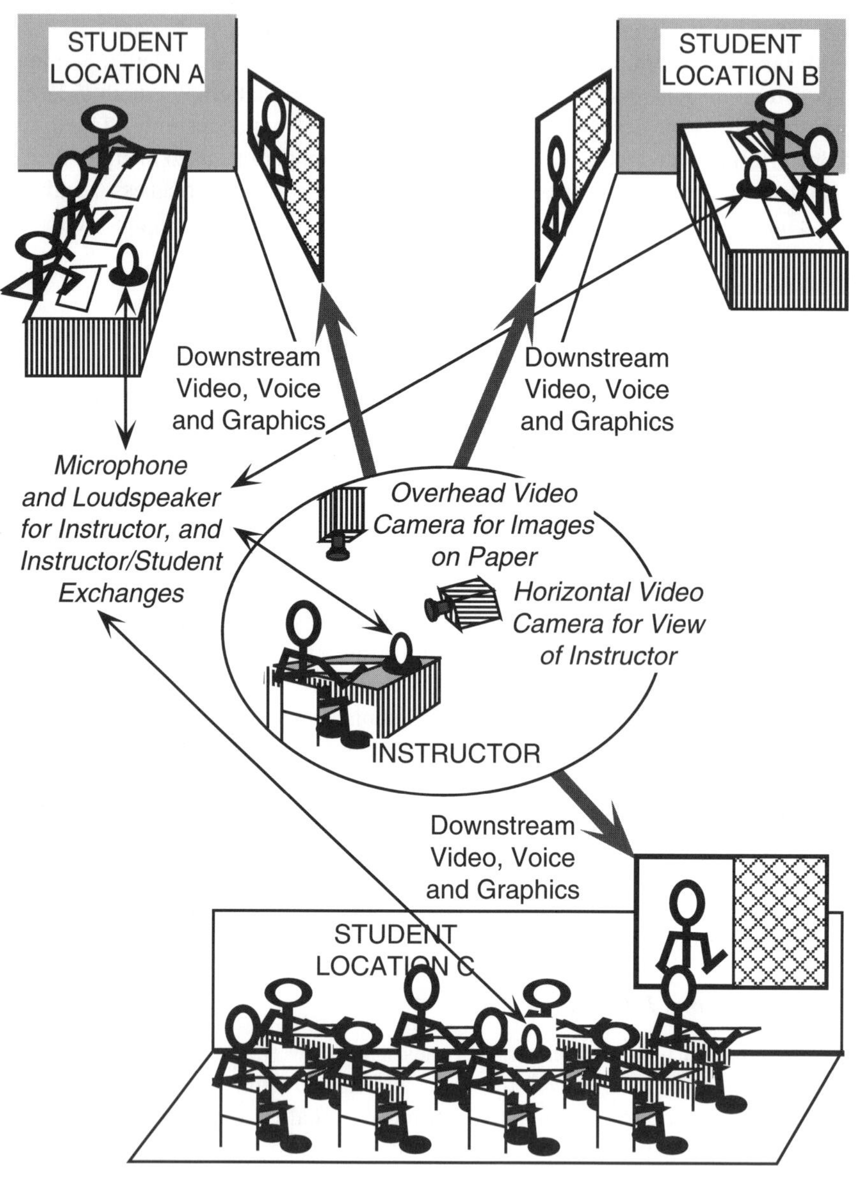

Figure 7.5 Distance-Learning Facility

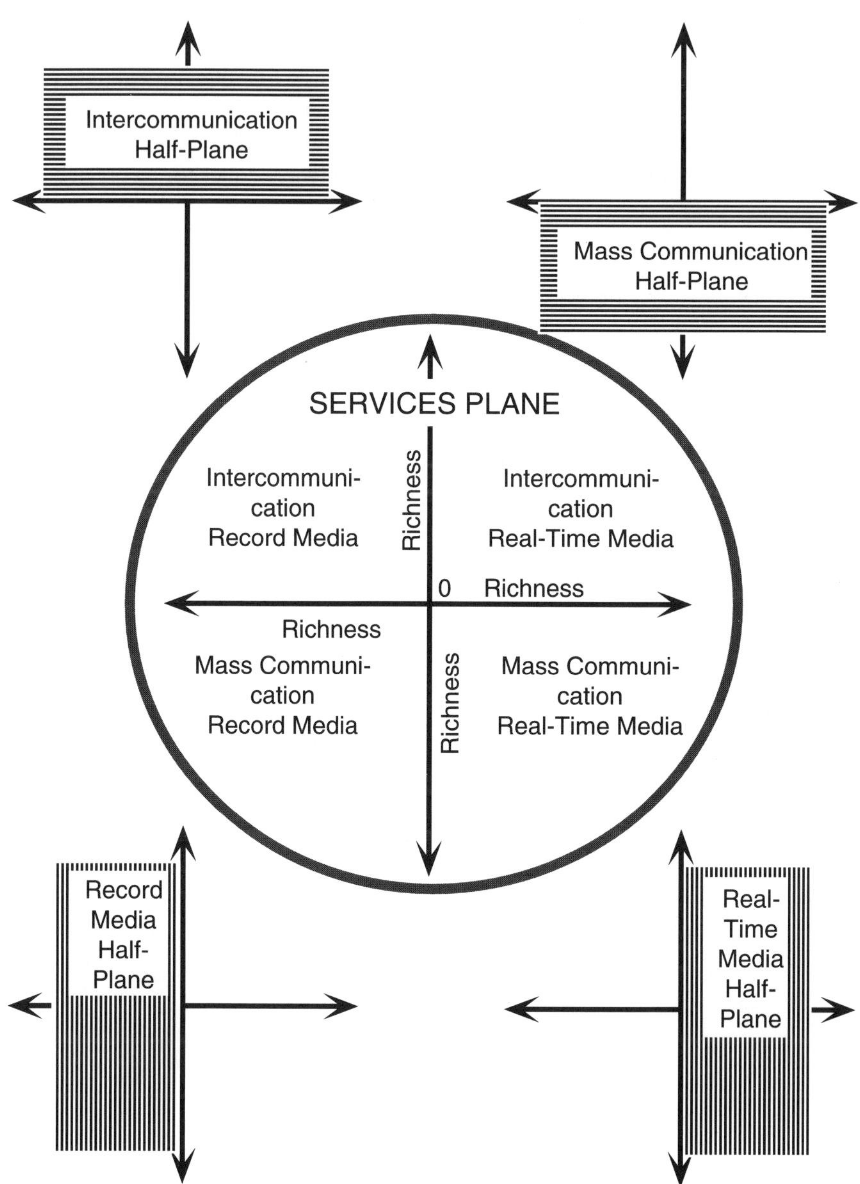

Figure 7.6 Development of Services Plane

(upper half) from mass communication (lower half). The four quadrants are used to segregate services according to the functions they perform. Further, if the axes are marked off in richness in both directions from the crossing point, and the horizontal axis represents the richness in the upstream direction, while the vertical axis represents the richness in the downstream direction, we can arrange communication services in order in the quadrants. This is shown in **Figure 7.7**. Because the richness of the connections in the downstream direction is always equal to, or greater than,

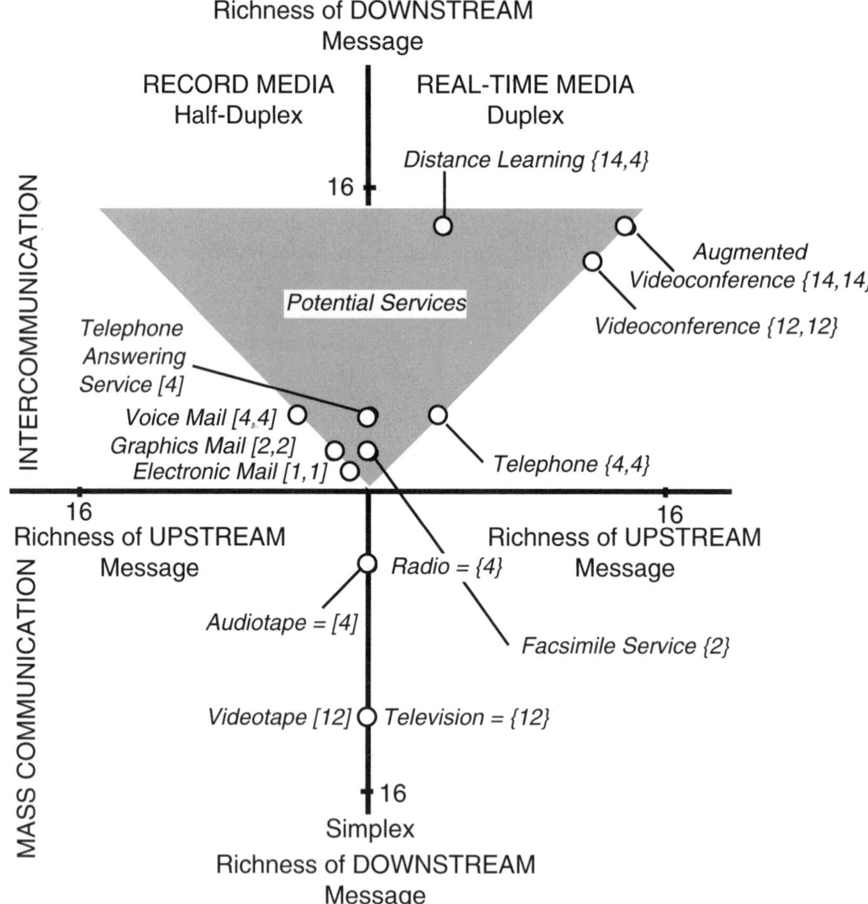

Figure 7.7 Services Plane

By definition, downstream is the direction in which the richer messages flow. Thus, services over half-duplex and duplex channels are limited to the triangular area in the intercommunication half-plane. Services over simplex channels are limited to the vertical axis of the mass communication half-plane.

the richness of the connections in the upstream direction (by definition, above), services can exist only in the upper 45° sector of the real-time media and intercommunication quadrant (upper right-hand side). Services that form the boundary line at 45° to the axes have the same richness in each direction; they share the maximum common information space. In the record media and intercommunication quadrant (upper left-hand side), the array of potential services follows a similar pattern—services exist only in the upper 45° sector. In the mass communication half-plane (lower half-plane), there can be no upstream messages and no common information space so that all services are located along the vertical axis.

(6) Notation

We shall employ a convenient shorthand to represent the position that a service occupies on the services plane. If they employ record media, intercommunication services will be represented by a combination of two numbers enclosed in square brackets, e.g., $[x,x]$, and, if they employ real-time media, in curly brackets, e.g., $\{x,x\}$. The numbers are the richness of the downstream message followed by the richness of the upstream message. Mass communication services are represented by a single number (denoting their richness) enclosed in the same styles of brackets, i.e., $[x]$ or $\{x\}$.

7.2.2 Communication by Telephone

(1) Communication Among Persons

Among persons, communication occurs on a physical and intellectual level. It is influenced by the emotional states of the participants and involves messages that are expressed in spoken words, texts, and images (stationary or moving). In addition, the parties exhibit a sense of purpose and recognize that they are participating in a continuing process. Furthermore, persons are endowed with freewill so that they may communicate selectively. For senders, freewill gives the freedom to attempt to initiate communication when it is convenient for them, and to seek to communicate with whoever appears to be able to satisfy their objectives. For receivers, it gives the right to refuse to communicate, and to treat senders' messages in different ways. While some messages demand immediate attention, others can be afforded less priority, some can be ignored completely, and some may be interrupted if the receiver loses interest or has other things to discuss.

(2) Using the Telephone

A significant fraction of all personal telecommunication is conducted over the telephone. We can divide the activity associated with its use into three stages:

- **Physical stage**: establishing a path along which signals can flow between users. Under the supervision of software, procedures initiated by the originating user execute this task. It requires compatible terminals and network equipment.

- **Procedural stage**: ensuring that the information is in a common language and is exchanged between the right parties. At this stage, the parties check each other's authority to communicate, and ensure that their information is exchanged in compatible formats. This task employs the routine intelligence of the users.
- **Intellectual stage**: managing the flow of information to create an understanding of the message. This stage is peculiar to person-and-person communication—it employs the users' intellectual capabilities to convert the information exchange into the messages that they wish to impart. Achieving a common understanding depends upon the intellectual processes employed by the users, and mutual comprehension is not always possible.

(3) Need for Judgment

For a telephone conversation to be completed successfully, there is need for a common understanding of the telecommunication process on the parts of the participants. Fortunately, the procedures are common knowledge, so that the calling and called parties have a high expectation of creating the connection and of engaging in a productive communication session. That the protocols are relatively simple, is due to the presence of intelligent beings at both ends of the connection. They exercise judgment—judgment to know that the connection is made, that the right person has been reached, when to speak and when to listen, whether the message is being received, and that the information provided has been understood. Of course, things can go wrong. The calling party may have the wrong number, or may have the right number but enter it incorrectly. The called party may answer in a tongue that the calling party does not understand, causing one of the parties to change languages, or to hang up. Before the calling party has completed the message, the called party may lose interest and terminate the call. The connection may be noisy, making it necessary for the parties to speak and to listen carefully, and to repeat parts of their conversation. These are all situations in which judgment must be exercised. They include the possibility of starting over from the beginning.

7.2.3 Communication by Video Camera

(1) Video Telephone

From time-to-time, video telephone service is suggested as a natural extension of voice telephone service that will benefit the users and exploit the rapidly improving, technical capabilities of public telephone networks. Requiring both a video signal and a voice signal, major stumbling blocks have been the large bandwidth required by the visual information, and agreement on a common video format. So that interested organizations can develop compatible equipment, ITU-T has standardized a coding technique for video telephone that provides communications quality video over 2, 64 kbits/s, digital channels—one for voice, and one for compressed video. Organizations such as AT&T, Northern Telecom, and NEC produce equipment that carries voice and video frames over a single voice channel. However, users' reactions to videotelephone service are mixed, and vary over time. After an initial fascination with the technology, continued use brings to the fore some of the limitations of the technique. Sitting in front of a camera attempting to maintain eye contact with the caller, is a discipline we are not used to. What we are used to is the telephone and the freedom to lounge comfortably while talking, and to dress, or not to dress, as we choose. Eventually, there comes a realization that videotelephone introduces restrictions that outweigh the advantages of being seen—at least for routine communications. Furthermore, there is no clear evidence that the addition of video to a telephone call provides more information than the callers can exchange by voice alone (see § 7.7.2(3)).

(2) Video Conferences

To accommodate the large screen requirements of video conferences, ITU-T has recommended coding techniques that provide full-color, full-motion, pictures and sound at 1.5 Mbits/s, and fractions thereof. In contrast to video telephone, video conferences are an acceptable mode of communication for many business purposes. We discuss them in § 7.6.5.

7.2.4 Communication by Data Terminal

(1) Communication Between Persons and Machines

When messages are exchanged between persons and computers, data processors, file-servers, and like machines, they are expressed in alphanumerics and graphics, and the communication session is governed by rules that have been programmed into the application. To communicate effectively, persons must abandon their natural mode of communication—free-flowing speech and gestures—and adopt the use of keyboards and video displays. With them, the user can select functions and enter commands and whatever data may be involved in the transaction. In return, the user receives text and/or graphics, and cryptic requests for further actions. Provided the user is well trained in the operation of the terminal, and understands the performance of the application being executed, such arrangements work well. They are the foundation of the knowledge-based activities being pursued by an increasing number of workers.

(2) Degree of Difficulty

Key to communication mediated by data terminals, is the software that executes the data exchange functions. Too complex, and it is likely to be slow and cumbersome; too simple, and it might not protect the user from choices that cause gridlock and loss of messages. Like other machine-based operations, communication by data terminal can be described as user-friendly, user-hostile or user-seductive; they are defined as follows:

- **User-friendly**: by most persons, the rules governing the operation of user-friendly services and equipment can be learned readily, and they cause little or no increase in anxiety level in users.
- **User-hostile**: by most persons, the rules governing the operation of user-hostile services and equipment are difficult to learn, and they cause a significant increase in anxiety level in users.
- **User-seductive**: most persons are fascinated by the performance of user-seductive services and equipment, and are eager to use them. They dispel anxiety and produce pleasure in the user.

Early attempts at machine-based communication are best described as user-hostile; to employ them required the user to have an ability to subordinate the need to communicate, to the need to present and handle messages exactly as the software demanded. With time, the importance of making them easy to use was recognized and contemporary, data terminal-mediated, communications systems exhibit a substantial degree of user friendliness. Applying the descriptors to a broader range of services and equipment, for many persons, telephone, radio, and television are user-seductive, and most VCRs are user-hostile.

7.3 DOCUMENT SERVICES

Some personal communication services are based on the transfer of information in the form of text and/or graphics that exists at the sender's station as physical documents or on video screens. At the receiver, the message appears as hard copy (i.e., on paper), or soft copy (i.e., on a video screen). The principal methods include:

- **Facsimile** {2}: a communication service that captures images from paper, transmits them to a distant site, and reconstructs them on paper. Facsimile is designated a real-time service (of richness 2) because the image is transmitted immediately as it is formed, and reproduced immediately as it is received.
- **Teletex** [1]: a communication service that transports character-based messages to remote sites where they are printed out or displayed on a video screen. Teletex is designated a record service (of richness 1) because the message is composed and stored until the receiver is available, then transmitted and stored again until the recipient is ready.

7.3.1 Facsimile

In a facsimile system, the item that is copied is a document that may contain typed and printed matter, handwritten notes, and images in the form of graphs, diagrams, photographs, etc. Most facsimile systems scan the original document to detect black, white, and gray-scale patterns, convert the patterns into analog or digital signals, transmit them over a telephone connection to a remote site, and produce a replica of the original. Times to complete the transmission range from minutes for older equipment to seconds for newer equipment. Other facsimile systems are able to code, transmit, and reproduce replicas of color documents. We discuss coding for facsimile transmission in § 4.3.3 and the component parts of a facsimile unit in § 10.2.3.

7.3.2 Teletex

In 1983, recognizing a demand to link sophisticated terminals, several telecommunications organizations inaugurated a service called Teletex to promote communication among electronic typewriters with storage capabilities, and communicating word processors. At a Teletex terminal, texts are composed, edited, and stored, and the terminal may be used for local functions independent of the transmission and reception of messages. The basic character set includes all letters, figures, and symbols used in Latin-based languages and the user may add national characters as required. Text to be transferred to another terminal is assembled in memory and transmitted automatically once a circuit has been established with the receiving terminal and a send instruction has been given by the receiver. In some respects, Teletex and a single node, computer-based message system (CBMS, see § 7.4.4) are at the extremes of the implementation spectrum of text mail services. A one node CBMS forces all communication through the central, message computer: Teletex allows each terminal to handle the forwarding and receipt of its own messages.

Teletex provides direct electronic document exchange between electronic typewriters, word processors, or personal computers equipped with sending and receiving storages and communications capabilities. Once a connection is completed, formatted documents are exchanged between the sending storage of the calling machine and the receiving storage of the called machine. Transmission speeds are 2400 bits/s, or higher, depending on the common capabilities of the machines involved and the channel chosen. Teletex employs an 8-bit code set for alphanumerics, graphics symbols, and control characters. In addition, the calling procedures require the exchange of 72-character reference information frames before the exchange of messages. The frames contain the called terminal identifier, the calling terminal identifier, date and time transmitted, and the document reference.

7.3.3 Mixed-Mode Documents

Mixed-mode documents contain fields of text (that can be handled efficiently by teletex), and fields of graphics (that can be handled efficiently by facsimile). To preserve the ability of the recipient to edit and process them, they are exchanged in blocks that contain characters, and blocks that contain bit-maps. **Figure 7.8** shows the communication procedures involved in facsimile, teletex, and mixed-mode document services. The presentation layer consists of two sublayers. The upper sublay-

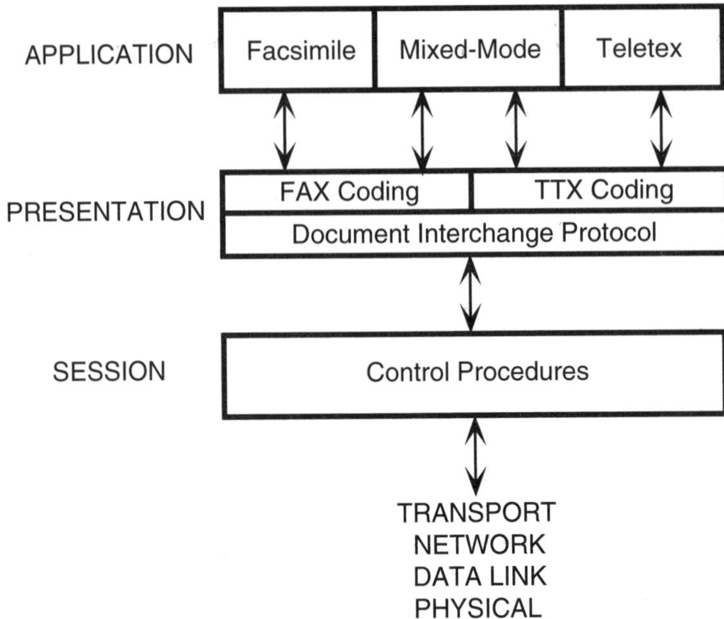

Figure 7.8 Teleservice Protocol Architecture
Mixed-mode documents contain discrete areas of bit-mapped and textual information. They are handled as Facsimile and Teletex.

er is concerned with the coding of the facsimile and teletex documents, or the coding of the text and graphics portions of a mixed-mode document. The lower sublayer is concerned with the transmission of the document structure. This is accomplished by describing the document as a set of logical objects and as a set of physical objects that conform to a specific layout. The document's content, structure, and attributes are represented as a series of protocol elements that are exchanged between the transmitter and receiver.

7.4 MESSAGE SERVICES

Electronic message services can serve many of the non-real-time needs of businesses and households. They have the advantage over voice telephone service in that no one needs to be there to answer, and over normal mail such that delivery is made at electronic speed. Electronic mail services facilitate non-real-time exchanges of text, graphics, or voice messages among persons; they transport messages point-to-point between persons, or distribute them within communities of interest. They are store-and-forward message services in that messages are stored to await requests for delivery from the intended recipients. They may be described as follows:

- **Text mail** [1,1]: composed at, and delivered by, data terminals, the text message may be stored in a central system of electronic mailboxes, or stored at one or other of the terminals serving the persons involved. When the intended recipients command, the messages are retrieved, forwarded (if necessary) and displayed on a video terminal, or printed.
- **Graphics mail** [2,2]: entered in facsimile machines, graphics messages are reconstructed on similar machines that serve the communicating parties. Transmission between senders and receivers is rapid; usually, the messages are reproduced on paper. It is the storage medium that preserves the messages until the recipients retrieve them.
- **Voice mail** [4,4]: recorded using telephone connections to voice storage devices (tapes or solid-state memories), voice messages are delivered on commands from the ad dressed station. Delivery may be by the storage devices directly (answering machines at the recipients' telephones, for instance), or over telephone connections initiated by the recipients.

7.4.1 Text Mail

Text mail is an intercommunication, record service of richness 1 with point-to-point (sometimes point-to-many), half-duplex topology. For many years, a text message service has been provided by communicating teletypewriters and teleprinters. Called *Telex* service, it began in the early 1900s, and developed into a worldwide text communication service with some two-million users. In its simplest form, the operator composed the message at the keyboard of the teletypewriter. This machine produced a strip of paper tape in which the message was encoded as punched holes. When ready, the operator contacted the intended receiving station and passed the punched paper tape through the reader on the teletypewriter. This produced a signal at some 30 to 50 bits/s in 5-bit Baudot code that mimicked the message entered by the operator. If direct communication was not possible, the operator sent the message to an intermediate station where a paper tape copy was made and used later to relay the message to its destination. Improved over the years by the introduction of automatic features, and later through the use of electronics, it still provides slow, but reliable, service. In some parts of the world, Telex service is more readily available, and more dependable, than telephone service. In many countries, Telex is being upgraded to Teletex.

7.4.2 Graphics Mail

Graphics mail is an intercommunication record service of richness 2 with point-to-point (sometimes point-to-many), half-duplex topology. For more than 100 years pictures have been sent over wires. In recent years, the desktop facsimile transceiver has become an important adjunct to business. Because ideographs are difficult to break down into a limited number of keystrokes for typewriter entry, graphics mail is particularly useful in countries where the written language employs them. In

Japan, for instance, graphics mail is much more important than text mail, and the Japanese have pioneered in facsimile communication systems. All the features of store-and-forward operation can be employed. Delayed delivery, multiaddress delivery, and mailbox delivery, are of most interest to users. Desktop facsimile units are commonplace in industrial and commercial organizations. Connections are made on an *ad hoc* basis over telephone lines, or other circuits as needs arise.

7.4.3 Voice Mail

Voice mail is an intercommunication record service of richness 4 with point-to-point (sometimes point-to-many), half-duplex topology. Voice mail records the spoken words of the senders, stores their messages, and delivers the sounds of their voices to the recipients when they request their messages. Also, voice mail may deliver general announcements, be used for message collection, furnish call-answering capability (both for no answer and line busy conditions) automatically deliver messages, and keep trying until the message is accepted, or refused, and record both sides of a telephone conversation. Control is exercised through the touchtone buttons on the telephone set. The system prompts both sender and receiver, asking them questions to elicit essential information that they key in through the touch pad. Voice mail services are offered by a number of North American organizations that route messages over telephone networks to central facilities where they are stored in electronic mailboxes until the recipients request they be forwarded, or they are handled under special arrangements made by the senders. In the public telephone network, advanced management support services incorporate versatile, artificial voice announcement and instruction capabilities, and electronic PBXs are available with voice message service facilities.

7.4.4 Computer-Based Message Systems

In a computer-based message system (CBMS), the originator enters a message on a data terminal connected through the public switched telephone network, a public packet network, or private connections, to a central message computer. It verifies the originator's password, provides editing support, assembles the message, interprets the address and other instructions, and stores the message in an electronic storage device. In most systems, the originator may send the same message to one person, a defined group of persons, or to all persons on the system. Also, the sender may request confirmation that the message was forwarded to the recipient, and keep a copy for file. Usually, the recipient is able to review the sender's name and the subject of the message before calling for the complete message. Each message can be deleted, filed, forwarded to someone else, answered, or processed in other ways.

(1) Local Resources

A CBMS can serve a localized community of interest. Many organizations have established internal systems of this kind. For multi-location organizations, if a single computer is employed, intra-location messages must be sent to the computer site and then returned; inter-location messages are sent to the central site, stored, and forwarded to their destination sites. Besides data terminals connected directly to the central message computer, messages can be composed, edited, formatted, inserted, and retrieved through other local computers using indigenous word-processing capability, or through stand-alone word-processing equipment. For long messages and reports, pre-processing of this kind reduces the connection time to the CBMS processor.

(2) Distributed Systems

As the user community grows, it is prudent to distribute the electronic message system over several node computers. This approach reduces the risk of losing the entire system if there is a catastrophe and may reduce communication charges. More sophisticated networks allocate part of each node computer to provide message and operational back up for another computer. In this way, system integrity can be guaranteed in spite of significant operational difficulties. Some of these options are diagrammed in **Figure 7.9**. Besides versatile message computers, a comprehensive text mail system requires up-to-date user directories and network information, including a complete listing of the procedures for inter-network working.

(3) Combined Networks

Telex, Teletex, and CBMS networks can be interconnected. Many CBMS providers make some provision for the receipt and dispatch of Telex and Teletex messages. One technique employs gateways that store-and-forward the messages and make protocol and frame conversions as required.

7.5 INFORMATION RETRIEVAL

At the beginning of the 1970s, the intercommunication and mass communication authorities in the United Kingdom demonstrated systems that allowed large numbers of users to call for text or graphics information and display it in color on the screens of modified television receivers. Using a common display format, one service employs the telephone network in an interactive way to search through databanks containing thousands of pages of information: the other service supplies a hundred or more pages in the vertical blanking interval (VBI) of a broadcast television signal (see § 3.3.1) without affecting the television service. The first service was called Viewdata, and the second is called Teletext. Together, the two services were known as Videotex, with Viewdata being described as wired-Videotex and Teletext described as broadcast-Videotex. Similar systems were quickly developed in France and Japan, and a little later in Canada. In 1983, CCITT recommended the substitution of the name Videotex for Viewdata. Accordingly, the two services are

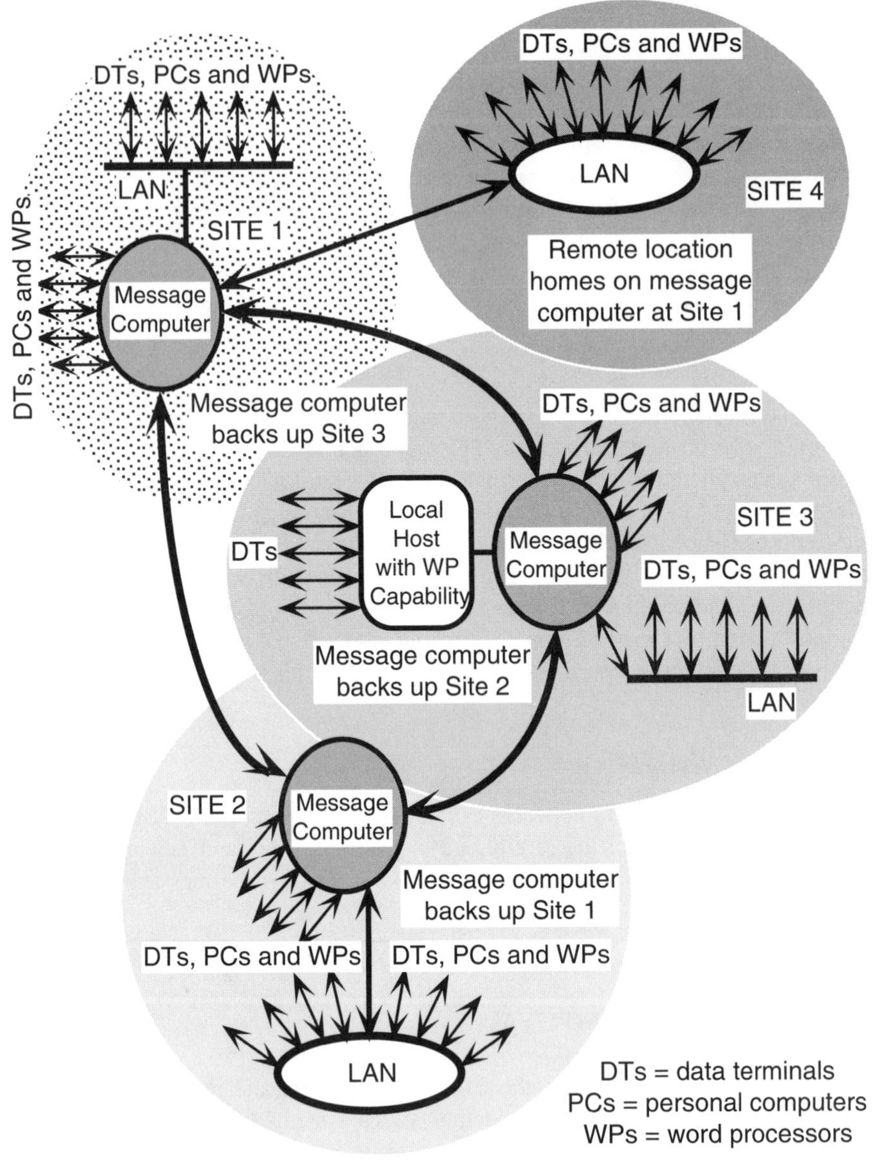

Figure 7.9 Multi-Site Computer-Based Message System

now referred to as *Videotex* and *Teletext*. In **Figure 7.10**, we show the basic facilities required for the two services.

7.5.1 Videotex—Electronic Encyclopedia

In its primary role as provider of detailed information to a large user community, Videotex [2,1] can substitute for any number of reference books, catalogs, and

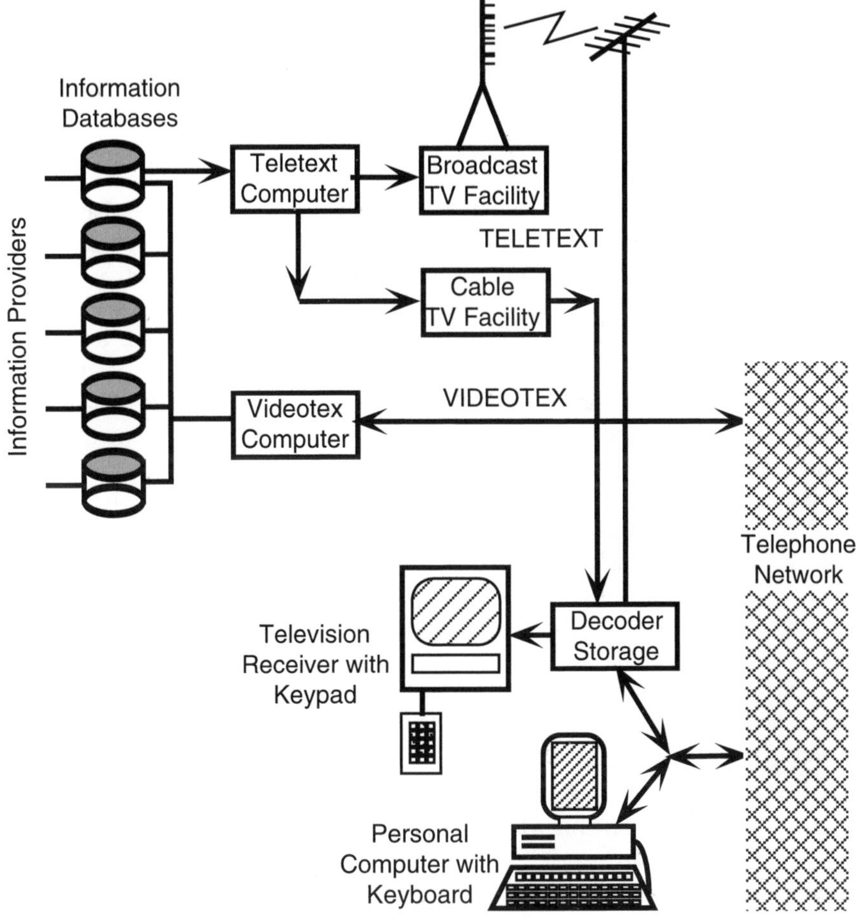

Figure 7.10 Videotex and Teletext Systems
Videotex provides encyclopedic information over the telephone network. Teletext provides the same limited amount of information to all users.

directories. Further, the information supplied can be up-to-date, providing immediate changes in price or schedule, or other quantities. For those who work with such data, the service provides an easy to use capability to access, review, and compare complex information before committing to a specific action. Because it is two-way and computer-controlled, videotex can support interactive applications. This is particularly true for customers who employ a personal computer as a terminal. For instance, shopping from home is possible—for merchandise, whose specification is called up, general features are displayed, availability indicated, price listed, order accepted, and charged to a credit card, or paid for by funds transfer—or for transportation, whose timetable is called up, a specific run selected, space availability indicated, reservation made and confirmed, and paid for. In well-developed public services, upwards of 1000 different information providers may contribute to a databank of several hundred thousand pages. Probably not all of it is available to casual users. Many providers restrict their information to special groups distinguished by their professional status or industry association, and willingness to enter into subscription arrangements.

To satisfy the greatest number of users, a videotex system must be easy to use (user-friendly), flexible (to accommodate the changing needs of the users), and responsive (to provide relevant information in a time that is competitive with the use of alternative sources of the same information). To some extent these qualities depend upon the hardware with which the system is implemented; to a much greater extent, they depend upon the software that controls access to, searching for, and display of, the information. Usually, videotex databases are tree-structured so that users can search them from the general to the specific. Except for the final page in the sequence, each page of search information is organized as a menu from which the relevant option can be selected by entering a number (from the keypad or keyboard). On first acquaintance, this type of search routine is seductive to many users; however, repeated use may bring frustration with the constant searching among the menus representing the branches of the information tree. For experienced users, object-oriented techniques must be employed to retain their attention. Perhaps the most popular are logical search procedures in which the users define the information domain sought in terms of keywords and logical relationships among them. To make full use of these methods requires a PC or an intelligent terminal.

7.5.2 Teletext—Electronic Magazine

In many ways, teletext [2] complements videotex and adapts its capabilities to a mass market. Thus, like television, teletext can be used by everyone at the same time without encountering a busy signal. In addition, it is cheaper to use since it requires no telephone connection. These advantages are obtained by restricting the number of pages of information to that which can be made available in a reasonable time—that is the time the user is willing to wait before turning to more familiar sources of information such as newspapers, guides, and magazines. The challenge for the service provider is to create an electronic magazine that contains information that will be seductive, causing users to prefer teletext to other sources of information.

The number of text characters or picture elements transmitted with each frame of the television picture controls the waiting time for page capture. In present broadcast systems, four lines (in the VBI) out of every 525 lines (in a complete picture) are used to carry mass teletext information. With four to six pages of text transmitted each second, the average waiting time for a particular page in a 100 page magazine will be about ten seconds. Dramatic decreases in search time can be achieved if all 525 lines are used. By dedicating an entire channel to teletext information, consulting a 100 page electronic magazine is virtually instantaneous; and consulting a 12,000 page electronic magazine would take no more time than present teletext. Several cable television systems devote one, or more, channels to teletext.

7.5.3 World System Teletext

Inserted within the envelope of the broadcast television signal, teletext signals must conform to limitations set by the broadcast spectrum. Further, the time to transmit a single page must be as short as possible. Thus, while teletext is compatible with videotex, it uses some of the capabilities in moderation. Slowly, the individual national systems have been brought together into common specifications by ITU-T and other bodies. The current World System Teletext specification is compatible with previous standards (which are subsets of it). In addition, it contains non-Latin alphabets, more colors, better graphics, and page linking. Page linking is a technique designed to reduce access time by acquiring in advance the pages most likely to be sought next by the user. Through the use of multi-page decoders, and information pages that contain the number(s) of the next likely pages, the receiver can search for them while the user digests the current screen. If this activity lasts for several seconds, the next screen has a high likelihood of being available in the decoder when it is called for. Other operating schemes are possible. For instance, one of the pages stored by the decoder can be the Current Index so that the user can return to it quickly.

7.5.4 Competition

(1) Other Sources of Information

The ready availability of more familiar, alternative reference sources, and the lack of a clear need that videotex and teletext can assuage better than anything else, is a deterrent to their expansion. For instance, when the user seeks information normally contained in the newspaper, more than a 15 to 20 second wait at the terminal—while the receiver is switched-on, warms up, and the connection is made (videotex), or the proper page is seized (teletext)—may be enough to discourage electronic searching in favor of leafing through the newspaper. However, such a wait is acceptable if the information can only be obtained by going to the public library.

(2) Technology Push or Market Pull?

To its supporters, videotex promises to become an electronic library, an electronic art gallery, and an electronic publishing medium. They believe it can make major contributions to improving the quality of life by reducing time spent shopping, allowing the conduct of personal business from home, making working at home possible for information workers, and providing knowledge and education. Similarly, teletext provides up-to-the-minute information of general interest to advertise, advise, and inform; and is a cost-effective way to distribute newsletters, market reports, and other information to which specialist groups are prepared to subscribe. To existing communications providers, videotex and teletext represent opportunities to provide new services over existing equipment. To entrepreneurs, the demand has yet to be verified and the return is uncertain. Videotex and teletext appear to be influenced more by technology *push* than market *pull.*

(3) PCs and Optical Discs

To North American users, videotex and teletext are largely unknown—television is for entertainment, not serious information. Those who need reference materials already have on-line searching capability using their personal computers, modems, and the many computer-based, information retrieval services now available to them. They can retrieve vast amounts of current and historical information, participate in noticeboard activities, and purchase catalog items and obtain tickets for travel or entertainment. Furthermore, the sophistication of the searching routines matches the computer abilities of the users. Compared to the dumb terminal approach of videotex, computer-based information services demonstrate a user-friendly approach that will be hard to beat. Other technical developments may also erode the demand for videotex. Optical discs make research materials available in the home, and on-line, library retrieval services make the world of information available to serious investigators.

(4) Multimedia Information Sources

As precursors of the electronic information age, Videotex and Teletext have shown that electronic encyclopedias and electronic magazines are possible and practical. However, future information services are likely to be provided by multimedia systems, i.e., systems that compose coherent inputs from messages of different richness. Streams representing text, graphics, audio, and video dimensions, will be stored, retrieved, and distributed on command to create a composite information field that suits the unique requirements of the user. The goal will be to come as close to face-to-face exchanges as possible. This includes the opportunity to manage emphasis and presentation so as to comprehend. The key to implementing interactive multimedia systems appears to lie in the domain of computer science, particularly those areas that have to do with the representation of objects and the management of abstractions. The dissemination of multimedia *documents* will depend on the ability of communications links to transport mixed signals from what may be several separately located sources to the user's location. The digital networks described in the later Chapters will be able to provide physical transport; some have suggested that the terminals and nodes may require new protocols to handle these situations.

7.6 ELECTRONIC MEETINGS

Meetings vary in purpose, size, and format. They vary in range from:

- **Large forums** in which papers are exchanged, graphics are used, and views and ideas are argued back and forth by persons whose ambitions and objectives may be at odds with each other, to...
- **Small gatherings** of like-minded persons in which the outcome depends solely on the persuasiveness of the spoken word, and the faith and trust each person has in the others.

In the United States, millions of meetings occur each working day. More than half of them last less than one hour, require no visual aids, and are between persons in the same building or the same facility. For some of the remaining meetings, participants must travel to the meeting site. In doing so, they deprive their home locations of their presence, use up time in traveling, and may encounter stressful situations that impair judgment (e.g., jet-lag, accidents, or terrorist activities). Insofar as electronic meetings provide an extended meeting space among remote sites, they can reduce the need for travel, save time, and contribute to the well-being of the individual. In short, electronic meetings should improve productivity.

7.6.1 Use of Electronic Media

Through the use of electronic media that provide an extended information space, meetings can be conducted so that those who participate are able to stay at their home sites, eliminating travel to share a common physical space. If a small number of sites are involved, the participants may interact freely over communication channels that are likely to be equally rich at all locations. If a large number of sites are involved, interaction must be controlled by a chairperson, and it is likely that one site is more important than the others. Under this circumstance, the channels from the more important site to the other sites may be richer than the channels from them to the primary site. The effect of limiting the dimensions of the exchange by introducing electronic media is not completely understood. What is known is that the degree of satisfaction with electronic meetings expressed by the participants depends directly upon these factors:

- The richness of the channels employed
- The degree of familiarity of the participants with each other.

The degree of satisfaction varies inversely with:

- The complexity of the topics discussed
- The number of persons and/or locations involved.

From these providers of electronic meeting services conclude:

- For routine discussions among a few peers who are well known to each other, good quality audio channels may suffice
- For unique discussions among several levels of management in several locations, full-color television may be insufficient
- For deliberations among technical persons, computer conferencing may be an ideal medium.

7.6.2 Range of Electronic Meetings

For intercommunication applications, the spectrum of electronic meetings extends from computer conferencing to video conferencing (that may be augmented by graphics mail). **Figure 7.11** shows the set of electronic meeting alternatives:

- Computer conferencing [1,1]
- Audio conferencing {4,4}, that may be augmented by text or graphics {5,5 or 6,6}
- Videoseminar {12,4 or 14,4}
- Video conferencing {12,12 or 14,14}.

As the number of participants and/or locations increases, each participant is less able to use the return channel(s) and the communication capabilities degenerate. The services plane can be used to visualize the migration of one service into another. For large numbers of participants the following effects occur:

- Computer conferencing [1,1] degenerates to a computer noticeboard [1]
- Audio conferencing {4,4} degenerates to radio {4}
- Video conferencing {12,12} and augmented videoconferencing {14,14} may degenerate to:
  - a videoseminar {12,4 or 14,4}, and eventually to television {12}
  - audio conferencing {4,4 or 6,6} and eventually to radio {4}.

7.6.3 Computer Conferences

Computer conferencing is an ideal technique with which to support activities directed to producing a report or position paper. Participants can compose segments of the report, criticize each other's contributions, have access to resources at their home locations, and ponder issues without off-site distractions. In fact, computer conferencing may be an ideal medium for developing consensus without being swayed by zealous proponents of one position or another. To implement it, each participant has a terminal that is connected to a central computer equipped with conferencing software. To reach consensus on an assigned topic, participants pass

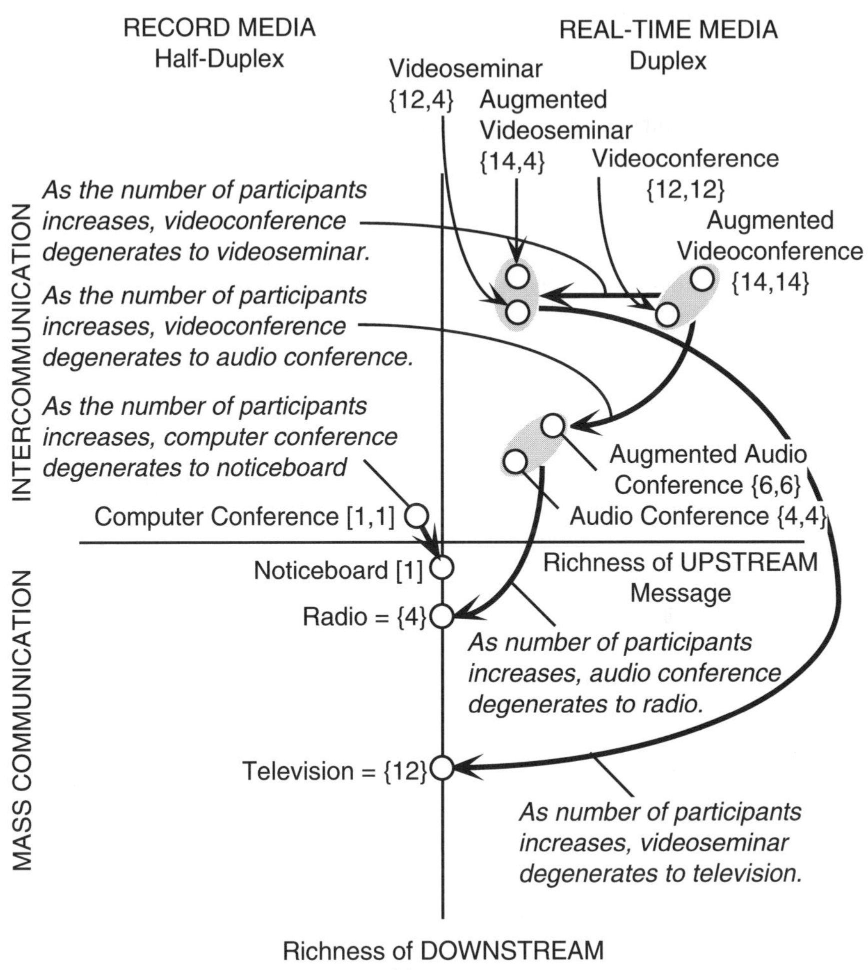

Figure 7.11 Spectrum of Electronic Meetings

messages to the entire group, a subset, or a single individual. The product of the conference activity is stored and made available to all participants as their work proceeds. The arrangement provides electronic interaction among a restricted group of individuals for a common purpose. The computer attends to the exchange of messages, the maintenance of files, and the assembly of the body of data that is the outcome of the computer conference. Like an electronic mail system, the participants do not have to be involved simultaneously. Unlike an electronic mail system, the participants share a common data space and message community that does not have to be defined each time, and can operate on common information without having to build it each time.

7.6.4 Audio Conferences

(1) Use of Telephone

In a relatively primitive form, audio conferencing has been available for a long time. For groups small enough to gather around a desk, it can be implemented between two or more locations using speakerphones, or special conference telephones can be used in small conference rooms. For those served by modern telephone switches, a three-way conference (using telephones) can be established unaided, and the conference can be extended one station at a time to encompass larger groups. Conferences of around six persons can be connected by the call originator, and, by calling each party in turn through the public switched network, larger groups of persons can be connected by an attendant. Very large groups of persons can call a *meet-me* bridge at a pre-arranged time for connection in a conference arrangement, or they may join a meeting already in progress. Usually, as a precaution against uninvited listeners, the connections are supervised by an attendant who has a list of those invited to participate. If privacy is important, the connection procedure may also include the use of a password. Up to 100 persons can be connected by a single bridge. Unless there is a prearranged protocol administered by a strong chairperson, in such large groups individual recognition and participation in the discussion is difficult, and most users must be content to listen.

(2) Sound Quality

For audio conferences, telephone connections are far from ideal. For one thing, the users hear the sum of whatever background noises are present on all the circuits. This includes both noises in the environment local to each user, as well as noise present on the circuits themselves. Tests have shown that relatively insensitive microphones worn by the users are better than one or two sensitive microphones that pick up sounds from an entire group at a particular conference location. Other tests have shown that the overall teleconferencing environment is improved by using a second channel to produce a stereo effect that assists the listeners to differentiate between persons who are speaking at the remote locations. Also, multichannel sound has been reported to improve the conference ambiance. Above all else, participants seem to value good sound quality.

(3) Augmentation

For many purposes, it is helpful to augment voice conferences with text or graphics. When equipment or functions need to be described, or the relationships among variables need to be illustrated, there is no substitute for a picture; and, when financial statements need to be analyzed, there is no substitute for properly prepared tables.

7.6.5 Video Conferences

The concept of a videoconference between two sites is shown in **Figure 7.12**. For multi-location conferences, full video coverage of all rooms at all times requires that each location must be connected to every other location by a separate video channel, i.e., for n locations, each location is connected to $(n - 1)$ other locations so that there must be $n(n - 1)$ video channels. Such an arrangement becomes expensive in a hurry! As an alternative, a video switch can be employed so that only the scene from the conference room in which the speaker of the moment is located is distributed to all other locations. This reduces the video connections to each location to one two-way connection, but limits the video-space of the participants; they are required to look at the speaker, and cannot look at others to judge reactions, etc. For conferences among a number of locations, some observers find it confusing to see different scenes as the conversation shifts from site to site. In some systems, the

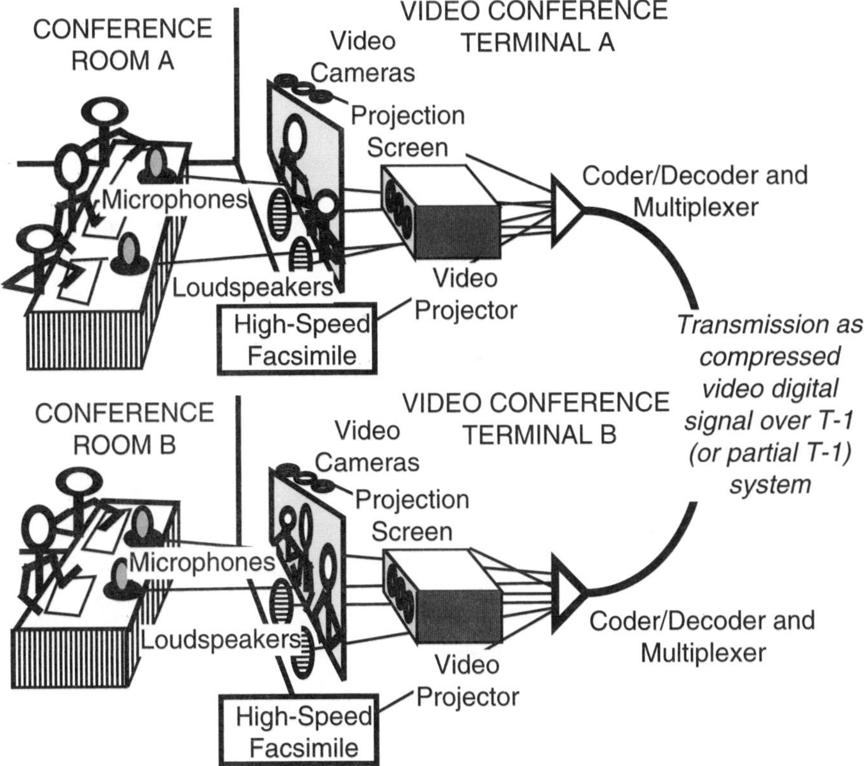

Figure 7.12 Video Conferencing

video switch is voice activated. So long as the conversation is orderly, this technique works reasonably well—but, if persons at different locations interrupt each other, particularly if they shout, the video may switch back and forth in a haphazard fashion. Alternatively, one speaker may capture the system by talking rapidly, and without pauses, so that no one else can speak. The convenience of voice activated switching can be defeated unless participants adopt a cool, calm, permissive style of conversation.

Often, when conferences take place among a large number of locations, one location is more important than the others—company headquarters for instance. Many meetings can be arranged using one-way video, audio, and graphics connections from headquarters to all sites, and an audio connection from each site to the chairperson at headquarters. This arrangement is similar to the distance-learning example shown in Figure 7.5. It is convenient when the primary site wishes to share information with the secondary sites, and wants the secondary sites to ask questions about, or to react to, the information imparted.

7.7 HUMAN FACTORS

Since the late 1960s, technical journals, professional magazines, and even daily newspapers have extolled the future promises of *wired* cities, *paperless* offices, and *electronic* cottages. Driven by technology, a burgeoning array of messaging and meeting services that seemed to fit many business needs have been implemented. Yet, with some notable exceptions, communication in many industries continues in much the same way it did 30 years ago. As a consequence, we are forced to conclude that there is more to personal communications than the availability of high-technology facilities. In fact, the innate desires, objectives, and cultural values of the users appear to play a significant role in determining the adoption and rate of growth of new telecommunication services.

7.7.1 Electronic Message Services

(1) All Services

Electronic message services have been readily accepted by scientific, professional, and commercial organizations. Usually, little difficulty is encountered in comprehending and employing them. Many users feel that they provide more effective, and more equal, communication with all levels of the operation, especially with those in remote locations. Electronic message services reduce the frustration associated with uncompleted telephone calls that result from line busy or party unavailable conditions. Satisfaction is gained from knowing that the message will be delivered at a time when the recipient is available. The store-and-forward feature is particularly effective when communication spans several time zones, or when colleagues work different hours.

(2) Text or Graphics versus Voice

Text mail and graphics mail can be read from a video screen or printed copy. Specific points, and their context, can be studied readily, and at length. Time can be taken to compose replies, and they can be edited several times before sending. Long messages can be handled a piece at a time so that they present no more difficulty to the originator or recipient than a long letter.

Voice mail has an important advantage over text mail and graphics mail—it requires no terminal other than a touchtone telephone. Voice mail can be sent and received from practically anywhere. What is more, voice mail delivers a message in the sender's voice with all the emphasis and color retained. However, if detailed consideration of specific points is required, the message must be listened to carefully, and in its entirety. Because they require considerable attention, long voice mail messages impose hardships on the originator and the recipient.

These points lead to the following conclusions:

- Text mail and graphics mail may be more suited to lengthy arguments, to position statements, and to technical matters
- Voice mail may be more suited to instructions and to short administrative messages
- For those more skilled in dictating than typing, voice mail is probably an easier service to employ than text mail
- For those more skilled in typing than dictating, text mail is probably an easier service to employ than voice mail.

Successful employment of any of these message systems requires that users monitor their mailboxes. No matter how user-friendly and speedy an electronic message service is, it will be to no avail if the messages are left unread or unheard.

(3) Saving Time and Increasing Participation

Saving time is the chief, direct contribution that electronic message services make to the business community. By eliminating the normal delay in letter mail, they accelerate the pace of decision-making. Another direct contribution is opening up participation in decision making to all who have something to contribute. Electronic mail can have a positive effect on productivity, provided those who receive it have other productive things to do to fill up the time they save, and all *electronic* participants diligently study each others' contributions. However, it is unlikely that the number of productive things to do will expand in sympathy with the use of electronic message services. Consequently, the diligent use of electronic mail services could result in a reduction of the need for so many professional and management persons. One of Parkinson's celebrated conclusions is that work expands to fill the time, and consume the resources allocated to it. With all users seeking to remain fully employed, electronic message services may generate more discussion, not less.

7.7.2 Electronic Meeting Services

(1) Purpose of Meetings

During the work day persons meet face-to-face to discuss sensitive, confidential, or personal problems; negotiate and reach agreements; solve problems; share information; or for any combinations of these reasons. In each case, the participants may belong to the same peer group, different peer groups, share a superior–subordinate relationship, or be total strangers.

(2) Observed Behavior

Many of these combinations have been studied in electronic meetings that use services of different richness. Analysis of the behavior of the participants leads to the following:

- Regional personnel perceive a greater need for, and utility of, electronic meeting services, than do headquarters persons
- There is a substantial difference between the measured and perceived effectiveness of electronic meeting services (most users believe that video is an important element—problem-solving tests do not bear this out)
- An electronic meeting imposes a degree of formality that reduces extraneous discussion and concentrates the attention of the participants on the subject at hand (meetings are more topic-oriented and less person-oriented)
- Acquainted persons perform more efficiently than strangers in an electronic meeting
- Acquainted persons perform more efficiently in an electronic meeting than in face-to-face meetings (some estimate as much as 30% shorter meetings)
- Groups in which no clear authority structure or previous acquaintance exists become more uncertain with regard to norms, values, and role expectations as the richness of the electronic meeting medium decreases
- So long as audio is included, both problem-solving and information sharing meetings are relatively insensitive to the total richness of the meeting service employed
- As the richness of the meeting service decreases, many persons find it easier to say no, or to argue for positions that they do not totally support
- For sensitive, confidential or personal discussions, and for meetings whose purpose is to reach an agreement, the degree of success is sensitive to the richness of the communications channel (face-to-face meetings are best for these purposes).

(3) Importance of Voice

In university psychology departments, extensive testing of the speed with which problems are solved has shown that face-to-face meetings provide no significant advantage over communication by voice alone. Then why have electronic meetings not substituted for face-to-face meetings and revolutionized the way in which business is conducted? Some reasons may be:

- **Inappropriate application of psychology department results**: tests in psychology departments are conducted in a cooperative environment with the participants seeking to overcome the limits placed on them. Usually, they have no alternative means available to accomplish the assigned end. In business, cooperation is not necessarily present, and the alternative to an electronic meeting is to travel to a *real* meeting—a mode that is familiar, well understood, and has worked in the past.
- **Imbalance of sensory channels**: audio conferencing has been described as similar to talking to people in a dark room. This may be something we are willing to do with one person (by telephone), but not with several. In fact the analogy of the dark room is not exact. Audio conferencing is an activity in which the participants talk to unseen persons while viewing their present surroundings. The field serving their ears, and the field serving their eyes are not the same. For more than a person-and-person conversation, the participants find visual cues helpful in separating the speakers and may become confused when none is available in the immediate vicinity. Calling persons together in a conference room matches sound and sight, and restores the coincidence of the inputs. This may be the reason many persons consider video to be an important component of an electronic meeting system intended for use in a variety of situations.
- **Richness of medium**: the richness of the medium affects the size of the group and the complexity of the subjects that can be handled effectively. With an audio-only link {4,4}, more than a few persons discussing a complicated or unfamiliar topic can result in chaos. Graphics equipment increases the richness somewhat and may help to sort out conflicts and partition the problems. Computer conferencing [1,1] may assist with understanding the complexity at the expense of slowing the communication down and introducing an unfriendly interface. The richness of full-motion video {12,12} will make person identification easier, restore the balance between eyes and ears, and should make it possible to handle larger groups. The richness of the medium employed has other effects:
  - *insincere support*. As the richness decreases, many persons find it easier to argue for positions they do not support. For this reason, for meetings in which the parties intend to discuss important issues, even full-motion video may not be entirely adequate.
  - *change in importance of participants*. As richness decreases the conference participants tend to equalize in importance. Thus, full-motion video allows easy identification of headquarter's persons by their manner, style, and dress. In an audio conference, none of the visible marks of rank is available. Attention is focused much more on eloquence, logic, and other

vocal skills. A computer conference may even reverse the ranking, for successful participation depends on easy familiarity with terminals and protocols. This is more likely to be found in working-level persons than in management-level participants.

(4) User Benefits

Those who use electronic meeting facilities report improvements in personal productivity and in the working environment. They cite decreases in the time away from work, greater job satisfaction, improved access to other persons, and more productive meetings. An important prerequisite appears to be a senior manager determined to break the face-to-face meeting habit. When used regularly, electronic meetings are shorter and more timely; more issues are raised and resolved, decisions can be made based on more correct information and less mis-information, and unproductive time is reduced. The results are said to be greater management visibility, improved communication throughout the corporation, improved morale, and increased cooperation.

7.7.3 Intra- and Inter-Organization Meetings

(1) Intra-Organization Meetings

Most of the applications of electronic meetings are reported to be within large institutions. This is not surprising since a community of interest is already formed, the resources may be easier to obtain, and the location of the terminals is already established (each of the separated facilities). In addition, in most large organizations, there is a sense of belonging that makes employees who have never met feel like unacquainted members of the same family rather than strangers. Conditions are such as to heighten the probability that an intra-organizational electronic meeting can be completed over relatively simple facilities (audio-only or augmented audio) to the satisfaction of all participants.

(2) Inter-Organization Meetings

But what of inter-organization electronic meetings? Inevitably, this implies more conferences among strangers, so that the richness of the medium must increase to have the same probability of satisfying all parties. More expensive facilities (full audio, video, and graphics capabilities) must be used. What is more, compatible terminal equipment is required so that one terminal can communicate with the other, and a switched video transmission network is necessary so that the expensive transmission facilities can be shared among a large number of users. Thus, inter-organization electronic meetings are more complicated and more expensive than intra-organization meetings. For this reason, they have been slower to develop.

7.7.4 Other Factors

(1) Travel and the Telephone

Already, the ubiquitous telephone is the major tool used in business and personal life to share information and to solve problems. In fact, we are likely to reach for a telephone without even considering travel as an alternative way to achieve our purposes. However, by expanding our information field, the same instrument stimulates our natural curiosity so that the simultaneous achievement of greater use of telecommunications and reduced travel, is probably impossible.

(2) Adequacy of Audio Conference

For groups of acquainted, compatible persons who wish to solve a problem, or to share information, an audio meeting link is probably as efficient as a face-to-face meeting. Therefore, if problem-solving or information-sharing is all such a meeting must achieve, it can be held using relatively simple electronic meeting facilities, and travel can be eliminated. Perhaps, such is the case for regular meetings of the kind at which project progress is reviewed or sales information is exchanged.

(3) Advantages of Electronic Meetings

However, meeting electronically is not a matter of having the equivalent of a face-to-face meeting over a communication link. Because intimate personal contact is lost, it is less than a face-to-face encounter; but it is also much more, for many persons can be present at each location so that a better outcome is achieved through having those that really know what they are doing participate in the decision making. Important matters need no longer be left to traveling generalists, but can be discussed to any depth within the capabilities of the staff resident at each terminal location.

(4) Complex Objectives

Many meetings have complex objectives—and very often the stated purpose for the meeting is an excuse for accomplishing something else. Some attendees have special reasons for traveling to the meeting site, just as those who wish to have an electronic meeting may have special reasons to stay home. Some of their reasons will be official; some may be personal but public; and some may be personal but private. These factors influence the decision to travel or to communicate much more than whether the objectives can be accomplished over electronic meeting facilities.

(5) Location of Meeting Rooms

The decision to travel or to communicate will be influenced by the location of the electronic meeting rooms. Within an organization, greater use of electronic meetings may occur between suburban locations and downtown headquarters, than between suburban locations. This is particularly true in older metropolitan complexes where going downtown means traffic snarls, uncertain parking, and the risk of vandalism. In this environment, travel problems exceed electronic meeting problems, and those working in the suburbs can be expected to use electronic meetings as often as the headquarters staff will allow them to. For the same reasons, a downtown public electronic meeting site will do little to encourage inter-organizational electronic meetings unless it is conveniently located, provides adequate support services, and is easy to use.

(6) When Travel is Perceived as Dangerous

Given the motivation of many individuals to travel (despite ready access to a telephone), and that much of this travel may be outside the traveler's own organization, it appears unlikely that the availability of electronic meeting facilities will produce a wholesale reduction in travel. For this to occur, travel must become a lot more difficult. A recent case-in-point is provided by Operations Desert Shield and Desert Storm in the Persian Gulf—international airline fares increased dramatically, and the threat of terrorism became very real for many companies and individuals. Newspapers reported a 60% increase in the use of limited-motion videoconferencing between international cities during these operations. Since then, the demand has fallen as travel has returned to normal.

8

Multiplexing

Because transmission costs increase with distance, on all but the shortest connections, the bearer carries as many channels as possible. Making this happen is the function of a *multiplexer*.

8.1 TYPES OF MULTIPLEXERS

Two basic techniques are used:

- **Frequency division multiplexing** (FDM): to each signal a fraction of the bandwidth of the bearer is assigned so that the multiplexed signals exist at the same time, at different frequencies, on the same bearer.
- **Time division multiplexing** (TDM): to each signal the entire bandwidth of the bearer is assigned for very short periods of time so that the multiplexed signals exist in series, at different times, on the same bearer. Multiplexing is:
  - *synchronous* if the timeslots are assigned to each signal in turn in a regular sequence paced by the system.
  - *asynchronous* or *statistical* if the timeslots are assigned to signals as required.

Figure 8.1 provides a conceptual view of FDM and synchronous TDM. The equipment that combines analog voice signals into an FDM or TDM signal is known as a *channel bank*. In the case of TDM: when transmitting, the channel bank performs quantization, coding, μ-law compression, and multiplexing; when receiv-

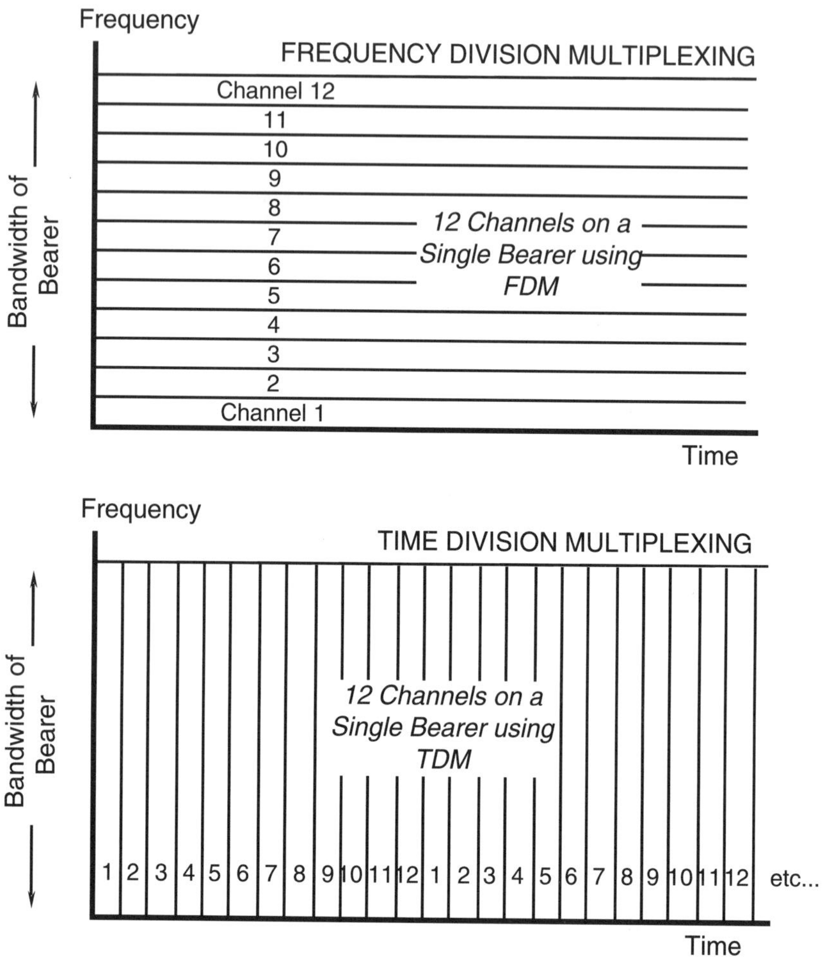

Figure 8.1 Comparison of FDM and TDM Techniques
In FDM, each signal is assigned a portion of the bandwidth of the bearer. The signals are sent simultaneously, in parallel. In TDM, each signal is assigned the complete bearer for a short time. The signals are sent in series.

ing, the channel bank performs demultiplexing, decoding, μ-law expansion, and reconstruction of the voice signals. Equipment that combines digital signals (voice or data) into a TDM signal is known as a *multiplexer*.

8.1.1 Frequency Division Multiplexing

For many years, frequency modulated (FM) microwave radio systems carried frequency division multiplexed voice signals in medium- and long-haul telephone transmission applications. Later, to increase the capacity of long-haul systems, single-sideband (SSB), amplitude-modulated (AM) microwave systems were employed. Now, the combination of optical fibers and digital techniques has displaced FDM in favor of TDM. Nevertheless, because of the amount of FDM equipment in-place, a significant fraction of the world's voice traffic will continue to use these techniques into the next century. **Figure 8.2** shows how 12 analog voice channels are combined in a *channel bank* to form a composite, multiplexed signal. To produce a double-sideband, suppressed carrier (DSBSC) signal (see § 2.3.2 (1)), each voice channel modulates a carrier chosen from the set 64, 68, 72,...108 kHz. When passed through a filter, only the lower sideband remains. Because each LSB spans a different frequency range, they may be added as shown at the bottom of Figure 8.2. This arrangement of channels is known as a *group*. **Figure 8.3** illustrates how groups are combined in steps to become part of a signal containing 600 (or more) channels. In turn, these signals are used to modulate carriers in radio relay systems. Of course, a complete transmission system consists of two multiplexing and demultiplexing arrangements—one for each direction of transmission.

8.1.2 Synchronous Time Division Multiplexing

Synchronous time division multiplexers combine *n* lower-speed, digital signals on higher-speed, single bearers. Whether signals are present or not, timeslots on the bearer are allocated to each input channel at a rate determined by the multiplexing equipment. To ensure that they are synchronized with other equipment in the network, a stable clock is incorporated in all major multiplexers (see § 9.1.5(3)). Demultiplexing restores the *n* lower-speed channels. Many time division multiplexers operate at the octet level:

- In **octet-level multiplexing**, octets from each input line are interleaved on the common bearer in time slots reserved for them. After each sequence of octets (one from each input line) has been serviced, the multiplexer adds a bit to define the frame. A delay of one octet time (i.e., the time required to transmit 8-bits at the input speed) occurs in the passage of each datastream from input line to multiplexed output line.

However, multiplexing is not limited to the octet-level process described above. Multiplexing occurs at the bit-level and the frame level:

- In **bit-level multiplexing**, bits from each input line are interleaved in turn on the common bearer in a regular fashion. Timing is more critical than at the octet-level, but throughput delay is reduced. In the public switched telephone network, bit-level multiplexing is used to build higher-speed multiplexed datastreams from lower-speed multiplexed datastreams (see § 8.2.3).
- In **frame-level multiplexing**, frames from each input line are interleaved on the common bearer. Frame-level multiplexing, which is likely to be asyn-

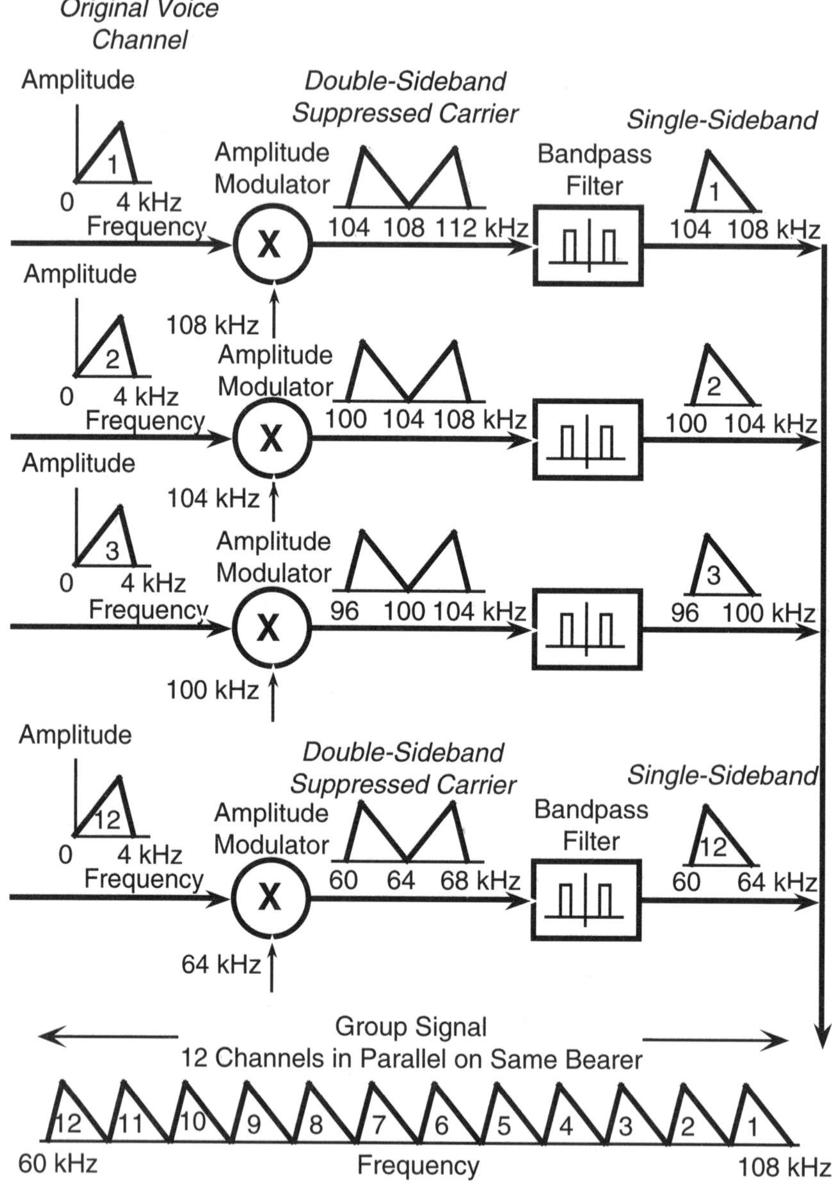

Figure 8.2 Frequency Division Multiplexing of Voice Channels
Twelve voice channels are frequency division multiplexed to produce 12 channels in parallel on one bearer.

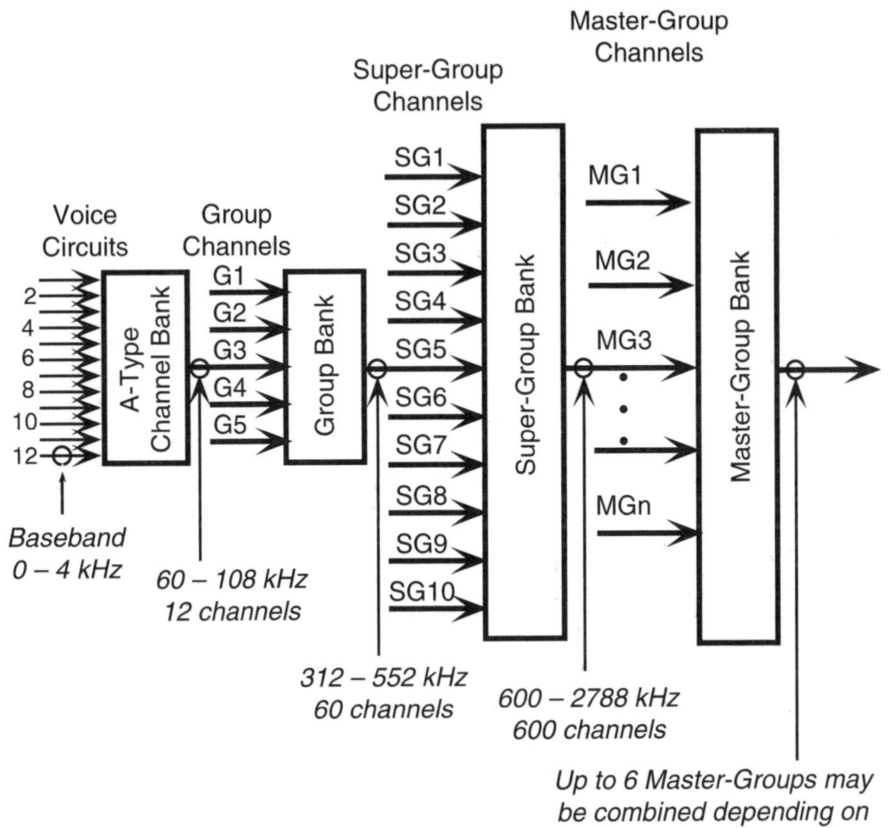

Figure 8.3 Frequency Division Multiplexing Hierarchy

chronous, is performed by data concentrators (see below) and packet switches (see § 9.2.1).

To avoid congestion, the speed of the multiplexed line must be equal to, or greater than, the sum of the speeds of the input channels, *plus* the speed of the framing channel. **Figure 8.4** illustrates the multiplexing and demultiplexing functions, and shows the time relationship among octets on the lower-speed lines, and octets in the datastream on the higher-speed line. In a dynamic process, the multiplexer collects octets from the input lines, reads them out at the higher-speed of the multiplexed line, and places each in a time slot that is allocated to it on a regular basis. If one or more of the incoming lines are inactive, one or more of the slots on the multiplexed line will be empty. After all time slots have been serviced once, the multiplexer adds a bit to define a frame. At the receiving end, the demultiplexer uses the framing bits to locate the beginning of a frame, collects the octets from the

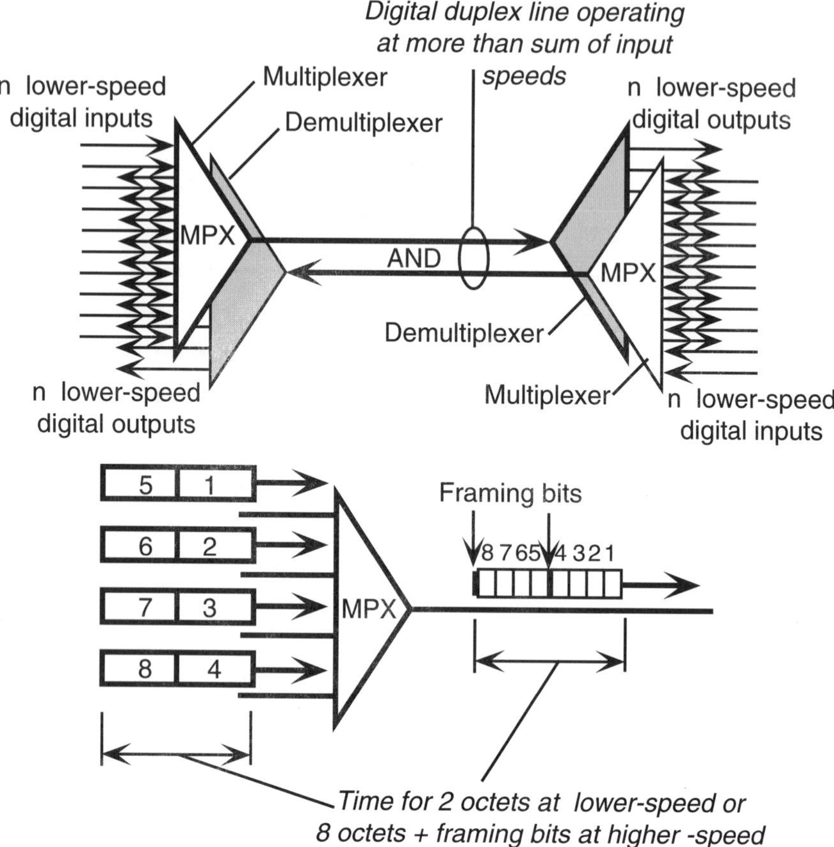

Figure 8.4 Time Division Multiplexing
In TDM, each signal is assigned the entire bearer for a short time.

multiplexed line and distributes them to the lower-speed output line corresponding to the input line from whence the octet came. Throughout the process, frame-level synchronization must be maintained so that the demultiplexer can assign the received octets to the correct output channels. For this reason, the process is called *synchronous* multiplexing.

8.1.3 Asynchronous or Statistical Time Division Multiplexing

(1) Bursty Traffic

In data communication, traffic may consist of bursts of data separated by relatively long periods of silence. An example is provided by a keyboard/video terminal that is connected to a host computer. During periods of input activity, characters may be generated at an average of three per second (corresponding to 30, 6-character words per minute) and transmitted asynchronously as bit-framed characters at 300 bits/s. Under this circumstance, the line is occupied only 10% of the time. Between periods of input activity there may be periods of thinking, during which nothing is transmitted, making the long-term average occupancy significantly less than 10%. If a cluster of terminals is involved, and a synchronous, time division multiplexer is employed to carry the datastreams to the host, only a small percentage of the time slots allocated to any terminal may be occupied. Obviously, this is a very inefficient use of the transmission facility.

The bursty nature of data transmissions has led to the development of a class of multiplexers known as statistical multiplexers (SMPXs), or asynchronous multiplexers. Only when data are available at an input do they allocate a timeslot on the multiplexed line to them. Arrival times at the multiplexer, not what input line they are on, determine the order in which octets are sequenced. As they are placed on the multiplexed line, the multiplexer adds an identifier so that the demultiplexer will know to which output line the octets are to go. To reconstruct the original datastreams, this information is vital—but, it adds overhead bits to the process. In principle, some, or all the input lines can operate faster than the multiplexed line provided that the average number of input bits per second, plus the average number of overhead bits per second added by the multiplexer, is less than the speed of the multiplexed line.

EXAMPLE 8.1 Calculate the Throughput of the Statistical Multiplexing Arrangement Shown in Figure 8.5

Assumptions: in **Figure 8.5**, 8-bit characters, framed by start and stop bits, are presented on input lines at 300 bits/s. During multiplexing, the framing bits are stripped off and 2 bits are added to each character to identify the input line from which it came. Thus, the characters consist of 10 bits throughout the multiplexing process.

Derivation: because the characters are generated irregularly, and transmitted asynchronously, the average bit rate on each input line is less than 300 bits/s. The multiplexed line runs at 600 bits/s. As each character arrives at the SMPX, the multiplexer allocates a time-slot to it and places it on the common bearer. If two, or more, characters arrive at the same time, the SMPX handles them in a consistent manner—we assume that it takes them in the sequence line #1, line # 2...line #4. The second character must wait while the first is regenerated at the higher speed and transferred to a time-slot on the multiplexed line. Under the circumstances shown in Figure 8.5, the occupancy of the multiplexed stream is approximately 80% (16 slots are filled out of a total of 20), and the throughput (i.e., information bits transmitted per second) is 480 bits/s. The situation in Figure 8.5 is not particularly efficient...

- The addition of 2-bit identifiers to 8-bit characters results in a coding efficiency of 8 ÷ 10, i.e., 80%.

STATISTICAL MULTIPLEXING: *The speed of the multiplexed line is twice the speed of the input lines, i.e. 600 bits/s. Framed characters are handled on a first come, first served basis. In the event of a tie, they are handled in input number sequence.*

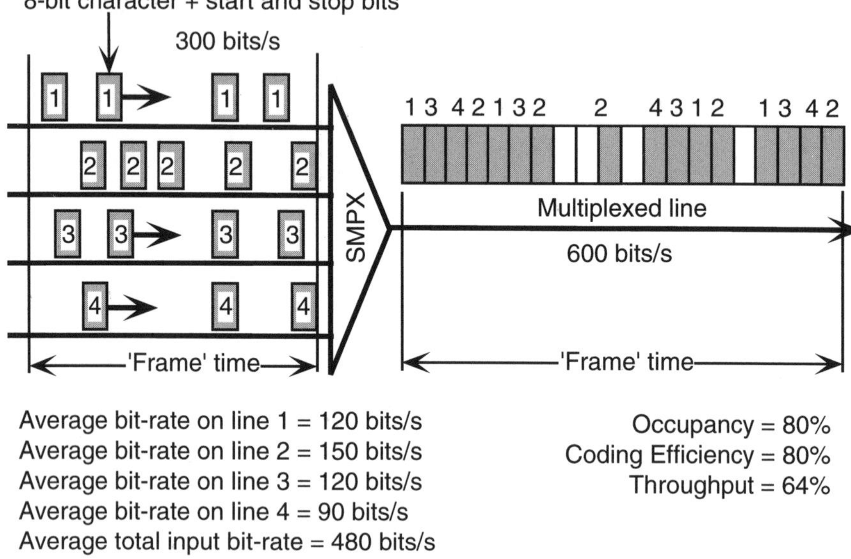

SYNCHRONOUS MULTIPLEXING: *the speed of the multiplexed line is four times the speed of the input lines, i.e. 1200 bits/s.*

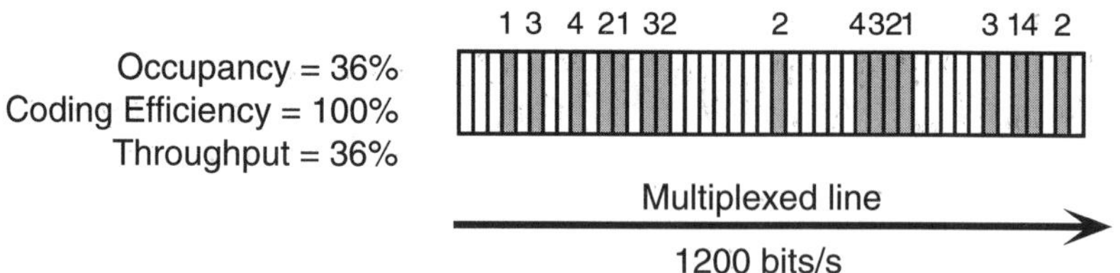

Figure 8.5 Statistical Multiplexing (Example 8.1) and Synchronous Multiplexing (Example 8.2)

- In conjunction with an occupancy of 80%, this produces a transmission efficiency of 64%.

EXAMPLE 8.2 Calculate the Throughput of a Synchronous Multiplexer when Handling the Traffic of Example 8.1.

Assumptions: the multiplexed line runs at four times the speed of an input line and any bits added to demarcate frames can be ignored.

Derivation: the speed of the common line will be 1200 bits/s. In a total of 44 timeslots, only 16 are occupied, thus the occupancy is 16 ÷ 44 = 36%. In synchronous multiplexing, there is no need for identifying bits so each character consists of 8 information bits. Thus the coding efficiency is 100%, and the transmission efficiency is 36%.

Comment: for the conditions postulated, statistical multiplexing performs almost 80% better than synchronous multiplexing. Further, if the multiplexed line speeds were the same, the SMPX would be capable of handling more lines of bursty data traffic than the TDMPX. Care should be taken not to extrapolate these results too far. They are true only for the particular situation postulated in Figure 8.5—and this includes the statistics of the input traffic.

EXAMPLE 8.3 Determine the Average Wait for Service at the Statistical Multiplexer shown in Figure 8.5.

Assumptions: a statistical multiplexer can be modeled as a M/D/1 queuing system. (Of several queuing models that have been formalized, M/D/1 most closely represents the multiplexing situation. M/D/1 is shorthand for a queuing system in which the arrival process is represented by a Poisson distribution, the service function is the same for all customers and there is a single server that handles arrivals on a first-come, first-served, basis.) **Figure 8.6** shows the M/D/1 queuing model of the SMPX shown in Figure 8.5.

Derivation: for an M/D/1 queue, the average time waiting for service is $\lambda/[2(\mu - \lambda)]$ seconds, where λ is the average number of framed characters arriving per second and μ is the average number of addressed characters transmitted per second.

Numerical Value: In Figure 8.5, λ = 48 framed characters per second (i.e., 480 bits/s) and μ = 60 addressed characters per second (i.e., 600 bits/s). Substituting in the expression introduced above, the average time waiting for service is 0.0333 seconds.

M/D/1 MODEL

Average arrival rate = 48 framed characters per second

Buffer

Service time = same for all arrivals

Serving rate = 60 addressed characters per second

Figure 8.6 M/D/1 Model

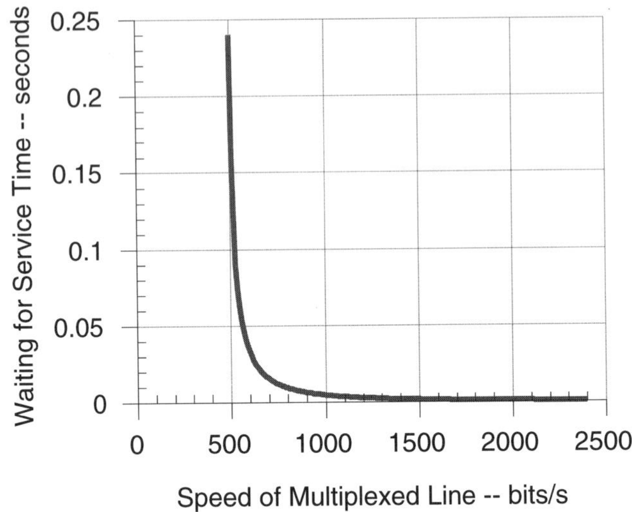

Figure 8.7 Waiting for Service Time

Extension: **Figure 8.7** plots values of the average time waiting for service against the speed of the multiplexer line. At 1200 bits/s, the average delay waiting for service is only 0.00222 seconds. As the speed of the multiplexer line slows toward 480 bits/s (the average total input bit-rate), the delay increases dramatically.

Comments: results derived from the M/D/1 model are relatively simple to obtain, and serve to illustrate the general behavior of a statistical multiplexer. However, the representation of the input process as a Poisson distribution (an infinite number of sources generating short messages at random) is a serious distortion of reality. This compromise is made to simplify the analysis.

EXAMPLE 8.4 Determine the Average Delay between Source and Destination Using Line 1 in Figure 8.5.

Assumptions: same as Example 8.3, and there is no propagation delay across the circuit.

Derivation: the framed characters are assembled in buffer storage at the input to the SMPX at the speed of line 1 (10 bits at 0.00333 seconds = 0.0333 seconds), and read out as addressed characters at the speed of the multiplexed line (10 bits at 0.00167 seconds = 0.0167 seconds) in a first in, first out (FIFO) manner. If the multiplexer is empty when a framed character appears on Line 1, it passes through with only the delays for assembly and read out (a total of 0.05 seconds). If the multiplexer is serving a framed character when the next one arrives, it must wait in the queue. The average waiting time was derived in Example 8.3.

Numerical Value: the total time for a framed character to pass from source to destination over the statistical multiplexer is:

- Time for source DTE to send character on Line 1 and assemble in buffer storage in SMPX = 0.0333 seconds
- Waiting for service time = 0.0333 seconds (see Ex. 8.3)
- Time to read from source SMPX on to line (i.e., service time) = 0.0167 seconds
- Time to read from destination SMPX to destination DTE = 0.0333 seconds
- Total Delay = 0.1167 seconds.

Comments: to send a framed character from source to destination without speed changing or multiplexing requires 0.0333 seconds. To send a framed character from source to destination using TDMPX (i.e., no wait for service time) requires 0.1167 - 0.0333 = 0.0834 seconds. To send a framed character from source to destination using SMPX requires 0.1167 seconds. Obviously, there is a tradeoff between transit time and transmission efficiency.

(2) Synchronous Transmission

SMPXs can be configured to use synchronous transmission on the common bearer. To do this, the offered traffic is organized into an SDLC-style frame in which the text portion contains the traffic sequence and their identifiers. Such a frame is shown in **Figure 8.8**. With four input lines, the multiplexer of Figure 8.5 can use a 2-bit code to identify them. In practical applications, these identifiers may be 4 or 5 bits long to allow 16 or 32 input lines to be handled at once.

(3) Data Concentrator

- **Data concentrator**: a frame buffering and multiplexing device that buffers frames received from several lower-speed lines until they can be sent to a central site on a single higher-speed line.

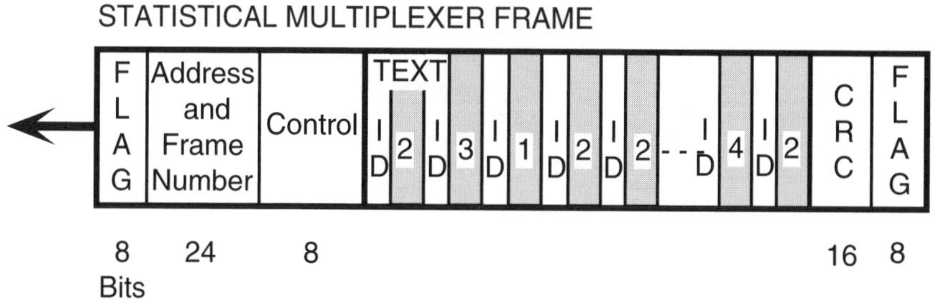

Figure 8.8 Statistical Multiplexer Frame for Synchronous Transmission

Data concentrators are likely to perform as primary sites with respect to inputs from DTE devices (polling, error checking, requesting resends, etc.) and as secondary sites with respect to the hosts they serve (responding to polls, resending frames, etc.). The principle of a data concentrator is shown in **Figure 8.9**.

8.1.4 Inverse Multiplexing

Earlier, we stated that a synchronous time division multiplexer combines *n* lower-speed, digital signals on a higher-speed bearer. In many ways, inverse multiplexing may be considered to be the reverse of this activity:

- **Inverse multiplexing**: creating a higher-speed, point-to-point, digital circuit from several independent, lower-speed digital channels that exist between the originating and terminating locations.

Inverse multiplexers are used to send composite, higher-speed signals (such as a 384 kbits/s videoconference signals) over lower-speed channels (such as 6 ×64 kbits/s channels). Employed when a channel that matches, or exceeds, the higher-speed signal is not available, or when the higher-speed signal must be sent to a location served by lower-speed, switched lines, the inverse multiplexer establishes the lower-speed connections, segments the higher-speed data stream and sends the data segments over the lower-speed lines to the destination. At the receiving location, an inverse demultiplexer receives the segments and reassembles them into the

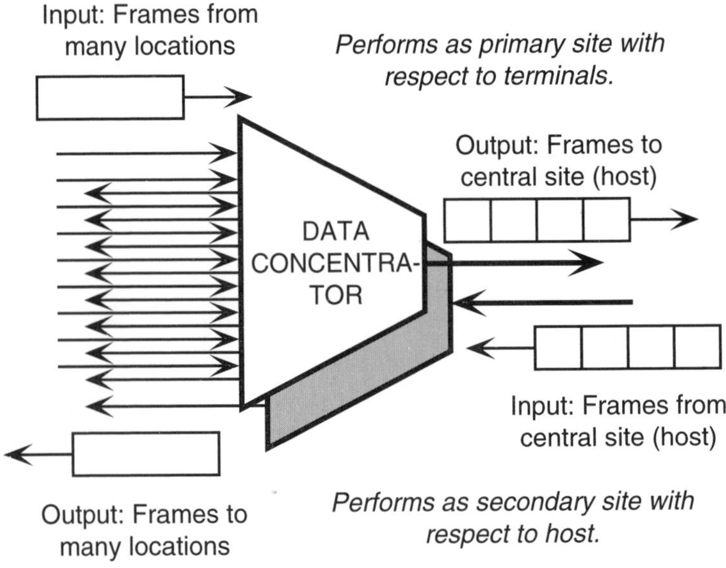

Figure 8.9 Data Concentrator

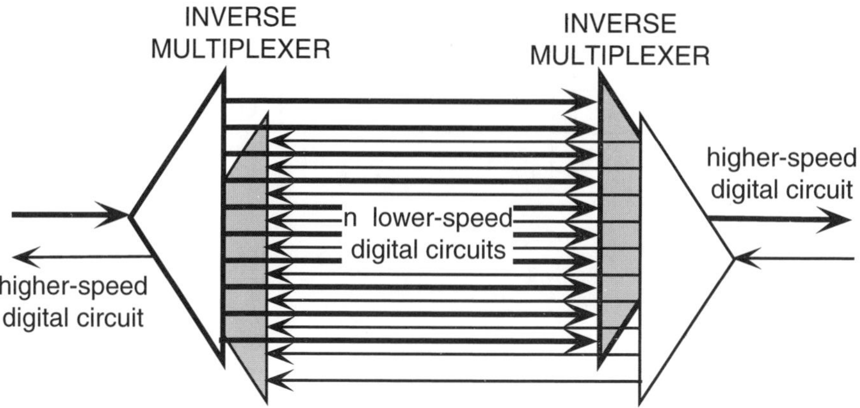

Figure 8.10 Inverse Multiplexer

higher-speed stream—an action that may require reordering some segments and compensating for different channel delays. The principle of inverse multiplexing is shown in **Figure 8.10**. The need for inverse multiplexing is likely to diminish as carriers deploy more high-speed switched services (see § 10.5.2(4)).

8.2 T-CARRIER MULTIPLEXING

T-carrier multiplexers produce composite signals that contain 24, 48, 96, 672, 2016, and 4032 digital voice signals. Also, provided certain restrictions are observed, they transport data signals. The operating speeds of T-carrier equipment extend from 1.544 Mbits/s to 274.176 Mbits/s in four levels. While there were good reasons for the development of the entire family, today, the first (T-1, 24 voice channels, 1.544 Mbits/s) and third (T-3, 672 voice channels, 44.736 Mbits/s) levels dominate the T-carrier environment. Accordingly, while we will describe all members, we will emphasize T-1 and T-3.

8.2.1 T-1

(1) Original System

In its original form, a T-1 system employs channel banks that process 24, analog voice signals to become 24, 64 kbits/s, PCM signals, and octet multiplexes them to produce a bit stream of 1.544 Mbits/s that is carried on conditioned, four wire circuits. Shown in **Figure 8.11** each frame contains 24 octets that represent samples from 24 voice signals and a single bit that is used for synchronizing (framing bit). In total, then, there are $(8 \times 24) + 1 = 193$ bits per frame. Because each voice signal is sampled 8000 times per second, the frame must repeat 8000 times per second. Thus, the output bit rate is $193 \times 8000 = 1.544$ Mbits/s. The T-1 line signal at the

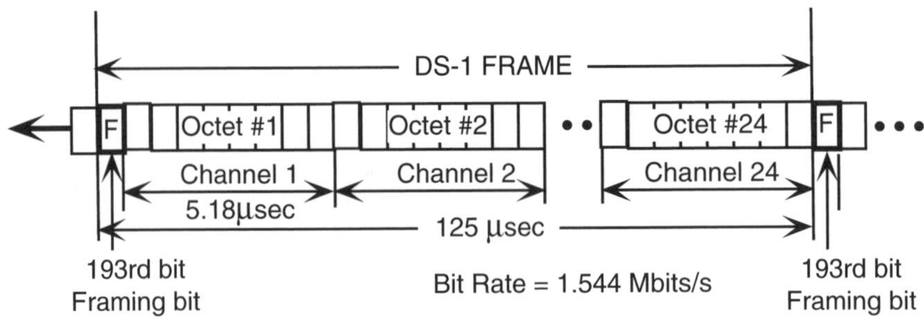

Figure 8.11 DS-1 Frame Format
Frame consists of 24 octets from 24 digital voice channels and a single framing bit.

output of the multiplexer is described as DS-1, i.e., digital signal level 1. The digital input signals are described as DS-0 signals, i.e., digital signals level 0.

Figure 8.12 shows the general arrangement of the components required to implement a traditional, copper-based, T-1 facility between a customer's premises and a carrier's central office. Digital signals are sent and received by the customer's digital PBX that includes equipment that performs the multiplexing and

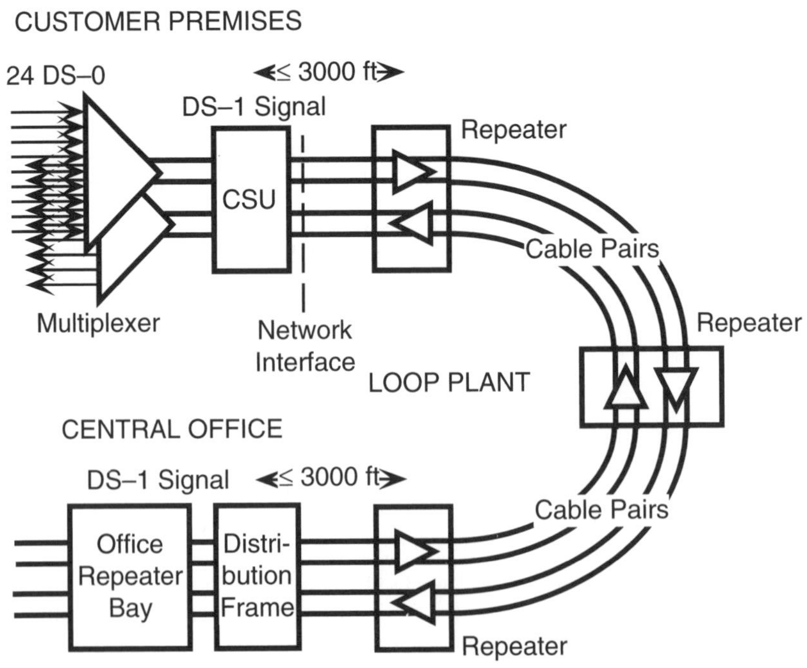

Figure 8.12 T-1 Facility

CSU (channel service unit, see § 5.1.1(2)) functions. Twenty-four DS-0 signals are multiplexed to produce a DS-1 signal on the network side of the CSU. Connections are made over two pairs of wires with intermediate repeaters spaced at intervals of approximately 6000 ft. The first and last repeaters in the T-1 line are spaced ≤ 3000 ft from the central office and the customer's premises. Because of the timing jitter that accumulates in long sequences of repeaters, the maximum distance over which T-1 is used is 50 miles. Ninety-five per cent of the time, T-1 facilities provide an error rate of 1-bit in error in 10^6 bits. Using 400 pairs, 200 T-1 systems can operate over a single, 900 pair cable. Newer technology has produced *repeaterless* T-1s that can operate up to 12,000 ft on two pairs, without intermediate repeaters. Each pair carries 784 kbits/s and employs echo cancellers to achieve duplex transmission over each pair. Called *high bit-rate digital subscriber line* (HDSL), the technology is aimed at providing local T-1 connections that do not require special engineering.

(2) DS-1 Signals

DS-1 employs alternate mark inversion (AMI, see § 4.5) signaling. Ones (1s) are signified by alternating positive and negative, raised-cosine pulses, and 0s are signified by zero level. Because they are represented by the absence of pulses, long runs of 0s may cause the repeaters to lose synchronization. To maintain reliable working, on average, the signal stream must contain at least one 1 in eight bits (12.5% 1s), and in no case can there be more than 15 consecutive 0s. The way in which this is achieved depends on whether the T-1 is employed exclusively in carrying voice, or in carrying data, or voice and data:

- **Voice only**: during active periods, the PCM quantizing procedure achieves the required *ones-density* automatically (see Figure 3.3); during idle periods, the transmitter sends the octets 00000010 (bit 7 forced) or 00011000 (bits 4 and 5 forced); and, in unassigned channels, the octet 11111111 is sent.
- **Data only**, or **voice and data**: in unassigned channels, the transmitter sends the octet 00011000. The 0 in the 8th slot identifies it as a code introduced by the carrier. In assigned channels, substituting an arbitrary code in the middle of a user's data stream is unacceptable. Instead, one technique forces bit 8 in all user data time slots to 1. As a result, the user data rate drops to 56 kbits/s—but the *data* can contain unlimited sequences of zeros. This is the technique employed in DDS (Digital Data Service, see § 13.3.1(2)). Another technique is tailored to providing a full 64 kbits/s *clear* channel. Phantom 1s are inserted into any sequence of eight 0s in such a way that the receiving equipment can detect and remove them. To do this, the pulse representing a phantom 1 is inserted in the same polarity as the previous (genuine) 1. This is a *violation* of the normal AMI sequence and is recognized by the system.

The most common scheme used to introduce bipolar variations is known as bipolar with 8 zeros substitution:

- **Bipolar with 8 zeros substitution** (B8ZS): in a sequence of 8 zeros, a bipolar violation is inserted at the 4th and 7th position. Each violation is followed by a 1 in the next position (5th and 8th position). The receiving equipment is programmed to accept the sequence 000V10V1 as 00000000 so that the operation is transparent to the user. The upper signal waveform in **Figure 8.13** shows the AMI format for the sequence...⇐ 011100000000011... The lower signal waveform shows the application of B8ZS. The sequence becomes...⇐ 0111000V10V1011...—a pattern that the receiver can detect and correct. Use of B8ZS removes the need to introduce special codes and other techniques to satisfy the ones-density requirement, and eliminates the no more than 15 zeros requirement in the user's data.

A scheme that does not employ bipolar violations to achieve a full 64 kbits/s clear channel is known as:

- **Zero-byte time slot interchange** (ZBTSI): the transmitter maintains a buffer that is filled by successive groups of four frames (96 octets) from the T-1 stream. In the buffer, octets with all zeros are removed and the remaining non-zero octets consolidated at the back of the buffer. The first octet space at the front of the buffer is filled with a 7-bit binary number that gives the location (in the sequence of 96 octets) of the first all zeros octet. The eighth bit in the octets is used to indicate whether there are more all zeros octets in the

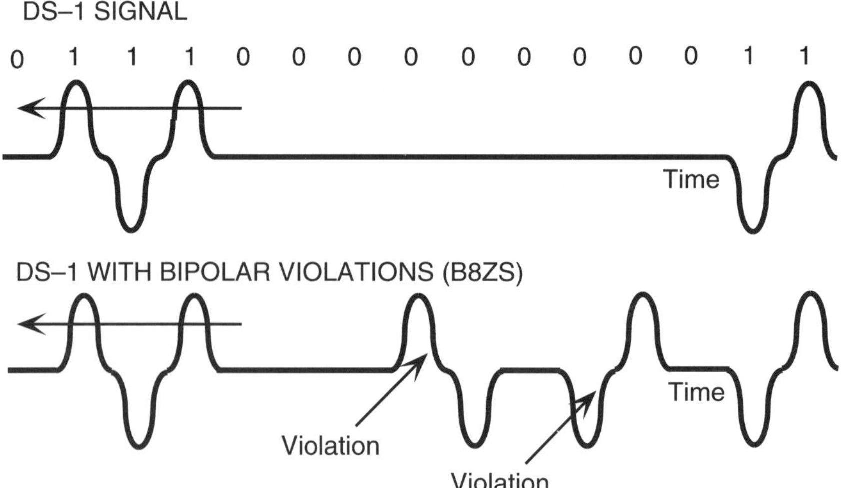

Figure 8.13 DS-1 B8ZS Signaling
Phantom 1s are inserted in any sequence of 8 zeros to preserve reliable working. On average, the stream must contain 12.5% 1s, and there can be no more than 15 consecutive zeros.

sequence. The second octet space is used to indicate the position of the second all zeros octet, and so on. (There are as many spaces as all zeros octets.) At the receiver, all zeros octets are reintroduced in the positions indicated. Because the processing delays the stream by 1.5 milliseconds, ZBTSI is less popular than B8ZS. Use of ZBTSI removes the need to introduce special codes and other techniques to satisfy the ones-density requirement, and eliminates the no more than 15 zeros requirement in the user's data.

(3) Framing and Channel Signaling—Voice Only

Over time, the framing of the DS-1 signal stream has evolved from the simple 193rd bit approach illustrated in Figure 8.11, to the use of sequences of framing bits to provide control patterns and error information. Two common arrangements include:

- **Superframe format** (SF): also called D4 superframe format, 12 DS-1 frames form a superframe that contains 2,316 bits and takes 1.5 milliseconds to transmit at the DS-1 rate. The 12 framing bits form a pattern (⇐ 100011011100) that is used for synchronization. In addition, they are divided into two sequences. The bits in the odd-numbered frames provide the sequence ⇐ 101010; it is used by the receiving equipment to establish, maintain, or re-establish frame boundaries. The bits in the even-numbered frames provide the sequence ⇐ 001110; it is used by the receiving equipment to identify frames 6 and 12. (Frame 6 occurs on a 1 preceded by three 0s—the first zero is the 12th bit from the previous frame—and frame 12 occurs on zero preceded by three 1s).
- **Extended superframe format** (ESF): 24 DS-1 frames form an extended superframe that contains 4,632 bits and takes 3 milliseconds to transmit at the DS-1 rate. The 24 framing bits are used for the following purposes:
  - synchronize terminal equipment (F-bits)
  - secure multiframe alignment (F-bits)
  - provide a 4000 bit/s data link (FDL, facility data link) that carries application-dependent information and is used to acquire historical facility performance data (D-bits)
  - provide means of checking the error performance of the system (C-bits).

Figures 8.14 and **8.15** present views of SF and ESF. Twelve 193-bit frames are stacked one on the other to produce the SF diagram, and 24, 193-bit frames are stacked to produce the ESF diagram. Seen in this way, overhead bits and user octets are arranged in columns that represent individual bit streams. To read the serial DS-1 stream, start in the left-hand upper corner, proceed along frame #1, return to the left-hand side and continue along frame #2, etc., until you reach the bottom, right-hand corner. The sum of the 24 octet bit streams (i.e., the 24 DS-0 signals) is called the *payload*.

SUPERFRAME

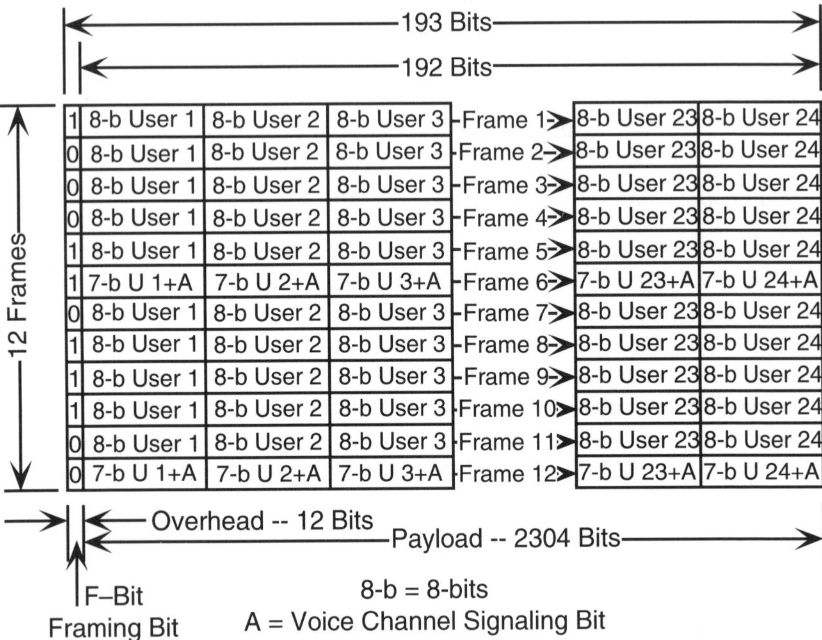

Figure 8.14 Superframe Layout

Twelve 193-bit DS-1 frames are stacked on one another. Seen in this way, overhead bits and user octets appear in columns that represent individual message streams. SF employs a 12-bit sequence of framing bits to provide synchronizing and frame alignment.

In frames 6 and 12 of SF, and frames 6, 12, 18, and 24 of ESF, bits are robbed from the user to provide a 1333 bits/s signaling channel for each voice channel. Used in pairs (SF and ESF), they can report four states, and used in quadruples (ESF only), they can report sixteen states. Messages include *idle*, *busy*, *ringing*, *no-ring*, and *loop open.*

(4) Framing and Channel Signaling—Data Only

For data only applications, the general formats of SF and ESF that have been described are used except that per channel, robbed bit signaling is eliminated in favor of common channel signaling. Customers have access to all but the 24th channel that is reserved for synchronizing and signaling chores (see below).

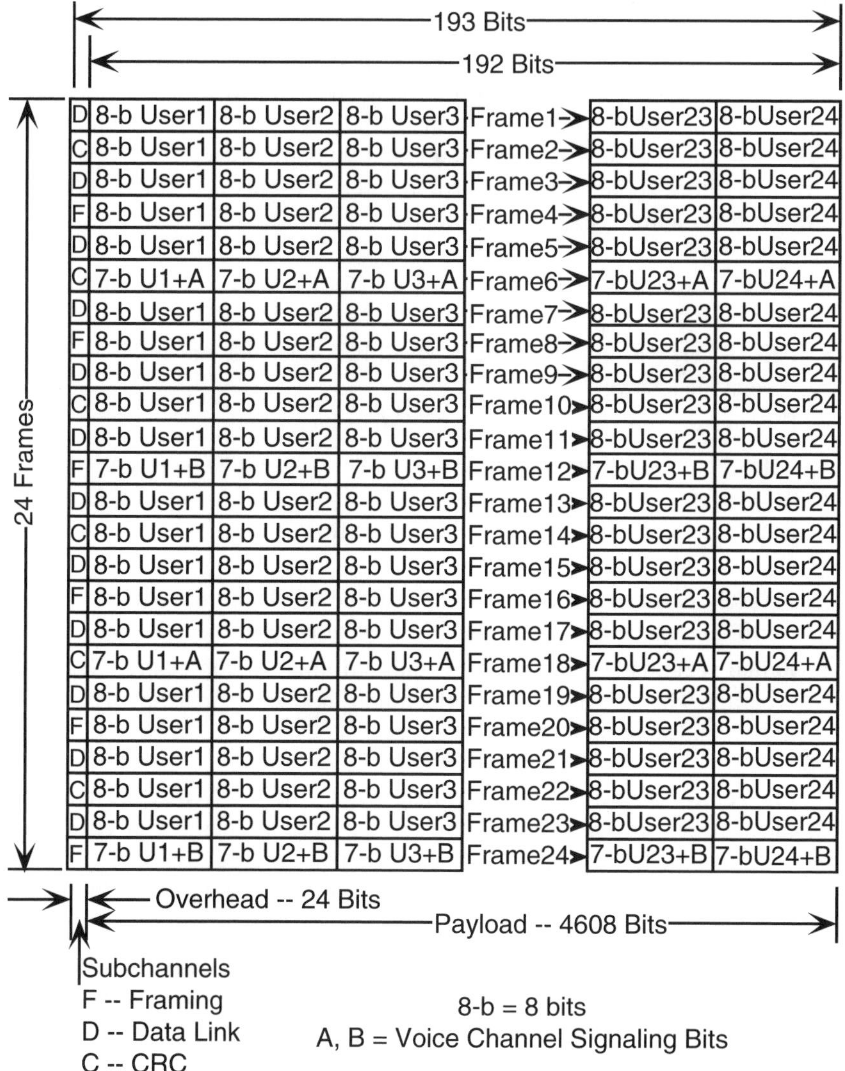

Figure 8.15 Extended Superframe Layout
Twenty-four frames are stacked together to make this diagram. ESF employs a 24-bit sequence of framing bits to provide synchronizing, frame alignment, facility data link, and error checking.

(5) ESF Error Checking

The error checking capability of ESF employs a 6-bit cyclic-redundancy check (CRC) process in which, at the sending end, the polynomial formed by the 4,632 superframe bits is divided by a 7-bit polynomial (1000011) to produce a 6-bit remainder (the frame check sequence, FCS). It is transmitted in the 193rd bit of frames 2, 6, 10, 14, 18, and 22. At the receiving end, the CRC calculation is repeated. If the FCS produced at the receiver matches the FCS received from the sender, no error was introduced by transmission. The CRC process and the bit patterns in the framing information are used to define a set of events that describe the error performance of the link. Some of them include:

- **Out-of-frame** (OOF) **event**: when two out of four consecutive framing bits are incorrect
- **CRC-6 error event**: when the received FCS does not match the FCS calculated at the receiver
- **ESF error event**: an OOF event, CRC-6 event, or both
- **Errored second** (ES): any second in which one, or more, CRC-6 events are detected
- **Bursty second** (BS): any second in which from 2 to 319 CRC-6 error events are detected
- **Severely-errored second** (SES): any second in which from 320 to 333 CRC-6 error events are detected, or one OOF event occurs
- **Failed seconds** (FS) **state**: state achieved after 10 consecutive SESs have occurred. The state remains in effect until the facility transmits 10 consecutive seconds that are not severely errored.

Figure 8.16 shows the general arrangement of a T-1 facility operating with ESF. Channel Service Units in the ESF facility store the number of error events detected. Using the facility data link, controllers poll the CSUs for data that is used to establish the level of performance being achieved. This is important information for billing and for scheduling maintenance.

8.2.2 Evolution of T-1 Systems

(1) Common Types

Originally developed to multiplex and transport voice signals, T-1 systems have evolved to general purpose, voice and data, digital multiplexing, and transmission systems that transport:

- Digital voice at 64, 32, or 16 kbits/s on DS-1 channels
- Digital data at 56 kbits/s or multiples of 64 kbits/s (i.e., n × 64 kbits/s, where n = 1, 2, etc.) on DS-1 channels

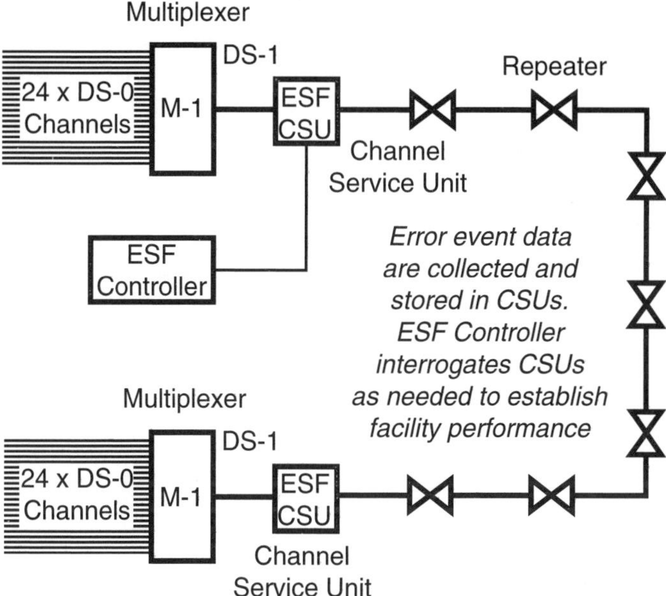

Figure 8.16 T-1 Facility With ESF Error Checking

- Sub-rate data at 2.4, 4.8, 9.6, or 19.2 kbits/s on DS-0 channels.

Accommodating these signals and signal speeds on DS-1 lines requires a range of multiplexers. Some of them include:

- **M24**: original T-1 system. Customer access = 24 voice channels. Per channel signaling uses 8th bit robbing in every 6th frame. Framing uses 8000 bits/s. Output = DS-1.
- **M44**: customer access = 44 voice channels that are converted to 44, 32 kbits/s voice channels. They occupy 22, DS-0 channels. The remaining 2, DS-0 channels are used for common channel signaling. Framing uses 8000 bits/s. Output = DS-1.
- **M48**: customer access = 48 voice channels that are converted to 48, 32 kbits/s channels. They occupy 24, DS-0 channels. Per channel signaling uses bit robbing. Framing uses 8000 bits/s. Output = DS-1.
- **M88**: customer access = 88 voice channels that are converted to 88, 16 kbits/s voice channels. They occupy 22, DS-0 channels. The remaining 2, DS-0 channels are used for common channel signaling. Framing uses 8000 bits/s. Output = DS-1.
- **M96**: customer access = 96 voice channels that are converted to 96, 16 kbits/s channels. They occupy 24, DS-0 channels. Per channel signaling uses bit robbing. Framing uses 8000 bits/s. Output = DS-1.

- **M24/64**: customer access = 23, 64 kbits/s data channels. Employs B8ZS, and 1, 64 kbits/s channel for data synchronization, alarms, and signaling. Framing uses 8000 bits/s. Output = DS-1. Known as *clear channel capability* (CCC) or *clear 64* (i.e., the customer has full use of the 64 kbits/s channel), it is used with ESF.
- **M24/56**: customer access = 23, 56 kbits/s data channels. Output = DS-1. The bit layout for a 56 kbits/s data superframe is shown in **Figure 8.17**. The 8th bits in the 23 user time slots are set to 1 to ensure meeting the ones-density requirement. The 24th time slot is used for data synchronizing, as an alarm channel for reporting failure in the outgoing direction (bit 6, the Y bit), and as an 8 kbits/s data channel for use by the digital transmission surveillance system (bit 7, the R bit) employed in DDS.

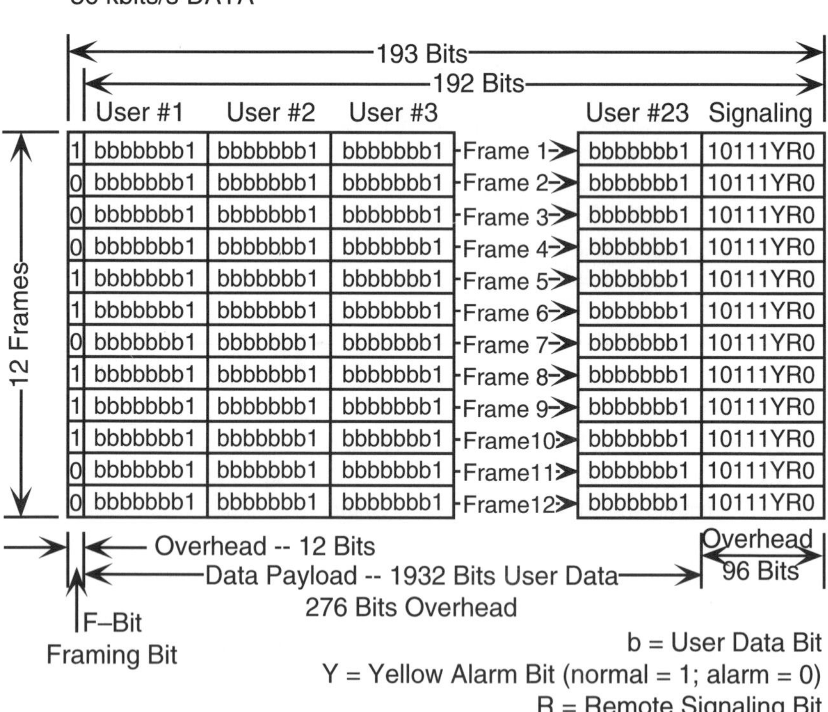

Figure 8.17 56 kbits/s Data Superframe Layout
The 8th bits in the 23 user time slots are set to one to meet the 1s-density requirement.

(2) Subrate Data

In DDS (see § 13.3.1(2)), subrate data (2.4, 4.8, 9.6, and 19.2 kbits/s) are multiplexed on DS-0 channels that are then multiplexed with other DS-0 channels on a DS-1 line:

- To connect a **single subrate station** to a M24/56 multiplexer, bytes are repeated as necessary to match the sender's speed to the speed of the DS-0 line. In this case, it is known as a DS-0A signal.
- To connect **several subrate stations** over the same DS-0 channel, subrate signals are multiplexed in a subrate multiplexer (SRM) to produce a stream of octets that contains a subrate synchronizing bit (bit #1) and six data bits (bits 2 through 7). The 8th bit is set to 1 to ensure meeting the 1s density requirement. Thus, the 64 kbits/s signal contains 48 kbits/s of customer subrate data, 8 kbits/s of subrate synchronizing and 8 kbits/s of ones. It is described as a DS-0B signal.

  Some typical SRMs include:
- **M5**: customer access = 5, 9.6 kbits/s data channels. Output = DS-0B.
- **M10**: customer access = 10, 4.8 kbits/s data channels. Output = DS-0B.
- **M20**: customer access = 20, 2.4 kbits/s data channels. Output = DS-0B.

SRMs are used in combination with M24/56 multiplexers to produce DS-1 signals. An example is a combination of 23, M5 multiplexers feeding an M24/56 multiplexer:

- 23 **M5 + M24/56**: customer access = 115, 9.6 kbits/s data channels to 23 M5 SRMs. Output from each M5 = DS-0B. Input to M24/56 multiplexer = 23, DS-0B. Output = DS-1. The bit layout for a 56 kbits/s, subrate data, superframe is shown in **Figure 8.18**.

(3) Unchannelized T-1

Some data are not organized in octets. For instance, the output of a video codec may be a continuous stream of bits composed of variable length code words. Also, a customer may combine subrate data in one continuous stream to improve on the 71.5% efficiency associated with the subrate data superframe in Figure 8.18. For transmission over a DS-1 line, the bitstream can be divided into segments of 192 bits to which framing bits are attached. Since neither the sending nor the receiving equipment makes use of the boundaries that define channels, the technique is described as *unchannelized*. Incompatible with public networks, unchannelized operation is confined to private connections. It is the user's responsibility to ensure that the bit stream complies with T-1 transmission requirements.

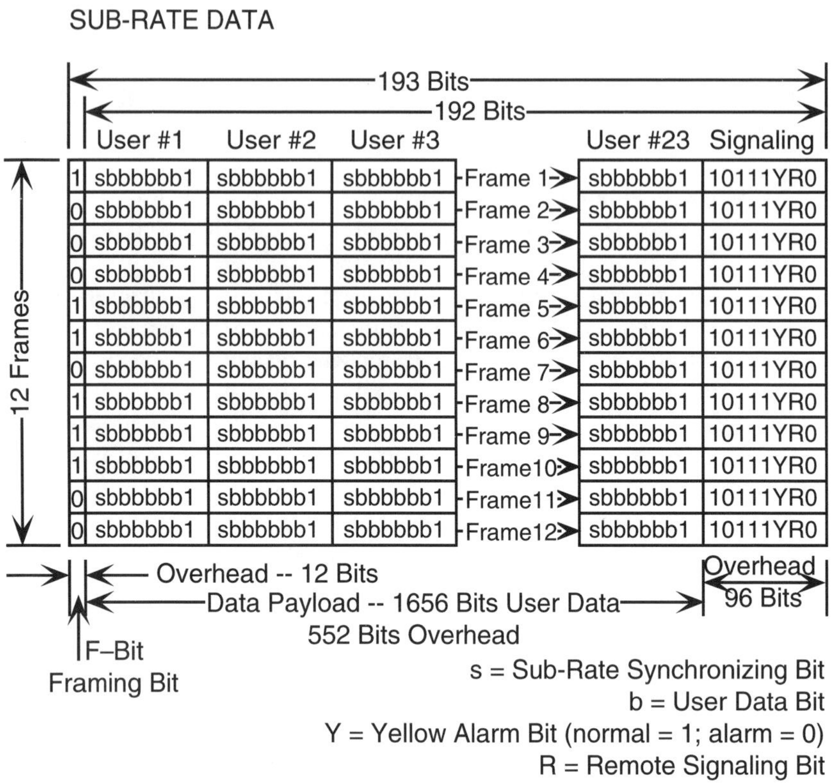

SUB-RATE DATA

193 Bits

192 Bits

F-Bit	User #1	User #2	User #3		User #23	Signaling
1	sbbbbbb1	sbbbbbb1	sbbbbbb1	Frame 1	sbbbbbb1	10111YR0
0	sbbbbbb1	sbbbbbb1	sbbbbbb1	Frame 2	sbbbbbb1	10111YR0
0	sbbbbbb1	sbbbbbb1	sbbbbbb1	Frame 3	sbbbbbb1	10111YR0
0	sbbbbbb1	sbbbbbb1	sbbbbbb1	Frame 4	sbbbbbb1	10111YR0
1	sbbbbbb1	sbbbbbb1	sbbbbbb1	Frame 5	sbbbbbb1	10111YR0
1	sbbbbbb1	sbbbbbb1	sbbbbbb1	Frame 6	sbbbbbb1	10111YR0
0	sbbbbbb1	sbbbbbb1	sbbbbbb1	Frame 7	sbbbbbb1	10111YR0
1	sbbbbbb1	sbbbbbb1	sbbbbbb1	Frame 8	sbbbbbb1	10111YR0
1	sbbbbbb1	sbbbbbb1	sbbbbbb1	Frame 9	sbbbbbb1	10111YR0
1	sbbbbbb1	sbbbbbb1	sbbbbbb1	Frame10	sbbbbbb1	10111YR0
0	sbbbbbb1	sbbbbbb1	sbbbbbb1	Frame11	sbbbbbb1	10111YR0
0	sbbbbbb1	sbbbbbb1	sbbbbbb1	Frame12	sbbbbbb1	10111YR0

12 Frames

Overhead -- 12 Bits

Data Payload -- 1656 Bits User Data
552 Bits Overhead

Overhead
96 Bits

F–Bit
Framing Bit

s = Sub-Rate Synchronizing Bit
b = User Data Bit
Y = Yellow Alarm Bit (normal = 1; alarm = 0)
R = Remote Signaling Bit

Figure 8.18 Sub-Rate Data Superframe Layout
Six bits in the 23 user time slots are used for subrate data. The 1st bit is employed for synchronizing the data and the 8th bit is set to one to meet the 1s-density requirement.

8.2.3 T-Carrier Hierarchy

(1) Hierarchy

Called the T-carrier hierarchy, the facilities shown in **Figure 8.19** combine DS-0 signals into progressively higher-speed streams. The four levels are designated T-1, T-2, T-3, and T-4, with an intermediate level, T-1C between T-1, and T-2, and another intermediate level, T-4NA between T-3 and T-4. In conformity with the DS-0, DS-1 nomenclature introduced for the input and output signals of T-1, the output signals are designated DS-1C, DS-2, DS-3, DS-4NA, and DS-4, and the multiplexers are designated:

- **M1**: combines 24, DS-0s into one DS-1, also known as M24, M24/56 and M24/64

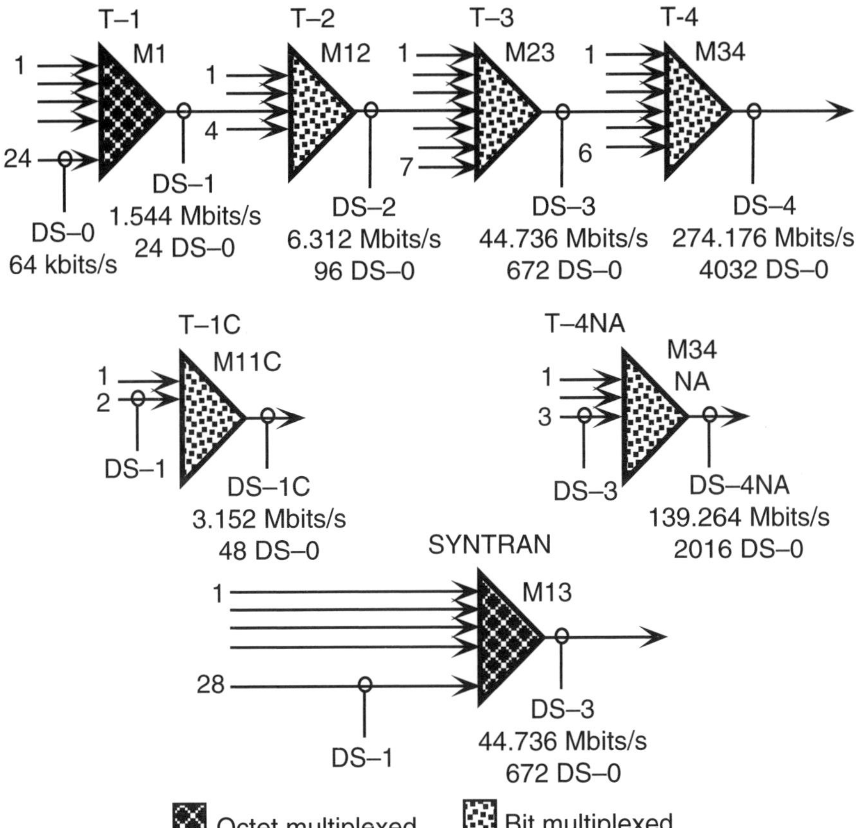

Figure 8.19 T-Carrier Hierarchy
T-carrier equipment combines DS-0 channels into progressively higher-speed data streams.

- **M1C**: combines 2, DS-1s into one DS-1C
- **M12**: combines 4, DS-1s into one DS-2
- **M13**: combines 28 DS-1s into one DS-3
- **M23**: combines 7, DS-2s into one DS-3
- **M34NA**: combines 3 DS-3s into one DS-4NA
- **M34**: combines 6 DS-3s into one DS-4.

For the multiplexing process to succeed, DS-0 octets must be formed in 125 μseconds and must arrive at the first level multiplexer (M1) in synchrony with octets from other sources so that they can be octet-multiplexed into DS-1 frames. In turn, individual frames of the DS-1 stream must be formed in 125 μseconds, must arrive at the second level multiplexer (M12) in approximate bit-synchronism with frames from other DS-1 sources, and their speeds must be equalized before being bit-multiplexed into a DS-2 frame. Proceeding to the third level, DS-2 frames must arrive at the third level multiplexer (M23) in approximate bit-synchronism with frames from other DS-2 sources, and their speeds must be equalized before being bit-multiplexed into a DS-3 frame. Etc... The activity is described as:

- **Plesiochronous multiplexing**: a multiplexing process in which two or more signals are generated at nominally the same digital rate and their significant instants occur at nominally the same time.

With signals homing on the multiplexers from many locations, maintaining frame or bit synchronism is no easy task. Nevertheless, so that each octet shall arrive at its destination without error, the beginning of each frame must be identified unambiguously, the octets they contain must be demultiplexed exactly, and the bits decoded perfectly. Providing satisfactory service to each of the 4032 users of a T-4 facility allows no room for errors.

(2) T-1C

T-1C converts a T-1 transmission link to one that transports twice the number of bits per second without changing the physical transmission arrangements—it operates over the same wire pairs with duobinary signaling (see § 4.7.5), the same repeater spacing (6000 ft), and the same limitation (≤ 50 repeaters). Using 300 pairs, 150, T-1C systems can operate over a single, 900 pair cable. This is a 50% increase over the performance of T-1. The DS-1C signal is formed by interleaving 2, DS-1 streams, bit by bit, and adding a control bit after every 52 bits (26 from each DS-1). To maintain the ones-density, one DS-1 stream is logically inverted before multiplexing. Also, because the input DS-1 streams are likely to be operating at slightly different speeds, extra pulses (stuffing pulses) are inserted in specific locations in each stream to equalize their rates and raise the composite to 3.09 Mbits/s. With the insertion of one control bit for every 52, DS-1 bits, the final speed is 3.152 Mbits/s. At the receiver, the stuffing bits are removed before demultiplexing the DS-1C stream into two DS-1 streams.

(3) T-2

In M12, the DS-2 signal is formed by interleaving 4, DS-1 streams, bit by bit, and adding a control bit after every 48 payload bits. To maintain the ones-density, bits in DS-1 streams #2 and #4 are logically inverted before multiplexing. Also, because the DS-1 streams are likely to be operating at slightly different speeds, stuffing pulses are inserted in specific time slots in each stream to equalize their rates and raise the composite to 6.183 Mbits/s. With the insertion of one control bit for every 48, DS-1 bits, the final speed is 6.312 Mbits/s. At the receiver, the stuffing bits are removed before demultiplexing the DS-2 stream into 4 DS-1 streams. Because T-2 requires special low-capacitance cable and repeaters spaced at approximately 3 mile intervals, it is not widely deployed.

(4) T-3

In M13, the DS-3 signal is formed from 28, DS-1 signals by bit-interleaving groups of four DS-1s to create 7, DS-2 signals as described above; in turn, the 7, DS-2 signals are bit-interleaved to create one DS-3 signal. In the conversion of 7, DS-2 to DS-3, a control bit is added for every 84 payload bits, and bits are stuffed in specific slots to equalize the speeds of the individual streams and achieve a DS-3 speed of 44.736 Mbits/s. The overhead functions associated with M13 consume 1.504 Mbits/s. DS-3 employs bipolar signaling and may be transmitted over coaxial cables, microwave radios, or optical fibers. DS-3 frame formats are known as traditional DS-3 (also called M23), C-bit parity, and SYNTRAN (also called synchronous M13). The first two are asynchronous techniques (i.e., actions are driven by the state of the signal stream); the third is a synchronous technique (i.e., actions occur in synchrony with an external timing source). They are described as follows:

- **Traditional DS-3**: the DS-3 signal is partitioned in frames containing 4,760 bits. Called M-frames, they consist of seven subframes of 680 bits each. The subframes are divided into eight blocks of 85 bits. In these blocks, the first bit is a control bit and the other 84 are payload bits. Each frame, then, includes 54 overhead bits that are used to perform frame alignment, subframe alignment, performance monitoring, alarm and status reporting, and other functions. **Figure 8.20** shows the format of a DS-3 frame and identifies some of the uses of the control bits. In the last block of the seven subframes, a single bit can be stuffed into one, some, or all of them, to adjust the speeds of the 7 DS-2 signals. In the subframe in which it occurs, the presence of a stuffing bit is indicated by setting the three C bits to 1. Because the DS-3 rate is not an integral multiple of the DS-1 rate (44.736 ÷ 1.544 = 28.974), and bits are stuffed as needed, individual DS-1 frames *slide* relative to the DS-3 frames. This makes locating all the bits belonging to a specific DS-1 input signal equal to demultiplexing the entire stream.
- **C-bit parity**: the format of the traditional DS-3 frame is preserved. However, the 7 DS-2 signals created in the multiplexing process are generated at exactly 6.306272 Mbits/s and a stuffing bit is inserted in every subframe. The result is a DS-3 signal at exactly 44.736 Mbits/s. Because they are not needed to report the presence or absence of stuffing bits, the 21 C bits can be used for other purposes. For instance, the three C bits in subframe 3 indicate the parity status of the preceding frame at the origination point of the stream. At the receiver, comparison of the locally calculated value (P1, P2) with this value indicates whether an error has occurred. Other bits are used to provide data links, alarms, and status information.

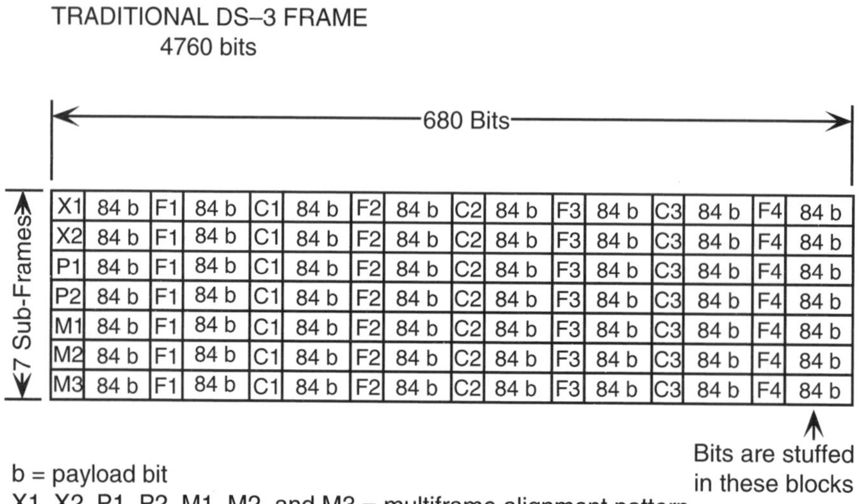

TRADITIONAL DS–3 FRAME
4760 bits

680 Bits

7 Sub-Frames

X1	84 b	F1	84 b	C1	84 b	F2	84 b	C2	84 b	F3	84 b	C3	84 b	F4	84 b
X2	84 b	F1	84 b	C1	84 b	F2	84 b	C2	84 b	F3	84 b	C3	84 b	F4	84 b
P1	84 b	F1	84 b	C1	84 b	F2	84 b	C2	84 b	F3	84 b	C3	84 b	F4	84 b
P2	84 b	F1	84 b	C1	84 b	F2	84 b	C2	84 b	F3	84 b	C3	84 b	F4	84 b
M1	84 b	F1	84 b	C1	84 b	F2	84 b	C2	84 b	F3	84 b	C3	84 b	F4	84 b
M2	84 b	F1	84 b	C1	84 b	F2	84 b	C2	84 b	F3	84 b	C3	84 b	F4	84 b
M3	84 b	F1	84 b	C1	84 b	F2	84 b	C2	84 b	F3	84 b	C3	84 b	F4	84 b

Bits are stuffed in these blocks

b = payload bit
X1, X2, P1, P2, M1, M2, and M3 = multiframe alignment pattern
P bits indicate parity values. P1 = P2 = 0 indicates that the sum of all payload bits in the previous frame is 0; P1 = P2 = 1 indicates the sum is 1
M bits are used for frame synchronization; M1 = 0, M2 = 1 and M3 = 0
F bits are used for subframe synchronization; F1 = 1, F2 = 0, F3 = 0 and F4 = 1
X bits are used for alarm and status functions
C bits identify presence of stuffing bit in block 8

Figure 8.20 Traditional DS-3 Frame Layout

- **SYNTRAN**: the format of the traditional DS-3 frame is restructured and 699 M frames are combined into a synchronous superframe so that DS-1 frames have fixed locations within the superframe (699 M-frames occupy 74.375 milliseconds—the time required for 595 DS-1 frames). The principal purpose is to provide simple, reliable access to all DS-0 and DS-1 signals contained in the SYNTRAN DS-3 signal. To do this, *octets* from 28, DS-1 channels of exactly 1.544 Mbits/s are multiplexed directly to DS-3. No stuffing bits are used—a circumstance that frees approximately 66 kbits per second for other overhead functions. Because each M-frame contains 588 octets, and a full set of frames (i.e., 28) from the 28 DS-1 input signals requires 672 octets, a 125 μs segment of the input signals occupies more than one M-frame. **Figure 8.21** shows how the 672 octets might occupy parts of two frames. The C bits, and the bandwidth no longer occupied by stuffing bits, are used for data link, CRC error checking, and other purposes.

Principally, traditional DS-3 is used by carriers in public networks and SYNTRAN DS-3 is used in private networks. It is likely that the newer technology and broader capabilities of SONET and SDH (see below) will make them the systems

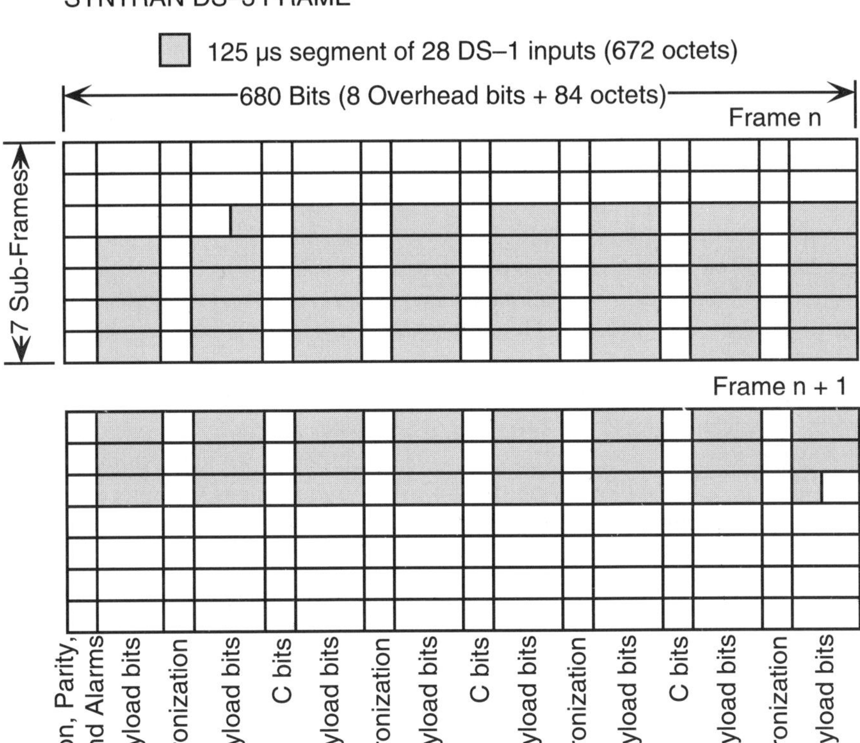

Figure 8.21 SYNTRAN DS-3 Frames

of choice for future applications that require the multiplexing and high-speed transport of both voice and data signals.

(5) T-4

In M34, 6, DS-3 signal streams are combined into a DS-4 stream that operates at 274.176 Mbits/s to create T-4, a high-density multiplexing system that is the natural extension of the T-carrier family. However, it is little more than a definition. An intermediate step, DS-4NA has been standardized as a 139.264 Mbits/s stream that is formed from 3, DS-3s. With SONET and SDH available, it is likely that neither of these systems will see much use in the future.

(6) Comparison of Levels in North America, Europe, and Japan

Five level multiplexer hierarchies have been standardized in Europe and Japan. **Figure 8.22** shows how the number of DS-0 signals that North American, European, and Japanese multiplexers carry varies with level and region. To make this comparison, M4NA has been placed at level 4 and M4 has been placed at level 5. At level 1, the three are synchronous, and in Japan and North America, the number of DS-0s they carry is the same (24). In Europe, the first level (E-1, 2.048 Mbits/s) multiplexes 32 DS-0 channels. Of these, 30, DS-0 channels carry customer traffic, and 2, DS-0 channels are used for supervision. At level 2, both Japanese and North American multiplexers carry 672 DS-0 channels. However, in Japan, octet-multiplexing is employed and the process is synchronous, while in North America, bit-multiplexing is employed and the process is asynchronous. In Europe, the second level bit multiplexes 128 DS-0s. At level 3, with the exception of SYNTRAN

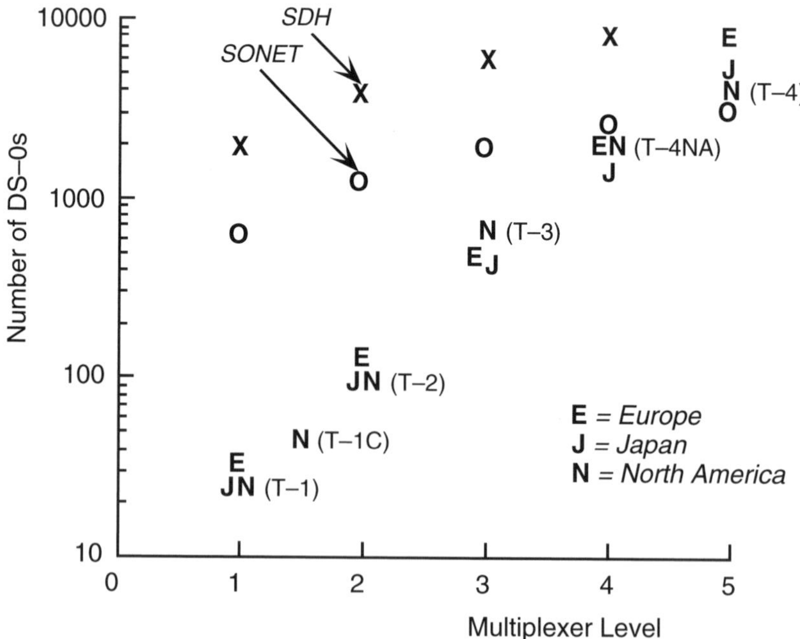

Figure 8.22 Comparison of North American, European, and Japanese Multiplexing Levels

M13, the multiplexers employ bit–multiplexing, carry different numbers of DS-0s, and are asynchronous. At levels 4 and 5, the multiplexers employ bit-multiplexing, carry different numbers of DS-0s, and are asynchronous. The lines at the top of the chart indicate the numbers of DS-0s multiplexed by different levels of SONET and SDH. They are discussed below.

8.3 SONET AND SDH

8.3.1 Synchronous Optical Network

In the United States, carriers are implementing synchronous digital networks that employ optical fiber bearers. Called synchronous optical networks (SONETs), each facility conforms to standards that define a set of optical interfaces for network transport. First promulgated as United States national standards by ANSI/ECSA, and now incorporated in ITU-T standards, they include electrical and optical speeds, and frame formats. As is to be expected of a more recent system that employs newer technologies, the features available in SONET make it superior to T-3 and T-4 systems. They include improved transport performance, ability to identify subsidiary streams with certainty, international connectivity without format conversion, and enhanced control and administrative functions.

(1) SONET Transport Speeds

SONET uses octet-multiplexing to create a higher-speed stream from lower-speed tributary signals:

- **Tributary signal**: a lower-speed signal that consists of a synchronized stream of octets.

  The electrical signal hierarchy has *n*-members…

- **Synchronous transport signal level 1** (STS-1): with a basic speed of 51.84 Mbits/s, STS-1 signals are designed to carry DS-3 (synchronous or asynchronous) signals, or a combination of DS-1, DS-1C, and DS-2 signals that is equivalent to DS-3.
- **Synchronous transport signal level N** (STS-N): with speeds that are multiples of STS-1, i.e., n ×51.84 Mbits/s (where n may assume any integer value between 1 and 255), STS-N signals are created by interleaving N STS-1 signals, octet-by-octet. For various reasons, the values N = 3, 12, 24, 48, and 96, are preferred.

Corresponding to the STS signal hierarchy, the optical signals transmitted over the fiber facility are:

- **Optical carrier level 1** (OC-1): the optical equivalent of STS-1
- **Optical carrier level N** (OC-N): the optical equivalent of STS-N.

(2) STS-1 Frame

To achieve compatibility with existing systems, synchronous transport signals are divided into fixed frames of 125 μs duration. An STS-1 frame consists of 6480 bits or 810 octets. Shown in **Figure 8.23**, the octets can be arranged in a frame of 9 rows of 90 columns (with one octet in each location) that are divided as follows:

- **Transport overhead**: 9 rows x3 columns (27 octets) are occupied by transport overhead (TOH) information made up of 9 octets (3 rows by 3 columns) of *section* overhead (SOH) and 18 octets (6 rows by 3 columns) of *line* overhead (LOH).
- **Synchronous payload envelope**: the remaining 9 rows ×87 columns (783 synchronous payload envelope (SPE). Of the 783, 9 octets (1 column) are reserved for *path* overhead (POH) information.
- **STS-1 payload**: the remaining 774 octets (9 rows x86 columns) are payload.

Some of the functions performed by the overhead streams include:

- **Section overhead**: defines and identifies frames, and monitors section errors and communication between section terminating equipment.
- **Line overhead**: locates first octet of SPE (pointer function), and monitors line errors and communication between line terminating equipment.
- **Path overhead**: verifies connection, and monitors path errors, receiver status, and communication between path terminating equipment.

The lower part of Figure 8.23 shows the physical meaning of section, line, and path. The STS-1 line is deployed between STS-1 multiplexers. It consists of several *sections* that are bounded by repeaters or the multiplexers that terminate the line. On the *line*, the STS-1 signal may contain several tributary signals. In the Figure we use the example of a DS-1 signal. A *path* for this signal exists between the terminating M1 multiplexers. There are as many paths as there are tributaries in the STS-1 signal. Overhead octets are included in the STS-1 frame to supervise and manage events associated with the transport of user data at the section, line, and path levels. **Figure 8.24** shows the organization of octets in columns 1, 2, and 3 of two STS-1 frames. Also, Figure 8.24 shows that, in a dynamic situation, the SPE does not necessarily start in column 4, nor must it remain within one frame. This has an impact on the position of the path overhead octets. Shown in the fourth column of the frame in Figure 8.23 (the first column of the SPE), path overhead information floats across the frame to the column in which the SPE starts. The important point is that one octet of POH occurs immediately before every 86 octets of payload.

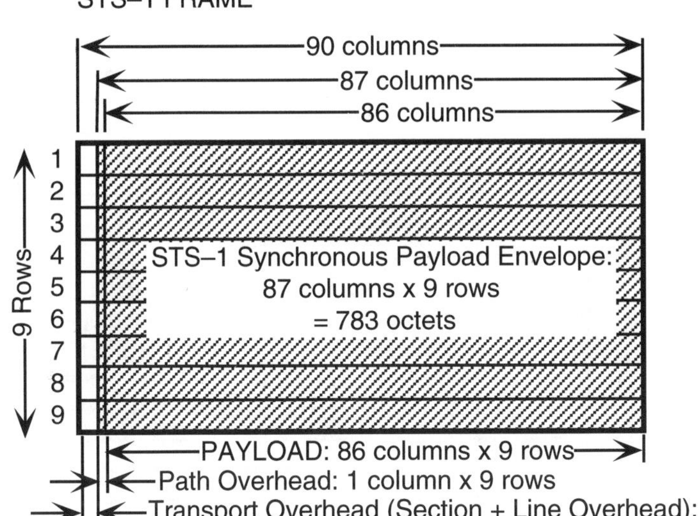

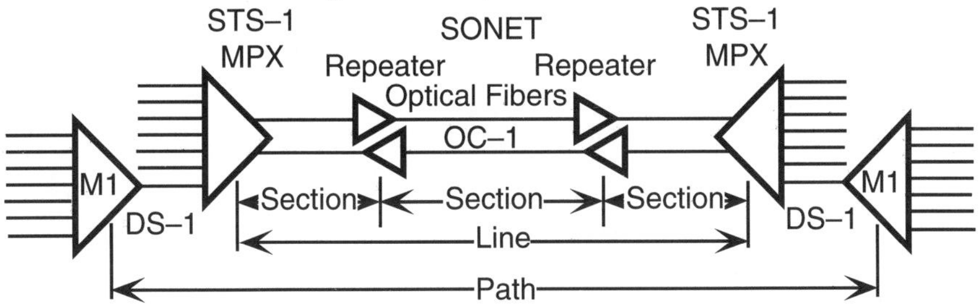

Figure 8.23 STS-1 Frame
At the first level, SONET employs a frame of 6480 bits that is repeated every 125μs.

Payloads whose sizes are less than an STS-1 payload (i.e., < 774 octets) are carried in special structures. Known as *virtual tributaries* (VTs), they consist of a fixed number of octets that occupy 9 rows ×*n* columns in the SPE. For instance, VT1.5, the virtual tributary for DS-1, consists of 27 octets (9 rows ×3 columns, equivalent to 1.728 Mbits/s). They are divided into 2 octets of VT overhead, 24 DS-0 octets (or 192 bits of unchannelized data), and, to preserve the information contained in SF or ESF framing sequences, 1 octet for framing. For European applications, VT2, the virtual tributary for E-1, consists of 36 octets (9 rows ×4 columns, equivalent to 2.304 Mbits/s). VT3, the virtual tributary for DS-1C, consists of 54 octets (9 rows ×6 columns), and VT6, the virtual tributary for DS-2, consists of 108 octets (9 rows ×12 columns). To make loading easier for payloads that

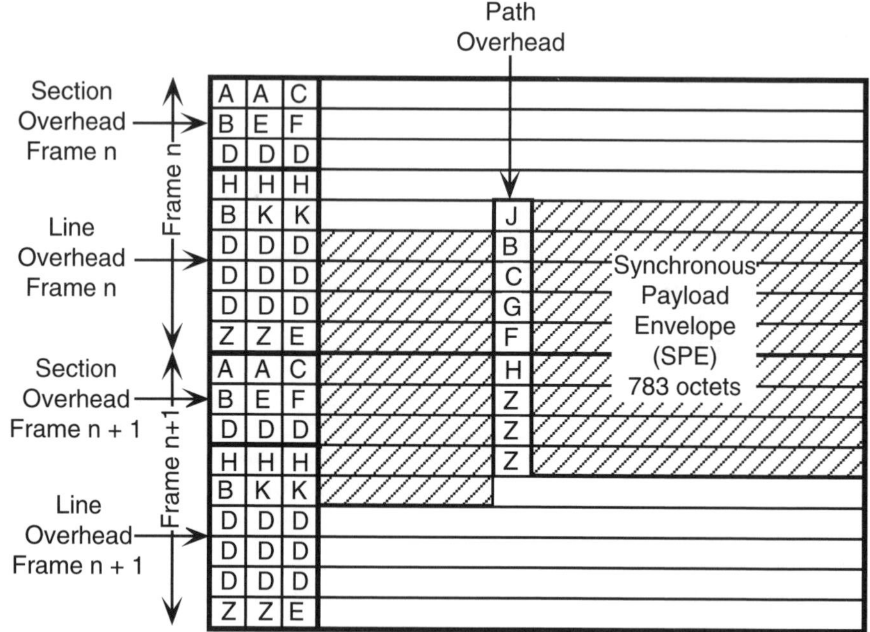

FUNCTIONS
A octets -- framing
B octets -- error monitoring
C octets -- identification
D octets -- data communication channels
E octets -- orderwire
F octets -- network uses
G octet -- receiver status
H octets -- pointer
J octet -- indicates path exists to receiver
K octets -- data communication channel
Z octets -- future growth

Figure 8.24 SPE Contained In Two STS-1 Frames

consist of a mix of VTs, the SPE can be divided into 7 equal compartments (9 rows ×12 columns) that contain 4 VT1.5s, or 3 VT2s, or 2VT3s, or 1 VT6 or a compatible combination of signals. For payloads of only one size VT, the VTs are allowed to float in the SPE. **Figure 8.25** shows the principle of loading a mix of virtual tributaries into the SPE of an STS-1 frame.

DS-3 signals are carried as STS-1 payloads. For SYNTRAN, it is a matter of mapping the signal (information octets + overhead bits) into the space available.

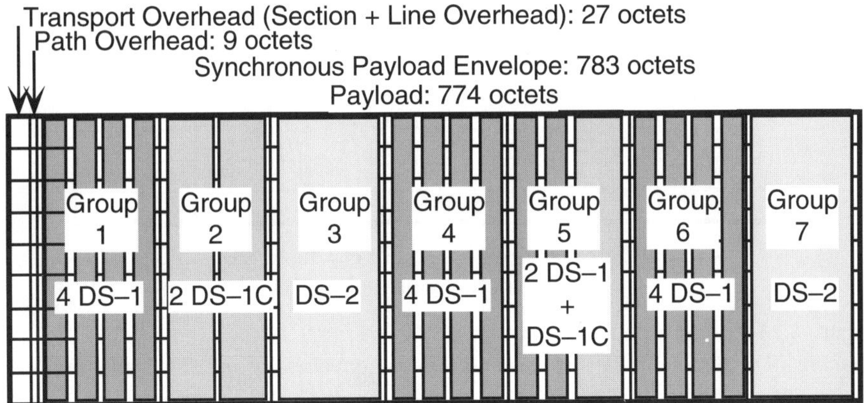

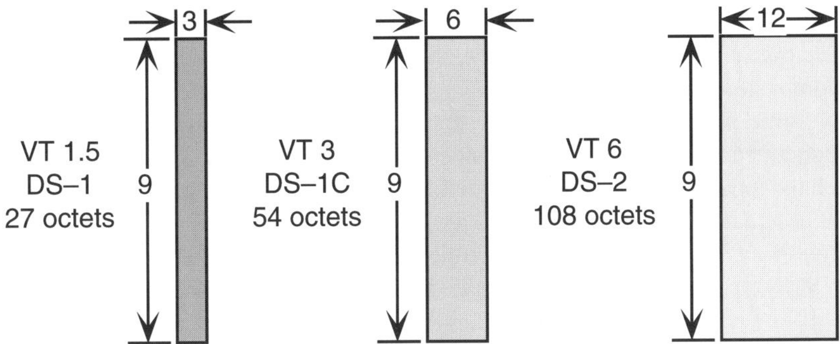

Figure 8.25 Virtual Tributaries In STS-1 Frame
Virtual tributaries are payloads (< 774 octets) that occupy 9 rows ×n columns. They are assembled in ways that optimize frame occupancy.

For traditional DS-3, the technique must compensate for timing variations due to the asynchronous nature of the signal. This is achieved by providing an asynchronous stuffing bit in each SPE row (in the Z octet row). In addition, fixed stuffing bits are added to fill the STS-1 frame. The arrangement is shown in **Figure 8.26**.

(3) STS-N Frame

Of 125 μs duration, an STS-N frame consists of N ×6480 bits arranged in 9 rows ×90N columns. STS-N signals are created by octet-multiplexing lower-speed STS signals. For example, 3, STS-1 signals are octet-multiplexed into an STS-3 signal,

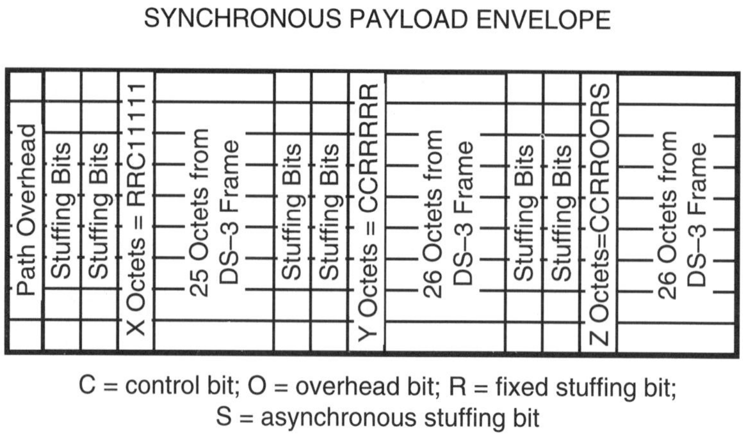

Path Overhead | Stuffing Bits | Stuffing Bits | X Octets = RRC11111 | 25 Octets from DS–3 Frame | Stuffing Bits | Stuffing Bits | Y Octets = CCRRRRR | 26 Octets from DS–3 Frame | Stuffing Bits | Stuffing Bits | Z Octets=CCRROORS | 26 Octets from DS–3 Frame

C = control bit; O = overhead bit; R = fixed stuffing bit;
S = asynchronous stuffing bit

Figure 8.26 Asynchronous DS-3 Frame in SPE
Placing a DS-3 frame in the synchronous payload envelope of an STS-1 frame requires the use of asynchronous stuffing bits (S) to compensate for timing variations and fixed stuffing bits (R) to fill the frame.

or 4, STS-3s are octet multiplexed into an STS-12 signal. **Figure 8.27** shows an outline of an STS-3 frame, and a multiplexing arrangement that will produce it. Before transmission at 155.52 Mbits/s, the entire frame is scrambled (see § 4.5.2) to randomize long sequences of zeros.

Some input signals are larger than STS-1—suppose such a signal is large enough to require an STS-N frame. Under this circumstance, the N STS-1 frames into which the input signal is mapped are defined as a *concatenated* structure, and the signal is designated STS-Nc. The frames are multiplexed and move through the network in the SPE of the STS-N frame as an entity. Octets H1 and H2 in the line overhead of each STS-1 frame are used to denote the concatenated condition. Because all the individual frames are part of the same signal and destined to take the same path, path overhead information is confined to the first STS-1 frame of the group.

8.3.2 Synchronous Digital Hierarchy (SDH)

To transport the broadband signals envisaged for BISDN (see § 10.7.1), ITU-T has defined a synchronous digital hierarchy. Compatible with existing facilities, SDH handles ISDN basic and primary rate signals [i.e., 64 kbits/s (DS-0) and 1.544 Mbits/s (DS-1) or 2.048 Mbits/s (E-1)], and SONET STS-N signals. SDH provides transport at these levels.

- **Synchronous transport module level 1** (STM-1): a basic speed of 155.52 Mbits/s, i.e., 3 ×STS-1
- **Synchronous transport module level N** (STM-N): speeds that are multiples of STM-1, i.e., N ×155.52 Mbits/s. STM-N frames are created by interleaving N STM-1 frames, octet-by-octet.

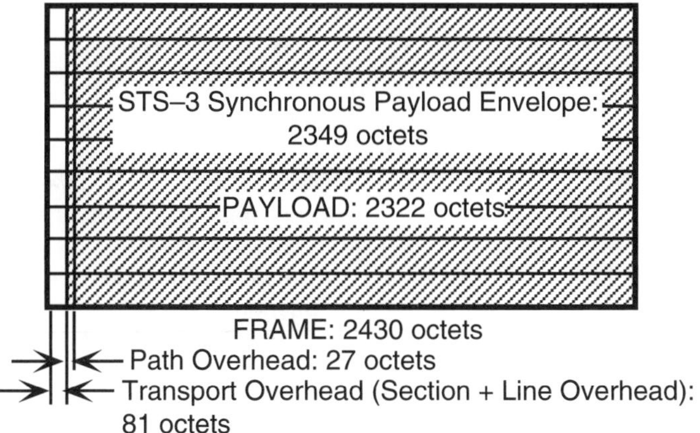

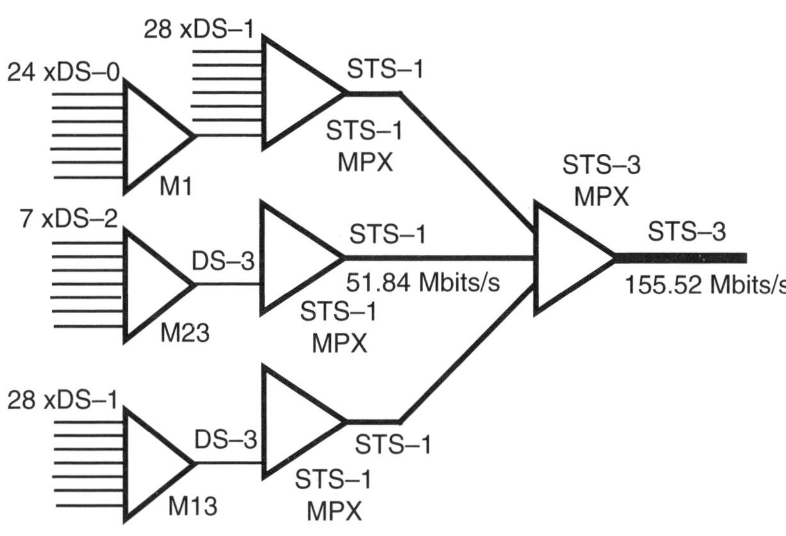

Figure 8.27 STS-3 Frame and Typical Multiplexing Arrangement

Table 8.1 compares levels and speeds in SDH and SONET.

(1) STM-1 Frame

The organization and structure of SDH are similar to SONET—unfortunately, some of the nomenclature is different. What is also different is that SDH is intended for use by microwave radio relays as well as optical fibers. SDH signals flow in fixed

Table 8.1 Comparison of SONET and SDH Transport Levels

RATE Mbits/s	SONET Electrical	SONET Optical	SYNCHRONOUS DIGITAL HIERARCHY	APPLICATIONS
51.84	STS-1	OC-1		*Feeder plant, distribution networks.*
155.52	STS-3	OC-3	STM-1	*Distribution networks, integrated digital loop carrier, thin interoffice routes, broadband UNI.*
622.08	STS-12	OC-12	STM-4	*Access rings, loop feeder, interoffice transport.*
1244.16	STS-24	OC-24	STM-8	*High traffic interoffice transport, interoffice backbone.*
2488.32	STS-48	OC-48	STM-16	*Interoffice rings, BISDN feeder.*
4976.64	STS-96	OC-96	STM-32	*Gigabit networks.*

length frames of 125 μs duration. An STM-1 frame consists of 19,440 bits that are transported at a speed of 155.52 Mbits/s. As shown in **Figure 8.28**, it consists of 9 rows of 270 columns with one octet in each location (2430 octets). In each frame, 9 rows ×9 columns (81 octets) are occupied by information related to the frame structure and information that facilitates frame transport from one node to the next (section overhead). The remaining 9 rows ×261 columns (2349 octets) are the payload. Made up of a limited set of structures that have a specific size and fixed relationship to each other, it contains user information, and information that defines the network path for each payload structure (path overhead). By way of illustration, in Figure 8.28 the payload consists of three equal structures called administrative units (see below). Compatibility with other network signals is achieved by:

- **DS-0**: transmitting frames every 125 μseconds.
- **DS-1** (North America and Japan): the 192 payload bits in a DS-1 frame and STM-related overhead bits (path overhead), are mapped into a 9 rows ×3 columns (27 octets) structure (24 octets for signals + 3 octets for overhead) that can be carried in the payload and delivered at a rate of 8000/s.
- **E-1** (Europe): an E-1 frame is mapped into 9 rows ×4 columns (36 octets, 32 octets for signals + 4 octets for overhead) that can be carried in the payload and delivered at a rate of 8000/s.
- **SONET**: the STS-1 SPE (783 octets) is exactly one-third of the STM-1 payload (2349 octets), and the speed of STM-1 (155.52 Mbits/s) is three times the speed of STS-1 (51.84 Mbits/s).

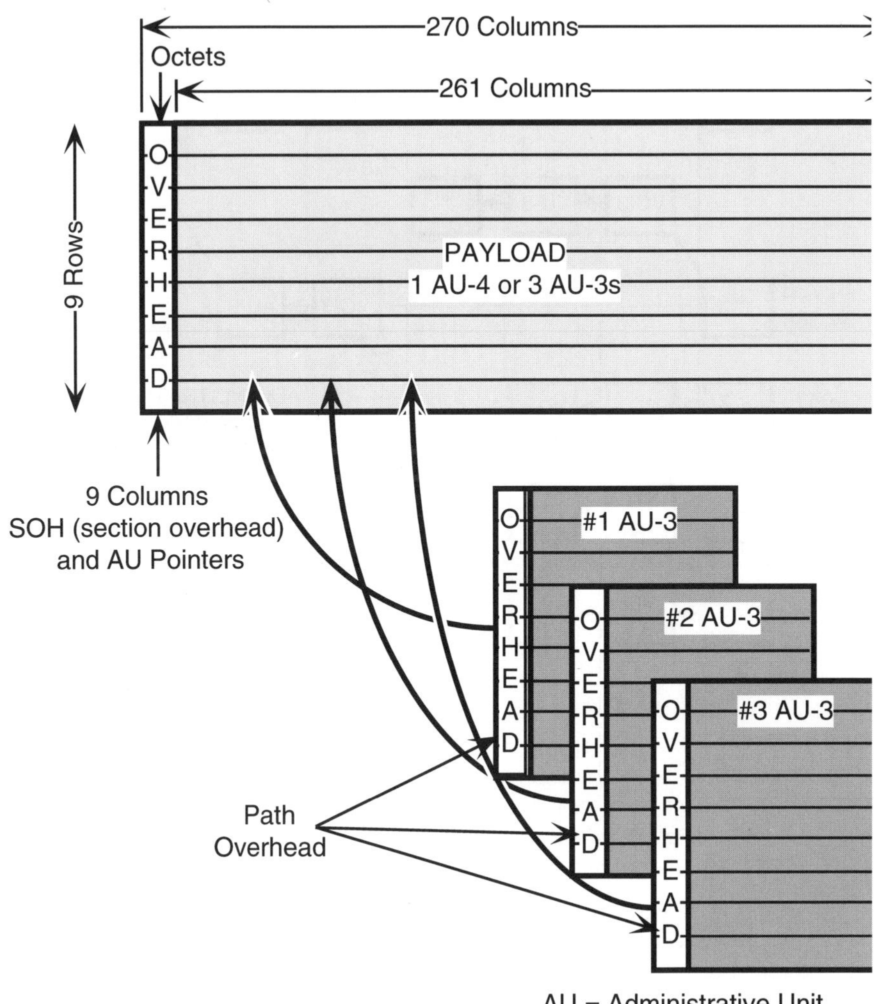

Figure 8.28 STM-1 Frame
At the first level, SDH employs a frame of 19440 bits that is repeated every 125μs.

(2) Creating STM-1

To create an STM-1 frame, a 125 μsecond segment of each *tributary* signal is mapped into a structure called a *container* whose size and designation (C-4, C-3, etc...) reflects the speed of the user's signal. For instance, if the tributary speed is 1.544 Mbits/s, the container (C-11) carries 24 octets, and if the speed is 44.736

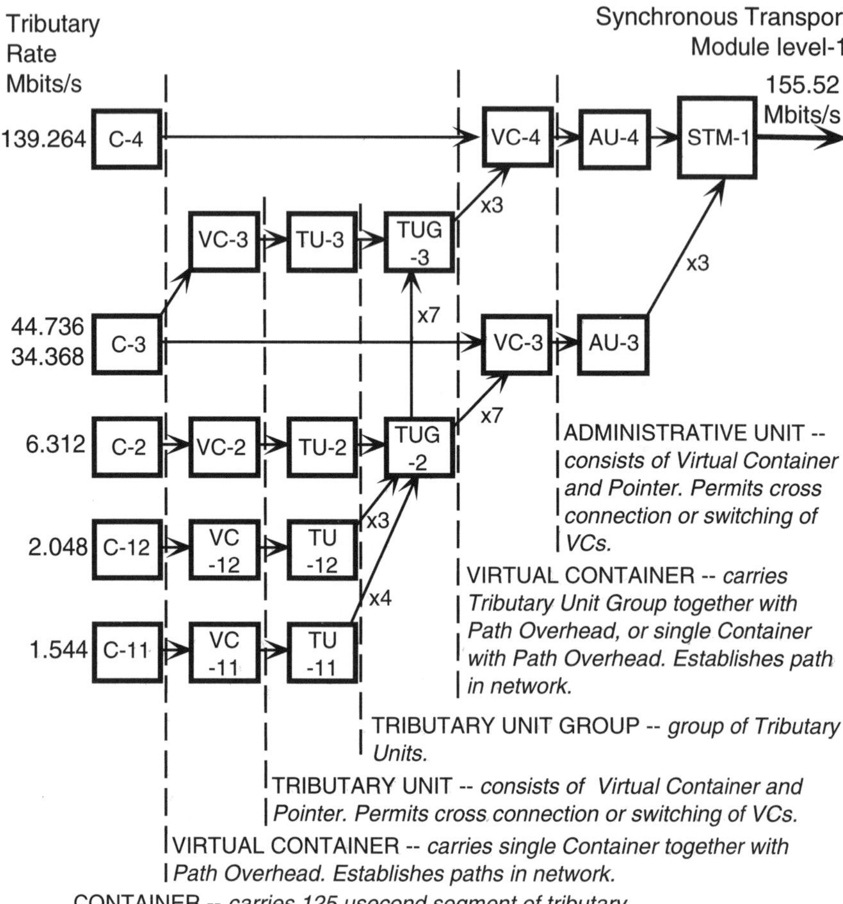

Figure 8.29 STM-1 Tributary Multiplexing Structure
The payload of an STM-1 frame can be made up from standard rate tributary signals.

Mbits/s, the container (C-3) carries 672 octets. As shown in **Figure 8.29**, the container is then joined to path overhead information to become a virtual container (VC). Next, to keep track of the start of the virtual container, pointer information is added, and the whole is identified as a tributary unit (TU). Continuing the multiplexing process, several TUs are collected together to form tributary unit groups (TUGs). In turn, several TUGs are loaded into another virtual container (at level 3 or level 4), more pointers are added to keep track of the segments, and administrative units (AUs) are formed. By ones (AU-4) or threes (AU-3), the AUs are formed into an STM-1 frame that is transported between nodes at 155.52 Mbits/s. Using the pointers to locate them, the segments of the payload are delivered to the receiving node.

9

Switching

Because it is not possible for users to have permanent connections to all those they might wish to call, communication networks consist of links interconnected by nodes. The nodes receive traffic from one link, create a temporary connection, and send it on another link so as to satisfy the users' immediate objectives. The work of the nodes is performed by switches. In this Chapter we describe the principles of the operation of circuit switches and packet switches including asynchronous transfer mode (ATM).

9.1 CIRCUIT SWITCHING

Circuit switches selectively establish and release connections between transmission facilities to provide dedicated paths for the exchange of messages between users. The paths are established before the information exchanges begin, and, until the users terminate the sessions, are maintained for their exclusive use. During each exchange, the only delay the parties will encounter is propagation delay. Before they began to converse, however, there was a network-related delay of several seconds while the caller entered the called number and the telecommunication facilities hunted for, and established, the required paths.

9.1.1 Hardware Model

To understand the performance of a digital end office, we need a model. **Figure 9.1** shows an arrangement that is typical of present day, Class 5, switching facilities

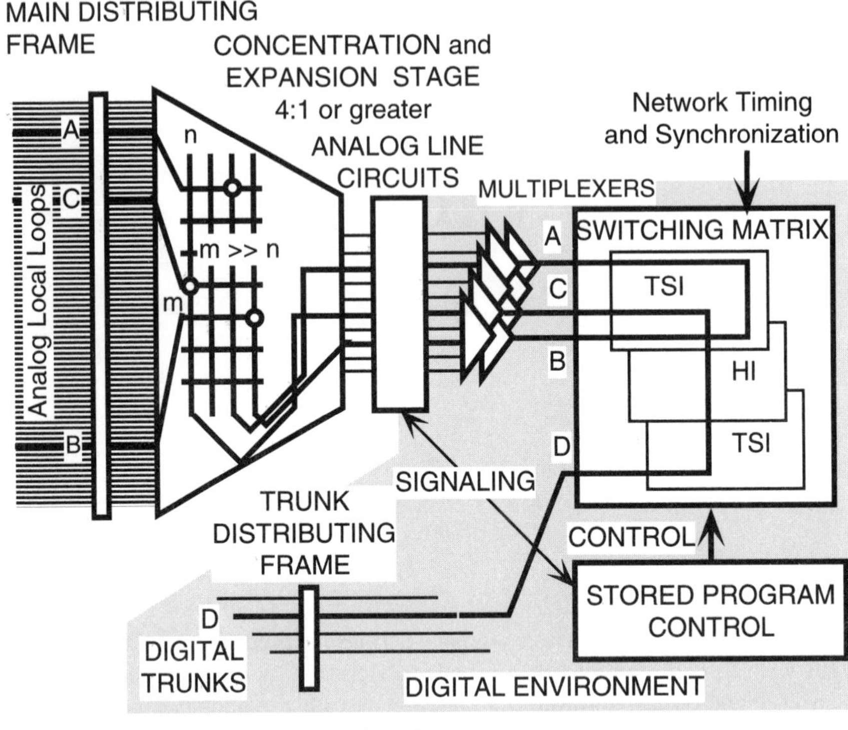

Figure 9.1 Elements of a Stored-Program Control Digital Circuit Switch
Increasingly, end office switches are digital machines. Most local loops are analog. On average, they are used 2% of the time.

that serve a customer community equipped with analog loops. The functions of the principal elements can be described as follows.

(1) Concentration and Expansion

At any specific time, the number of active loops is likely to be much less than the total number of loops installed in an exchange area. Accordingly, the number of circuits switched simultaneously is less than the total number of lines supported by the office. A concentration stage is employed to connect the active loops to the input ports of the switching stages, and to disconnect the arrangement as soon as the communication is over, to make the capability of the switching stages available to other circuits. In reverse, the concentration stage acts as an expansion stage to connect active lines from the switching stages to the loops that complete the calls. Figure 9.1 shows the concept of a concentration and expansion stage in the environment of a digital end office. The specific implementation involves a matrix in which connections can be made between horizontal and vertical members by energizing crosspoints. The number of horizontal links (m) corresponds to the number of loop appearances at the distributing frame; the number of vertical links (n) corresponds to the number of channels that can be handled simultaneously by the switching matrix ($m >> n$). In practice, to improve reliability and permit easy expansion of the system, the large matrix serving all $m \times n$ possible connections is subdivided into a set of identical, smaller matrices that each perform a fraction of the task.

(2) Analog Line Circuit

To be processed by a digital end office, analog signals must be converted to digital signals. An analog line circuit that performs this conversion, and the functions associated with powering, ringing, and testing the local loop, is shown in **Figure 9.2**. The subscriber line interface circuit (SLIC) is responsible for the 2-to-4 wire conversion function of the hybrid, for signaling detection, and for battery feed. In the PCM codec, the analog voice signal is converted to a 64 kbits/s, companded, PCM, unipolar signal for transmission to the multiplexer. The reverse process is employed to produce analog voice from 64 kbits/s, companded, PCM, unipolar signals received from the demultiplexer. Since most of the terminations in an end office are conventional, analog loops, the cost of the line circuits becomes a significant frac-

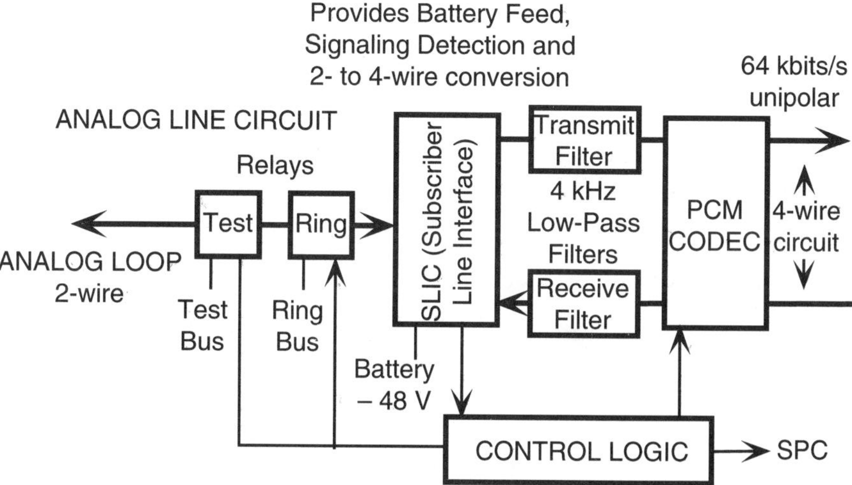

Figure 9.2 Analog Line Circuit
Makes the connection between analog loop and digital switch.

tion of the total cost of the switching system. The cost is increased by the need for them to withstand lightning surges and power line crosses, and to interface with a variety of station equipment.

(3) Multiplexer

Standard techniques are employed to multiplex 24 voice channels in a digital stream at 1.544 Mbits/s and demultiplex the 1.544 Mbits/s stream to 24 voice channels. The multiplexed streams are applied to the switching matrix shown in Figure 9.1.

(4) Switching Matrix

In the switching matrix, connections are made so that a circuit is completed between the calling and called parties. The stored program control seeks out a suitable combination of time-slots and multiplexed highways to perform this task, and instructs the hardware elements to employ them for the duration of the communication session. Usually, the switching matrix consists of sections devoted to time-slot interchanging and highway interchanging.

(5) Time-Slot Interchange Process

Time-slot interchangers are known as time switching stages and designated by the symbol T. The principle of time-slot interchanging is shown in **Figure 9.3**. Incoming frames on a multiplexed highway are read into buffer memories; one memory handles odd-numbered frames, the other memory handles even-numbered frames. While one frame is read into one buffer, the preceding frame is being read out of the other buffer. As shown in Figure 9.3, frames are read in as a sequence of 24 octets—one octet from each of 24 voice channels. Once the memory is filled, the octets are read out in a sequence defined by the stored program control. The result is an outgoing frame on the same multiplexed highway that contains the same octets as the incoming frame. However, the outgoing frame is delayed by one frame time, and the octets are in a different order. For small switching systems in which it is possible to multiplex all active lines on one highway, a single TSI stage is all that is required to interconnect the users.

(6) Highway Interchange Process

In addition to time-slot interchangers, larger systems employ highway interchangers. Described as time-shared, space-division switching that employs a space array, highway interchangers are known as space-switching stages and designated by the symbol S. The principle of highway interchanging is shown in **Figure 9.4**. Incoming highways are represented as the horizontal lines in a two-dimensional matrix; outgoing highways are the vertical lines. Incoming and outgoing highways are interconnected by gates that are activated by enabling signals (E). Thus, E(1,1) operates the gate that connects incoming highway (horizontal) 1 to outgoing high-

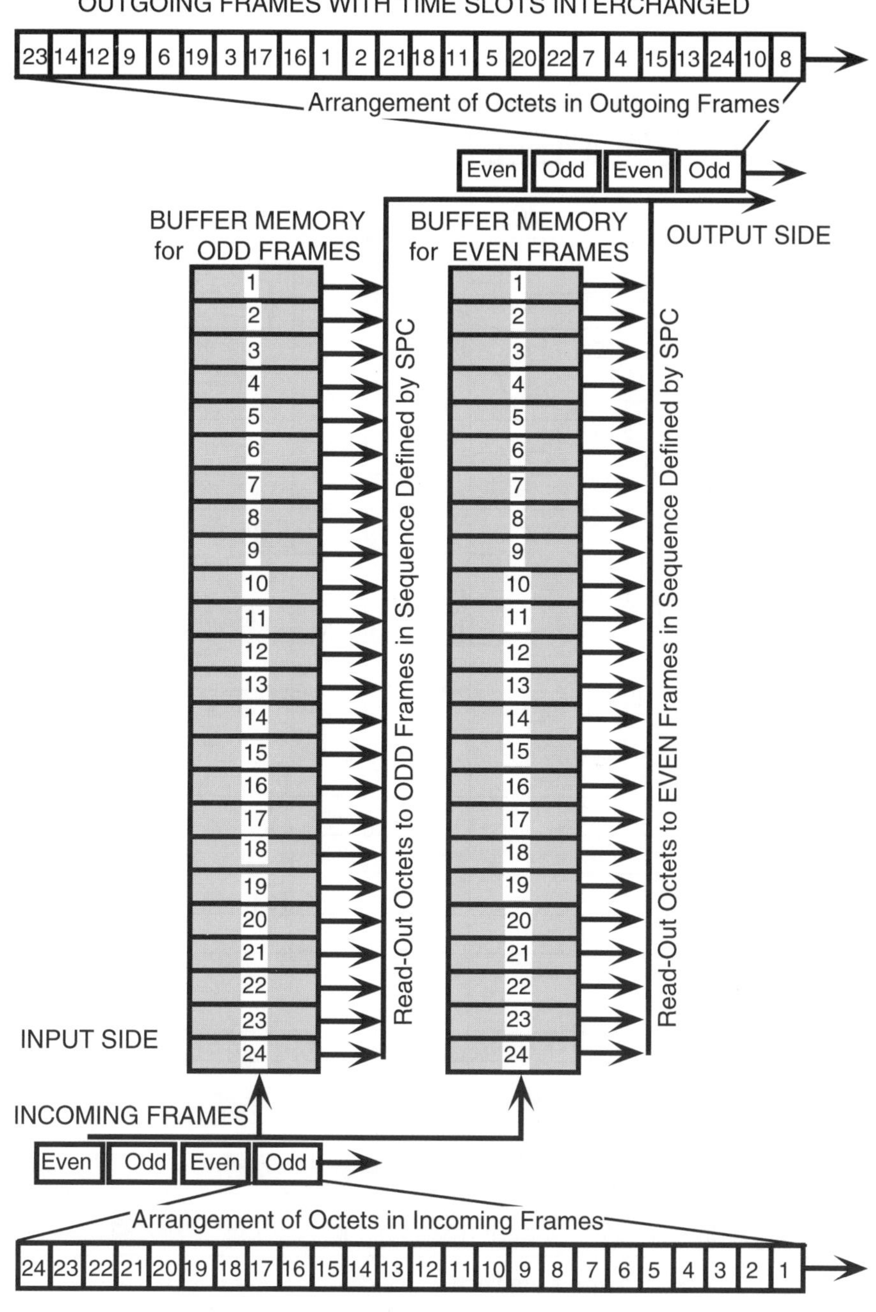

Figure 9.3 Principle of Time–Slot Interchanger
Time-slots are interchanged without changing the highway.

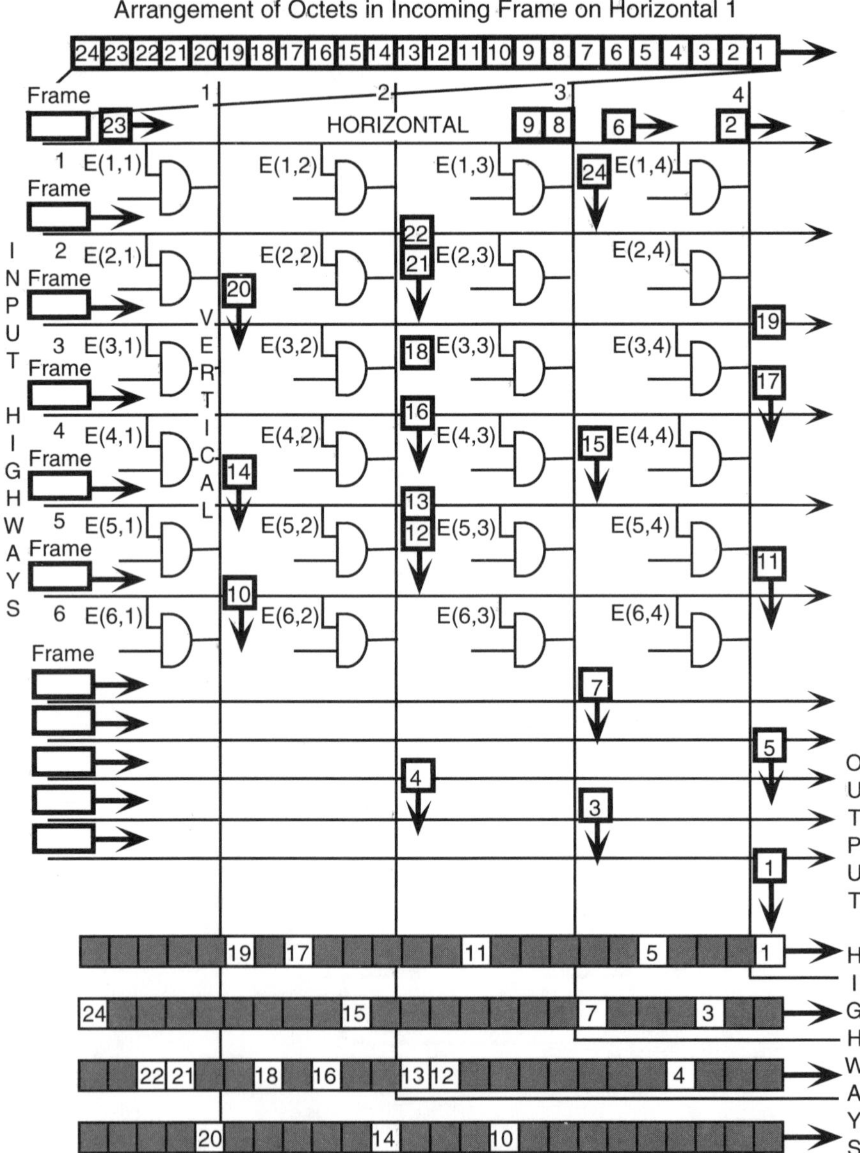

Figure 9.4 Principle of Highway Interchanger
Highways are interchanged without changing the time-slot.

way (vertical) 1. The enable signal is a pulse of octet length that is synchronized with the incoming octets to switch an octet-at-a-time on to different highways. Figure 9.4 shows the decomposition of an incoming frame of 24 octets on highway 1 into octets that are switched to different outgoing highways. Because there is no octet or frame buffering, there is no delay through the interchanger, and the octets leave the HI in the same order, and at the same time, as they entered; however, they are on different highways. The enabling signals are created in response to instructions from the SPC.

(7) Timing

For TSIs and HIs to operate successfully, the entire digital environment must perform in synchrony. It is vital that the processes change state at the exact instant that one octet ends and another octet begins. Achieving this condition requires that all operations be governed by a highly stable, digital clock and that adjustment procedures be included in those components in which delays occur.

(8) Time-Space-Time Switching Stages

To outfit practical, wire-center size, switches, combinations of time and space switches are used. A Time-Space-Time array is shown at the top of **Figure 9.5**. In this architecture, the space stage interconnects the time stage serving the calling party with the time stage serving the called party. The time stages rearrange the order of octets from the channels that serve the customer loops to match slots available on the internal channels associated with the space stage. The stored-program control does not associate the incoming and outgoing time-slots with any particular space stage time-slot, and any free slot may be chosen to create a connection. The path-hunting process is a matter of finding an available time-slot at each end of the space array. In the circumstance that a pair cannot be found, blocking ensues.

(9) Space-Time-Space Switching Stages

A Space-Time-Space array is shown at the bottom of Figure 9.5. In this architecture, the time stage interconnects the space stage serving the calling party with the space stage serving the called party. Because the calling and called parties define the external time slots (input of calling party and output to called party), the time stages must interchange them to create the connection. Flexibility is provided by the ability to choose among any of the TSIs that can perform the required slot interchange. In the circumstance that a TSI with the necessary capability cannot be found, blocking ensues. TST and STS networks can be designed with identical call-carrying capacities and blocking probabilities.

9.1.2 Software Model

The software that drives switching machines must exhibit a robustness, speed of operation, reliability, and extendibility that is several levels better than its data pro-

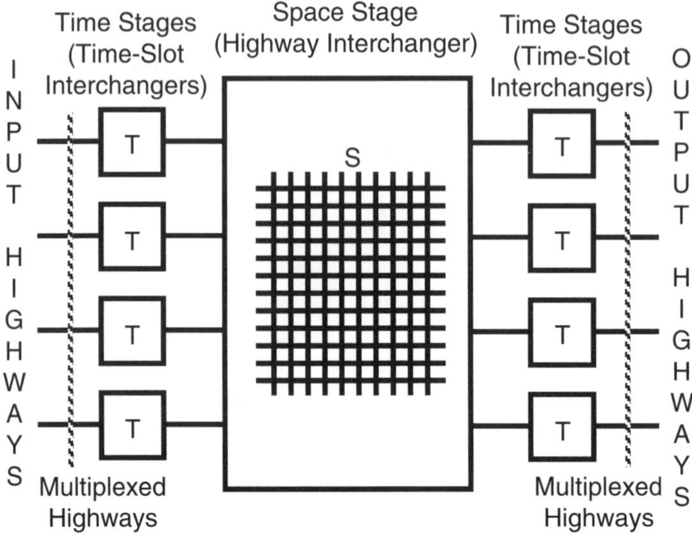

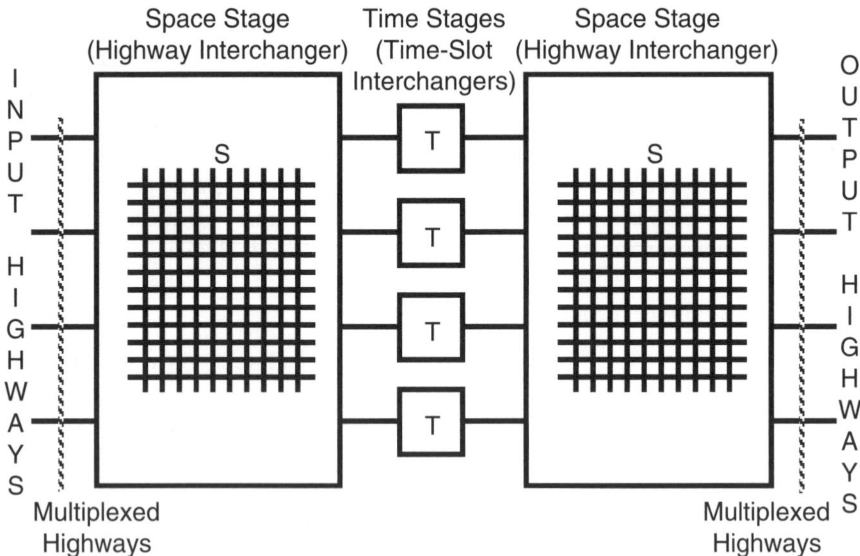

Figure 9.5 TST and STS Architecture

cessing cousins. Without such properties, it will be impossible to handle the real-time requirements of a continuous stream of call requests and circuit allocations and accommodate undefined future services. Switching software can be divided into two parts:

- **Generic program**: a general-purpose software package installed in all machines, it provides instructions that perform call processing, system maintenance, and system administration.
- **Office database**: collection of data that is unique to a specific switching location. The information on attached equipment and customers that it contains is used by the generic program to perform specific functions in the environment of a particular office.

(1) Call Processing

Call processing software initiates, maintains, and terminates a message path between calling and called parties. It consists of six basic steps:

- **Activity detection**: approximately every 100 milliseconds, scan programs test each customer line for signs of call origination, and each terminating trunk for signs of an incoming call. When the scan program detects a customer going off-hook, it stores the equipment number of the originating line (LENO, i.e., line equipment number—originating) in a service request buffer. At regular intervals, the executive control program reads the line equipment numbers requesting service and passes them to the origination program.
- **Number interpretation**: the origination program uses the equipment number to search the office database to identify the characteristics of the calling station (directory number, dial-pulse, or touchtone signaling, etc.), creates a call record in temporary storage, locates an idle digit receiver that matches the signaling characteristics of the station, finds a path through the network between the calling loop and the digit receiver, and connects them. Completion of this sequence is followed by return of dial-tone to the calling station and initiation of the digit scan program that scans the digit receiver for dialed digits, and collects them in a digit buffer. At this point, the executive control program invokes the digit analysis program to identify and validate the called numbers. When satisfied that the number has been received correctly and that it is a valid directory number, the program enters it in the call record.
- **Ringing and answering**: the termination program uses the directory number to search the office database to obtain information on the line equipment number of the loop serving the terminating station (LENT), and any associated routing instructions. Then, the terminating line is checked to determine it is not busy, and network paths are found between LENO and LENT and between the ringing circuit and the terminating line. If the terminating loop is not busy, and a network path is available between the loops serving the calling and called parties, ringing voltage is applied to the terminating loop and ringback is applied to the originating loop.
- **Talking**: the scan program is activated to detect an answer (going off-hook) from the called station, or a disconnect (going on-hook) from the calling sta-

tion. When an answer is detected, the executive control program disconnects ringing, establishes the talking path, notes the start of call in the call record, and releases the resources that have been used to establish the call.

- **Terminating**: the scan program maintains surveillance of the connection to detect a disconnect signal from one, or the other, party. At that time, disconnect programs return the resources to general availability.
- **Charging**: based on the call record, a charge is computed by the billing program (if appropriate).

(2) System Maintenance and Administration

Approximately 80% of the generic program is concerned with maintenance and administration functions. Among others, the functions are:

- **System maintenance functions**: designed to maintain the correct operation of the switching system, they include:
  - fault recognition and recovery procedures that detect improper hardware and software operation and enable the system to regain normal operation
  - congestion control and management techniques that enable the system to cope with traffic loads close to the ultimate capacity for which the system was designed
  - audit routines that appraise the use of memory and other resources and recover unused assets
  - recovery procedures that restore the system to normal operation when it encounters a major, system-affecting problem. They include reloading the entire generic program and office database from permanent storage, as well as procedures that contain less drastic options.
- **System administration functions**: designed to allow the operating company to manage the switching system as a network resource, they include:
  - collection of traffic engineering statistical information
  - office database change capability
  - verification of contents of system memory locations and office database entries.

(3) Office Database

The generic program is customized to each end-office environment by the office database. It contains two types of data:

- **Customer data**: customer-related information such as:
  - line equipment numbers for the loops that connect customers
  - directory number to line equipment number conversion
  - features available to each customer
  - identification of trunks available to remote offices

  - call-routing information
  - call-billing information.
- **Equipment data**: defines the hardware configuration of the end-office, including:
  - number of equipment frames, and number of circuits within each frame
  - switching network configuration
  - addresses of frames, circuits, and switching network components
  - size of memory, and other information relevant to processor capabilities and performance.

The uniqueness of each end-office is reflected in the size and variety of the entries in the office database. In designing its structure, care must be taken to preserve the independence of the generic program interface. Usually, this is accomplished by requiring entry through a master head table that resides in a fixed location in memory, and by structuring the database as a hierarchical array of lists accessed from this table.

9.1.3 Digital Cross-Connect Systems

- **Digital cross-connect system** (DCS): a computer-based facility that reroutes channels contained in synchronous transmission systems without demultiplexing.

DCSs are not switches in the sense that they set up and tear down circuits on a call-by-call basis. Used in relatively large digital trunking situations, DCSs perform as the digital equivalents of analog distribution frames. Connections are loaded into the connection map memory to meet the average needs of the network, and connections are changed by entering new information from a keyboard. In some systems, several maps may be stored in memory so that the channel distribution can be changed easily in response to recurring situations. Like a distribution frame, DCS permits reconfiguration of circuits; unlike a distribution frame, DCS operates within the multiplexed stream and performs the task of rerouting in an electronic matrix. Time-slot by time-slot, DCSs remap DS-0 channels among T-1 lines without demultiplexing. All channels must carry the same type of traffic and employ the same format. For instance, DS-1 lines operating with B8ZS, cannot connect with DS-1 lines using ZBTSI; DS-1 lines containing 56 kbits/s data channels cannot connect with DS-1 lines containing clear-64 channels, etc. DCSs are time-slot manipulators, not format changers. For this reason, DCSs are not applied to asynchronous T-2 and T-3 systems. Finding a specific DS-1 signal that is bit-multiplexed in a traditional DS-3 signal requires demultiplexing the stream. However, DCS can be applied to synchronous DS-3 signals and to SONET streams.

Figure 9.6 shows a simple application of DCS. T-1 carrier systems containing inter-LATA traffic originating from two exchanges, home on a digital cross-

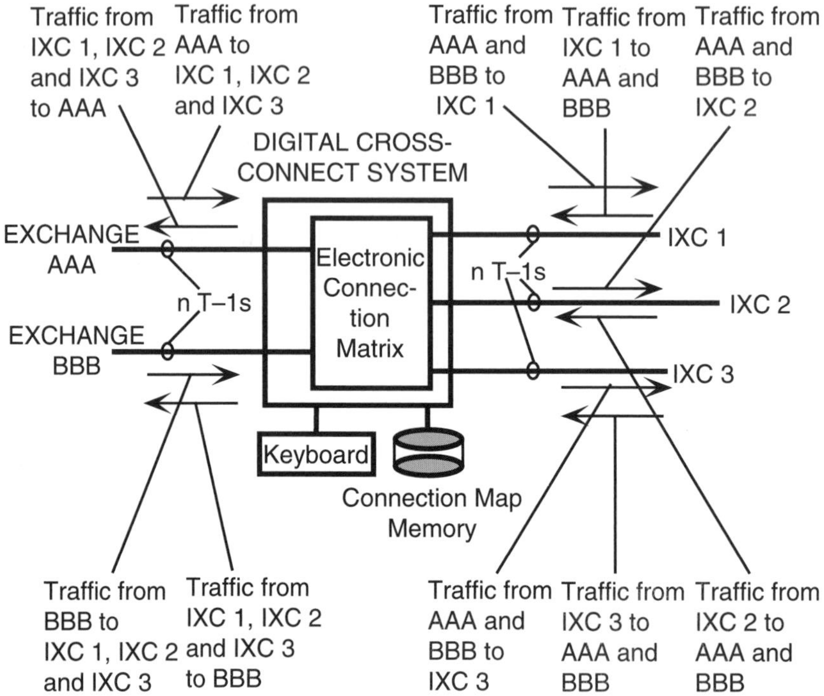

Figure 9.6 Digital Cross-Connect System
A DCS performs as the digital equivalent of an analog distributing frame. Connections are changed by entering relevant information at the keyboard.

connect system. On leaving the originating exchanges (AAA and BBB), the traffic is segregated in time-slot groups assigned to specific IXCs. At the DCS, traffic in the incoming time-slots is moved to outgoing time-slots in the multiplexed lines that serve the intended IXC. Traffic from the IXCs is routed to the intended terminating exchange in a similar manner. As each IXC's long-term traffic patterns change, the connection map is modified to accommodate them.

9.1.4 Synchronization

The deployment of digital switching and transmission facilities throughout public and private networks requires that all the transmission paths connected to each of the digital switches deliver signals that are timed to satisfy the requirements of switch operations, and are sequenced so that information exchanges occur between the correct pairs of senders and receivers. Conceptually, this is not a difficult task for one switch in isolation. However, when thousands of communicating public and private network switches are involved, digital traffic is passed between millions of pairs of users over transmission paths that vary significantly in length (and therefore propagation delay), and through equipment that may vary significantly in performance (and therefore introduce jitter and other uncertainties). Despite these perturbations, activities associated with each node and associated equipment must occur in intervals that begin at the same instant. This *must* be so if frames are to be reorganized and rerouted in a reliable manner. To do this requires that all nodes have access to sources that can report time in exactly 125 μ second intervals—the basic unit of time established by the sampling rate of digital voice. The practical situation calls for a network solution.

(1) Types of Synchronization

When digital signals are transmitted over synchronous systems, synchronization must exist at the frame, time-slot and bit levels:

- **Frame synchronization** is necessary so that the sender and receiver can agree on the beginning and ending of a group of associated bits, and proceed to identify the time-slots within the frame
- **Time-slot synchronization** is necessary so that the slots allocated to each sender/receiver pair can be identified correctly.
- **Bit synchronization** is necessary so that the receiver will reconstruct the same bit sequence as the sender transmitted.

(2) Hierarchical Synchronization

Practical network synchronizing techniques employ a hierarchical network approach. The levels are represented in **Figure 9.7**. They include:

- **Source nodes**: one, or more, network nodes serve as timing reference points. Known as source nodes, they are equipped with the most stable clocks, are in communication with receiver nodes equipped with less stable clocks, and, in large networks with more than one source node, each source node is connected to at least two other source nodes for coordination and security.
- **Level 1 receiver nodes**: the receiver nodes connected to the source nodes are equipped with less stable clocks. They employ the source nodes as timing references to synchronize (discipline) their clocks.
- **Level 2 receiver nodes**: these receiver nodes are connected to level 1 nodes. They employ level 1 nodes as timing references to discipline their clocks; and so on.

The accuracy of these clocks is expressed in terms of their frequency stability, and the number of *slips* they permit:

- **Slip**: event that occurs when a timing error causes a time-slot to be repeated or deleted.

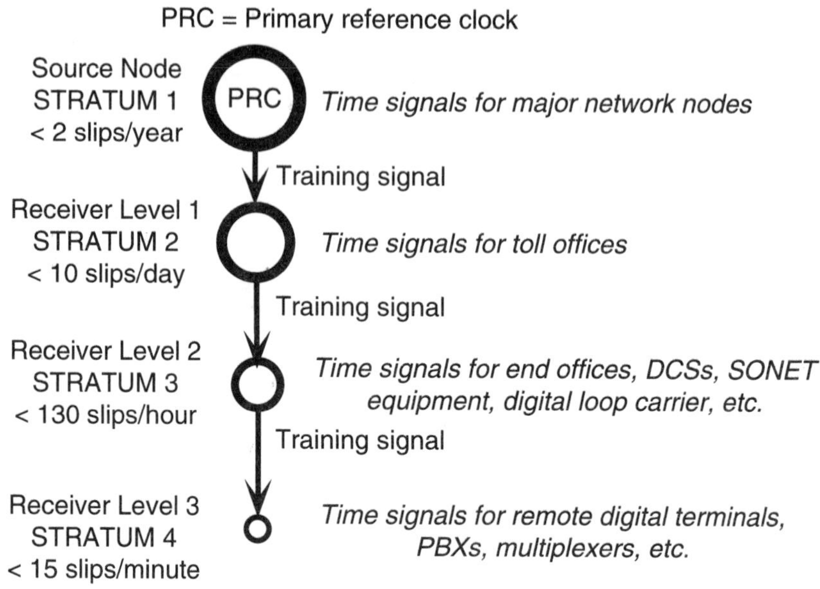

Figure 9.7 Timing Nodes for Network Synchronization

As more links are implemented by optical fiber circuits, the dominant source of errors is becoming the accuracy of the clocks that preserve the 125 μsecond rhythm of the network.

(3) AT&T Switched Network Synchronization

Through a network of 16 source nodes equipped with primary reference clocks (PRCs), AT&T provides timing and synchronizing services to its own digital network, to the digital networks of many local-exchange carriers, and to private digital networks. The clock hierarchy is as follows:

- **Stratum 1**: this group of 16, free-running, PRCs consists of an ensemble of primary or secondary atomic standards that achieves a frequency accuracy of better than 1×10^{-11}, and permits no more than two slips per year. Fourteen of the clocks are located in the continental network; the others are located in Puerto Rico and Hawaii.
- **Stratum 2**: disciplined by the PRCs, these clocks are located in toll offices. They achieve a frequency accuracy of better than 1.6×10^{-8}, and permit no more than ten slips per day.

- **Stratum 3**: disciplined by clocks in Stratum 2, these clocks are located in local networks. They achieve a frequency accuracy of better than 4.6×10^{-6}, and permit no more than 130 slips per hour.
- **Stratum 4**: disciplined by clocks in Stratum 3, these clocks are located in private networks. They achieve a frequency accuracy of better than 3.2×10^{-5}, and permit no more than 15 slips per minute.

9.2 PACKET AND CELL SWITCHING

9.2.1 Packet Switching

Taking advantage of the bursty nature of data communication activities, packet switches statistically multiplex packets of data (see § 6.2.2) on network links.

- **Packet switch**: a network node that receives packets, performs error and congestion control, may store the packets briefly, and sends them on toward their destination over routes determined by information carried in the packet, by information stored at the node, and by the state of the network.

A packet switched network has been described as a network of queues. Each packet switch consists of controllers, databases, and queuing elements designed to make the receipt, holding, and forwarding of packets as rapid and as reliable as possible. In **Figure 9.8**, we show the principle of a packet switching node. Each port consists of buffers in which packets are queued to await processing according to data they carry and data stored in the switch. The network map provides the node with information on the state of the network; routing information is contained in a database; and port status is available to all controllers.

Classical packet switching employs the Balanced Link Access Procedure (LAP-B, see § 6.1.4) at layer 2. Over the links between terminals and network, LAP-B functions as follows:

- Recognizes flags (to define frame limits)
- Executes bit stuffing (to achieve bit-transparency)
- On the transmit side, generates frame check sequences (FCSs, see § 4.2.2(8))
- On the receive side, confirms FCSs
- On the receive side, acknowledges receipt of frames and requests retransmission to correct errored frames (go-back-n or selective repeat ARQ, see § 4.2.3(1))
- On the transmit side, repeats unacknowledged frames
- Performs flow control
- Using logical channel numbers, multiplexes packets over the links

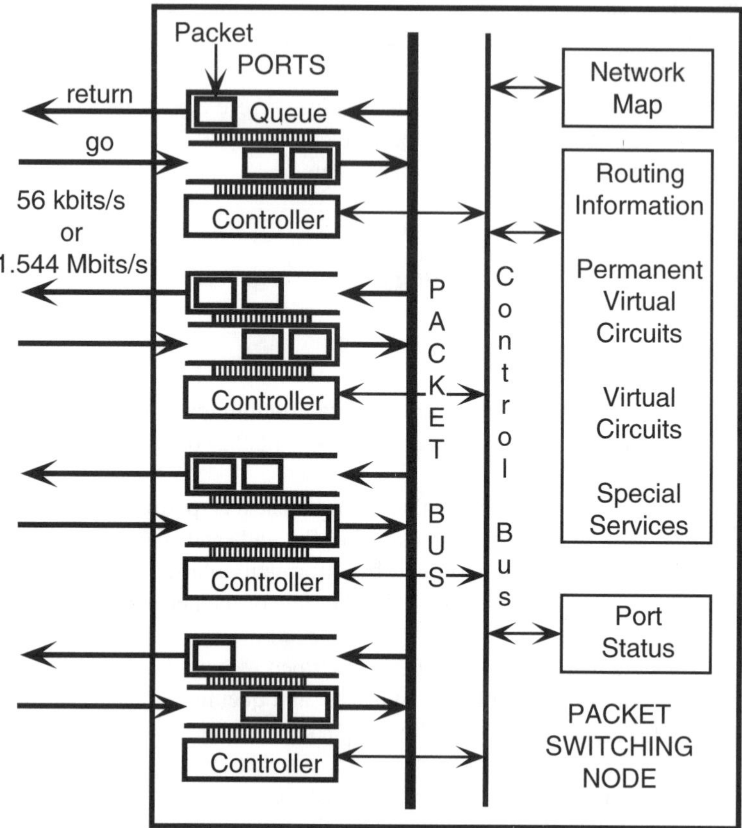

Figure 9.8 Principle of Packet Switching Node
Each port buffers incoming packets and routes them in accordance with information they contain.

The decision to forward a packet over one link or another (a layer 3 action) involves the processing of information carried in the three octets that comprise the packet header (see Figure 6.6). For instance:

- If it is a *call request packet*, forwarding will depend on its destination and the best link available to facilitate migration toward this destination at the time the packet is received. The number of the link selected, and the logical channel number carried by the packet will be stored in the switch directory (routing information). The presence of a call request packet implies a connection-oriented service is being used and establishes a virtual circuit.
- If it is an *information packet* carrying a logical channel number that *is* stored in the directory, it is part of a connection-oriented service over a virtual cir-

cuit. It will be routed over the same link as the call request packet that preceded it.

- If it is an *information packet* carrying a logical channel number *not* stored in the directory, it is part of a connectionless service. It will be forwarded over the best available link without regard to the way any other packet has been forwarded.
- If it is an *information packet* that carries a logical channel number that is associated with a permanent virtual circuit, it will be afforded priority treatment over the assigned link.

These examples allow us to distinguish two methods of packet routing:

- **Distributed routing**: the node decides which path the packet shall take to its destination on the basis of information about traffic conditions and equipment status (network map, port status). Each packet is routed without regard to packets that may have preceded it. The node's decision is based solely on the situation existing at the time of packet arrival.
- **Centralized routing**: a primary (and perhaps an alternate) path is (are) dedicated to a pair of stations at the time of need. For the sequence of packets that constitutes a block of user's information, the same primary (or alternate) *en-route* resources are used by each packet. Information on all possible routes that can be assigned is stored in a database (routing information). Routes established in this manner are known as *virtual circuits*.

In addition, circuits comparable to private lines are employed:

- **Permanent virtual circuit**: a virtual connection that is permanently assigned between two stations. Fixed logical channel numbers are used by the sending DTE and the receiving DCE. Because it is a permanent (virtual) connection, no connect or disconnect packets are required to set up the circuit, and information packets may be transmitted at will by the originating DTE.

We have more to say on X.25 packet nodes, and their integration into networks in § 12.1.1.

9.2.2 Frame Relay

- **Frame relay**: a technique that transports variable length frames across a modified X.25 interface at speeds faster than conventional X.25 services provide.

The frame relay UNI (user-network interface, also known as FRI, frame relay interface) contains a data link layer and physical layer only. The packet layer of X.25 is eliminated so that user information streams up to 8192 octets can be transmitted intact. Individual frame addresses (for statistical multiplexing) are added to the data link layer (X.25-2). Error detection (but not correction) is performed, and corrupted frames are discarded (without requesting retransmission). No call set up mechanism exists. As a result, frame relay supports permanent virtual circuit operation only. Nevertheless, it is an important technique for the rapid transport of long streams of user data between specific nodes in private networks. The sender uses a 10-bit data link connection identifier (DLCI) to specify the permanent virtual circuit that connects to the receiving equipment. At the network node, as soon as the FR nodal processor reads the DLCI, it prepares to forward the frame to the next processor. The frame may be kept intact, or fragmented into cells. If cell relay is employed across the NNI (node-network interface), the nodal processor handles the frame in cell-size segments that are routed using local virtual circuit identifiers (VCIs). As soon as a segment is ready for transmission, it is forwarded to the next node. At the nodal processor that serves the receiver, the segments are reassembled and the frame is passed to the receiver. Congestion control is implemented by bits in the address octets. They are used by a nodal processor that is approaching capacity to notify the sending and receiving processors of that fact. We pursue this topic further in § 12.1.3.

9.2.3 Fast Packet Switching

- **Fast packet switching** (FPS): techniques for transporting data, voice, or video messages in a fixed length packet.

FPS techniques do *not* support data transport across an X.25 interface at speeds faster than conventional X.25 switches! Rather, they support the transport of fixed length cells (packets) that can contain data, voice, or video information. FPS employs statistical multiplexing across the NNI. FPS techniques are *fast* because the cells can be routed using hardware that recognizes the destination address, and error control is applied only to the entire message, not to each cell link-by-link (i.e., error control is left to the transport layer). While several FPS techniques have been developed by individual switch vendors, ATM (asynchronous transfer mode) is now the preferred technique.

9.2.4 Asynchronous Transfer Mode

- **Asynchronous transfer mode** (ATM): a high-speed (155 and 622 Mbits/s), fast packet switching technique employing short, fixed length cells that are statistically multiplexed over virtual connections. Each cell consists of a 5 octet header and a 48 octet payload (data, voice, or video).

The short cell reduces delay and jitter, and the fixed length permits synchronous operation of the switch fabric. Before information is accepted from a source, a virtual circuit must be established to the receiver. If resources are not available, the request for service is refused.

(1) ATM Switch Components

Figure 9.9 shows the major components of an ATM switch. The switching matrix, or switching *fabric* as it is called, operates synchronously. The ports receive or send bits at speeds of 155 or 622 Mbits/s (STS-3 and STS-12, or STM-1 and STM-4). The components perform the following tasks:

- **Input port controllers**: manage the input streams; among other things, they:
  - may buffer incoming cells
  - align cells for switching
  - on the basis of information in the headers, identify output ports and establish a path across the fabric for the cells, or attach self-routing labels to guide the cells across the switching fabric.
- **Output port controllers**: manage the output streams; among other things, they:
  - strip off self-routing labels
  - may buffer cells awaiting transmission
  - align cells and transmit them to the next node.
- **Switching fabric**: routes incoming cells to the correct output ports. The structure of the switch fabric varies with the size of the switch. We describe some of the possibilities below.
- **Control processor**: directs the operation of the switch. To achieve very high-speed operation, as many of the logical and control operations as possible are

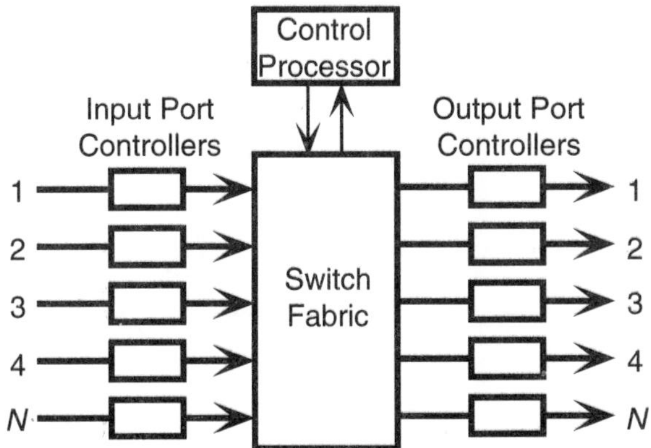

Figure 9.9 Principal Components of ATM Switch
The switching fabric operates synchronously. The ports receive and send 53 octet cells at 155 or 622 Mbits/s.

performed in hardware. The control processor contains a network database (with virtual circuit and other information), and performs system maintenance and administration tasks.

(2) Switching Fabrics

For *small* ATM switches—the sort that are deployed in private networks—a common time-divided structure is employed as the switching fabric. Since every incoming cell occupies the fabric for a short time during which it is available to all outgoing ports, the structure can be used for point-to-point and point-to-multipoint (multicasting) operation. Two popular structures are:

- **Time-divided bus**: all incoming cells flow over the shared bus. Its speed determines the maximum speed of the switch, and access to the bus is granted to each input port, in turn. Operation has been reported at speeds over 20 million cells per second with this type of fabric. The concept is shown in **Figure 9.10** and is similar to the packet switch shown in Figure 9.8.
- **Multiplexing and storage**: each cycle of incoming cells occupies the common storage for a slot time. They may be read out in any order to the output ports. Operation has been reported at speeds over 10 million cells per second with this type of fabric. The concept is shown in Figure 9.10. In principle this technique is similar to the time-slot interchanger shown in Figure 9.3.

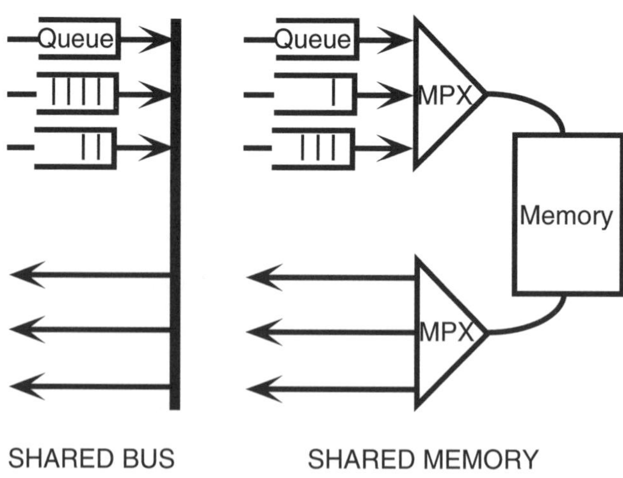

Figure 9.10 Time-Divided Switch Fabrics
Small ATM switches employ a common time-divided structure as the switching fabric.

For *large* ATM switches—the sort likely to be deployed in public networks—the switching fabric consists of elementary switching units (switching elements) that are connected in a specific manner. The network is implemented with solid-state switching elements to provide one, or several, non-blocking paths between ports. At regular intervals (known as *slot* times) the switch fabric transfers one cell from as many of the input controllers as have received cells to the appropriate output controllers. Paths across the network are selected when calls are set up, or the cells are *self-routing*. In this technique, the input port controller adds a tag to each arriving cell that identifies the port to which it must be delivered. The switching elements in the fabric route the cells by inspecting their tags. If a number of cells, destined for the same output port, arrive simultaneously at different input ports, the output processor must be fast enough to handle them all in one slot time, or some must be delayed or discarded. Several structures are of interest for the large switch application. Dividing them into single path and multiple path fabrics, we will describe some of them briefly:

- **Single path switching fabrics**: a unique path is available between each input/output pair:
  - *fully interconnected mesh*. Each input port is permanently connected to all output ports. Because it increases in complexity in proportion to the product of the number of input and output ports (i.e., N^2), this network design is not economical for very large switches. However, fully interconnected mesh structures can be used in moderate size, non-blocking, self-routing switch networks, and in switching elements that form other fabrics.
  - *crossbar* or *matrix*. The connection between individual input/output pairs is made by closing a gate to connect the input horizontal to the output vertical. The arrangement is similar to the matrix of the highway interchanger shown in Figure 9.4. Because it increases in complexity as the product of the number of input and output ports (i.e., N^2), this network is not economical for very large switches. However, crossbar and matrix structures can be used in moderate size, non-blocking, self-routing switch networks, and in switching elements that form other fabrics.
  - *Banyan*. A family of self-routing networks composed of 2×2 switching elements. With a complexity that increases as $N \log N$, it is better suited to large networks—but it is a blocking network. If switching elements larger than 2×2 are employed, performance is improved but blocking remains a problem. However, if the incoming cells are sorted in the order of their destination ports and presented on adjacent network inputs, the network is non-blocking, provided that multiple inputs are not destined for the same output at the same time.
  - *Batcher-Banyan*. A network that sorts incoming cells on the basis of their output port sequence (Batcher), connected by a *perfect* shuffle, interconnection network to a self-routing network of 2x2 switching elements (Banyan). The addition of the Batcher network significantly increases the

complexity of the whole. A Batcher-Banyan network is shown in **Figure 9.11**. To overcome the problem of multiple inputs seeking the same output port at the same time, all but one of the interfering inputs must be buffered, otherwise delayed, or destroyed.

- **Multiple path switching fabrics**: several paths are available between each input/output pair:
  - *augmented Banyan*. Adding switching stages to a simple Banyan network produces a multiple path, self-routing fabric.
  - *Benes*. A particular case of an augmented Banyan network.
  - *Clos*. A non-blocking arrangement of switching elements and links that has long been used in telephone switches. It may be the technique of choice for very large networks.
  - *parallel switching fabrics*. Single path or multiple path switching fabrics can be arranged in parallel to create multiple path networks. They have the advantage of redundancy so that the failure of a plane is inconvenient, but not catastrophic. Also, as described below, they can be connected so that more than one cell can be delivered to an output port at the same time.

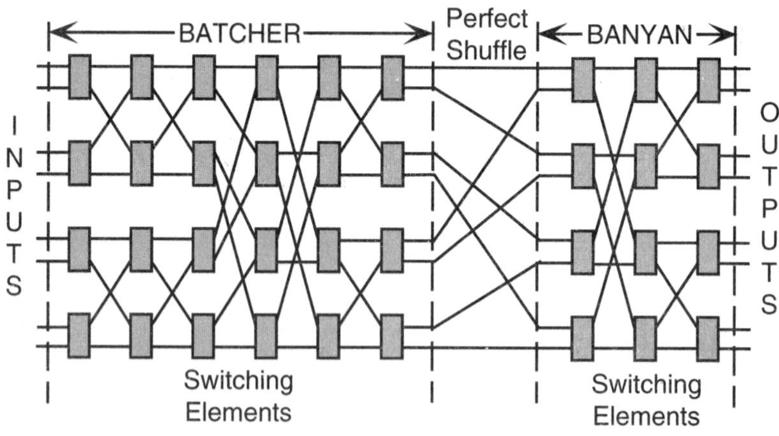

Figure 9.11 Batcher-Banyan Switching Fabric

(3) Performance Requirements

As many as a dozen ATM switch designs are being pursued by major research and development organizations. In common, they are seeking to fulfill the following requirements:

- **Information rates**: varying from as little as 100 bits/s to more than 100 Mbits/s, inputs may be synchronous or asynchronous.
- **Multiple receivers**: to implement electronic mail or video distribution services (for example), one-to-many, one-to-all, and many-to-many connections are required. Implementing these multicasting and broadcasting services requires a *copy* capability within the switching fabric.
- **Blocking**: before a message is sent, resources must be available to complete the transfer. If they are not, the call attempt will be refused (blocked). If the switching fabric is non-blocking, temporary congestion can be accommodated by queues at the input and output. If it is blocking, temporary congestion may require internal queues.
- **Delay**: when queues are present, network transit time depends on the cell arrival time distribution, the number of cells waiting for service and the service time. The difference in cell delay between light traffic conditions and congested conditions can be large and will affect channel throughput significantly. Under such conditions priority handling may be necessary for synchronous traffic.
- **Cell loss**: when traffic to a port exceeds the capability of whatever queues are available, cells are lost. Also, when traffic is misrouted, i.e., delivered to the wrong port, the cells are lost. Service integrity requires that cell loss be ≤ 1 cell in 10^8 cells.

(4) ATM Switch

Figure 9.12 is a high-level diagram of an ATM switching fabric that incorporates a sorting network, parallel routing networks, and several queues. Queues of cells may be formed at each input port (to buffer incoming cells), at the input to the sorting stage (to buffer arrays of cells that will be routed each slot time), at the outputs of the routing stages (to accommodate several cells seeking the same output port), and in the recirculation buffer (to accommodate those cells seeking the same outputs that cannot find room in the output buffers). The sorting network is likely to be a Batcher network, and the routing networks will be self-routing and based on a Banyan, or augmented Banyan network. Not all of the buffers shown will be implemented in every design. Those cells that cannot be buffered or delayed are destroyed. We pursue applications of ATM and cell relay in §s 10.7.2 and 13.3.2.

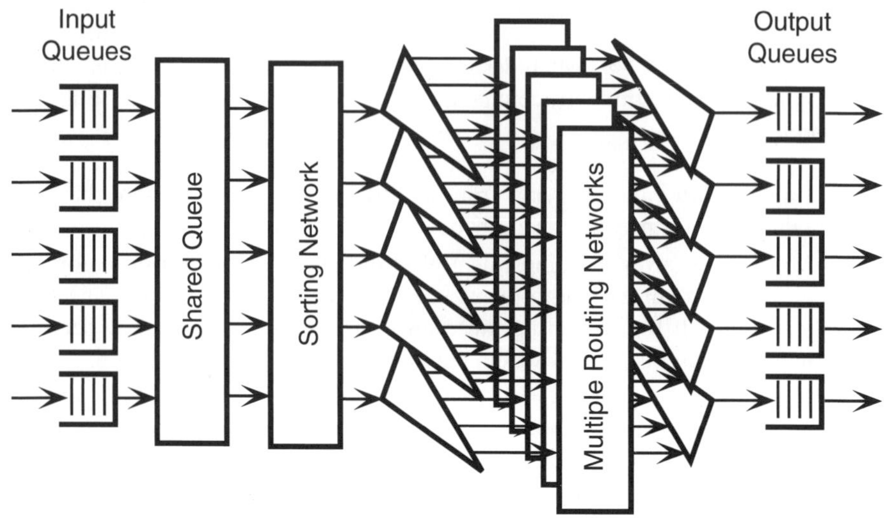

Figure 9.12 Principle of Self-Routing ATM Switch
High-level diagram of ATM switch consisting of queues, sorting network, and parallel routing networks.

10

Public Switched Telephone Network

As soon as the first installations demonstrated the feasibility of real-time communication between separated persons, the demand for telephone service created the telephone *exchange*. Staffed by human operators, it became a daunting assembly of plugboards and patch cables that were connected on-demand to provide talking paths between subscribers. Eventually, automatic switches were perfected and the *public switched telephone network* (PSTN) was born. In this Chapter, we describe the modern-day PSTN and take a look at the broadband future.

10.1 SIGNALING

Described as the command and control component of a telecommunication network, signaling is the common thread that unites all the companies participating in the PSTN, and their customers. It is the name given to the exchanges of specialized messages that inform the users, and the facilities, of what actions to take to set up, maintain, terminate, and bill a call.

10.1.1 Types of Signaling

For convenience, we distinguish between two types of signaling:

- **Station signaling**: the exchange of signaling messages over local loops that connect the users' stations to the serving switch; station signaling is limited to an exchange area

- **Inter-office signaling**: the exchange of signaling messages between switches that serve the routes between calling and called parties; inter-office signaling occurs between the switches that are involved in completing a call within the LATA, or completing a call through IXC facilities and those of an LEC in another LATA.

  Signaling messages are of four kinds:
- **Alerting**: indicating a request for service (such as the caller going off-hook, or the terminating office ringing the called party)
- **Supervising**: providing call status information so that the users and/or the intermediate facilities can set up, maintain, and terminate the call (such as a station busy signal or a ringback signal)
- **Controlling**: providing tones, announcements, and other information related to the call (such as the tone that indicates ready to receive credit card number, and announcements such as *number changed to…*, or *number no longer in service*)
- **Addressing**: providing the called number, and other routing-associated information [such as 1 + (area code) + (7-digit number)].

10.1.2 Telephone Numbering Plan

So that persons may telephone others around the world, ITU-T maintains a universal numbering plan in which customer stations are identified by subscriber codes, exchange codes, area codes, and country codes. The range of digits in each of these codes is shown in **Table 10.1**. However, national telephone authorities have modified it to accommodate local numbering plans and no country employs all thirteen digits. For instance, the country code for the United States is 1 (one-digit, not three-digits); the country code for Brazil is 55 (two-digits, not 3-digits) and the area code for Sao Paulo is 11 (two-digits, not 3-digits); and, while the country code for Ireland is 353 (3-digits), the area code for Dublin is 1 (one-digit, not 3-digits) and the area code for Galway is 91 (two-digits, not 3-digits). Some general principles apply to the codes:

- **Country code**: ITU-T divides the world into nine areas numbered 1 through 9. Countries have country codes beginning with the digit assigned to their

Table 10.1 Universal Numbering Plan Codes

COUNTRY CODE		AREA CODE		EXCHANGE CODE		SUBSCRIBER CODE
XXX	–	NXX	–	NXX	–	XXXX

where, X = any digit and N = digits 2 through 9.

area. Countries with few telephones are assigned two additional digits. Countries with more telephones are assigned one additional digit. For Canada, United States, and certain Caribbean countries—countries that participate in the North American Numbering Plan (NANP)—their country code is 1, the digit that identifies the geographic area North America. Breakdown by subarea in the NANP area is accomplished by 3-digit area codes.

- **Area code**: North America is divided into subareas that are assigned 3-digit area codes. With the restriction that the first digit is one of eight digits, 800 codes are possible. Some of them (500, 800, and 900, for example) are assigned to special services.
- **Exchange code**: identifies specific switching facilities located in a central office within the area defined by an area code. Many central offices handle more than one exchange code.
- **Subscriber code:** identifies an individual station line that connects the subscriber to the serving switching facilities.

In North America, long-distance (i.e., interLATA) calls are placed by dialing 1 + area code + exchange code + subscriber code. Dialing 1 as the first digit indicates to the serving switch that an interLATA call is about to be placed and prepares it to access the long-distance network. The digit 1 is the network access code, it is not part of the number that identifies the called station, nor is it the NANP area country code; that is only used when placing a call from another area of the world.

10.1.3 Signaling Arrangements

Signaling is described in several, non-exclusive ways:

- **Associated**: when signaling messages are sent over the same route (transmission and switching facilities) as the information that is exchanged between the parties
- **Disassociated**: when signaling messages are sent over a route different from the route taken by the information that is exchanged between the parties
- **Common-channel**: when signaling messages relating to many circuits are transported over a single channel by addressed messages
- **In-band**: when the associated signals are analog tones that exist within the bandwidth used for the information exchange
- **Out-of-band**: when the associated signals are analog tones that do not occupy the same bandwidth as the information exchange.

In the case of digital message signals, in-band and out-of-band are replaced by the terms:

- **In-slot:** when the associated digital signals use a bit position contained within each digital message octet (this is known as *bit-robbing*)
- **Out-of-slot**: when the associated digital signals use a time-slot assigned exclusively for signaling.

Some of the features of a signaling network are shown in **Figure 10.1**. In the United States, signaling between switching nodes employs disassociated, common channel signaling. The signaling data is encapsulated within a frame addressed to the terminating switch and sent to a *signal transfer point* (STP) that serves many switching centers. Here, the message is routed directly to the terminating switch, or directly to the STP that serves the terminating switch. The combination of signaling data links between switching centers and STP, and the interconnected STPs, forms a network. Called the *common channel signaling network* (CCSN), it is separate from the one that handles the users' information exchange. To increase the reliability of the signaling network, STPs are duplicated. Also, they are fully interconnected so that a signaling message need only pass through two pairs of STPs—the duplicated STPs serving the originating switch and the duplicated STPs serving the terminating switch.

The deployment of a common-channel signaling network has led to the development of specialized services. Databases used for these purposes reside in *network control points* (NCPs) that are connected to signal transfer points. Signaling intended to use these capabilities is called *direct*-signaling. We develop further information on signaling networks when we discuss integrated services digital networks and intelligent networks later in this Chapter.

10.2 TELEPHONE INSTRUMENTS

10.2.1 DTMF Telephone

The elements of a dual-tone, multifrequency (DTMF) telephone are shown in **Figure 10.2**. It consists of the following major components:

- *Ringer*, or alerting device
- *Switch,* known as the hook switch, that closes when the handset is removed from its cradle
- *Touchpad* that generates bursts of two frequencies when a button is pressed to enter a digit
- *Microphone* contained in the handset into which the user speaks
- *Earphone* contained in the handset to which the user listens
- *Transformer* (known as the *hybrid* or balance transformer) contained in the base that separates the user's talking signal from the signal to which the user is listening.

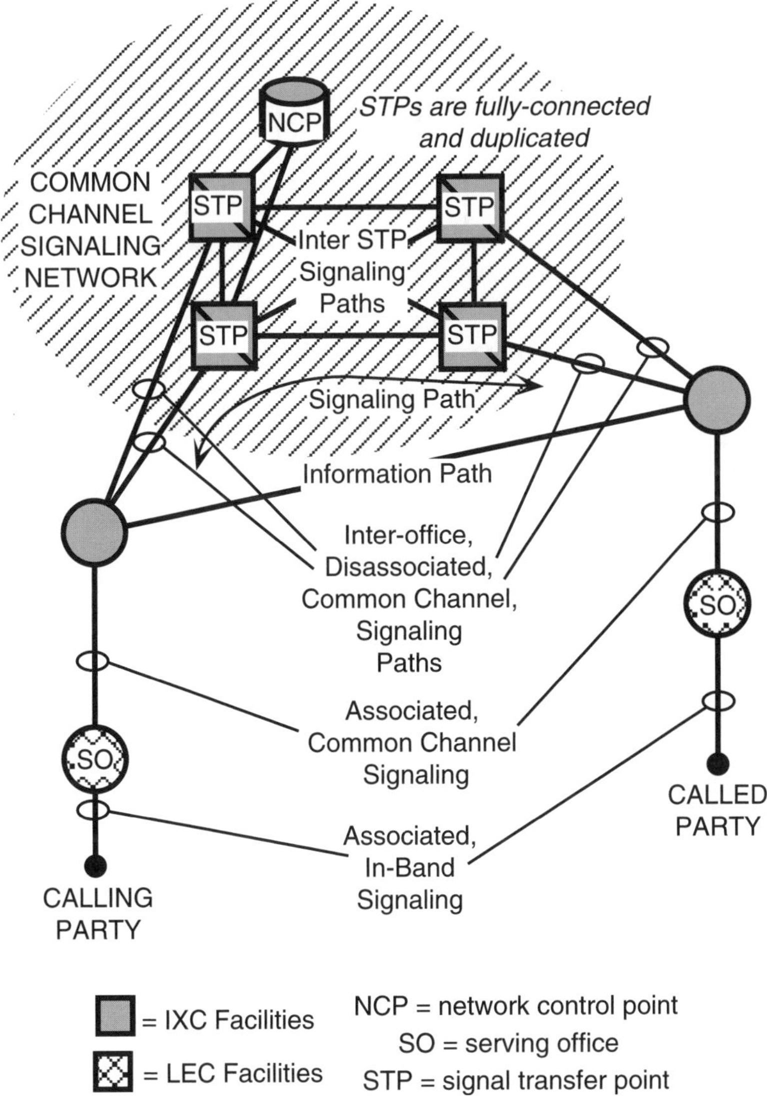

Figure 10.1 Elements of Signaling Network
From calling or called party to the appropriate serving office, signaling is in-band, and over the voice circuit. From SO to IXC switch, signaling is likely to be on a common channel in the same transmission facilities that will carry the message traffic. Across the IXC network, signaling is carried in a separate network. The information path is not set up unless a circuit is available from calling party to called party.

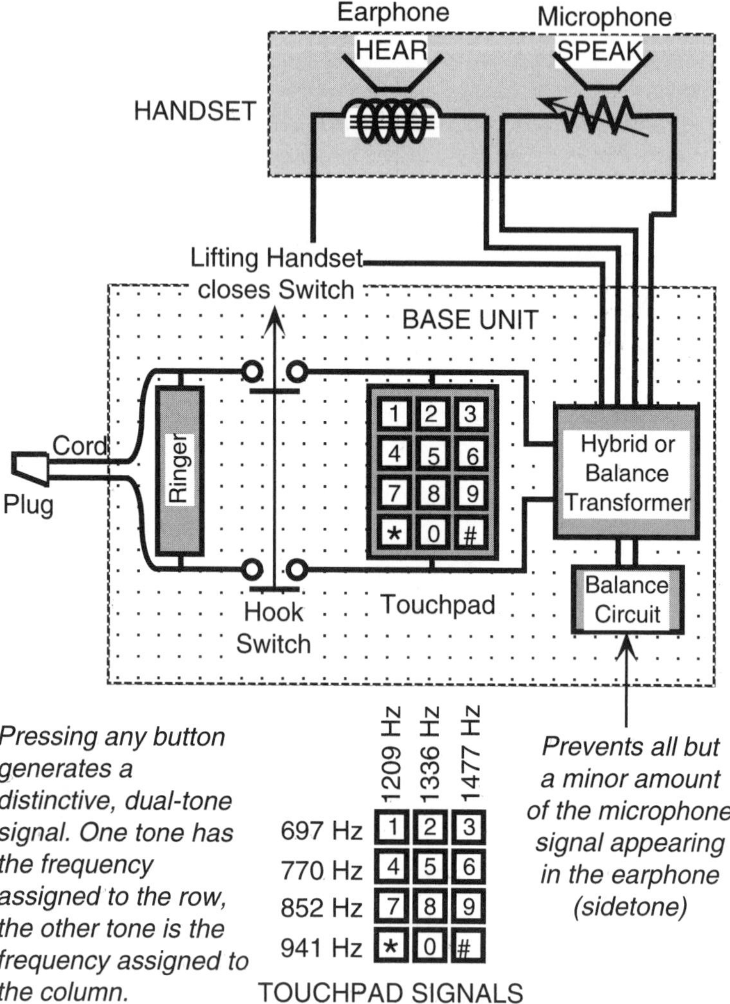

Figure 10.2 Dual-Tone Multifrequency Telephone

Additional electrical components complete the circuit.

10.2.2 Using the Telephone

To make a telephone call, the telephones serving the calling and called parties must be connected together so that signals can flow between them. Establishing such a connection is a function of the telephone network. For the sake of simplicity, consider the case of persons whose telephones are connected to the same end office so that their call is a *local* call; the major events are illustrated in **Figure 10.3**.

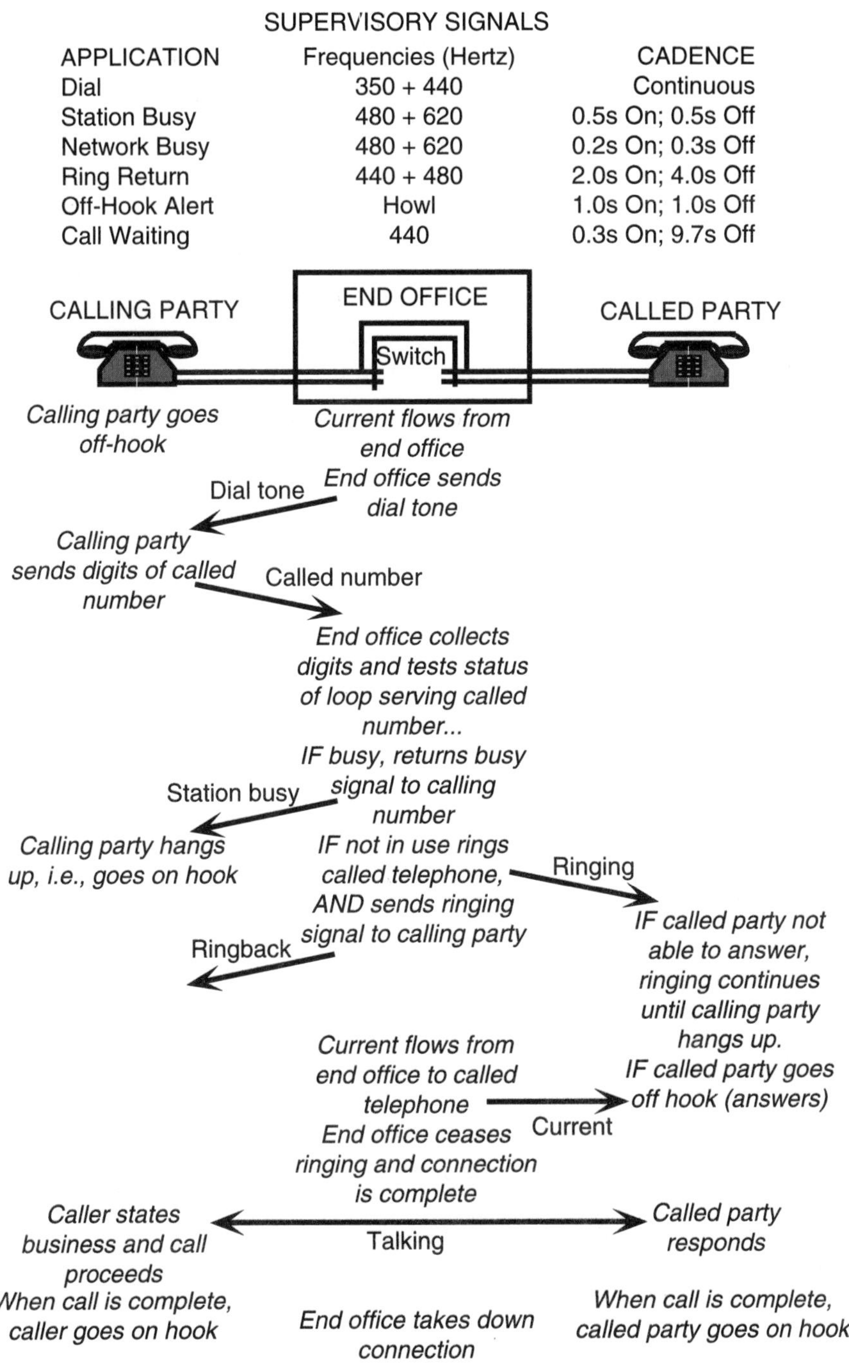

Figure 10.3 Establishing, Maintaining, and Terminating a Telephone Call

To initiate a call, the caller lifts the handset, thereby causing the hook switch to close and permitting current to flow from the end office. This action is known as going *off-hook*. On detecting the current, equipment at the end office connects the calling loop to a tone receiver and sends dial tone to indicate the office is ready to receive the called number. On hearing dial tone, the caller touches the buttons corresponding to the digits of the called number, an action that sends pairs of tones to the signaling receiver at the end office. The digit information is encoded as a combination of two tones selected from seven tones in the range 697 to 1477 Hz; because the range of frequencies falls within the voiceband it is in-band signaling, and users of DTMF telephones can hear the digit tones while keying. When the logic attached to the tone receiver is satisfied that a complete number has been received, equipment in the end office determines the status of the loop serving the called number. If the called loop is in use on another call, equipment at the end office returns a busy signal to the caller to indicate that the call cannot be completed at the moment. If it is not in use, a ringing voltage is applied to it. At the same time, a ringback signal is applied to the calling loop so that the caller is informed that the called number is being rung.

By applying a voltage to the pair of wires serving the called telephone, equipment at the end office activates the bell, or alerting device. Lifting the handset of the ringing telephone, the person answering the call closes the hook switch, thereby disconnecting the ringer, and connecting the microphone and earphone circuits to the line. The change in impedance that this produces indicates to the end office that the call has been answered and that the connection is made. Accordingly, equipment at the end office ceases to provide the ringing voltage to the called party and ringback signal to the calling party. The voice of the answering party activates the microphone and sends a signal along the line to the caller. Often, the microphone contains carbon granules that are compressed by the pressures of the sounds emanating from the speaker's mouth; this causes the resistance of the device to vary in sympathy with the sound pressure; in turn, the varying resistance modulates the current flowing from the end office. Because the microphone is connected at the center of the balance (or hybrid) transformer, most of the signal it produces is canceled out and does not reach the earphone. The small fraction of the signal that does, produces a sound known as *sidetone*. At the proper level, it provides a balance to the sound entering the user's other ear over the air path between ear and mouth. The result is a natural feel to the conversation. The voice signal from the caller is connected directly to the answerer's earphone through the transformer windings.

When the parties have completed their conversation, they hang up, i.e., they go *on-hook*. This action opens the hook switches on the calling and called telephones, interrupting the currents flowing from the end office and indicating a disconnect. This signal triggers a series of events that return the loops to idle status ready to participate in other calls. If only one party hangs up, the end office waits for approximately 15 seconds before applying dial tone to the off-hook telephone. If the user of the off-hook telephone does not dial a number in a reasonable time, the end office assumes the user forgot to hang up and places an off-hook alert signal on the loop. After a while, if the user still fails to hang up, the end office ceases attempts to get the user's attention, and the loop is left open and silent.

Because the wire connections between the end office and each telephone may be subjected to high voltages due to power line crosses and lightning, the telephone set must be able to withstand a wide range of interfering voltages. The telephone company provides a measure of protection against lightning and other destructive surges by attaching arrestors and fuses at the network interface point.

10.2.3 Facsimile Machines

Many users of the telephone network employ facsimile machines to transfer images. In § 4.3.3, we describe some compression schemes used in coding the image, and in § 7.3.1, we discuss some aspects of facsimile service. **Figure 10.4** shows the essential components of a digital facsimile connection. At the sender, page size documents are passed under a scanner where the images are converted to signals that are quantized, compressed, and transmitted as a digital signal stream.

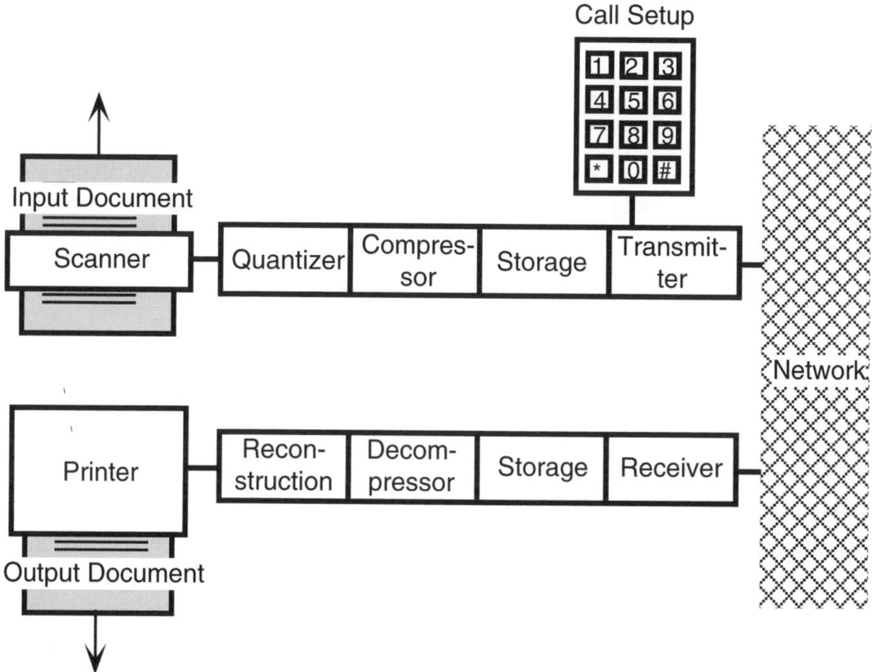

Figure 10.4 Digital Facsimile Machine
To increase the speed at which pages are transferred, the datastream is encoded and compressed.

At the receiver, the images are reconstructed by printing them dot by dot on a sheet of paper using the information provided by the digital signal stream, and the understanding that the reconstruction process is the inverse of the signal processing that was employed by the sender. A letter-size document that is scanned with a horizontal and vertical resolution of 200 dots per inch (dpi) could generate 3.2 million bits. Transmitting this amount of data over a 56 kbits/s, data channel, takes approximately one minute. To achieve faster service, data compression techniques are used. Speeds of up to 20 pages a minute are possible when the page is scanned at resolutions of 200 dpi × 200 dpi, and speeds of around ten pages a minute are possible for 300 dpi × 300 dpi resolutions. ITU-T defines four classes of facsimile machines. Some of their features are:

- **Group 1 Facsimile**: scans documents at a resolution of approximately 100 lines per inch. Uses frequency modulation to transmit a page in four to six minutes.
- **Group 2 Facsimile**: essentially the same as Group 1, but includes signal compression techniques to speed up delivery to two to three minutes.
- **Group 3 Facsimile**: scans black and white documents at 200 × 100, or 200 × 200 dpi. Scanning generates digital signals that are subjected to variable length coding (modified Huffman or modified Read coding, see § 4.3.3) to reduce redundant information. Transmits a page in less than one minute over modem-terminated, telephone line connections operating at speeds from 2400 to 9600 bits/s.
- **Group 4 Facsimile**: a black and white, digital system that employs variable length coding (advanced, modified Read coding) to reduce redundant information. Transmits a page in less than ten seconds over 64 kbits/s, digital connections. The group is divided into three classes according to combinations of horizontal and vertical resolutions that range from 200 dpi to 400 dpi.

10.3 TRANSMISSION FACILITIES

10.3.1 Outside Plant

(1) Cables and Twisted Pairs

The result of many years of evolution in telephone cables and outside plant construction techniques, the signal transport network that serves the telecommunication needs of an established community contains a mixture of materials and installation methods. They reflect the longevity of telephone practices and the relatively slow-speed with which new technologies diffuse through a system in which it is important to recover the investment in in-place equipment. For many years, telephones have been connected to the local switch by means of pairs of wires called *customer loops*; today, most telephones still are attached in this manner. Each loop is fashioned by twisting together a pair of plastic insulated, copper wires. To limit electrical interference (crosstalk) between them, neighboring *twisted* pairs are constructed with different pitches (twist lengths). The pairs are stranded in units of up to 100 pairs, and cabled into plastic-sheathed cores. Common cable sizes range from 6 pairs to 2700 pairs. Cables are laid in ducts under roadways and sidewalks, plowed into the earth, or suspended from telephone poles. The local loop cable plant is divided into three parts:

- **Feeder cables**: the backbone of the local loop network, feeder cables connect the distributing frame in the serving office to the serving area interfaces
- **Distribution cables**: they provide connections from serving area interfaces to service access points close to the user's premises
- **Drop cables**: they connect from service access points (such as pedestals, or other distribution points) to the customer's premises.

As far as possible, users are grouped into convenient areas that contain an interface (or cross-connect) frame. Called the *serving* area, it is shown in **Figure 10.5**. On the switch side of the serving area interface (SAI), connections are made by means of feeder cables (wire pairs or optical fibers). On the subscriber side of the SAI, connections are made through distribution cables to a pedestal. The final connection from the pedestal to the premises is made by a drop cable. All the facilities situated between the distribution frame of the serving office and the customers' premises constitute *outside plant* (OSP).

(2) Two-Wire and Four-Wire Connections

The terms two-wire and four-wire reflect the nature of the original telephone plant. Most connections in the OSP are *two-wire* (2W), i.e., the user's terminal is connected to the CO by a pair of wires that constitutes a duplex circuit and carries information in both directions. In some situations (DDS, digital data service, for instance), a duplex circuit requires two pairs of wires, one for each direction of transmission. This is called a *four-wire* (4W) connection. Four-wire connections must be employed when the distance is great enough to require repeaters or when the existence of two signals on the same bearer will cause them to interfere. As a consequence, 4W connections are employed exclusively on the network side of the CO. The 2W/4W transition is made on the subscriber line interface in the switch (see § 9.1.1(2)) in the CO. Also, a 2W/4W transition is made in the telephone instrument. A *hybrid* transformer is used to derive the talking and speaking connections (2 ×2W) for the microphone and earphone from the 2W circuit to the central office (see Figure 10.2). Today, even though many channels are carried by radio and optical fiber, they are still described as two-wire or four-wire connections.

(3) Echo

In an electrical circuit, it is important to ensure that each segment appears as an infinite line to the signal entering it. If this is not so, signal energy will be reflected

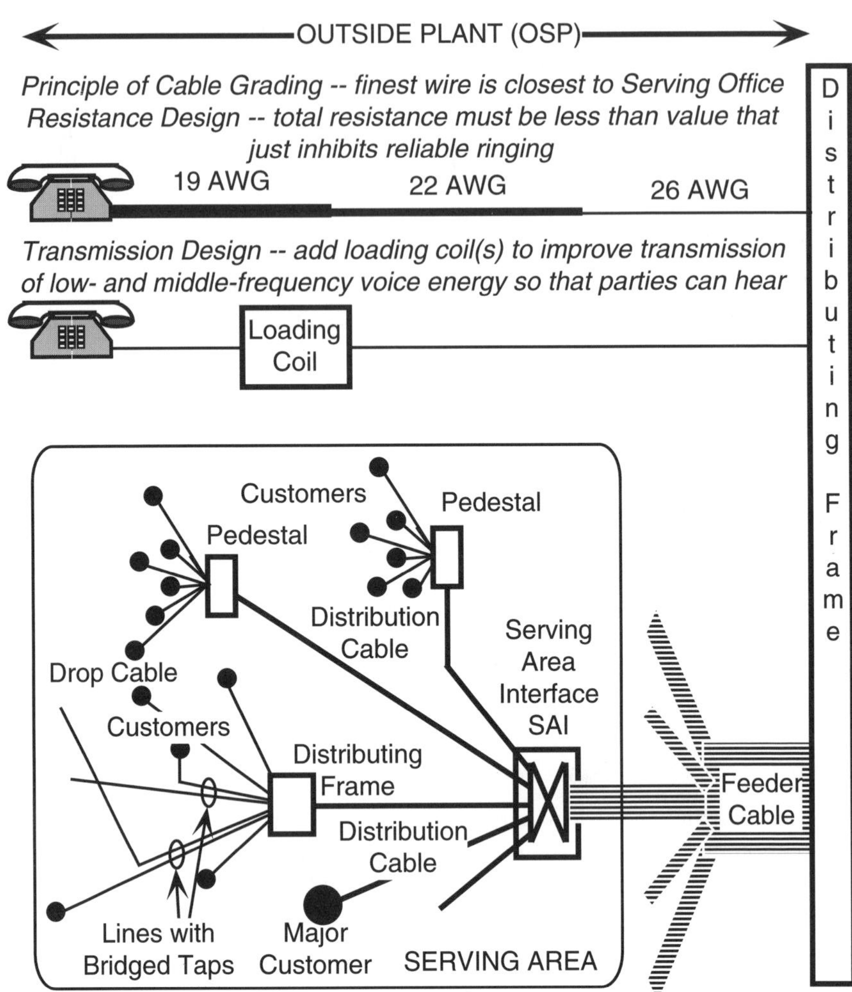

Figure 10.5 Wire Facilities in Outside Plant

back to the source in proportion to the impedance mismatch that exists. 2W/4W transitions represent a significant mismatch, and are a major cause of reflections. Under certain circumstances, the reflected energy creates an echo that is discernible and annoying to the sender. Subjective testing has established that echoes are less likely to annoy speakers when the delays are short, and more likely to annoy speakers as the delays become longer. To alleviate the annoyance, echoes must be attenuated; the longer the delay, the greater the attenuation that is required. Echo conditions are at their worst when the energy is reflected from the far end of a connection that includes significant propagation delay. Fortunately, electronic devices known as *echo suppressers* are available. At the sending end, they create an inverted echo signal that cancels the signal received from the impedance mismatch point.

(4) Bridged Taps

In areas subject to population fluctuations, installation practices permitted idle pairs to be re-used without disconnecting them from their original, but now inactive, terminating points. The new connection was bridged to an old one, leaving an active wire pair serving the new user attached to a portion of an inactive wire pair that served the former user. Such *bridged taps* (BTs) impose additional electrical loading on the loop circuits and cause the high-frequency performance to deteriorate. The overall effect depends upon the lengths of the active and inactive loops, the quality of the physical connections, and the proximity of the inactive loops to interfering sources that produce noise. With a growing demand for high performance circuits, bridged taps are no longer made, and those that remain in the loop plant are being eliminated as opportunities to do so become available.

(5) Resistance Design

Close to the switch, the feeder cables contain many wire pairs. As a consequence, to conserve duct space, and reduce cost, they employ small diameter wires (26 AWG). In the outlying feeder cable sections the number of wire pairs decreases and the diameter of the individual wire pairs may be increased (to 22 and 19 AWG, for example). The combination of wire sizes is chosen so that the total resistance of the loop does not exceed 1100 to 1300 ohms, the value that allows sufficient current to flow from the serving office battery for ringing. Known as *resistance design*, the technique ensures reliable signaling to customers situated at a considerable distance from the switch (30,000 to 50,000 feet).

(6) Transmission Design

Since there is no point to being able to signal if the parties cannot understand each other, inductance is added to long loops to reduce the voice band attenuation of the connection. Known as *transmission design*, the technique inserts loading coils at fixed intervals to improve the transmission of low- and middle-frequency voice energy. A loop must satisfy both resistance and transmission design parameters to provide satisfactory service to the user. As digital loop carrier equipment (see below) is installed and individual physical loops are replaced by channels contained in multiplexed streams, the concepts of resistance and transmission design become obsolete.

(7) Outside Plant Productivity

Customer loops and terminating telephones are the least productive telecommunication facilities. On average, residential facilities are in use about one-half hour each day, and business telephones may be used as much as one-hour each working day. Further, while most pairs may be assigned to specific customer stations, there is a fraction of the plant that is idle, waiting to be assigned to meet future customer requirements. Because OSP represents up to one-third of the total capital investment in an exchange area, increasing the intensity of use of these assets is important to local exchange companies. A strategy to increase the productivity of the OSP is to shorten loop lengths by moving switching closer to the customer. This creates a requirement for remote switching units, and for short-distance multiplexing to carry the traffic to the serving office. Other strategies introduce concentration and multiplexing at appropriate points where they can serve a community of customers. Known as *pair-gain* techniques, they replace a larger number of underutilized loops by a smaller number of loops that are in use more often. Pair-gain techniques are readily implemented in digital circuits and form the basis for modern digital end offices in which a large base unit switch is supported by remote switching units and digital loop carrier equipment.

(8) Penetration of Digital Electronics and Optical Fibers

Fine wire, bridged taps, and loading coils have the primary objective of providing a local network that will support voice frequency communications at a reasonable cost. For all but low-speed applications, these techniques are harmful to the carriage of digital signals. To pass digital voice (DS-0), the loops must be *groomed*, i.e., loading coils and bridged taps must be removed. As station growth takes place, a growing amount of digital electronics are being placed in the analog loop plant to expand the number of channels without adding new bearers. As demand for digital channels picks up, feeder and distribution cables are being replaced by optical fibers and digital loop carrier equipment. They are the technologies of choice for new construction, particularly new construction for business customers. Wider deployment depends on price reductions consequent on increasing volumes and the stimulation of demand for digital services in the home.

(9) Digital Loop Carrier

- **Digital loop carrier** (DLC): a carrier system that time-division multiplexes subscriber loops and operates between the user's premises and the serving central office. Also known as *digital channel banks* (D-channel banks) and *subscriber loop carrier* (SLC) systems, DLCs provide remote monitoring capabilities, perform concentration, and permit more channels to be implemented over existing OSP. Because they operate in the loop plant, DLCs provide standard support functions for the user's terminal, i.e., battery feed, overvoltage protection, ringing, supervision, coding and filtering, hybrid and testing (known by the acronym BORSCHT).

DLCs permit the use of digital technology in the feeder and distribution plant and newer equipment includes terminations for optical fibers. In principle, they transport multiplexed digital voice and data signals between central office terminals

(COTs) located at the serving office, and remote terminals (RTs) that provide digital services to a distinct geographic area known as a *carrier serving area* (CSA). As shown in **Figure 10.6**, each COT and RT will handle four T-1 facilities (or one T-2 facility) to provide 96, DS-0 channels. To serve a major concentration of business subscribers, 28 T-1 facilities (a T-3 facility) may be multiplexed on optical fiber to provide 672 DS-0 channels. They are fed through 7 COT/RT pairs. In some digital switches, the COT is incorporated within the switching electronics.

Contemporary digital loop carrier systems are:

- **D4-channel bank**: multiplexes 48 or 96 individual voice or data channels over DS-1, DS-1C, or DS-2 facilities.
- **SLC-96**: multiplexes 96 individual voice or data channels over four , DS-1 facilities or a DS-2 facility. Includes the option to concentrate 48 channels on each DS-1 facility.
- **D5-channel bank**: multiplexes 96 individual voice or data channels over four, DS-1 facilities or a DS-2 facility. Includes a common control subsystem that manages up to 20 units, and supports fiber connections to the users' premises.

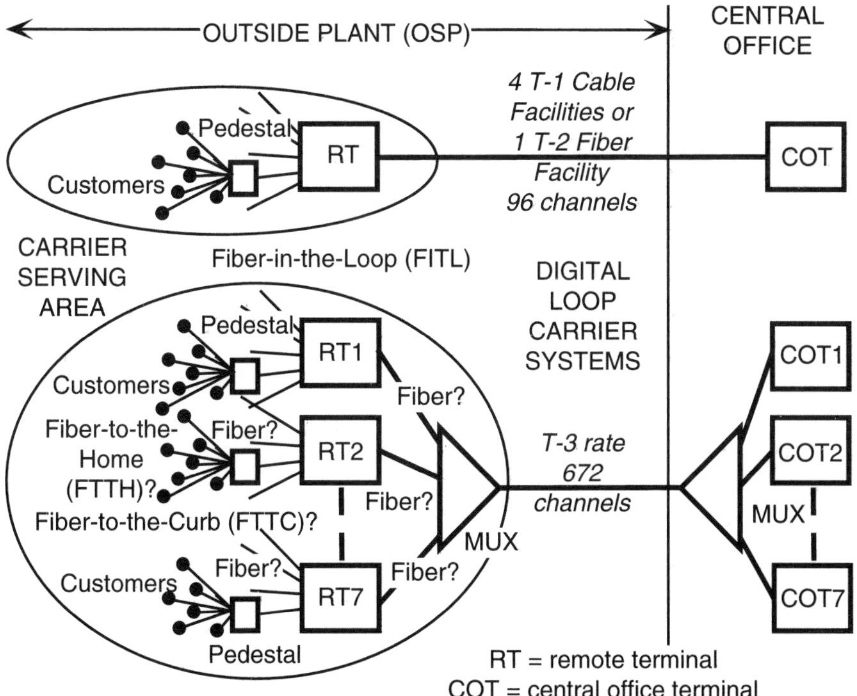

Figure 10.6 Digital Loop Carrier Facilities

- **Automated digital terminal equipment system** (ADTS): multiplexes individual voice and data channels over DS-1, DS-2, or DS-3 facilities. Provides improved monitoring and maintenance functions, routing capability, clear-64 and ESF capability, fiber connections to the users' premises, and supports SONET connections.

(10) 2B1Q Digital Subscriber Carrier

By coding two bits as one quaternary symbol (an action known as 2B1Q; the quaternary symbol assumes 1 of 4 amplitudes), it is possible to provide 160 kbits/s, duplex connections over two-wire copper loops up to 18000 feet. In the signal streams, 144 kbits/s are available for message transport, 12 kbits/s are used for framing and synchronization, 2 kbits/s are used for CRC error checking and 2 kbits/s are used for an operations data channel. Such connections are required to implement the basic rate interface for ISDN (see § 10.5.1). Extensions of the technology are used to provide non-ISDN-based services:

- **Digital added main line** (DAML): implements a second circuit over a single pair of wires for circuit expansion purposes.
- **Universal digital channel** (UDC): implements six digital circuits on a single pair of wires.

  Examples of the use of 2B1Q-configured DLCs are given in **Figure 10.7**.

(11) Loop Plant Topology

The traditional loop plant shown in Figure 10.5 can be described as a double *star*. The feeder cables centered on the SCO constitute one star and the distribution cables centered on the remote ends of the feeder cables constitute the other. Because it contains no electronic devices, the arrangement is described as passive. The introduction of DLCs has created an *active* double star configuration in which the number of circuits is no longer dependent on the availability of individual twisted pairs, and the continuing deployment of optical fibers in the loop, makes other arrangements possible. For instance, the function of the distribution plant can be performed by a bus or a ring, leading to *star bus* and *star ring* configurations, and the feeder plant may eventually become a second ring, leading to a *ring ring* topology. These possibilities are illustrated in **Figure 10.8**. To date, most of the optical fibers in the local loop have been placed there in support of digital loop carrier applications. With this equipment deployed in increasing numbers, extending optical fibers to pedestals (this is known as *fiber-to-the-curb*, FTTC), becomes possible and will make 64 kbits/s digital circuits (and multiples, i.e., $n \times 64$) available to just about all who want them. However, demand for interactive, broadband services is insufficient to justify widespread deployment to the home (this is known as *fiber-to-the-home*, FTTH). It will occur when the cost of optical technology is at, or below, the cost of maintaining and replacing today's copper systems. As always, when there is competition between technologies, advances are made on both sides.

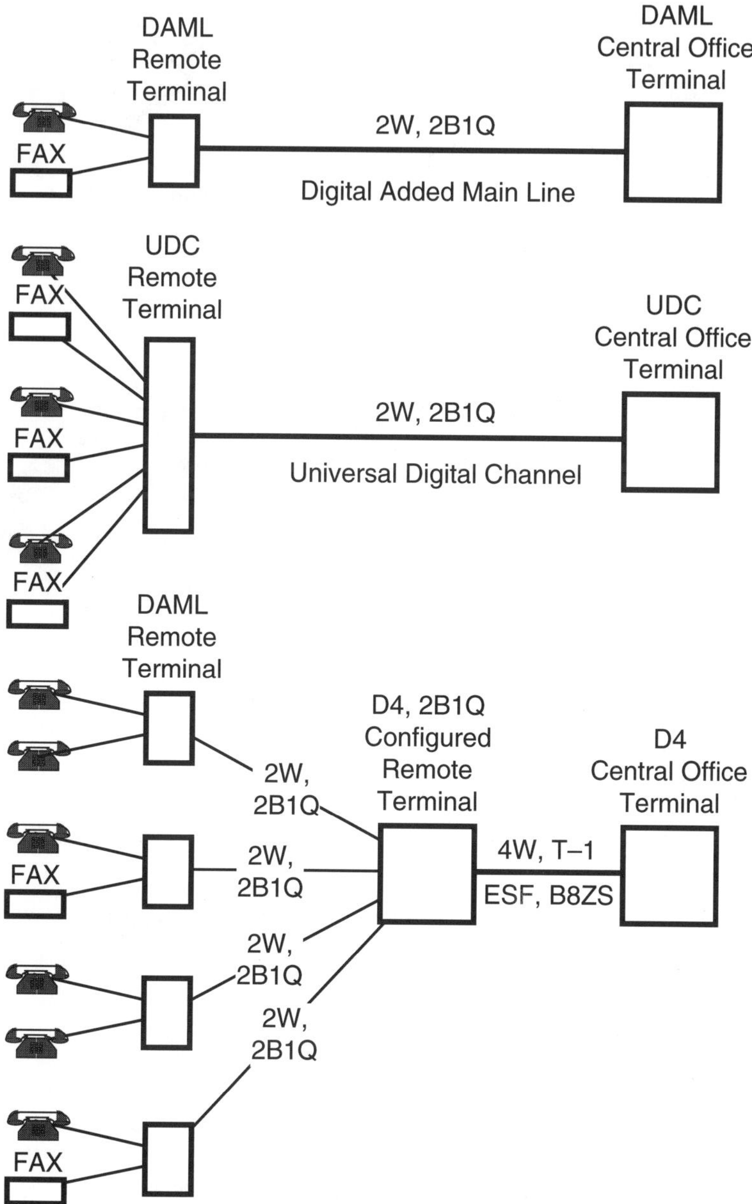

Figure 10.7 2B1Q Digital Loop Carrier Applications
Using two-binary, one-quaternary (2B1Q) coding, it is possible to provide 160 kbits/s, duplex operation on two-wire copper loops to 18000 feet. DAML implements an additional circuit over a pair of wires for circuit expansion. UDC implements 6 digital circuits. In the bottom diagram, a D4 channel bank supports several 2B1Q connections.

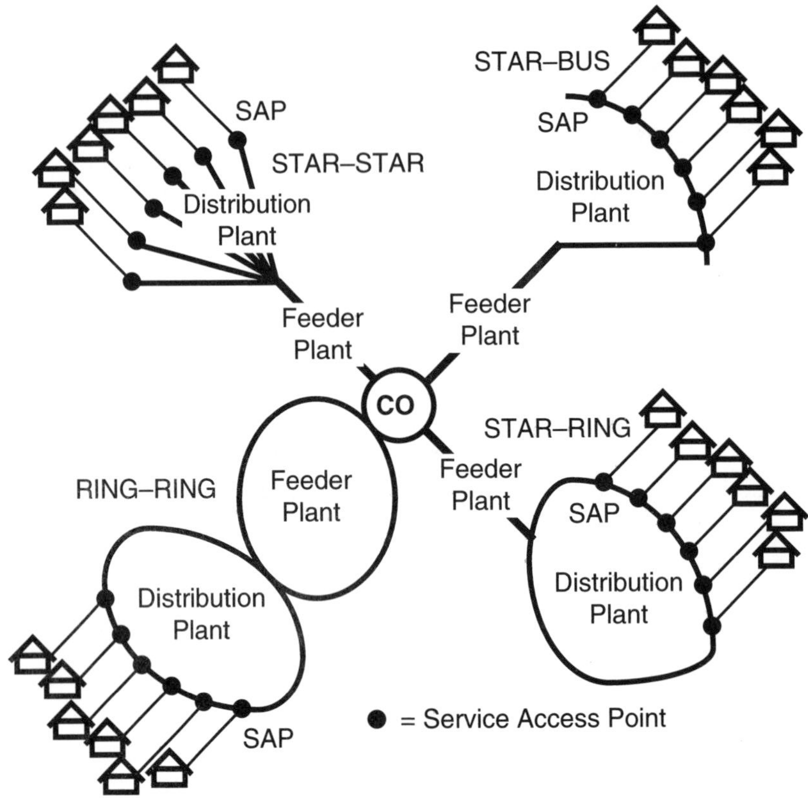

Figure 10.8 Evolution of Local Loop Topology

The combination of very-large-scale integrated circuits and copper pairs will be a hard one to replace.

(12) Video Dial Tone

An emerging technology shows promise of delivering high bit-rate digital information to households using existing cable plant. Called ADSL (asymmetric digital subscriber line), it transmits 1.5 Mbits/s video channels encoded with MPEG digital video compression standards (see § 3.4.5(2)). ADSL makes it possible for the LEC to provide switched television service to telephone subscribers. **Figure 10.9** shows the principle of *video dial tone*. Three levels of service are possible:

- **ADSL-1**: supports one MPEG encoded video channel, plus voice and low-speed command return channel. Operates over non-loaded loops up to 18000 feet.
- **ADSL-2**: supports two MPEG encoded video channels, plus voice and low-speed command return channels. Operates over non-loaded loops up to 12000 feet.

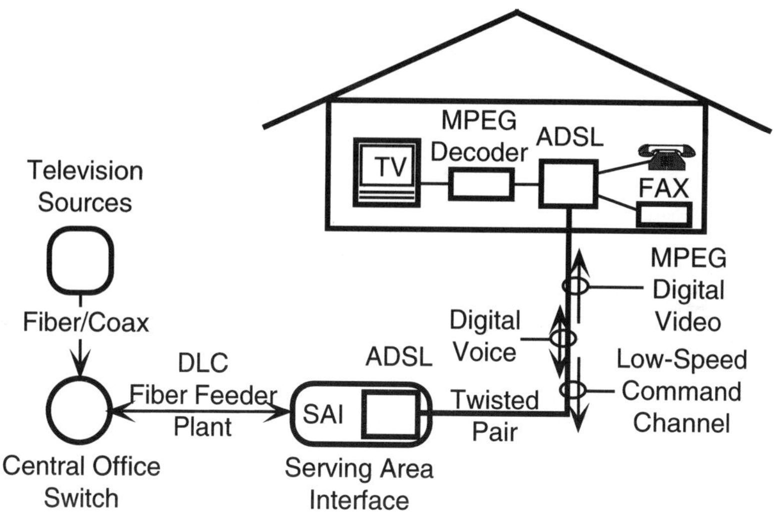

Figure 10.9 Video Dial Tone
ADSL makes it possible for the LEC to provide switched television service to telephone customers.

- **ADSL-3**: supports up to four MPEG encoded video channels, or other high bit-rate channels, plus voice and low-speed command return channels. Operates over distribution cables up to 6000 feet.

For all three versions, fibers are employed to transport the digital signals to the ADSL interfaces at the SAI.

(13) Voice, Data, and Television

A majority of households in the United States are connected to two networks—a switched network over which they receive local voice and data transport services from the serving LEC, and a non-switched network over which they receive television from the serving CATV company. With the continuing evolution of the regulatory environment, the possibility of combining the two networks is being explored. No easy task, it requires the reconciliation of dissimilar technologies—the distribution of television signals over cable is an analog, FDM-based activity, while the distribution of interactive voice and data signals is a digital, TDM-based activity. In the foreseeable future, they must remain so, for a consumer-oriented television service must be compatible with existing receivers, and local voice and data transport must be compatible with digital distribution facilities. The result is a hybrid network that contains both optical fiber (a point-to-point, digital bearer) and coaxial cable (a point-to-many, analog bearer). **Figure 10.10** shows a *fiber-and-coax* net-

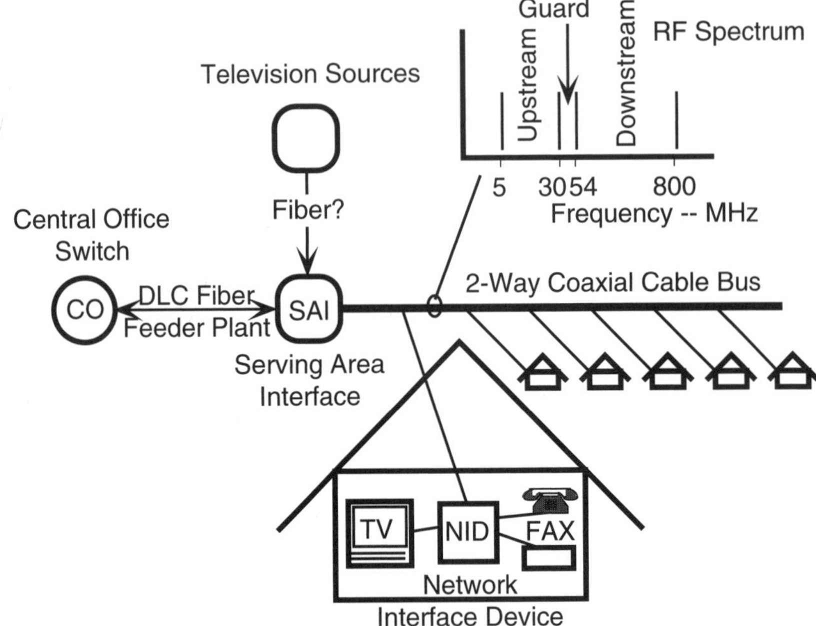

Figure 10.10 Fiber and Coax Network
The combination permits telephone service to be provided over the cables used for television distribution.

work that provides duplex voice and data circuits and simplex broadband channels. It employs optical fibers for signal transport to local distribution points (analogous to serving area interface points) where the digital, TDM signal streams are converted to an FDM stream and distributed to individual homes over shared coaxial cable. In each home, an interface device separates television from voice and data. Two-way operation over cable is achieved by allocating a portion of the spectrum for upstream signals. In several systems it is the frequency band 5 MHz to 30 MHz. In this circumstance, the band 30 MHz to 54 MHz serves as a guard band, and downstream signals occupy the cable passband above 54 MHz. Duplex voice and data circuits are carried between the SAI and the residence over coaxial cable; between the serving switch and the SAI, they are carried on optical fiber, digital loop carrier circuits.

10.3.2 Network Facilities

Within networks, traffic is carried between nodes (switches) over multiplexed routes that employ digital techniques to transport DS-0, DS-1, and DS-3 signals. The physical facilities employ twisted wire pairs in specially constructed cables, microwave radio or optical fiber lightguide bearers. Which is used depends on the number of channels, and on the cost of competing installations, although optical fibers are clearly the preferred technology.

(1) Twisted Pair Transmission Systems

Twenty-four and 48 PCM voice channels (i.e., T-1 carrier at 1.544 Mbits/s and T-1C carrier at 3.152 Mbits/s) are transmitted on wire pairs contained in multipair cables installed between local exchange facilities. Regenerators are spaced at intervals of approximately 6000 feet. In older systems, in-slot signaling is used so that the channel speed for data is reduced to 56 kbits/s. In newer systems, common-channel signaling and B8ZS coding (see § 8.2.1(2)) are employed to restore the channel speed to 64 kbits/s.

(2) Digital Microwave Radio Transmission Systems

Introduced between New York City and Boston in 1947, microwave radio relay systems quickly expanded across the continent to provide long-distance routes that would support direct distance dialing. The first systems employed frequency modulation (FM) and carried frequency division multiplexed (FDM), single sideband, suppressed-carrier voice channels (see § 2.3.2). Later, the capacity of existing systems was improved by converting the carrier modulation to single-sideband amplitude-modulation (SSB-AM). More recently, with the introduction of robust integrated circuits, the cost per channel-mile of digital radio systems has dropped below FM/FDM and SSB-AM/FDM radio systems for most routes. Coupled with the rapid adoption of digital switching, in-place analog microwave radio equipment is being replaced by digital radio—or optical fibers. To facilitate an orderly changeover, digital microwave radios employ the same frequencies and channel widths as analog microwave radios. Thus, in the 4, 6, and 11 GHz common carrier bands, channel spacings are 20, 29.65, and 40 MHz respectively. Commonly, 1344 voice circuits (2 ×DS-3 = 90 Mbits/s) are carried in each channel. Future plans call for the use of digital radios that conform to SDH (synchronous digital hierarchy, see § 8.3.3). It will be possible to transport 8 or 16 STM-1 streams at 6 GHz, and 24 STM-1 streams at 11 GHz (STM-1 = 155.52 Mbits/s). We describe some digital modulation schemes in § 4.7.

(3) Optical Fiber Systems

As a replacement for individual 12-, 24-, and 48-channel wire pair transmission systems, optical fiber systems are not economical. However, the 150 or 200 working systems contained in a 900-pair cable (200 × T-1 = 4800 two-way voice circuits: 150 × T-1C = 7200 two-way voice circuits), can be replaced by 12 or 24 fibers operating at 44.736 Mbits/s (FT3, 672 voice channels). Over most distances encountered within a LATA, all-optical fiber transmission systems can be installed without span repeaters. This improves network reliability and reduces operating costs. Further, in areas in which underground ducts are used, the substitution of fiber cables for copper cables frees significant space for future expansion. Where they can be used, optical fiber systems are the medium of choice for route expansion and for new installations. In the United States, individual carriers are implementing synchronous digital networks (SONETs) that employ optical fiber transmission media exclusively (see § 8.3.1).

(4) SONET Architecture

SONET makes use of multiplexers, digital cross connects, and other terminals, to provide point-to-point, hub and ring connections. Some basic configurations are shown in **Figure 10.11**. Originating and terminating signals are electrical and members of the digital signal hierarchy (i.e., DS-0, DS-1 and DS-3). Add and drop multiplexers (ADMs) are used to create a multidrop, point-to-point connection, and a ring connection. A digital cross-connect system (DCS) that cross-connects at the virtual tributary levels, is used to create a distribution hub. Because the individual fiber connections carry large numbers of channels, an important consideration for all of these configurations is the provision of alternate routes that can be invoked rapidly in the case of a cable cut or node failure. Automatic protection switching (APS) permits any one of up to 14 working lines to switch to a single back up line. In addition, SONET permits bi-directional rings in which the second fiber acts as back up for the first. In the event of a failure in the primary line, the back up fiber is used to bridge the gap by sending the frames around the ring in the reverse direction until they reach the other side of the fault. This action is described as *self-healing*.

10.4 LOCAL AND LONG-DISTANCE NETWORKS

10.4.1 Local Networks

(1) Central Office, Telephone Exchange, and Exchange Area

In simpler times, the caller picked up the earpiece, cranked the generator handle on the telephone instrument to call *central*, spoke with an attendant in the central office, and waited for the connection to the called party to be completed manually. As telephone facilities evolved, a switching machine was installed in the central office and became the focus of telephone activity in a given area. Called the *telephone exchange*, its customers defined an *exchange area*. They were connected to the exchange by loops—the dominant component of the outside plant. To complete a network that spanned the United States, switches were arrayed in a hierarchy of five levels. The telephone exchange was served by a Class 5 switch. The other switching levels were Class 4—Toll Center switch, Class 3—Primary Center switch, Class 2—Sectional Center switch, and Class 1—Regional Center switch. With the separation of long-distance transport from local transport, the upper layers have been abandoned in favor of fully interconnected mesh networks of large digital switches (for example, see § 10.4.2(1)).

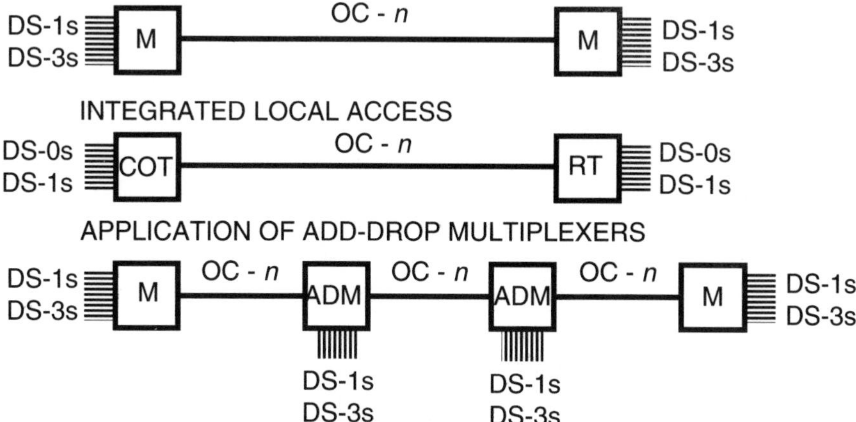

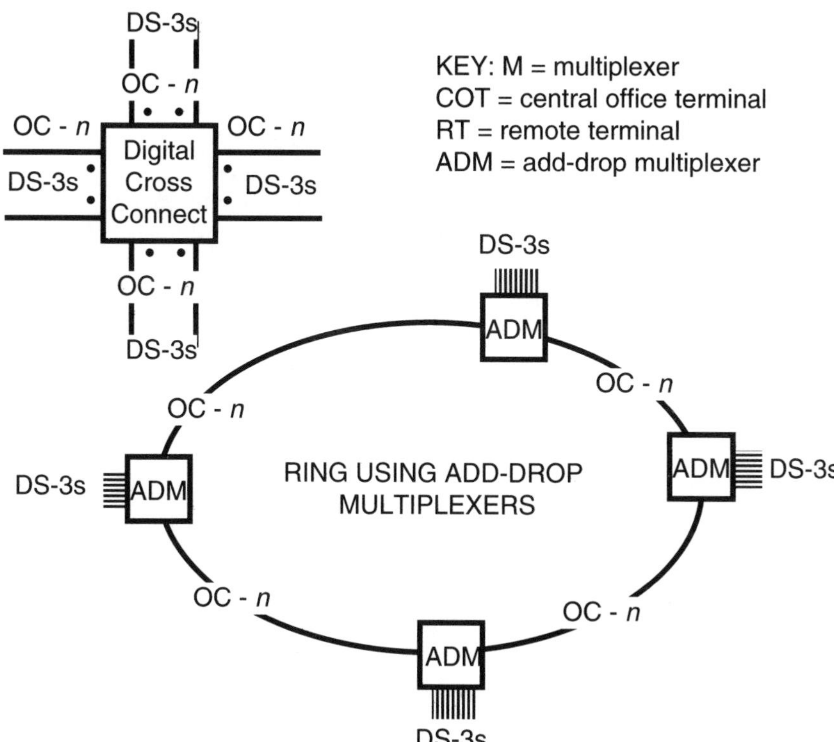

Figure 10.11 Basic SONET Configurations
Synchronous Optical Network (SONET) is an all-optical fiber transmission system that transfers octets at a minimum speed of 55 Mbits/s.

(2) End Office, Serving Office, and Wire Center

Digital switches and other electronic equipment have created a different telephone exchange. The changes are reflected in current names:

- **End office**: a facility that contains the lowest node in the hierarchy of switches that comprise the network. It is the network interface with the caller—connecting the caller's loop to the called party's loop if they share the same end office, or connecting the caller's loop to another switch if they do not. While the switch may serve a single exchange code, it is more likely to serve several. It is still called a Class 5 switch.
- **Serving office**: also known as serving central office (SCO). The end office that serves a particular user.
- **Wire center**: a physical location at which many cables come together to be served by a switching facility that spans several exchange codes. They may be wire pairs contained in feeder cables that carry analog voice from individual telephones, or optical fibers that carry digital, multiplexed traffic from remote switching units or pair-gain equipment in the exchange area—or from other network nodes.

Figure 10.12 provides a summary view of a modern end office that serves an urban area with business and residential customers. The office contains a digital switching machine that supports remote switch modules, a digital cross-connect system, and SONET and other multiplexer interfaces. The majority of circuits are provided over feeder cables to voice and low-speed data customers. Some customers use T-1 or T-2 digital loop carrier systems for voice or high-speed data, and other customers employ private lines that transit the office. The end office connects to other end offices, to a mobile telephone switching office (see § 14.2.2), and to IXC POPs for long-distance connections.

- **Points-of-presence**: the facility interface between the LEC and the IXCs is known as the point-of-presence (POP). A POP must be established on each trunk path that connects switches that belong to the local and long-distance carriers. On one side of the POP, the LEC is responsible for service; on the other side, the IXC is responsible.

Digital, local switches are designed to perform in all of these situations. They contain different complements of hardware and software modules that customize their performance to the environment in which they are to operate.

(3) Inter- and Intra- LATA Traffic Arrangements

To collect inter-LATA traffic from several exchange areas, the LEC may employ a tandem switch. In this role, it is known as an *access* tandem (AT). Also, it may provide alternate connections between exchange areas in the LATA and be the princi-

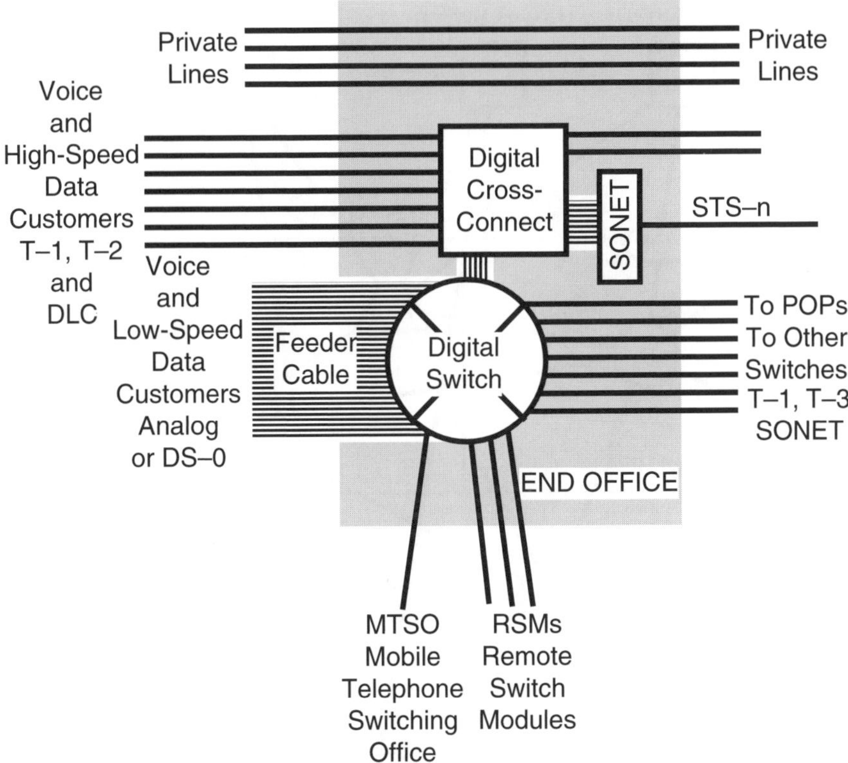

Figure 10.12 Overview of End Office
A typical end office includes a digital switch that mediates voice and low-speed data circuits between the feeder cable of the local loop, multiplexed inter-office connections, mobile telephone circuits, and remote switching units. A digital cross-connect reorders multiplexed voice and high-speed data circuits, for passage over network facilities. The office serves as a convenient location for terminating and testing private lines.

pal repository of resources for the *intelligent* network (see below). **Figure 10.13** shows some typical exchange areas and the inter-office connections that are required to serve the local and long-distance needs of the users. We describe them briefly here:

- **Exchange area A**: a single end office (that may include several exchange codes) serves the needs of users in Exchange Area A. Intra-LATA, inter-exchange traffic is sent directly to the end offices in Exchange areas B and C. Inter-LATA traffic directed to IXC (A) is trunked directly from the end office to IXC(A)'s toll switch. Long-distance traffic intended for other IXCs is routed through an access tandem switch to the individual carrier's switches. At

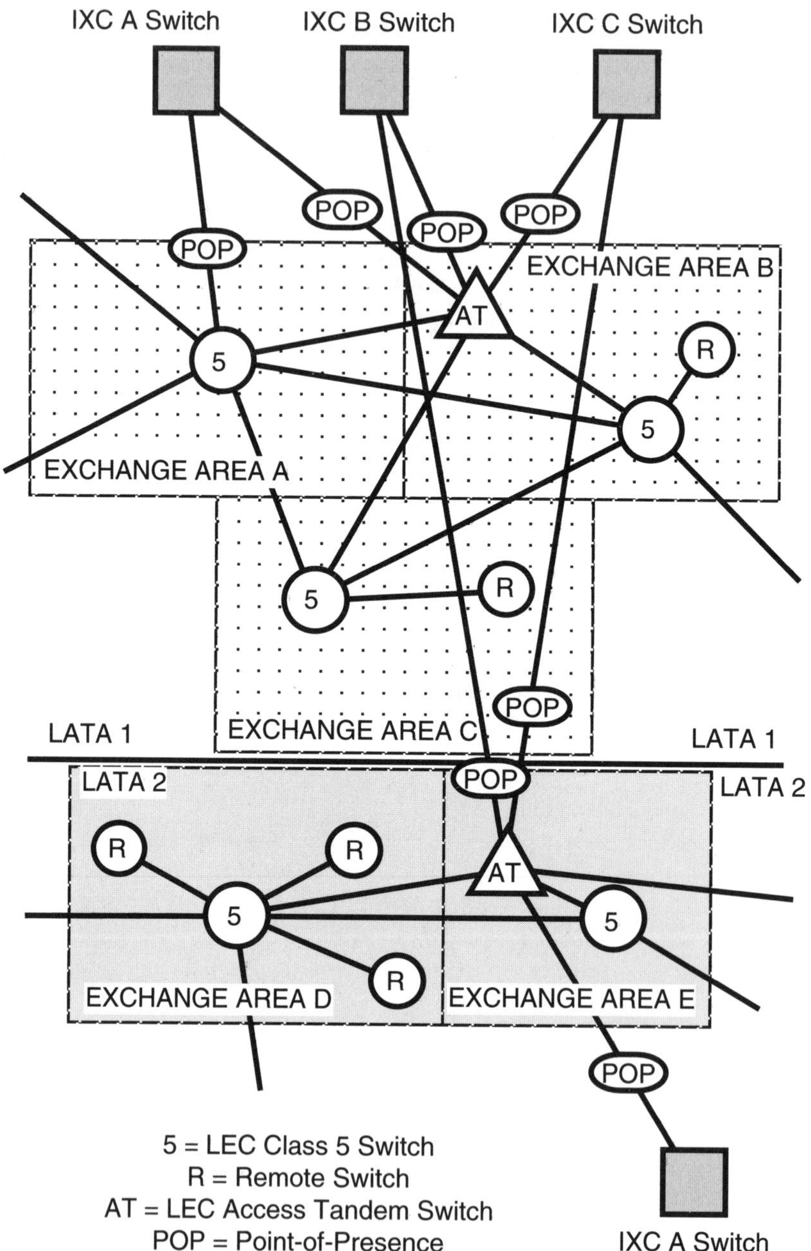

Figure 10.13 Local Switching Arrangements
Access tandems direct LEC inter-LATA traffic to IXC switches.

times of heavy demand, traffic intended for IXC (A) is carried through this switch.

- **Exchange area B**: an end office with a remote switching module serves the needs of users in Exchange Area B. Intra-LATA, inter-exchange traffic is sent directly to the end offices in Exchange areas A and C. Inter-LATA traffic intended for any IXC is trunked to the access tandem switch.
- **Exchange area C**: an end office with a remote switching module serves the needs of users in Exchange Area C. Intra-LATA, inter-exchange traffic is sent directly to the end offices in Exchange areas A and B. Inter-LATA traffic intended for any IXC is trunked to the access tandem switch.
- **Exchange area D**: an end office with three remote modules serves the needs of users in Exchange Area D. Intra-LATA, inter-exchange traffic is sent directly to the end office in Exchange Area E. Inter-LATA traffic intended for any IXC is trunked to the access tandem switch in LATA 2. Traffic between Exchange Areas A, B, or C and Exchange Area D is carried through a toll switch operated by one of the IXCs.
- **Exchange area E**: a single end office serves the needs of users in Exchange Area E. Intra-LATA, inter-exchange traffic is sent directly to the end office in Exchange Area D. Inter-LATA traffic intended for any IXC is trunked to the access tandem switch in LATA 2. Traffic between Exchange Areas A, B, or C and Exchange Area E is carried through a toll switch operated by one of the IXCs.

10.4.2 Long-Distance Networks

(1) AT&T's Network

To serve the long-distance communications needs of the continental United States, AT&T employs a network of over 100 switching machines, over 100 Network Control Point (NCP) databases, and 20 Signal Transfer Points (STPs). **Figure 10.14** characterizes the scope of AT&T's Network. Virtually all the switching machines are fully interconnected. There are approximately 6500 point-to-point, multiplexed links in the network. Signaling employs Common Channel Signaling System No. 7 (SS7). In § 9.1.5(3), we noted that AT&T maintains 16 Primary Reference Clocks (PRCs) that provide precise timing at critical locations in the network. Their output is used to *discipline* a hierarchical network of clocks with lesser accuracies that synchronize digital equipment at all levels of the network. A Timing Monitor System (TMS) verifies that this is done satisfactorily.

(2) Dynamic Nonhierarchical Routing

Calls are routed by a technique known as dynamic nonhierarchical routing (DNHR); they proceed from originating to terminating switches directly, or through a single intermediate switch so that over 100 routing options are available between

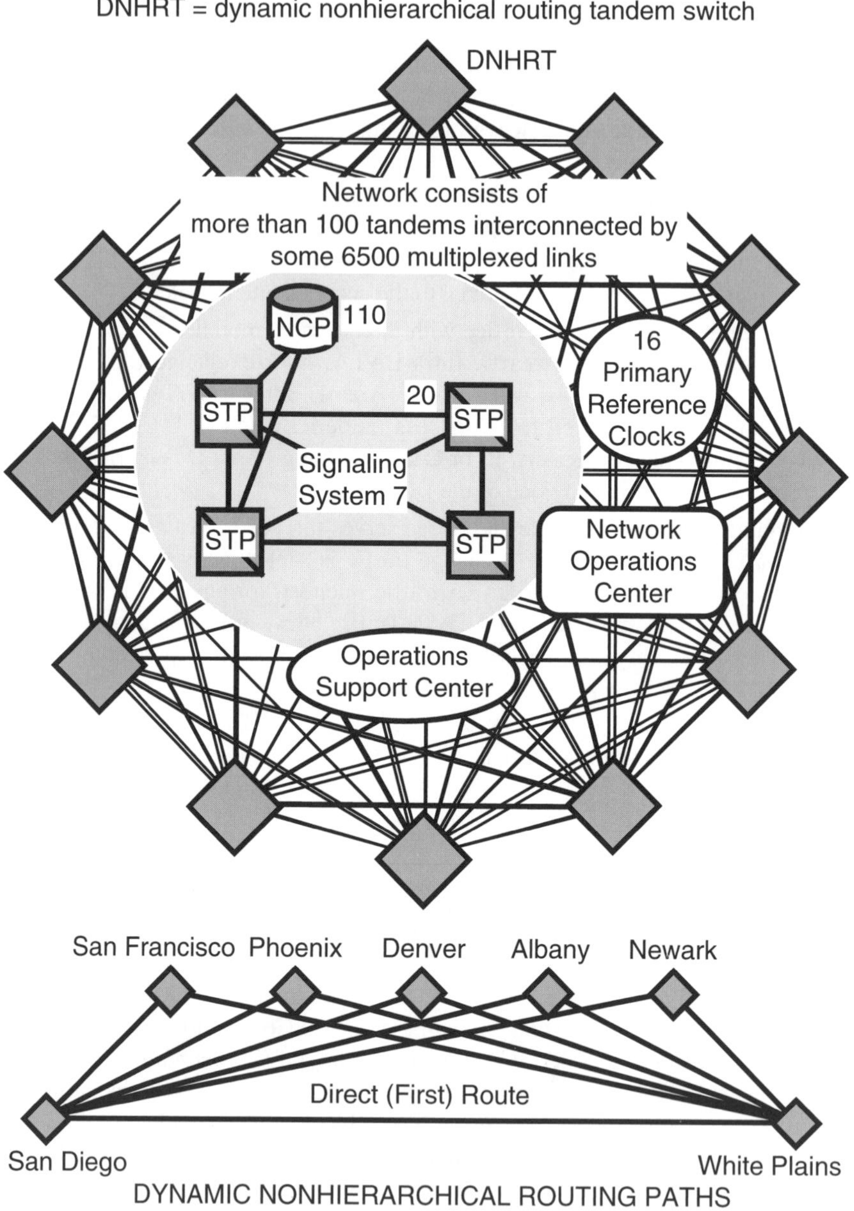

Figure 10.14 Overview of AT&T's Network and Dynamic Nonhierarchical Routing
In a DNHR network, calls are completed directly between the tandem serving the calling party and the tandem serving the called party. If this cannot be done, the calls are routed through one other tandem.

each pair of switches. No hierarchy of switching points is involved—hence the description *non-hierarchical routing*. Routing patterns vary by hour of day and day of week. At the bottom of Figure 10.14, we show examples of DNHR between switches located in San Diego and White Plains. The first route attempted is the direct San Diego to White Plains connection. The next attempt might be through the switch located in Albany. A third route could be through the switch located in Phoenix—and so on. Which is used will depend on the overall state of the network (possible restrictions due to failure or maintenance problems), the time of day (is it busy hour in Albany or Denver?), and day of week (is it a workday, a weekend, or Mothers' Day?). In no case does a call employ more than one intermediate switching point.

(3) Routing Controls

The possibility of selecting any one of a large number of routes to carry calls between switch pairs can result in an increase in the average number of circuits required to complete a call, and reduce the network capacity available for other calls. Under overload conditions, automatic expansive routing techniques may force the network into a congested, inefficient state, and lead to a decline in overall call-carrying capacity. (If all calls are completed through an intermediate switch, twice as many network links are involved as when all calls are first routed.) To forestall this condition, protective routing techniques are employed. They assess the probability of completion of calls attempted during periods of high use and favor those that are judged likely to be completed. Specific examples of protective routing techniques include:

- **Selective trunk reservation** (STR): activated by a reservation level placed on individual trunk groups, STR controls two link calls more stringently than single-link traffic. When the number of idle trunks in a particular group drops below a specified threshold, the idle trunks remaining are used preferentially to complete single-link (first-routed) calls. In cases of significant, widespread network congestion, alternate routed calls seeking to use the trunk group resources are rejected.
- **Hard-to-reach** (HTR): the ratio of the number of calls accepted by, and the total number of calls presented to, a specific switch, in a given time period, is called the answer-to-bid ratio (ABR). It is computed by analyzing call completions. When the ABR of a switch drops below a given threshold, it is classified as hard-to-reach (HTR). On subsequent bids for connection, the originating switch is informed over SS7 of the HTR status of the destination switch. By building a list of such information, individual switches reduce the number of times they request connection through a congested switch.
- **Selective dynamic overload** (SDO): relieves congestion at an overloaded switch by sending an instruction to all switches connected to it to reduce the volume of their requests for service for a fixed time interval.

To make these strategies work requires a centralized network management authority with the tools needed to detect the onset of congestion, and the real-time ability to tune the decision thresholds used in protective routing schemes on the basis of the latest performance data. AT&T employs a centralized Network Operations Center for this purpose.

10.5 INTEGRATED SERVICES DIGITAL NETWORK

After many years of study by ITU-T (and its predecessor organizations), PTTs are implementing an all-digital network. Called Integrated Services Digital Networks (ISDNs), they are switched, digital networks that provide a range of voice, data, and image transport services through standard, multi-purpose, user-network interfaces. ISDNs are expected to be linked into a worldwide digital network based on 64 kbits/s clear channels. In North America, ISDN deployment is moving more slowly. In part, this is due to the sophistication of the existing public networks, and the availability of competing service providers. In part, it is because digital transmission and digital switching by themselves do not make a network that is attractive to sophisticated users. What does is the exploitation of the capabilities of common channel signaling in a software driven, digital switching and transmission environment. Accordingly, North American carriers have concentrated on Intelligent Networks (INs). At the risk of oversimplification, ISDN can be considered to be a *provider*-driven version of an all-digital network with powerful, software-based capabilities, and IN can be considered to be the *market*-driven version. In the future, it is likely that the two approaches will come together in some fashion since interest in INs is developing in Europe and Japan, and interest in ISDNs is growing in North America. In this Section, we describe ISDNs and in Section 10.6, we describe INs.

10.5.1 Reference Configuration

A formal user-network interface (UNI) has been defined for ISDN. It is shown in the top, left-hand corner of **Figure 10.15**. Each block represents the physical implementation of certain functions:

- **Network termination 1** (NT1): the digital interface point between the network and the customer's equipment. It facilitates functions principally related to the physical and electrical termination of the network at the customer's premises.
- **Network termination 2** (NT2): includes enhanced functions associated with the data link layer and the network layer. It facilitates protocol handling, switching, concentrating, and multiplexing functions associated with PBXs, cluster controllers, local networks, and multiplexers.

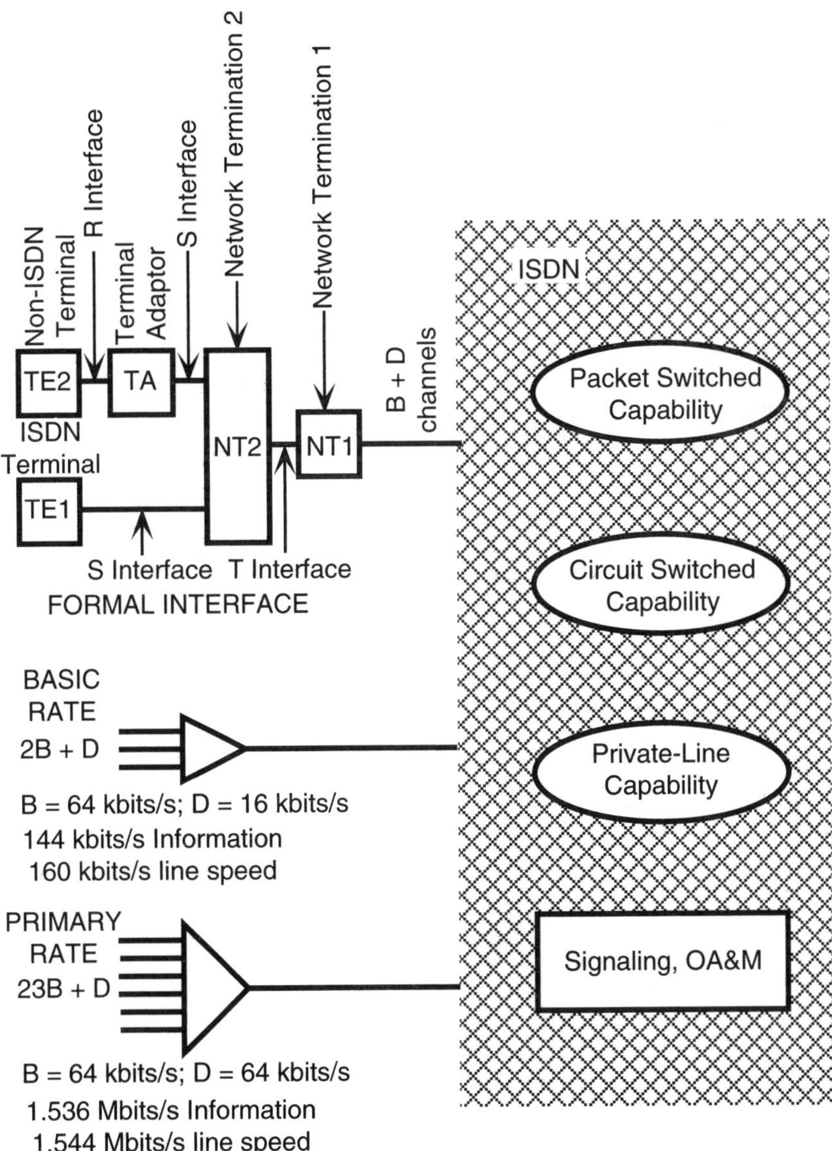

Figure 10.15 ISDN Architecture

- **Subscriber terminal 1** (TE1): terminal that is fully compliant with ISDN requirements. Can connect directly to NT1. Includes digital telephone, data terminal, and similar equipment.
- **Subscriber terminal 2** (TE2): terminal that does not fully comply with ISDN requirements. Must connect to NT1 through a terminal adapter.
- **Terminal adapter** (TA): provides the interface and protocol conversions needed for TE2 to connect to NT1.

ISDN supports two interfaces that connect customer premises equipment to the network. Shown in the lower left-hand quadrant of Figure 10.15, they consist of bearer channels (B-channels) and delta channels (D-channels):

- **Primary rate interface** (PRI): 23 B-channels + 1 D-channel. In North America, the PRI is supplied over a single connection that operates at 1.544 Mbits/s. This is the standard T-1 rate; it contains 1.536 Mbits/s of user and signaling data, and 8 kbits/s for framing the datastream.
- **Basic rate interface** (BRI): 2 B-channels + 1 D-channel. In North America, the BRI is supplied over a single connection that operates at 160 kbits/s. This speed contains 144 kbits/s of user and signaling data, and 16 kbits/s for overhead functions.

These arrangements are service independent, can carry voice, data, and image traffic simultaneously, and can support both switched and private line services. Moreover, data can be transferred in packet or in circuit mode.

10.5.2 ISDN Services

(1) Information Channels

The B-channel supports digital transmission at 64 kbits/s, is circuit switched, and employs user-specified protocols. If desired, subrates of 2.4, 4.8, 8, 16, and 32 kbits/s may be multiplexed on to the channel. Other information channels, called H-channels (for higher) have been defined; they are H0, at 384 kbits/s, H11, at 1536 kbits/s (T-1 user's information rate), and H12 at 1920 kbits/s. The last channel is used on networks that employ ETSI standards where the first level (E-1) multiplexing rate is 2.048 Mbits/s.

(2) Signaling and Supervision Channel

The D-channel supports a packet mode, ISDN protocol for call-control signaling, user-to-user signaling, and packet data. In addition, it is used for operations, administration, and maintenance messages. In the BRI, the speed of the D-channel is 16 kbits/s; in the PRI the speed is 64 kbits/s. When there is no signaling or network-related traffic, the entire channel can be used for customer data.

(3) Bearer Services

Information services available at the user-network interface are called bearer services. They are defined by information transfer mode (circuit or packet), information transfer rate (bits/s), and information transfer capability (speech, audio, etc.).

- **Switched circuit mode**: services include,
  - 64 kbits/s data
  - 64 kbits/s PCM voice
  - 3.1 kHz audio band data.
- **Private line circuit mode**: switched circuit mode services and, in addition,
  - 384 kbits/s data
  - 1536 kbits/s data
  - 1920 kbits/s in Europe.
- **Packet mode services**: employ X.25 protocol,
  - B-channel supports 64 kbits/s synchronous data transfer
  - D-channel supports 16 kbits/s synchronous data transfer.

(4) Switched Digital Services

For applications that require point-to-point, limited time connections to one of several sites, carriers have introduced switched digital services:

- **SW56**: 56 kbits/s switched digital service; this is switched DDS.
- **SW64**: 64 kbits/s switched digital service; this is switched DS-0 or ISDN B channel.
- **SW384**: 384 kbits/s switched digital service; this is ISDN H0.
- **SW1536**: 1536 kbits/s switched digital service; this is ISDN H11.

  SW64, SW384, and SW1536 are supported by ISDN-configured switches.

10.5.3 Signaling System #7

ISDN uses Signaling System #7 (SS7). The network structure and protocol architecture of SS7 for public network applications are shown in **Figure 10.16**. The architecture is a quad structure consisting of signaling end points (SEPs) and signal transfer points (STPs). The STPs are responsible for routing signaling messages from one SEP to another. To provide added reliability, each SEP is connected to the SS7 network by duplicate STPs.

(1) Signaling Network

The signaling network consists of nodes and links that connect Signaling End Points (SEPs) by high-speed data links to a network of Signal Transfer Points (STPs). The SEPs can be Switching Points (SPs), Service Switching Points (SSPs),

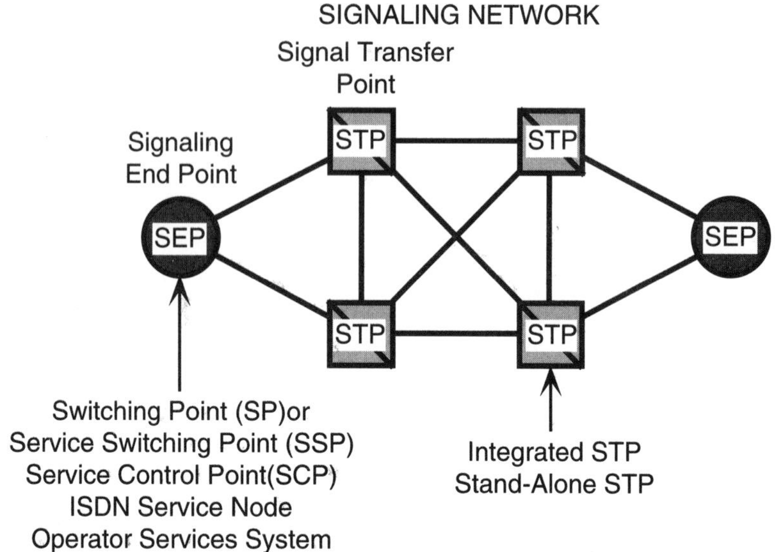

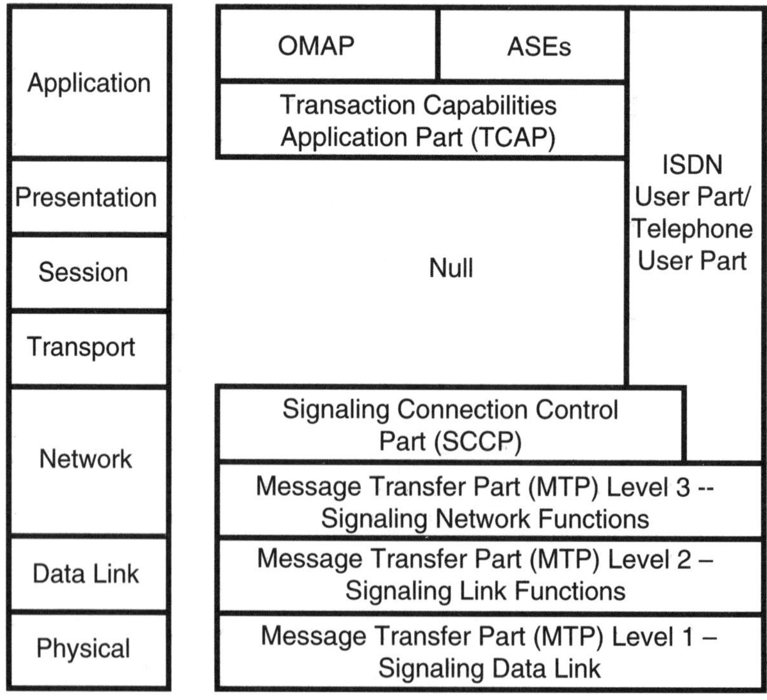

Figure 10.16 SS7 Architecture

Service Control Points (SCPs), ISDN Service Nodes (SNs), or Operator Services Systems (OSSs). The major functions these entities perform are:

- **Switching point** (SP): an end office (or other) switch at which the customer gains access to the network. Using common-channel signaling, the switch provides normal call switching functions—establish, maintain, and disconnect.
- **Service switching point** (SSP): embedded in the SP, the SSP processes calls that require remote database translations. The SSP recognizes special call types, communicates with the Service Control Point (SCP), and handles the calls according to SCP instructions.
- **Service control point** (SCP): provides database and call processing procedures for special calls.
- **ISDN service node**: supports provision of ISDN bearer services and supplementary services (see below).
- **Operator services system**: provides operator assistance, particularly directory assistance services.
- **Signal transfer point** (STP): performs as a link concentrator and message switcher to interconnect SEPs. STPs are stand-alone packet switches. For convenience, they may occupy the same site as other network equipment, including circuit switching nodes.
- **Integrated STP**: when a signaling point has an STP capability and also provides OSI Transport layer functions, it is said to have an integrated STP functionality.
- **Stand-alone STP**: when a signaling point provides only STP capability, or STP and SCCP capabilities, it is commonly called a stand-alone STP.

With the exception of STPs, these entities are implemented in software at the circuit switching nodes.

(2) Protocol Layers

SS7 protocols consist of a Network Services Part (NSP) that corresponds to layers 1, 2, and 3 of the OSI model, a Transaction Capabilities Part (TCP), and an Operations, Administration and Maintenance Part (OMAP). Together, the last two parts correspond to layer 7 of the OSI model. Because of the structure of the signaling environment, procedures that correspond to the Presentation, Session, and Transport layers of the OSI model are unnecessary—in the terms of the model maker, they are *null*. Each of these parts can be described as follows:

- **Message transfer part** (MTP): provides reliable transport of signaling information across the signaling network. This includes the capability to respond to network and system failures to ensure that reliable transfer is maintained. Briefly, the functions performed by the three levels are:

- *Signaling Data Link.* The physical layer, the signaling data link consists of two similar data channels operating in opposite directions so as to provide full-duplex signaling at rates from 4.8 kbits/s to 56/64 kbits/s. In North America, all signaling links are 56 kbits/s; in many countries they are 64 kbits/s.
- *Signaling Link.* The data link layer, the signaling link transfers signaling messages between two signaling points in variable length messages. Known as *signal units* (SUs), the formats of message signal units (MSUs), link status signal units (LSSUs), and fill in signal units (FISUs) are shown in **Figure 10.17**. Error detection is accomplished through the use of a 16-bit CRC. Two forms of error correction are used. The *basic method* employs go-back-N techniques; a method known as *preventive cyclic retransmission* (PRC) uses forward error correction and is employed with satellites and other long propagation delay links. A *signal unit error rate monitor* (SUERM) is employed to monitor the error performance of the

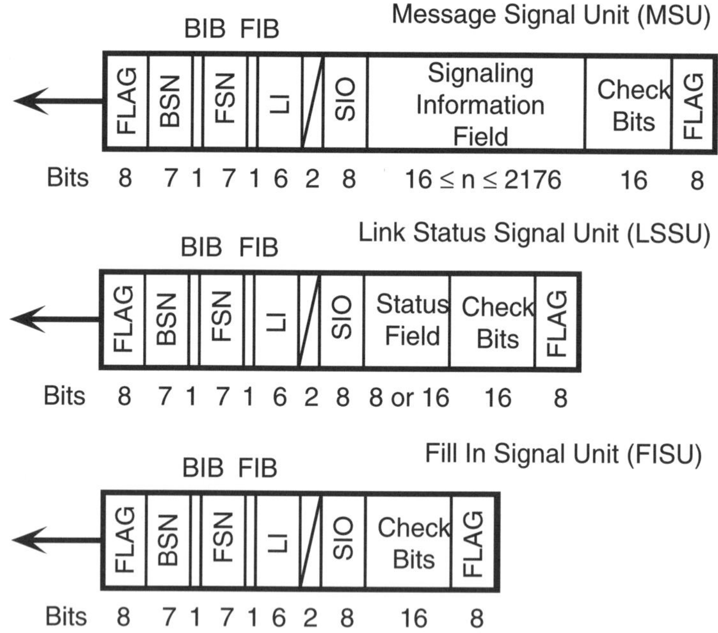

Figure 10.17 SS7 Signal Unit Formats

link and determine if it is of sufficient quality to remain in service. Flow control is accomplished through the use of LSSUs. The congested receiver notifies the sender of its condition with an LSSU that indicates busy. At the same time, it ceases to acknowledge incoming signal units. When there is no message traffic, FISUs are sent so as to sustain the error monitoring task and keep up to date with link performance.

- *Signaling Network Functions.* The lower half of the network layer, the signaling network functions level performs signal unit handling and management tasks. They include routing and recovery from failures.

- **Signaling connection control part** (SCCP): the upper-half of the network layer, SCCP enhances the services of the MTP to provide full OSI network layer capabilities. In particular, SCCP extends MTP's addressing and routing capabilities and provides connectionless (CL-SCCP) and connection-oriented (CO-SCCP) classes of services.
- **Transaction capabilities part** (TCAP): provides a set of tools in a connectionless environment that can be used by an application at one node to invoke execution of a procedure at another node, and exchange the results. TCAP consists of two sublayers—the component sublayer and the transaction sublayer. The component sublayer is concerned with requests for action at the remote end, or the provision of data in response to such a request. The transaction sublayer is concerned with the exchange of messages that contain these components.
- **Operations, maintenance, and administration part** (OMAP): provides the application protocols (ASEs, Application Service Elements) and procedures required to monitor, coordinate, and control all the network resources needed to make communication based on SS7 possible.
- **ISDN user part/telephone user part** (UP/TUP): represents the location of model layers invoked when the network employs the transport capabilities of the MTP and SCCP to provide call-related services to the user. UP/TUP controls the circuit switched connections between the originating and terminating stations. Flow control may be exercised at the UP level. When a sender receives a busy LSSU from a congested receiver, the MTP passes the information to the UP/TUP. The UP reduces the traffic to the congested receiver in steps until a normal condition is restored. TUP represents an interim arrangement employed between carriers until ISDNUP is defined and stabilized.

10.6 INTELLIGENT NETWORKS

ISDN originated with European PTTs seeking to provide networks that could carry voice and data. For several years the concept was not well-supported. What progress was made was made in technology, and *technology push* not *market pull* was the dominant motivation. Eventually, the digital network described above was defined and is enjoying limited implementation. In the meantime, some countries permitted alternative providers to be established to assuage the demands of business users for data communication. In the United States, local exchange carriers, equipment vendors, enhanced services suppliers, and large users bypassed the pure ISDN development in favor of an architecture that uses standard network interfaces to provide new services in a multi-vendor environment. Based on ISDN, and called *intelligent* network (IN), it has the following features:

- Distributes call processing capabilities across multiple network modules
- Employs reusable network capabilities to create a new feature or service
- Employs standard protocols across network interfaces.

IN is an evolving set of capabilities that permits the creation of custom services to suit specific customer needs. Mostly market-driven, IN is a major success for telecommunications providers.

10.6.1 IN Architecture

(1) Functional Components (FCs)

Unique to IN architecture, functional components (FCs) are elementary call processing commands that direct internal resources to implement specific IN services. Combined in different ways, they create a range of network services and features. Some functional component categories are:

- **Transfer control**: transfers control of service logic processing between IN modules.
- **Connection control**: requests SSP (services switching point) to complete any permitted connection between users, or between users and network resources.
- **Network participant interaction**: invokes specific resources in order to interact with users, e.g., speech synthesizer or DTMF digit receiver.
- **Network information management**: on request, permits an SLI (service logic interpreter) to augment, modify, or remove information from other entities so as to change service limits or regime.
- **Processing**: assigns resources to perform actions that satisfy IN service requests.
- **Information collection**: enables collection of network usage data from network modules.

Building network services on these service-independent, functional components ensures that the network will not be dominated by the characteristics of a particular service.

(2) IN Modules

Through the interaction of specially equipped switching systems (end offices and/or local tandems) with systems specifically programmed to provide the logic (service logic) needed to support new services, network operators can introduce them in response to market demands. **Figure 10.18** shows the physical and logical systems

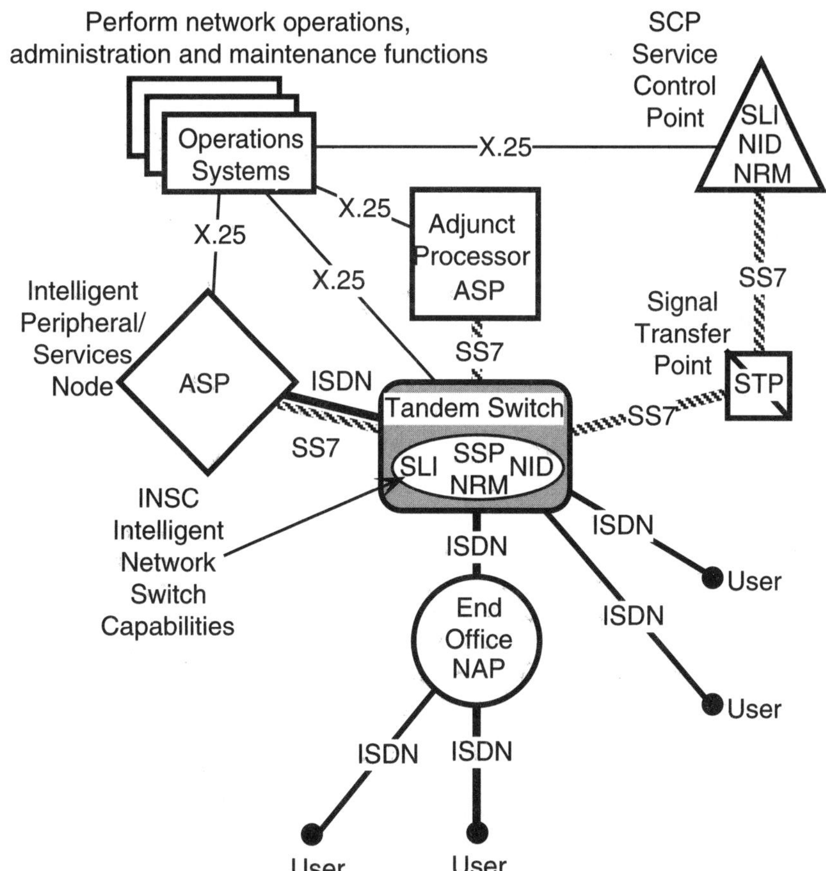

Figure 10.18 Intelligent Network Architecture

In IN, services are supported by a selection of modules resident in switches and other network equipment. Customer circuits are completed over ISDN circuits and signaling between IN resources uses SS7.

that constitute the modules of an intelligent network. The service logic—logic that defines the operation of specific network features or service applications—is contained in service logic programs (SLPs) that reside at service control points (SCPs), adjunct processors (APs), intelligent peripherals (IPs), and service nodes (SNs), or within the network switches themselves (SSPs). In more detail, the important elements of Figure 10.18 are:

- **Service control point** (SCP): consists of processor-based service logic interpreter (SLI), network information databases (NIDs), signaling network interface, and network resource manager (NRM):
  - *service logic interpreter* (SLI). Executes SLPs and handles exchanges between IN modules
  - *network information database* (NID). Contains customer and network information
  - *network resource manager* (NRM). Determines which IN module provides resources to continue call processing. May give routing instructions to establish connection to IN module.

Fully interconnected to IN switches through STPs, SCPs support distributed (network-wide) services. They provide database information and processing procedures for IN calls, and invoke service logic programs in response to messages from IN switching systems. For services such as 800 calling, SCPs may be service-specific, i.e., they only support 800 calling.

- **Signal transfer point** (STP): a packet switch that provides routing for signaling messages between IN equipped switches and SCPs.
- **Intelligent network switch capabilities** (INSC): identify calls associated with intelligent network services, formulate requests for call-processing instructions from service logic elements, and implement instructions; provide IN services to subtending users and lower level switches equipped as network access points. These capabilities are supported by SLI, NID, and NRM systems that perform the operations described above, and by *service switching points* (SSPs) that contain reference call models. They permit recognition of IN call-handling requirements at trigger points, query IN modules for processing instructions, and implement them. SSP software works with modules in SCPs, SNs, APs, and IPs, to provide advanced call processing capabilities within the switch. Intelligent network switch capabilities are deployed at access tandems and selected end-offices. (In Figure 10.18, we show deployment at a tandem switch.) As the demand for specific services grows, control migrates from the associated SCP to the logic modules in the IN equipped switch.
- **Network access point** (NAP): switches (such as the end office in Figure 10.18) not equipped with IN switching capabilities may be equipped as net-

work access points. NAP capabilities permit access to a limited set of IN services by users served by the switches. The NAP routes the call to an IN switching system, identifies the user, and requests IN treatment.

- **Adjunct processor** (AP): contains an adjunct service point (ASP) that responds to requests for service processing. Directly connected to an IN switching system (to assure timely communication), the adjunct processor supports services that require rapid response to users' actions. Examples are services that control provision of dial tone, or that may require rapid reconfiguration of call connections.
- **Intelligent peripheral** (IP): contains an adjunct service point (ASP) that responds to requests for service processing. Provides and manages resources such as speech recognition, voice synthesis, announcements, and digit collection equipment as required by IN service logic.
- **Service node** (SN): provides service logic and implementing resources in a single network entity. Communicates directly with a single IN-equipped switching system using ISDN basic, or primary rate interfaces. Contains adjunct service point (ASP) that responds to requests for service processing. Provides access to service logic programs that support a specific service, or group of services, or a full range of service functions (including attendant services). Allows network interaction with user including collecting dialed digits, receiving spoken inputs, and providing customized announcements.
- **Operations systems**: perform network operations functions such as resource administration, surveillance, testing, traffic management, and data collection.
- **Network interfaces**: using conventional analog or ISDN channels for access, they support standard protocols:
  - *SS7 Transaction Capabilities Application Part* (TCAP). The application layer protocol used between switching systems and SCPs or adjunct processors.
  - *ISDN interfaces*. Used between switching systems and intelligent peripherals or service nodes.
  - *X.25-based protocols*. Used for interfaces between network components and operations systems.

IN calls are identified at the originating SSP by means of a *trigger* table. When a quantity associated with the call matches a quantity carried by the table, information in the table is used to generate a request for instructions from the SLI designated by the NRM (the SLI may be in the switch, or in the SCP). Normal call processing is suspended until the SSP receives the reply. The SLI will respond with instructions for the SSP to execute a series of FCs that provide the service requested. IN architecture is designed to be flexible. Because the same FCs can be performed in different modules, configurations can be customized to suit traffic and performance requirements.

10.6.2 IN Services

IN services are provided by manipulating connections between nodes and addressable entities—such as other nodes or a user's station—anywhere in the network. For instance, three-way calling, a service available among extensions served by a PBX or Centrex, can be extended to remote stations using IN capabilities. During a conversation between party A in New York and party B in Chicago, suppose they wish to consult party C in Atlanta. The major steps to create the three-way call are:

- **Three-way calling**: a flashhook signal from A triggers the SSP serving A to request instructions from the assigned SLI for a three-way connection. As a result:
  - B is split from A temporarily and placed on hold
  - A receives dial tone and enters the directory number for C
  - The SCP provides routing information so that A can be joined to C
  - When A has established contact with C, A gives a second flashhook signal
  - The serving SSP joins B to A and C to produce a three-way talking state.

When A and B have consulted with C, A gives another flashhook signal, and the connection reverts to a two-way talking state between A and B. Other examples of IN services include the following:

- **800 database service**: provides toll-free service to the caller (the 800 service subscriber purchases service by serving area bands and is billed for all costs incurred). The associated SCP has information on how to handle the dialed 800 number; it can vary by time of day, day of week, originating number, designated IXC, and other items. 800 database service is offered by both LECs (intra–LATA) and IXCs (inter–LATA).
- **Network ACD**: enables subscribers to control the destination to which 1-800... calls are routed to match peak loads with availability of attendants and provide relief for overflow from congested automatic call distributers (ACDs). The associated SCP maintains a knowledge of the idle/busy status of a stored list of terminations and directs overflow traffic to a network address on the basis of instructions provided by the subscriber.
- **Area-wide Centrex**: based on an SCP that is connected to all switches within a LATA, provides PBX-like operations from designated stations across the LATA.
- **Custom local area signaling services** (CLASS, also known as TouchStar): a group of switch-controlled, common channel signaling features that use existing customer lines to provide end users with call management capabilities. Based on the ability to know the calling party's number (CPN) from messages on the signaling channel, they include:

- *automatic call back* (also known as call return): Allows the customer to automatically call the last incoming caller.
- *selective call forwarding* (also known as preferred call forwarding): Allows a customer to forward calls to another station based on the CPN.
- *automatic recall* (also known as repeat dialing): Allows the customer to recall the last party called.
- *distinctive ringing* (also known as call selector): Allows a customer to assign a distinctive ringing pattern to incoming calls on the basis of the CPN.
- *selective call rejection* (also known as call block): Allows a customer to reject calls on the basis of the CPN.
- *customer originated trace* (also known as call tracing): Allows the terminating party to request a trace of the last call received.
- *calling number delivery* (also known as caller ID): Allows specially equipped terminating station to display CPN. The number is sent during ringing.
- *calling name delivery*: Allows specially equipped terminating station to display the directory listing for the CPN. Directory information is sent during ringing.
- *bulk calling line identification* (also known as call tracking): Allows PBX and Centrex customers to receive call-related information on incoming calls.
- *voice messaging*: Enables a subscriber to route incoming calls to a remote voice storage system.
- *call waiting*: Provides an audible alert to a party that another caller is seeking contact. By using the hookswitch, the terminating party can place the original party on hold, and speak to the interrupting caller. A second hookswitch signal restores the original connection.
- *call waiting identification*: Presents the directory listing (and number) when the call waiting alert is given.

• **Alternate billing service**: through the use of a line information database, enables the calling party to bill a call to a number other than the called number (i.e., calling card or third-party number).

• **Mobile communication services**: IN can be extended to mobile cellular communications, paging, and other personal communications systems. Using SS7 for call control and database transactions, services such as selective call forwarding can be provided to these subscribers (see § 14.2.3(3)).

Through the use of functional components and the exploitation of common-channel signaling, these services will evolve as social and business environments change.

10.6.3 International Standards

(1) Objectives

Regional IN activity in Australia, Japan, North America, and Europe has resulted in ITU-T efforts directed to international IN standards. The objectives are to develop an architecture that, among other things, will accomplish the following:

- Be applicable to all telecommunications networks (i.e., PSTNs, narrowband-ISDNs, broadband-ISDNs, packet switched data networks, and mobile networks)
- Enable service providers to define services independent of service-specific developments by equipment suppliers
- Evolve to reflect implementation experiences, new technological opportunities, and market evolution.

Figure 10.19 shows the IN Conceptual Model developed by ITU-T working groups. The four planes include:

- **Service plane**: describes IN services and service features (SFs) from the user's perspective. Does not describe implementation. May include management services.

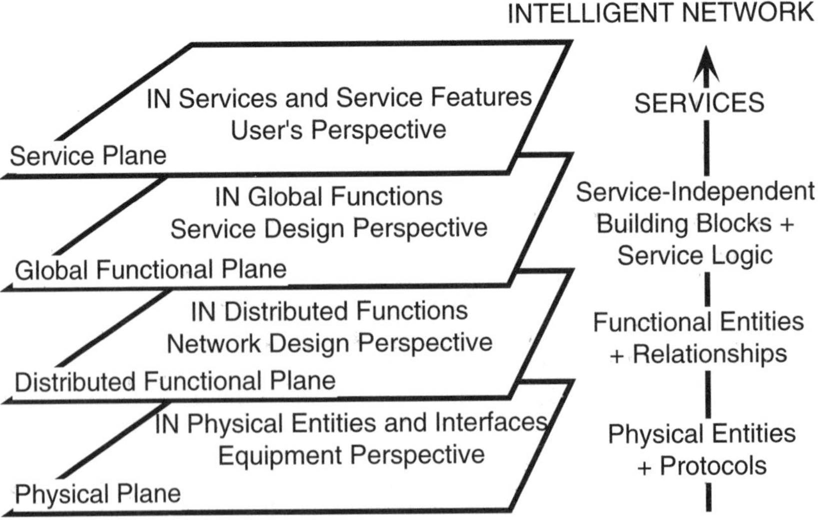

Figure 10.19 Model of Intelligent Network

- **Global functional plane**: describes IN functions from a service design perspective that treats the network as a single entity. Does not describe distribution of IN functions in the network. Contains call-processing model and service independent building blocks (SIBs). SIBs are combined with service logic (SL) to produce services in the service plane.
- **Distributed functional plane**: describes IN distributed functions from a network design perspective. Describes the IN network in terms of functional entities (FEs, i.e., units of network functionality) and the relationships (i.e., the information that flows) between them. Does not describe how the functionality is implemented and deployed. Sequences of functional entity actions and associated information flows are used to create SIBs in the global functional plane.
- **Physical plane**: describes the types of physical entities, the functions they implement, and the protocols by which they communicate. The combination of physical entities and protocols are used to create functional entities and relationships in the distributed functional plane.

(2) IN Capability Set 1

In IN Capability Set 1 (CS-1), ITU-T addresses the first stage of IN evolution. CS-1 includes services that are of high commercial value and can be implemented without significant impact on existing installations. They are *Type A* services, that is to say they are single-user, single-ended, single point-of-control and single bearer-services. Invoked by a single user, they carry no requirement for end-to-end messaging or control, and are provided over the same medium as the communication. Some examples are call forwarding, 1-800... calling, abbreviated dialing, customized announcement, custom ringing, virtual private networks (VPNs), universal personal telecommunications (UPT), etc. VPN services provide a subscriber with private network capabilities (such as private numbering plan and dialing restrictions) over the PSTN. UPT services provide a subscriber with a unique personal number that can be used across any number of networks and with various access arrangements. *Type B* services, that is to say services that can be invoked at any point during the call and may require end-to-end messaging and control, will follow in other Capability Sets.

(3) Distributed Functional Plane

Figure 10.20 shows the architecture of the distributed functional plane in CS-1. It consists of the following functional entities:

- **Call control agent function** (CCAF): provides end-user access to IN call processing over analog lines, ISDN BRI and PRI, analog multifrequency signaling circuits, and SS7 interfaces.

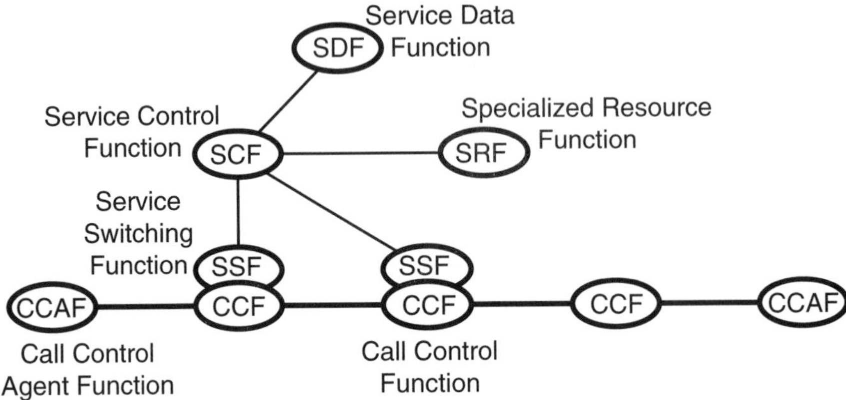

Figure 10.20 Distributed Functional Plane Architecture
IN Capability Set-1 creates services by passing call activity to CCF/SSF for processing. SSF passes the request to an SCF. The SCF invokes the appropriate service logic and interacts with SSF, SRDF, and SDF to provide the service.

- **Call control function** (CCF): provides call connection, maintenance, and termination.
- **Service switching function** (SSF): recognizes calls requiring IN features. Interacts with call processing and service logic to provide the services requested.
- **Service control function** (SCF): contains service logic that controls the implementation of IN services. Interfaces with service switching function, specialized resource function, and service data function.
- **Specialized resource function** (SRF): controls resources used by end-user to access network (e.g., DTMF receivers, protocol conversion, announcements).
- **Service data function** (SDF): provides access to service-related data.

For a call that requests IN service, the process can be described as follows. When the CCAF receives call activity from an end-user (DTMF digits, for instance), it passes it to the CCF/SSF for processing. If they constitute a request for IN service, SSF passes the request to an SCF. The SCF invokes the appropriate service logic and interacts with SSF, SRF, and SDF as necessary to provide the service requested by the end-user.

(4) Physical Plane

Figure 10.21 shows the architecture of the physical plane in CS-1. It consists of physical entities (PEs) that implement functional entities (FEs). Allowing for the difference in nomenclature, it is similar to Figure 10.18.

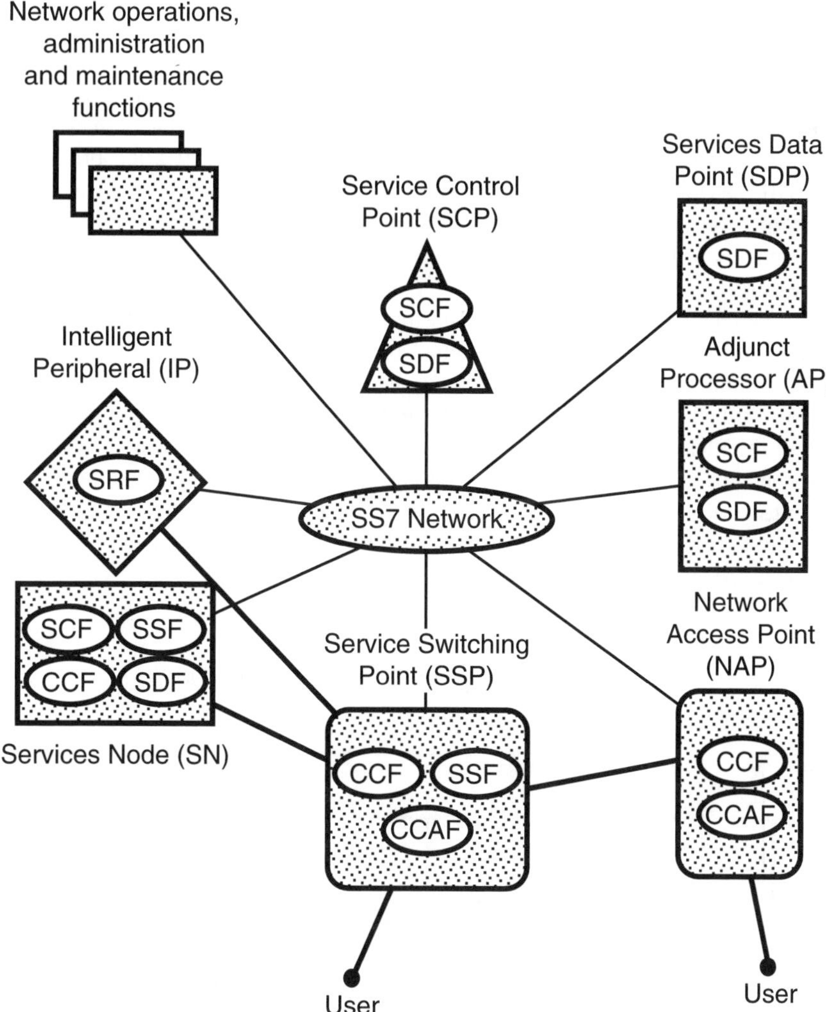

Figure 10.21 Physical Plane Architecture
The physical plane in IN Capability Set-1 consists of physical entities and network entities. Allowing for the difference in nomenclature, it is similar to Figure 10.18.

10.7 BROADBAND ISDN

10.7.1 Higher-Speed Network

Already, as a stand-alone concept, ISDN is obsolete. So long in development, it has been surpassed by the requirements of a future information age in which switched video, high-resolution graphics, television distribution, and highways between super-computers, will be normal communication requirements. Indeed, both ISDN and Intelligent Network are less than equal to the task. Based on a 64 kbits/s channel speed, and intended to handle voice and data up to the primary rate (1.544 Mbits/s), they are *narrowband* networks. To satisfy the communication needs of the future, a higher speed, supplementary network is needed. Called broadband ISDN (BISDN), it is expected to distribute information to subscribers at rates up to approximately 600 Mbits/s, and to support interactive, switched services up to approximately 150 Mbits/s. **Figure 10.22** shows a wide range of communication services arrayed according to channel speed, and the time that they are likely to occupy the channel. Thus, television and music are services that generate continuous signals that are likely to fill a channel for an extended period of time. On the other hand, data and voice services occupy the channel less intensely, and can share channels. ISDN covers the segment that extends to channel speeds equal to the speed of the PRI, and to channel occupancies of $< 50\%$. BISDN covers the segment that extends from channel speeds of 64 kbits/s to 600 Mbits/s and channel occupancies up to 100%. Not intended to replace (narrowband) ISDN, BISDN supplements and extends the range of services that can be offered by carriers. ITU-T divides broadband ISDN services into two major categories:

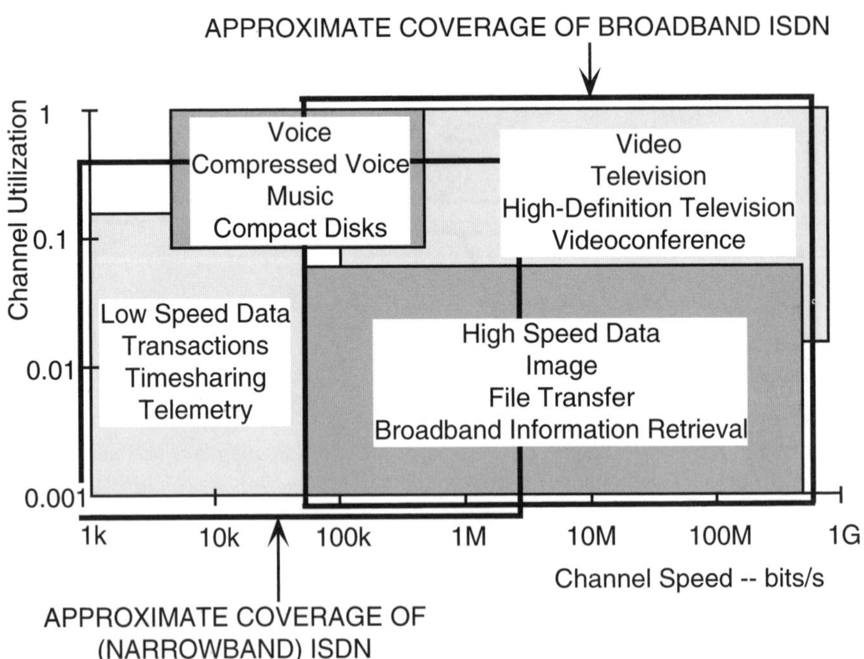

Figure 10.22 Service Coverage of Narrowband ISDN and Broadband ISDN

- **Interactive broadband services**: those in which there is an exchange of video and/or high-speed data information at rates between 64 kbits/s and 150 Mbits/s. They are divided into three types of services:
  - *conversational services*: They provide the means for real-time, broadband dialog between two users, or a user and a service provider. Important services in this category are video telephony and videoconference.
  - *messaging services*: They provide user-to-user, store-and-forward, broadband communication. An important service in this category may be video mail, i.e., storing a video message until the recipient requests to see it.
  - *retrieval services*: They provide the ability to retrieve broadband information from remote centers. An important service in this category may be broadband videotex, i.e., on request, the user is able to receive a mixture of sound, video, and data that describes a topic of interest.
- **Distributed broadband services**: those in which the information transfer is one-way, from distributor to customer at rates up to 600 Mbits/s. They are further divided into two categories:
  - *broadcast services*: They distribute broadband information without individual presentation control. A common example is broadcast television.
  - *custom services*: They distribute broadband information with individual presentation control. A common example is the electronic delivery of movies, or similar items, on request, i.e., pay-per-view television service.

In countries in which governments provide and control electronic communication, BISDN is a natural extension of the PTT's role. In the United States, it remains to be seen whether BISDN will become reality. Broadcast, broadband services are already provided to households by independent television broadcasters and cable television operators. Will carriers and cable system operators form alliances to transport data, voice, and television services to homes? Will they build what some are calling an *information highway* to every residence? Will they be part of the *national information infrastructure* (NII)? Only time will tell—but proposals for *wired cities* and a *wired nation* have been around since the late 1960s without result.

Figure 10.23 shows basic user interfaces and network access arrangements for ISDN and BISDN. On the subscriber side, the maximum digital rate for BISDN is at least 100 times the digital rate for ISDN, making the use of optical fiber necessary. On the network side, for transport, BISDN will use optical fibers or digital microwave radios that employ SONET and SDH multiplexing, and, for switching, BISDN will employ ATM.

10.7.2 ATM Bearer Service

ATM was introduced in § 9.2.4. It is a high-speed, fast packet switching (routing) technique employing short, fixed length cells that are statistically multiplexed over

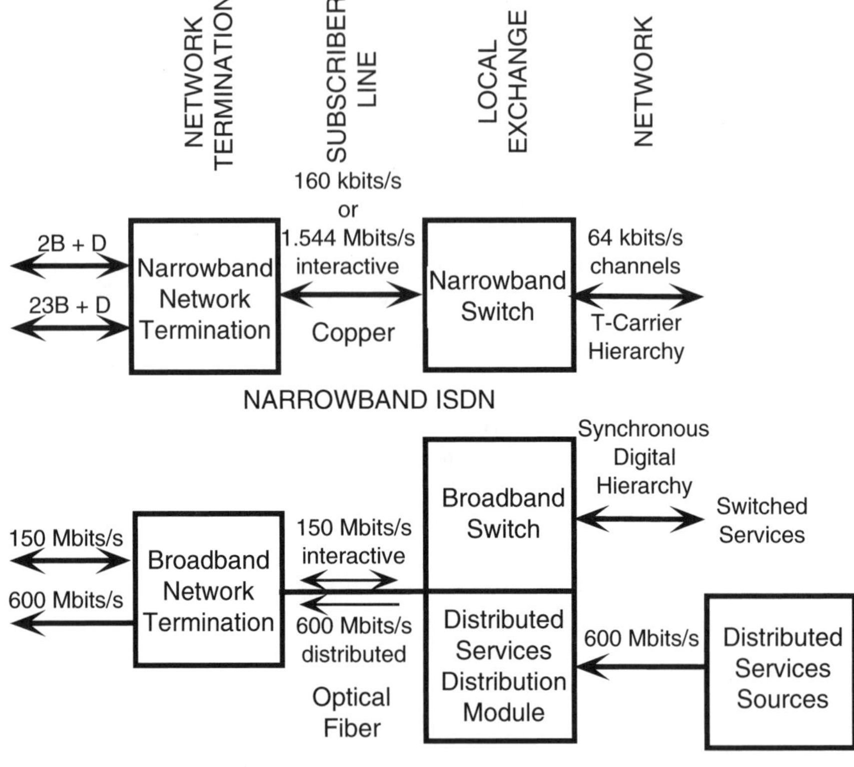

Figure 10.23 Interfaces and Access Arrangements

virtual connections. To accommodate large data packages, the protocol stack includes an adaptation layer.

(1) ATM Cell

Figure 10.24 shows the arrangement of an ATM cell. It consists of 48 octets of payload and 5 octets of header information. The headers used at the UNI and NNI are different, as follows:

- **ATM UNI header**: consists of the following:
  - 4-bit generic flow control (GFC) field that is used to assist in controlling the flow of traffic at the UNI.
  - 24-bit connection identifier that explicitly associates each cell with a specific virtual channel on a physical link. Consisting of two parts, a 16-bit virtual channel identifier (VCI) and an 8-bit virtual path identifier (VPI),

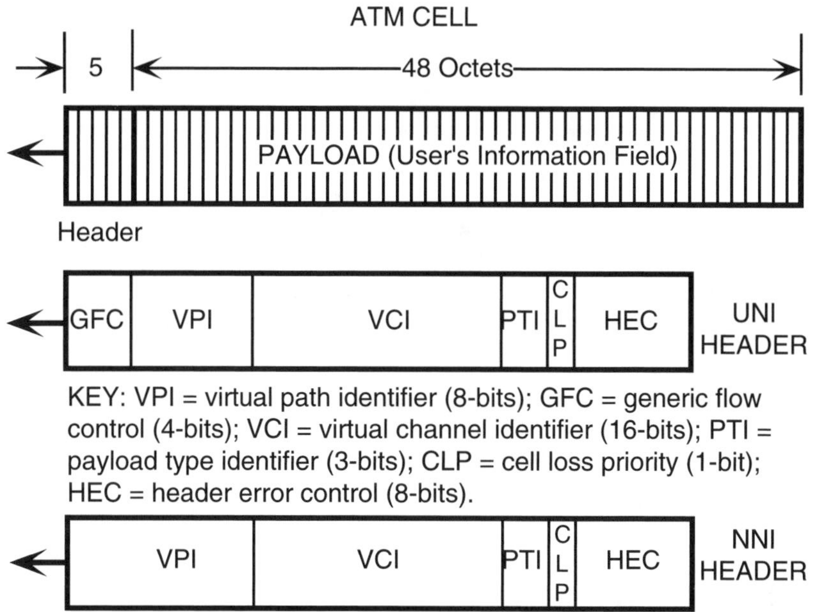

Figure 10.24 ATM Cell Format

the connection identifier is used in multiplexing, switching, and routing cells through the network. A connection identifier is assigned when the connection is established, and is retained for the duration of the connection.

- 3-bit payload type identifier (PTI) that indicates whether the cell contains upper-layer header information or user data.
- l-bit cell loss priority (CLP) field used to identify cells that can be discarded if the network has to discard cells to alleviate congestion.
- 8-bit header error control (HEC) that is used for error detection and correction of the header.

- **ATM NNI header**: similar to UNI except that the GFC field is replaced by four additional VPI bits to make the VPI field 12-bits at the NNI.

(2) Protocol Layers

Figure 10.25 shows the protocol layers for ATM bearer service. They are:

- **Physical layer**: the physical layer defines the transport of digital signals over multiplexed connections in a synchronous digital network. The bearer is likely to be optical fiber (SONET) or digital microwave radio (SDH).
- **ATM layer**: multiplexes and demultiplexes cells belonging to different connections identified by unique virtual channel identifiers (VCIs) and/or virtual

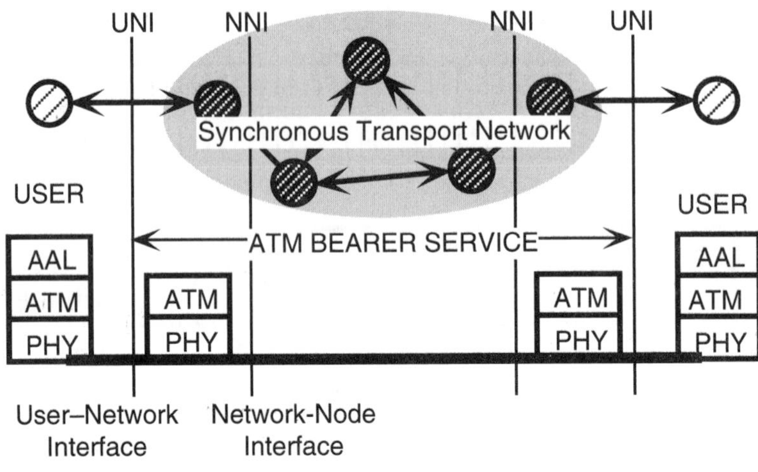

ATM PROTOCOL LAYERS

User Layer	User application layer.
ATM Adaptation Layer	Adapts user information to fixed cell, ATM environment
ATM Layer	Multiplexes cells, translates connection identifier for switching cells, passes information to AAL without processing, performs flow control at UNI.
Physical Layer	Provides physical transport of digital signals on synchronous digital network.

TYPES OF ATM ADAPTATION LAYER

AAL Type	1	2	3/4	5
Bit Rate	Con-stant	Variable		
Connection Mode	CO		CO CL	CO

CO = connection-oriented CL = connectionless

Figure 10.25 ATM Bearer Service
Designed to route BISDN traffic over SONET (or SDH) facilities using ATM nodes.

path identifiers (VPIs). Translates VCI and VPI at switches and cross-connection equipment. Passes cell *payload* (48 octets) to and from the ATM adaptation layer without error control or similar processing. Performs flow control at UNI. Implements connection-oriented and connectionless services.

- **ATM adaptation layer** (AAL): when sending, AAL receives different classes of information from higher layers in service data units (SDUs) and adapts them to occupy fixed length, ATM cells. When receiving, AAL receives data in 48 octet payloads from ATM layer and reassembles it to fit user's service data unit requirements. AAL-specific information passed to peer AALs (information field length and sequence numbers, for instance) is contained in the payload not the cell header. AAL is divided into two sublayers, *segmentation and reassembly* (SAR) sublayer and *convergence sublayer* (CS). Four classes of services are supported by AAL:
  - *Class A*: A time relationship exists between sender and receiver, the bit-rate is constant, and the service is connection-oriented (e.g., voice or fixed-rate video).
  - *Class B*: A time relation exists between sender and receiver, the bit-rate is variable and the service is connection-oriented (e.g., variable bit-rate audio and video).
  - *Class C*: No time relation exists between sender and receiver, the bit-rate is variable and the service is connection-oriented (e.g., connection-oriented data).
  - *Class D*: No time relation exists between sender and receiver, the bit-rate is variable and the service is connectionless (e.g., datagram, SMDS).

(3) Types of AAL

Five types of AAL layers have been defined. As a practical matter, types 3 and 4 have been combined so that AAL types 1, 2, 3/4, and 5 are in use. Shown in Figure 10.25 they can be described as follows:

- **AAL type 1**: supports constant bit-rate, connection-oriented services that have specific timing and delay requirements (voice and video, for instance). Provides segmentation and reassembly, may detect lost or errored information and recovers from simple errors. Error indications such as loss of timing, flow problems and corrupted user information are passed to the appropriate management entity.
- **AAL type 2**: supports connection-oriented, variable bit-rate services (video and audio, for instance). Provides segmentation and reassembly, and detection and recovery from cell loss or wrong delivery. The variable bit-rate complicates timing recovery and some cells may not be filled.
- **AAL type 3/4**: an all-purpose layer type that supports connection-oriented and connectionless variable bit-rate data services. Two operating modes are defined:

- *message mode*: Each service data unit (SDU) is transported in one interface data unit (IDU). Employs cyclic redundancy checking and sequence numbers. Lost or corrupted units may be repeated. Flow control is optional.
- *streaming mode*: Variable length SDUs are transported in several IDUs that may be separated in time.

- **AAL type 5**: supports connection-oriented, variable bit-rate, bursty, data services, on a best effort basis. Performs error detection, but does not pursue error recovery. AAL type 5 is essentially a connection-oriented only, type 3/4 layer. AAL 5 is also known as the *simple and efficient* layer (SEAL).

(4) CS/SAR Functions

By breaking them up in to small cells at the sending point, and reassembling them at the receiving point, convergence, segmentation, and reassembly are activities that facilitate the transport of large data packages. **Figure 10.26** shows the principle. The information to be transferred is contained in the service data unit (SDU) that can be up to 65535 octets long. In the header (and possibly a trailer), it includes information germane to the application. The SDU is divided up into 44 octet segments that become frames in the convergence sublayer (CS). The CS adds a header and trailer to form CS PDUs. They are encapsulated in 48 octet frames in the segmentation and reassembly sublayer (SAR). The SAR adds a header and a trailer. From the SAR, the frames are passed to the ATM layer where they are joined to 5 octet headers to become the 53 octet cells of Figure 10.24. The headers and trailers added in the type 3/4 ATM Adaptation Layer contain the following information:

- **Convergence sublayer**: in the header:
  - *common part indicator*: 1 octet used to set units for subsequent fields in the header and trailer (e.g., buffer allocation size).
  - *beginning tag*: 1 octet that is duplicated in the trailer. Incremented frame by frame, *Etag* provides association between header and trailer at the receiver.
  - *buffer allocation size*: 2 octets that inform the receiver of the maximum size of the buffer required to handle the payload. *BASize* is greater in streaming mode than in message mode.
- **Convergence sublayer**: in the trailer:
  - *alignment*: 1 octet of 0s used to provide bit alignment for the trailer.
  - *ending tag*: 1 octet that is duplicated in the header.
  - *length*: 2 octets that provide the length of the payload. Some frames are not filled with payload data, and must be buffered to size. See Figure 10.26.
- **Segmentation and reassembly sublayer**: in the header:

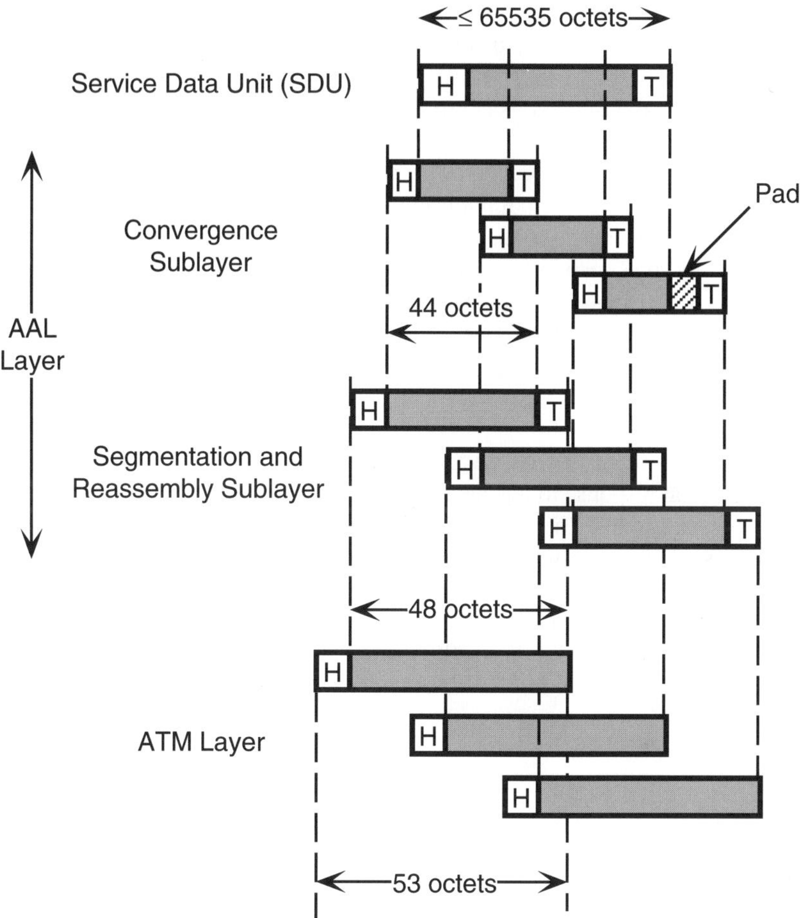

Figure 10.26 Principle of Convergence, Segmentation, and Reassembly in the ATM Adaptation Layer

Large system data units are segmented and formed into cells at the sending point. At the receiving point they are reassembled into the original SDU.

- *segment type*: 2 bits that indicate whether segment is first, continuation, or last in the string, or a single segment.
- *sequence number*: 4 bits that are incremented segment by segment. Used to identify lost or inserted segments.
- *multiplexing identifier*: 10 bits that allow multiplexing several user connections on a single network connection. For instance, *MID* is used to manage the interconnection of several terminals through a LAN gateway to servers on an ATM network.

- **Segmentation and reassembly layer**: in the trailer:
  - *length indicator*: 6 bits that provide the length of the payload.
  - *frame check sequence*: 10-bit remainder from cyclic redundancy check calculation. Used to detect errors in segment.

Headers and trailers associated with other types of AALs differ in detail.

CS/SAR functions are common to several kinds of packet switching. In §s 12.1.3 and 12.1.4 we describe their use in frame relay and switched multi-megabit data services (SMDS) networks, in § 13.3.2 we describe a private frame relay network in which cell relay is used inside the network (NNI), and a mixed network that includes an ATM backbone (see Figure 13.18).

10.7.3 Promise of B-ISDN

Over the years, virtually every advance in the network capabilities of the carriers has been made in response to business needs. Later, the same capabilities have been adapted to residential uses. What BISDN has encouraged is the development of new, high-speed technology. SONET and SDH are being incorporated into carrier's networks, and ATM is a viable, high-speed switching technology. In § 12.1.4 and § 13.3.2, we describe ways in which they are used in the public and private networks that support commerce and industry. It is here that immediate revenues can be generated, and the ideas for an *information superhighway* can be put to the test. When fully worked out, it will extend to residences.

11

Regional Networks

In this Chapter, we address some of the networks that make distributed data processing a reality. Using them, it is possible to collect data, create information, and distribute it within offices, over floors of buildings, in entire buildings, across campuses, and throughout urban cores.

11.1 LOCAL AREA NETWORKS

The pervasive deployment of microcomputers, and the burgeoning uses of desk-top processing, have encouraged the development of local area networks:

- **Local area networks**: non-switched, data communication networks that employ a common bearer to interconnect information processing devices within a limited area so as to serve groups of users (who share the resources of their trade). The abilities of LANs to serve the needs of diverse communities of users for specialized data processing is provided by employing the following:
  - a common bearer shared by all stations
  - a precise channel access procedure
  - a high operating speed.

In order to satisfy the needs of many sources, LANs employ megabit speeds; 1, 4, 10, and 16 Mbits/s are common, and 100 Mbits/s is available. The individual stations they connect may be in close proximity to each other, scattered over one floor of a multi-story building, distributed throughout the entire building, or situated on a commercial or academic campus. A typical network may be used to link as many as a hundred users together, as well as file servers, printers, hosts, and the like.

11.1.1 Implementation Options

(1) Baseband vs. Wideband

A basic difference among LANs is the way in which signals are carried on the common bearer:

- **Broadband** (or wideband): signals are carried on parallel channels over coaxial cable using frequency division multiplexing (FDM, see § 8.1.1). The technique is similar to CATV. Devices on the network are assigned to channels they share with other devices. In large systems, two cables may be used to increase the number of channels and/or segregate the channels by class of application. Like other quasi-analog techniques, broadband LANs find limited use today (e.g., in a video-intensive environment). They are being replaced by high-rate digital networks.
- **Baseband** (or narrowband): digital signals are carried directly on the common bearer. In a baseband network, a limiting factor is the data rate of the common channel. It affects the rate at which frames can be transported, and the distance over which they are sent (see § 4.5.4). Because of this, many baseband networks limit the total number of stations, the distance between stations, and the overall length of the bearer. Baseband systems are less expensive to buy, and easier to implement and maintain than broadband systems. In addition, because they are digital, they can take advantage of a continuing stream of device developments that are improving speeds and decreasing costs. They are the dominant networks for today's data applications.

(2) Bus vs. Ring

Another basic difference among LANs is the configuration of the common bearer. It may assume the form of a bus or a ring:

- **Bus**: all stations are connected to the same bearer—the *bus*—which is open ended. The loss of a station should have no effect on the others. Failure of the bus, however, will affect all stations, and may cripple the network. In a bus configuration, each station is connected to all other stations and a protocol is required to ensure that only one station sends at a time (medium access control protocol, MAC).

- **Ring**: each station is connected to two other stations so that there is a continuous, single-thread connection among all stations. A station receives messages from the station that precedes it on the ring, and sends messages to the station that follows it on the ring. In turn, the stations receive permission to transmit messages through the receipt of a *token*. The failure of a link or a station will affect all stations, and will degrade the performance of the network.

(3) Contention vs. Authorization

Yet another basic difference among LANs is the way in which access is gained to the bearer so that a station can transmit data without interference. It is obtained by contention, or by authorization:

- **Contention**: when using a contention technique, each station competes for transmission resources. They may:
  - send when ready (ALOHA)
  - send when ready *and* the bus is not occupied (carrier-sense, multiple access—CSMA)
  - send when ready *and* the bus is not occupied *and* listen for a collision (carrier-sense, multiple access, collision detection—CSMA/CD).

Under conditions of light traffic, the probability of a collision between two messages is low, and contention works well. Under conditions of heavy traffic, the probability of a collision between two messages is high, and so much time is spent in re-sending messages that almost nothing gets through.

- **Authorization**: each station sends when ready *and* authorized. Usually, authorization is granted through possession of a token—a data message that is passed from station to station. Operating in this manner, the possibility of a collision is zero; however, the speed of operation is slowed by the need to pass the token between stations and for stations that have nothing to send to handle the token. Under conditions of light traffic, token-passing is slower than contention. Under conditions of heavy traffic, token-passing is faster and more reliable than contention.

(4) Practical Topologies

Rings or busses are convenient descriptors. They have an intellectual appeal that makes the differences between LANs readily understood. Also, they have an historical appeal. Until the mid-1980s, all local-area networks were laid out in some fashion that mimicked these basic figures and the common highway ran to within a short distance of every user. With the development of better technologies, however, the LAN function has gradually been concentrated at a *hub*. Here, the LAN is constructed in miniature, and cables (and twisted pairs) connect to the users' terminals. The result is facility wiring that exhibits a star-like topology, and a central point that performs all of the functions that used to be performed by a distributed LAN. As you read the remainder of this Chapter, remember the concepts are bus and ring, but the reality may be a hub in a wiring closet to which all users are connected. Under this circumstance, the media access routines all take place on a back plane (or equivalent) that supports individual connections to users.

11.1.2 IEEE Standards

The Institute of Electrical and Electronics Engineers (IEEE) has pioneered in the development of standards for local-area networks. As a result of their work, four LAN specifications have been developed. Their scope is illustrated in **Table 11.1**. In addition, IEEE Committee 802.6 has developed standards for metropolitan-area

Table 11.1 Local Area Network Standards

OSI Layers	IEEE 802 Layers	Local Area Network Implementations			
Data Link Layer	Logical Link Control LLC	IEEE 802.2 Unacknowledged connectionless service Connection-mode service Acknowledged connectionless service			
	Medium Access Control MAC	IEEE 802.3 CSMA/CD	IEEE 802.4 Token-bus	IEEE 802.5 Token-ring	FDDI Token-ring
Physical Layer	Physical Layer	Baseband Coaxial cable 10 Mbits/s Baseband Twisted pair 1 Mbits/s Broadband Coaxial cable 10 Mbits/s	Broadband Coaxial cable 1, 5 and 10 Mbits/s Carrierband Coaxial cable 1, 5 and 10 Mbits/s	Baseband Twisted pair 1, 4 or 16 Mbits/s	Baseband Optical fiber 100 Mbits/s
		←BUS→		←RING→	

networks (MANs, described in § 11.5). The IEEE standards are organized around the layer hierarchy shown in **Figure 11.1**. In the IEEE LAN model, layer #2 of the OSI model is divided into the Logical Link Control (LLC) sublayer and the Medium Access Control (MAC) sublayer:

- **Logical link control sublayer**: defines the format and functions of the protocol data unit (PDU) that is passed between service access points (SAPs) in the source and destination stations.
- **Medium access control sublayer:** defines the format and functions of headers and trailers that are added to PDUs so that entire frames can travel between source and destination addresses over particular style LANs.

Because the decision to use a specific LAN defines the routing, station addressing, error control, and flow control techniques that will be used locally, only the LLC/MAC layer and the physical layer are involved in message transfers across it.

(1) Protocol Data Unit Format

Based on HDLC, IEEE 802.2—Logical Link Control (LLC), specifies the format in which data will be exchanged across the medium and how the stations will be addressed. The term Protocol Data Unit (PDU) is used to identify the frames that are employed. Their format is shown in **Figure 11.2**. The addresses displayed are service access point addresses, not physical device addresses. The addresses of the sending and receiving devices are included in the MAC header. Further, the control field may be one or two bytes long, depending on the function the PDU performs. The I (information) format is used for numbered information frame transfers

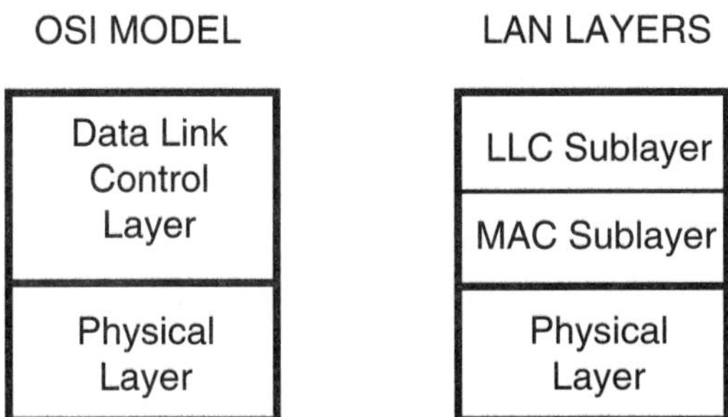

Figure 11.1 Comparison of Layers in OSI Model and IEEE 802 LAN Model
IEEE divides the Data Link Layer into two sublayers. Logical Link Control Sublayer defines the format of the protocol data unit (PDU). Medium Access Control Sublayer defines information that is added to the PDU to facilitate transfer using a specific access technique.

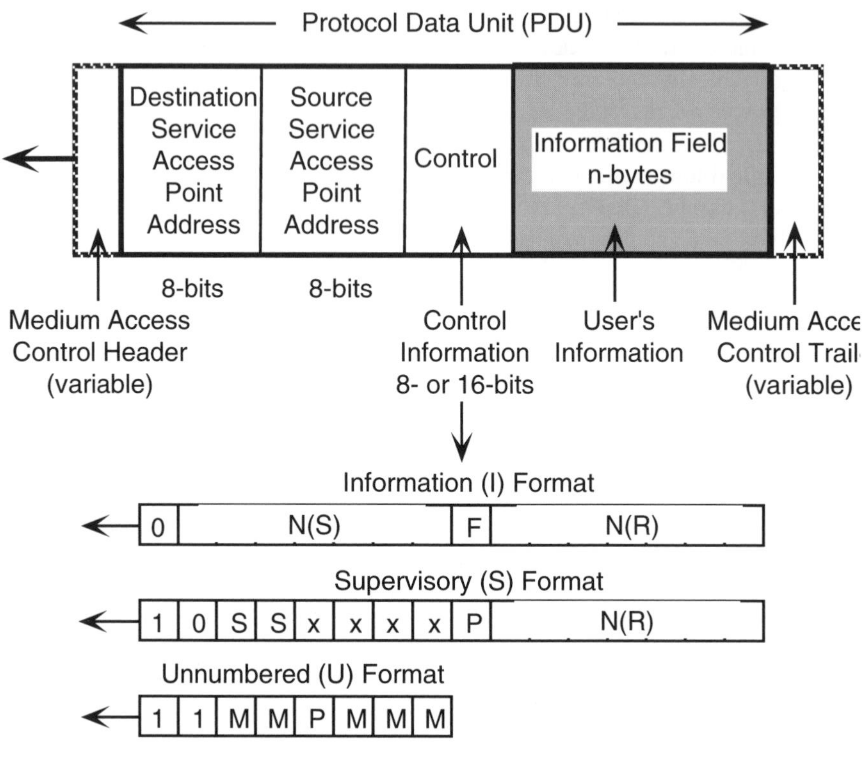

Figure 11.2 LLC Protocol Data Unit

between SAPs. The S (supervisory) format is used for numbered supervisory functions—as in SDLC, it contains no information field. The U (unnumbered) format is used for unnumbered supervisory functions and unnumbered information transfers.

(2) Classes of Service

Three classes of services are provided to stations attached to a LAN that employs LLC:

- **Connection-oriented service**: a logical connection is set up between originating and terminating stations. Acknowledgments, error and flow controls, and other features, are employed to ensure reliable data transfer between stations.

- **Acknowledged connectionless service**: messages are acknowledged but a logical connection is not established. This technique is used for large applications in which the overhead associated with connection-oriented service would make the operation too slow, yet it is important to know that the message was received.
- **Unacknowledged connectionless service**: messages are not acknowledged by the receiving station. Error control and flow control are not employed. This is a simple style of communication that is used in applications where the occasional loss or corruption of a PDU can be corrected by other processes.

11.2 BUS-CONNECTED NETWORKS

11.2.1 IEEE 802.3—CSMA/CD Bus

(1) CSMA/CD Operation

Based on *Ethernet*, a LAN pioneered by Xerox Corporation, and developed further by a team consisting of Xerox, Digital Equipment Corporation, and Intel Corporation, IEEE 802.3 describes bus-based LANs that employ carrier sense, multiple access, collision detection techniques. The principle of operation is shown in **Figure 11.3**. Key to successful data transfer is the continual monitoring of activity on the bus by each DTE. This is necessary for the following reasons:

- Each station will receive messages addressed to it
- Periods of no bus activity can be detected
- During transmission by the station, the appearance of another signal can be detected and a collision condition declared.

When a period of no activity on the bus is detected, the station with something to send may send it. Once this station has begun transmission, the other stations should detect the activity and withhold their own messages. However, if two, or more, stations begin to transmit at the same time, they will interfere with each other, a collision will be declared, and they must try again at a later time. To avoid a further collision, IEEE 802.3 defines a back-off algorithm for use by colliding stations to calculate some random time interval during which they will refrain from attempting to re-transmit their messages. In a production environment, satisfactory performance is obtained when offered traffic (see § 16.1.1(4)) is ≤ 10% of the bus capacity.

(2) CSMA/CD Frame Format

Details of the frame employed in CSMA/CD operation are given in **Figure 11.4**. The medium access control (MAC) header consists of the following:

- A 7-octet preamble used to establish bit synchronization

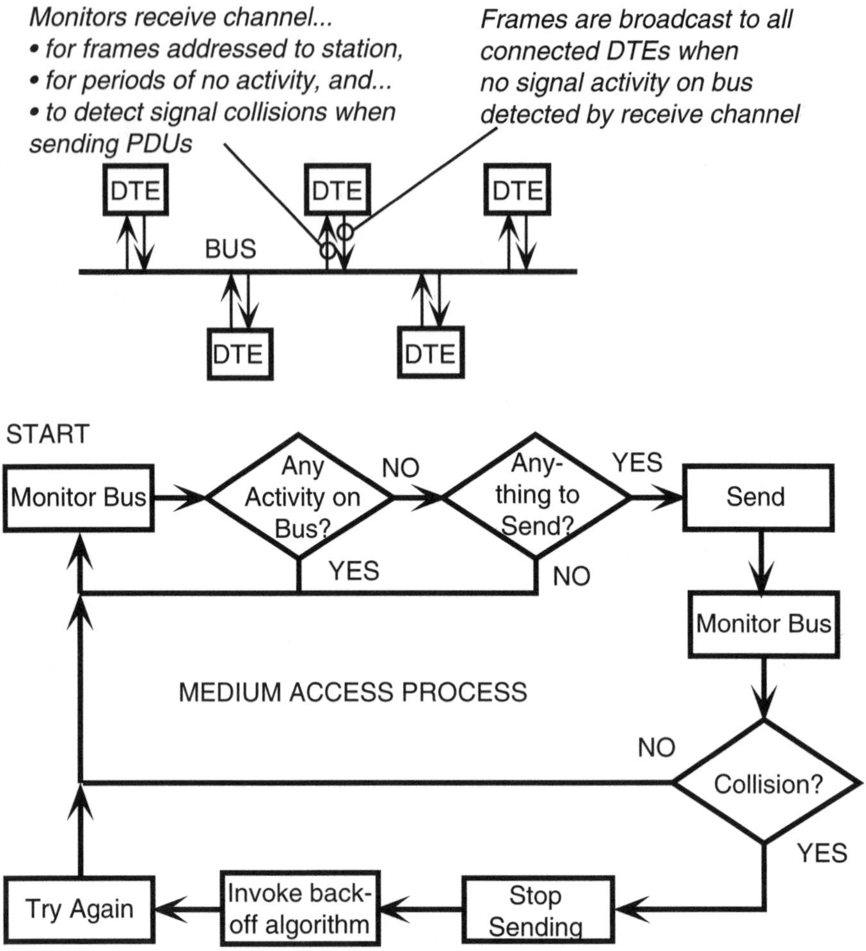

Monitors receive channel...
• for frames addressed to station,
• for periods of no activity, and...
• to detect signal collisions when sending PDUs

Frames are broadcast to all connected DTEs when no signal activity on bus detected by receive channel

DTE
DTE
DTE
BUS
DTE
DTE

START
Monitor Bus
Any Activity on Bus?
NO
YES
Any-thing to Send?
YES
NO
Send
Monitor Bus
MEDIUM ACCESS PROCESS
Collision?
NO
YES
Stop Sending
Invoke back-off algorithm
Try Again

Figure 11.3 CSMA/CD Medium Access Technique

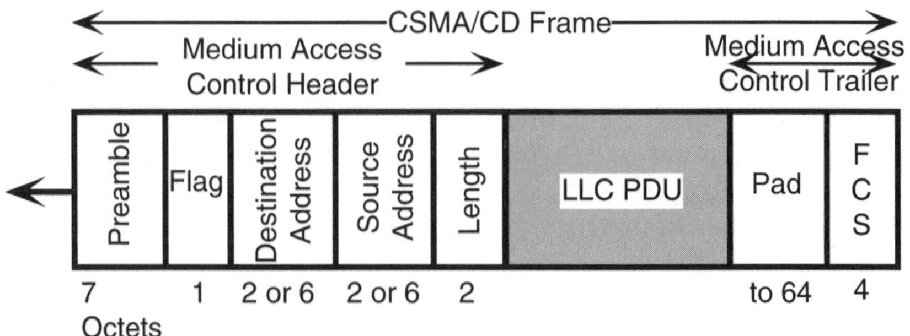

Figure 11.4 CSMA/CD Frame Format

- A 1-octet flag that locates the beginning of the data fields
- A 2- or 6-octet network address that identifies the terminating station
- A 2- or 6-octet network address that identifies the originating station
- A 2-octet length field that contains the number of octets in the logical link protocol data field that follows.

The two network addresses must be the same length (2 octets or 6 octets); they identify the physical units that originate and terminate the communication.

The MAC trailer consists of a 4-octet frame check sequence (FCS) and, if necessary, a field of octets that fills the frame out to a minimum total length of 64 octets. The 4-octet FCS is created through the use of a 33-bit generating function (see § 4.2.2(8)). A minimum frame length of 64 octets is required to stabilize the operation of the LAN.

(3) CSMA/CD Physical Layer

IEEE 802.3 is implemented in two ways:

- A 0.40-inch diameter coaxial cable bearer and Manchester signaling (Ethernet)
- A 0.25-inch diameter coaxial cable bearer and Manchester signaling (Cheapernet).

Cheapernet is less expensive, but, compared to Ethernet, has limited speed and distance capabilities.

11.2.2 IEEE 802.4—Token-Passing Bus

(1) Token-Passing

On a token-passing bus operating under IEEE 802.4, each station with something to send transmits only when permission to transmit has been received. Permission is conveyed by means of a token that is passed from station to station in a sequence. On a bus, all of the stations are assigned a position in the token-passing sequence, and the final station returns the token to the beginning station to provide the ability to repeat the process. Each station must maintain a record of its neighbors in the sequence, and the addition of stations to the bus will require modifications to the sequence. The basic medium access technique is shown in **Figure 11.5**.

(2) What Constitutes a Token on a Bus?

The token is an addressed supervisory frame that grants access to the bearer for a specific period during which the addressed station may send frames. At the end of the period, the station with the token generates a token frame addressed to the next station in the pre-arranged sequence. The time for which the station is in possession

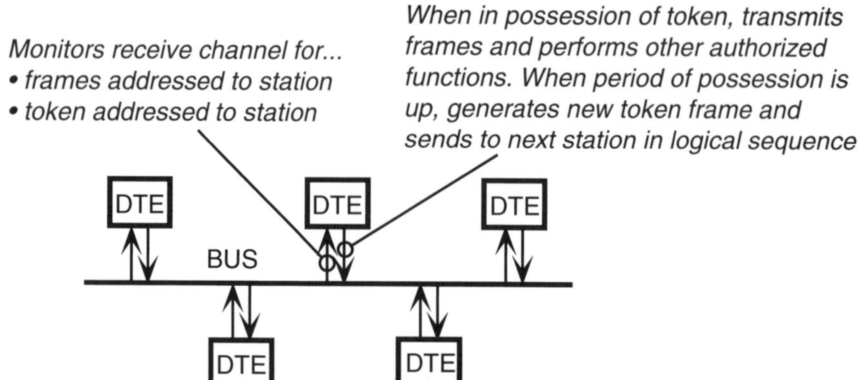

Figure 11.5 Token-Passing Bus Medium Access Technique

of the token can be adjusted so as to give stations that generate more traffic more time to transmit material than stations that generate less traffic. As the total traffic increases, it becomes important to tune the system to optimize the flow of data.

(3) Need for a Master Station

What if a station fails? It will not be able to regenerate the token frame when its turn comes and operations will cease. What if a station erroneously addresses the token to itself? If it is a one-time fault, probably no harm is done. However, if it is a permanent fault, something must be done to break the transmission monopoly the station enjoys. What if the token-passing station addresses the token frame not to the next in sequence, but to some other station? One, or more, of the stations will be cut out of the sequence and will not receive permission to transmit. And so on… A bus is prone to many conditions that can defeat the smooth operation of the token-passing strategy. For this reason, one station monitors the operation of the entire bus and is endowed with the ability to detect and correct whatever irregularities arise.

(4) Token-Passing Bus Frame Format

Details of the frame employed in token-passing bus operations are given in **Figure 11.6**. The medium access control header consists of:

- A 1-octet, or more, preamble used to establish bit synchronization
- A 1-octet flag that locates the beginning of the data fields
- A 1-octet frame control field that defines the frame as one of two frames:
  - *MAC* frame, in which case the data field contains information concerning token management or recovery from abnormal conditions

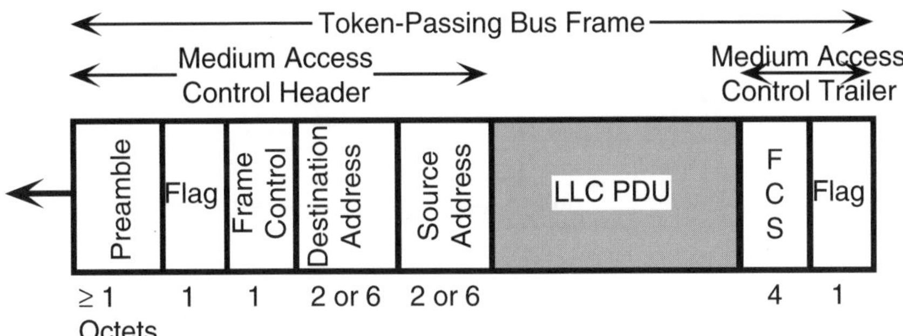

Figure 11.6 Token-Passing Bus Frame Format

- *user's data* frame, in which case the data field contains the information assembled by the logical link control sublayer

- A 2- or 6-octet network address that identifies the destination (terminating) station
- A 2- or 6-octet network address that identifies the source (originating) station.

The MAC trailer consists of a 4-octet frame check sequence (FCS) and a flag.

(5) Token-Passing Bus Physical Layer

IEEE 802.4 defines a broadband LAN that operates over a coaxial cable bearer and employs a modulated carrier for the transmission of information. Using frequency modulation, one technique provides a data rate of 1 Mbits/s over a single-channel; using phase modulation, another technique achieves data rates of 5 Mbits/s or 10 Mbits/s over a single-channel; using frequency division multiplexing, a third technique supports multiple channels that can operate at 1, 5, and 10 Mbits/s.

11.2.3 Personal Computer Networks

(1) PC Network

PC Network (developed by IBM and Sytek) is a bus-oriented network that can connect up to 72 stations (PCs) within a 1000 feet radius and support message exchanges at 2 Mbits/s. Network Adapter Cards (NACs) installed in each PC convert the digital signals into a frequency-modulated carrier at 50.75 MHz that is sent over a coaxial cable to a Network Translator Unit (NTU). At the NTU, the signal is converted to 219 MHz, and broadcast to all stations. A basic PC Network is shown in **Figure 11.7**. Each NAC arranges data into packets in an IEEE 802-like format for transmission, computes a 32-bit FCS for error detection and correction, detects collisions, and, on receipt of a good message unit, reclaims the data from the frame and presents it to the PC. NACs contain a unique 48-bit address that is used as the

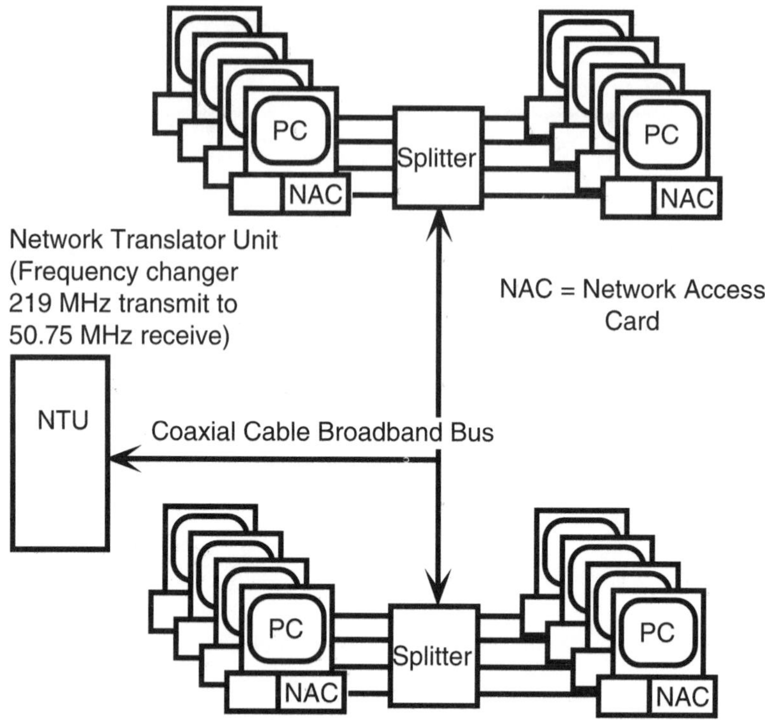

Figure 11.7 Principle of PC Network

destination or originating address, and IBM's NETBIOS (Network Basic Input/Output System) firmware that is responsible for managing all network traffic to and from the particular station.

(2) PC-to-Mainframe Connections

Data can be transferred between PCs by connecting them to a mainframe, sending all messages to it, and empowering the mainframe to distribute them to the appropriate PC. Installing a 3270-emulation card so that each PC appears to the mainframe as a 3270 terminal, will make the full capability of an SNA network available to the PC, i.e., the mainframe can be remote, and the PC can communicate directly with PCs (emulating 3270 terminals) in other domains. Through the use of emulation expansion boards, a PC can serve as a cluster controller for a group of PCs. Some of these possibilities are shown in **Figure 11.8**.

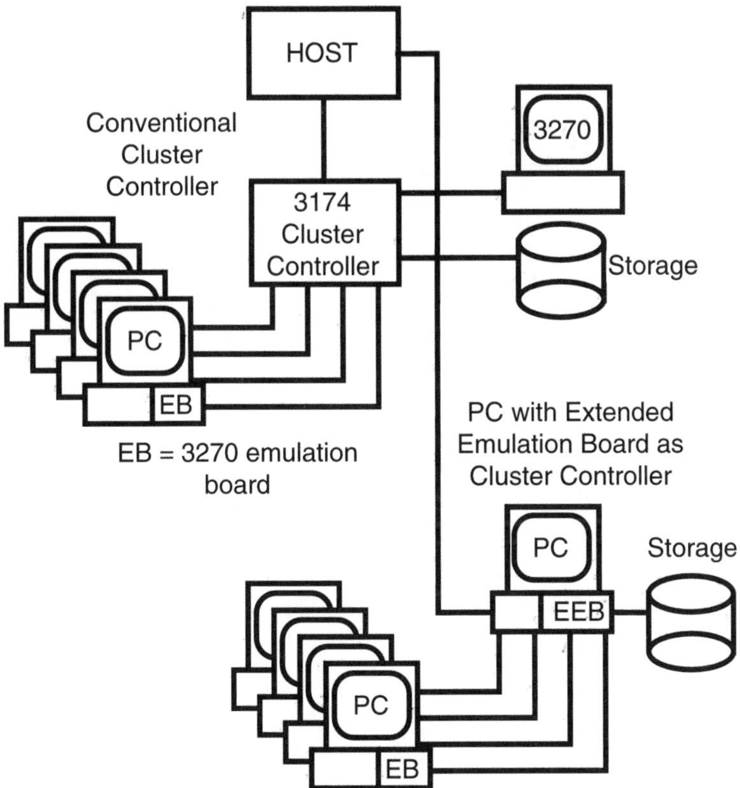

Figure 11.8 PC-To-Mainframe Connections

11.3 RING-CONNECTED NETWORKS

11.3.1 IEEE 802.5—Token-Ring

(1) Token-Ring

In a token-ring LAN operating under IEEE 802.5, each station is connected to two others so as to form a single-thread loop that connects all of the stations. Data is transferred around the ring from station to station; each station:

- Receives the data stream from the station preceding it on the ring
- May add to it
- Regenerates it
- Passes the data stream along to the station next on the physical ring.

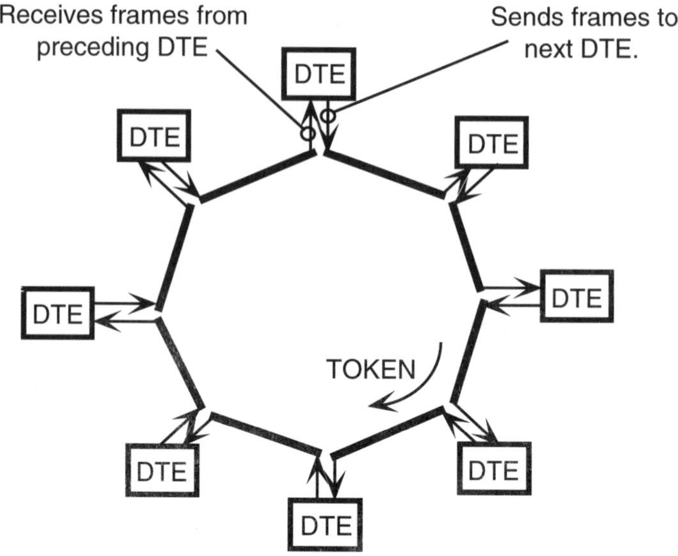

Figure 11.9 Principle of Token-Ring LAN

The principle of token-ring operation is shown in **Figure 11.9**. It is continued in **Figure 11.10** where some of the more important decisions are sketched.

(2) What Constitutes a Token in a Token-Ring?

In a token-ring network, the token available frame (empty token) consists of three octets, a start flag, an access control octet, and an end flag. An empty token is shown in **Figure 11.11**. The access octet contains:

- Three **priority bits** (PPP): they identify the level of priority a station must have to seize the token.
- A **Token bit** (T): it gives the token status. If it is 0, the token may be seized by a station that has sufficient priority. If it is 1, the token has been seized by another station and the frame is in use.
- A **monitor bit** (M): it is used to detect unclaimed frames.
- Three **reservation bits** (RRR): they provide a mechanism for lower priority devices to request the opportunity to transmit.

(3) Basic Operations

Some of the basic operations involved in the functioning of a token-ring LAN include:

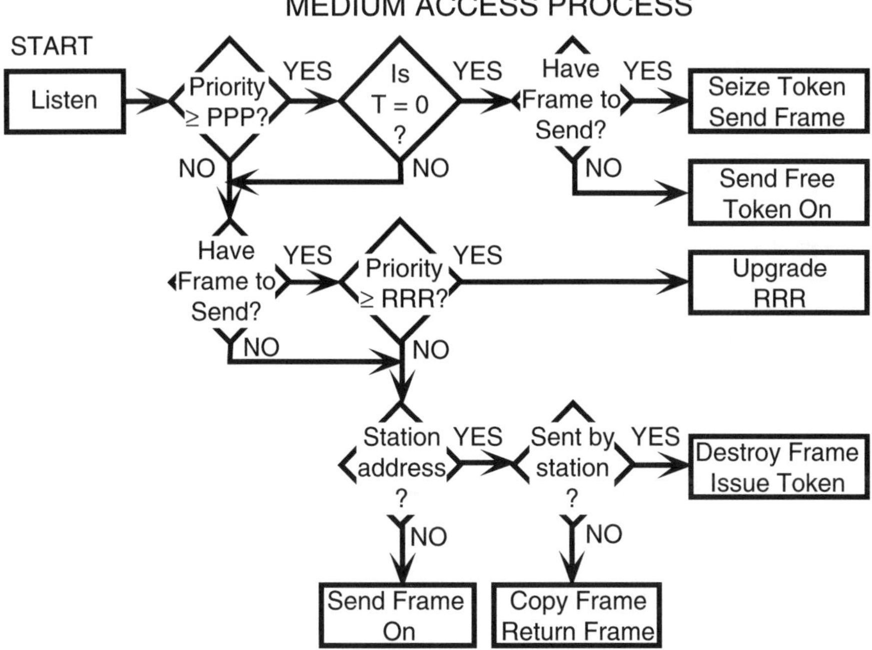

Figure 11.10 Token-Ring Medium Access Technique

- **Seizing the token**: when a station with data to transmit has priority equal to, or higher than the priority bits in an empty token frame, it seizes the token, changes the token bit to 1, adds data to it, and transmits the datastream to the next station on the ring. The formation of a transmission frame from an empty token frame is illustrated in Figure 11.11.
- **Receiving and acknowledging a frame**: when a station receives a filled frame that is addressed to it, it copies the information, reverses the source and destination addresses so that the frame will return to the originating station, adds an acknowledgment to the trailer, and sends the frame on to the next station.
- **Regenerating an empty token**: when the returned frame (above) is received by the originating station, it recognizes that it is one that was sent some time ago. It destroys the frame and regenerates an empty token frame in which T = 0. Before sending this frame to the next station, it replaces the priority bits in the access control octet with the priority expressed in the reservation bit space.
- **Using reservation bits**: when a station with something to send receives an empty token, it examines the priority level. If the token priority is greater than the priority the station enjoys, the station next examines the priority

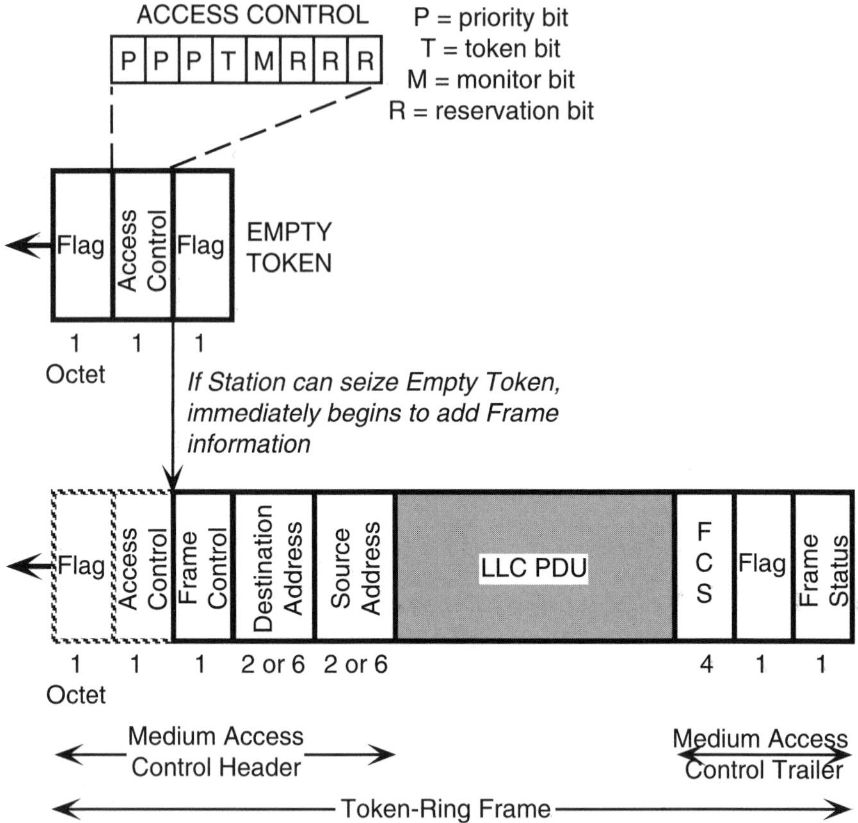

Figure 11.11 Token-Ring Frame Format

number contained in the reservation field. If it is less than the station priority, the station upgrades the R-bits to reflect its level of priority. In this way, the station protects the next empty token from being seized by a station with lower priority. However, it cannot prevent a station with a higher level of priority from upgrading the R-bits further. Nor can it prevent a station with the same priority, but located earlier in the sequence from seizing the next empty token.

- **Using the monitor bit**: when a station fails, messages addressed to it are likely to continue to circulate around the ring. To prevent this, one station is designated to monitor traffic. When filled frames pass the monitoring station, it sets the M-bit to 1; then, if a frame passes the monitor station with M = 1 the station knows it has made a complete circuit without being received and processed. The monitoring station destroys the frame and generates an empty token.

- **Using priority levels**: with three P-bits, eight levels of priority can be used. By assigning different levels to each station (or class of station) on the network, individual opportunities to send can be made to match the amount of traffic they have and the delay they can tolerate. Proper assignment of priority levels is key to the successful operation of a token-ring LAN. Adjusting each station's priority level so that all have an opportunity to transmit their information is a formidable challenge.

(4) Station/Ring Interface

Because each message is handled by all stations in the ring, it is important that the messages pass through each access point as quickly as possible. At the same time, the receiving station must interrogate each bit, recognize particular bits (T- and M-bit), and bit-patterns (PPP, RRR and destination address, for instance), and take action as necessary. When an empty token frame is first received, the flag is used to synchronize the station receiver and identify the beginning of the control octet. The next octet, the control octet, contains priority level, token status, and reservation status. If the station has something to send, it compares the priority bits with its own priority level. If the station has a higher priority than the three P-bits received, the next bit is crucial:

- If it is 0, the station is receiving an empty token (T = 0), and can seize it. It sets the T-bit to 1, and, as soon as the reservation bits have cleared, proceeds to transmit the frame control octet and the remainder of the message.
- If it is 1, the station is receiving a filled token (T = 1). There is no opportunity to send.

With a filled token, the next matter of interest is whether the PDU is addressed to the station. If it is, and

- It is not the PDU the station just sent, it copies the data it contains, performs a cyclic redundancy check, reverses the source and destination addresses, adds a frame status octet, and transmits the PDU to the next station, *or...*
- It is the PDU the station just sent, it generates an empty token frame, transfers the R-bits to the P-bit field, sets the R-bits to zero, and transmits the empty token to the next station. If the frame status octet indicates retransmission is required, the station waits until the next time it can seize an empty token to do this.

(5) Physical Medium

IEEE 802.5 is strongly supported by IBM, and is the basis for the IBM token-ring product line. IBM's cabling system uses twisted pairs with Manchester signaling at speeds of 1, 4, and 16 Mbits/s.

11.3.2 Fiber Distributed Data Interface Network

(1) Configuration

Intended for use within a campus, Fiber Distributed Data Interface (FDDI) Network provides the backbone for a distributed computing environment. Standardized by ANSI, FDDI is a dual-ring, optical fiber network that connects up to 500 stations in a ring of $\leq$ 100 kilometers. With multimode fibers, point-to-point links are limited to $\leq$ 2 kilometers; using single-mode fibers, individual links can be extended up to 50 kilometers. Data is passed at a speed of 100 Mbits/s using a token-passing protocol. 100 Mbits/s is defined by ANSI as the base data rate. It can be increased in increments of 6.144 Mbits/s. FDDI employs 4B/5B coding (i.e., 4-bit binary substituted by 5-bit binary—four signal bits are represented by a 5-bit symbol). Thus, the signaling rate for 100 Mbits/s is 125 Mbits/s. FDDI frames (up to 4500 octets) can be transported over SONET connections as STS-3c modules (STS-3c = concatenated payloads of 2322 octets, see § 8.3.1(3)).

Figure 11.12 shows the concept of an FDDI network. Stations are connected to the fiber rings by channel interface units (CIUs). Normally, CIUs receive frames from their upstream neighbors, and send frames to their downstream neighbors over the primary fibers. Should a link in the ring be interrupted, the gap is bridged by using the stand-by fibers in the manner shown in the lower diagram. Although conceived as a fiber-based network, versions of FDDI are available that employ shielded or unshielded twisted pairs. They can be used in offices and buildings that do not require long distances between stations (meters not kilometers), and are designed to employ existing wiring.

(2) Operation

Permission to send is derived from possession of a token. This message is circulated in the ring and can be seized by a station that has a frame to send. Possession of the token permits the station to send for a specific period. When the period is up, or the station has no more frames to send, the station generates a new token and sends it on to its downstream neighbor. The format of the token and information frames is shown in **Figure 11.13**. For all frames, the CIU performs an error check by computing the frame check sequence. If an error is detected, it sets the error symbol in the frame status field. For frames that are not addressed to one of the logical entities supported by the CIU, it sends the frame forward to its downstream neighbor. If the frame is addressed to one of the logical entities supported by the CIU, it copies the relevant information, forwards it to the logical entity addressed, sets the address recognized bit, and sends the frame on to its downstream neighbor to complete a circuit of the ring. On arriving back at its source, the originating station notes that the frame has (has not) been received correctly and removes the frame from the ring. If retransmission is required, it occurs next time the station has the token.

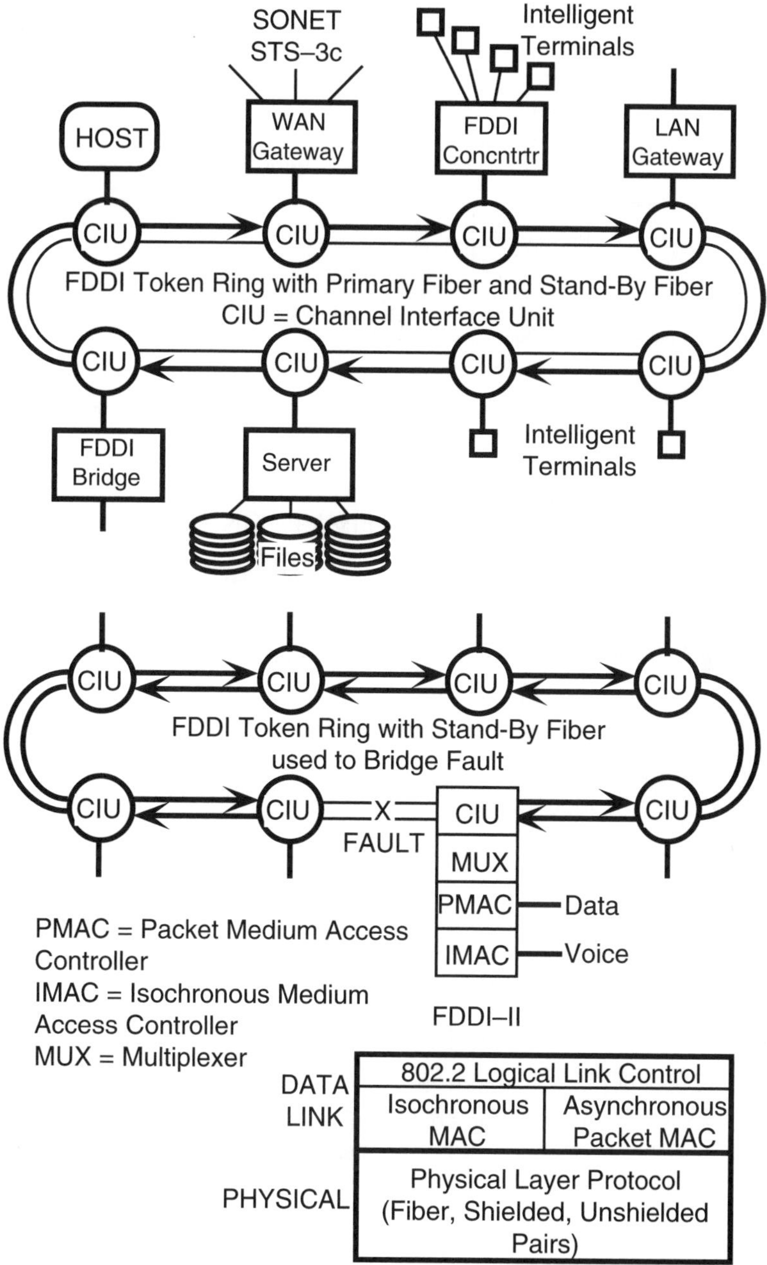

Figure 11.12 FDDI Network

In the lower diagram, two separate points are illustrated. 1) When a fault occurs, operation is maintained through the use of the stand-by fiber. 2) In FDDI-II, the channel interface unit is augmented with a multiplexer and controllers to carry both voice and data.

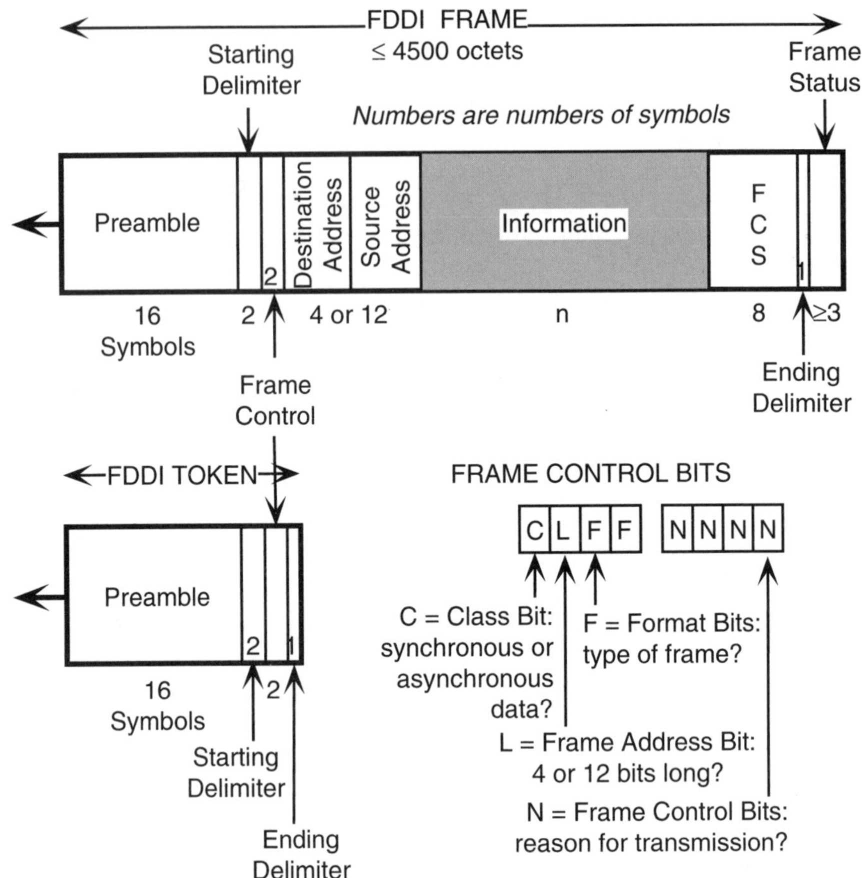

Figure 11.13 FDDI Frame Format

(3) FDDI-II and FFOL

FDDI is a data-only network. With the appearance of techniques such as SONET and ATM that handle asynchronous packet data and isochronous voice and video, FDDI has been enhanced to provide voice and data transport. Called FDDI-II, it includes a synchronous, two-way transmission capability. To achieve this state, each node is equipped with two medium access controllers. One, known as PMAC (packet medium access controller) supports data applications. The other, known as IMAC (isochronous medium access controller) supports voice applications. Asynchronous data or synchronous voice signals are carried in standard FDDI frames. FDDI-II operates in two modes:

- **Basic mode**: the same as FDDI, i.e., data only
- **Hybrid mode**: asynchronous data and synchronous voice are carried in FDDI frames. When receiving, the mixed frames are demultiplexed and directed to the appropriate MAC. When sending, the frames are directed to the multiplexer by their respective controllers for transportation over the ring.

In the lower diagram of Figure 11.12, we show the physical arrangement of MUX, PMAC, and IMAC, and the data link and physical layers for FDDI-II. The combination of multiplexer and IMAC is known as the hybrid ring control (HRC). Another extension of FDDI is known as FFOL (FDDI Follow-On LAN). As the name suggests, work is already underway to address a gigabit version of FDDI that will be compatible with higher-speed SONET levels (STS-48, 2.488 Gbits/s, for instance) and BISDN. FFOL will be compatible with FDDI and FDDI-II, and will transport the full range of signals required by BISDN.

11.4 LAN INTERCONNECTIONS

(1) Need to Interconnect LANs

Most often, LANs are employed to connect stations in a building or a building complex. Within the same local area there may be other LANs to which interconnection is necessary to share the same databases, or for other purposes. What is more, there may be other LANs in other buildings remote from the complex that require interconnection, and some of them may be a considerable distance away. As the effectiveness and efficiency of LANs improves, there are more and more reasons to want to interconnect them with wide-area networks so that the pockets of information activity that make up the modern corporation can communicate on any level. To achieve transparent information exchanges between users anywhere, differences between the networks must be reconciled.

(2) Problems Encountered in Connecting Networks

From the discussion of Ethernet, Token-bus, and Token-ring LANs, it is obvious that two networks can differ greatly in access methods, frame sizes, frame formats, sequence numbering techniques, addressing conventions, and other items. In connecting two networks, the management information that originates in one network, must be processable in the network in which it terminates, and in any intermediate networks that it traverses. Also, the user's data must be understandable in the network in which it terminates. Differences between networks can be organized on the basis of the OSI model layer in which they occur:

- Differences in the electrical and mechanical properties of the communication medium (e.g., cable vs. twisted pair, connectors, signaling methods, etc.) are differences in the implementation of the *physical* layers of the two networks.
- Differences in the types of frame information and the significance of individual bit positions (e.g., synchronizing patterns and flags, control bits, sequence numbering, error control, flow control, etc.) are differences in the implementation of the *data link* layers.
- Differences in segment sizes and routing techniques (e.g., centralized vs. distributed) are differences in the implementation of the *network* layers.
- Differences in message flow control and message segmentation are differences in the implementation of the *transport* layers.
- Differences in line discipline, failure recovery strategies, and communication management techniques are differences in the implementation of the *session* layers.
- Differences in codes used and encryption and compression techniques employed are differences in the implementation of the *presentation* layers.

(3) Connecting Devices

The device used to effect interconnection depends on the highest layer in which differences exist:

- **Repeater**: if the differences are limited to the implementation of the physical layers, repeaters—devices that link two connections together and amplify and regenerate signals—are used.
- **Bridge**: if the differences reach no higher than the implementation of the data link layers, bridges—devices that link identical networks (i.e. Ethernet to Ethernet) at the data link level—are used.
- **Router**: if the differences reach no higher than the implementation of the network layers, routers—devices that modify segment sizes, route frames between users on non-identical, but the same type of networks (i.e. Token-Ring to Token-Ring) and connect at the network level—are used.
- **Gateway**: if the differences involve the implementation of layers above the network layer, gateways—devices that link networks that are not of the same type (i.e. Ethernet to Token Ring, or LANs to SNA or X.25 networks) and connect at whatever layer is required—are used.

As a generic description, we may call all four of these devices *relays*—they connect two networks and move information from one to the other. Examples of the use of repeaters, bridges, routers, and gateways in connecting one LAN to another, are given in **Figure 11.14**. Their application must be worked out in exact detail for each specific combination of networks. Not only must normal operations be supported, but abnormal operations, including recovery from failures, must be assured. An example of the interconnection of two dissimilar, geographically separated LANs by a packet network is given in **Figure 11.15**. Each host is assumed capable of direct connection to their respective LANs, and TCP and IP are used as the transport and network protocols. The hosts generate frames that are matched to the requirements of an Ethernet LAN and a Token-Ring LAN. At the gateways, the frames are converted to frames that are compatible with the packet network.

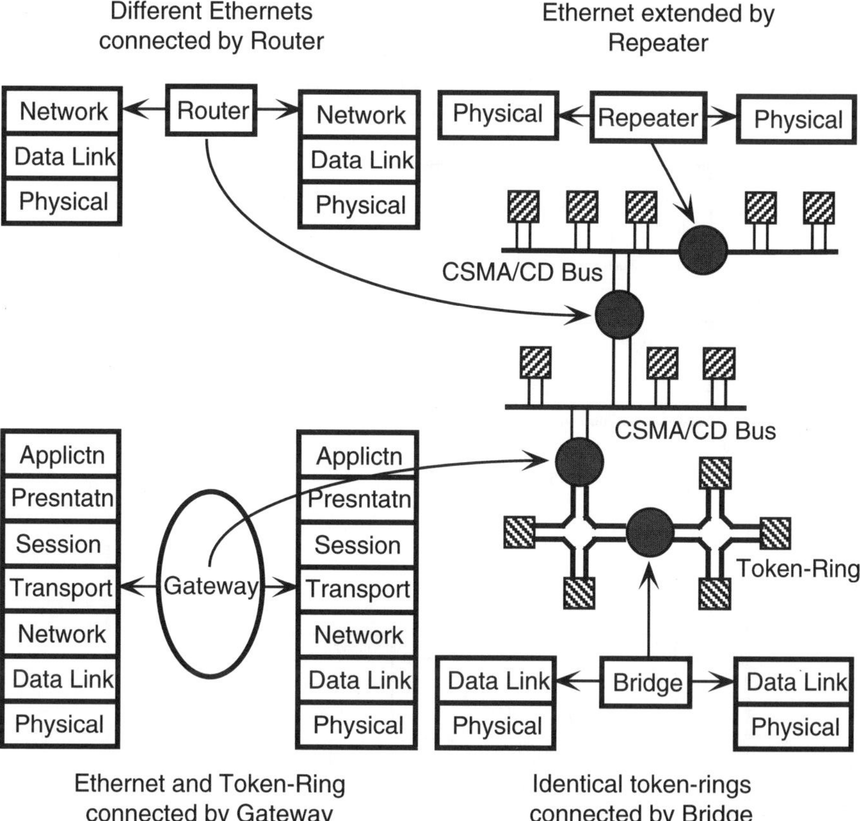

Figure 11.14 LAN Interconnection Techniques
Repeaters are used to connect the same LANs at the Physical Layer. Bridges are used to connect similar LANs at the Data Link Layer. Routers are used to connect similar LANs at the Network Layer. Gateways are used to connect dissimilar LANs at the Transport Layer.

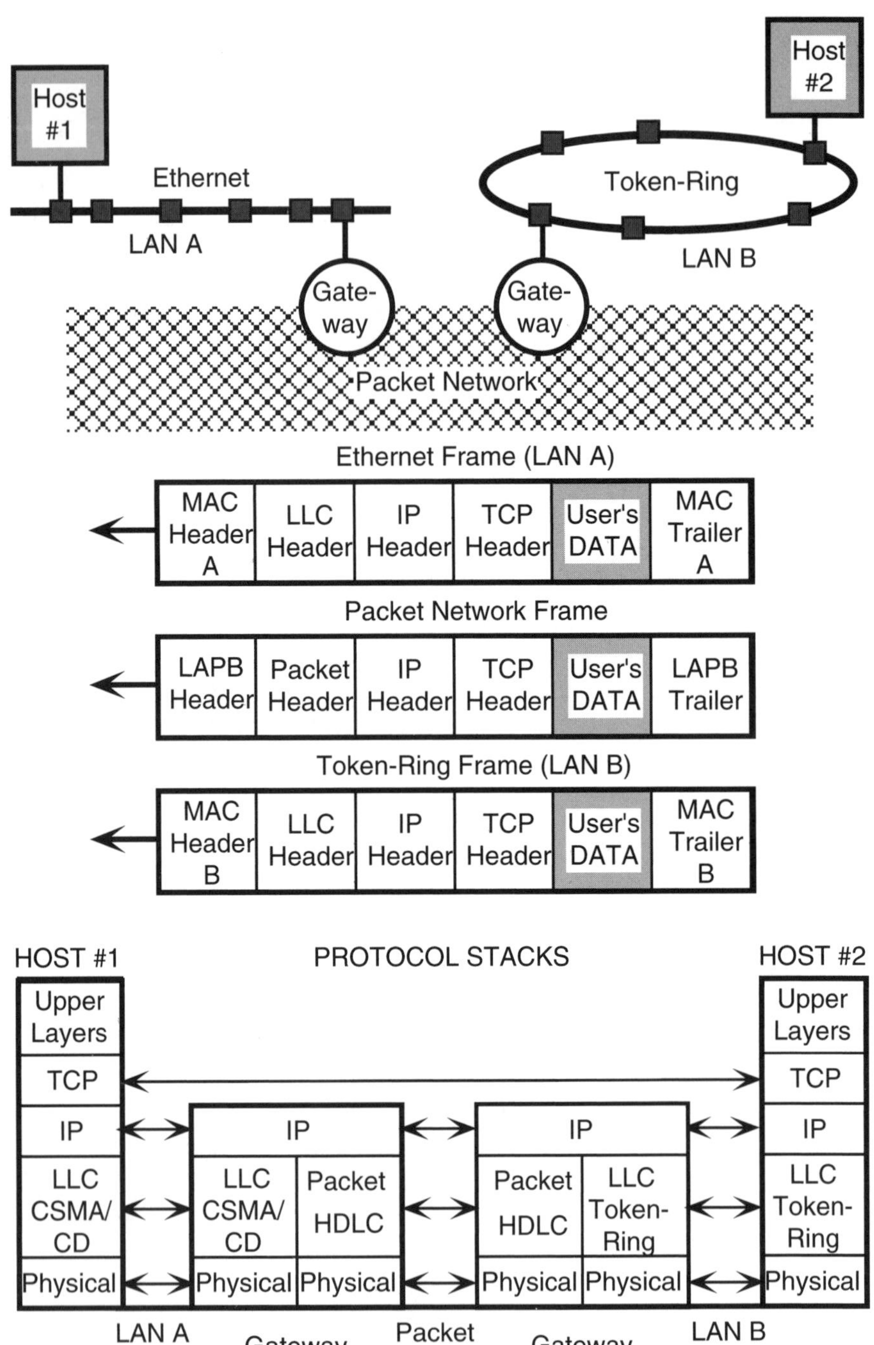

Figure 11.15 Connecting an Ethernet LAN to a Token-Ring LAN by Means of a Packet Network

11.5 METROPOLITAN-AREA NETWORKS

Metropolitan-Area Networks (MANs) are facilities that span urban cores, metropolitan areas, or major industrial campuses. Connecting the premises of many organizations, they provide the opportunity to employ public, switched, high-speed facilities to transport data, voice, and video. MANs cover a significantly larger area than local-area networks. Described by IEEE 802.6, they employ a *distributed-queue, dual-bus* (DQDB) access arrangement. **Figure 11.16** shows the dual-bus network topology in both open-bus and looped-bus versions. Each node serves as the collection point for local traffic and is connected to its partners by point-to-point connections that are operated to provide a transmission channel in each direction. Labeled *A* and *B*, these channels transmit data in fixed-size units. Called *slots*, they consist of 53 octets (52 octet payload + 1 octet header). Speeds on the busses range from 44.736 Mbits/s (DS-3) over coaxial cable or fiber, to 155.52 Mbits/s (STS-3, STM-1) over single-mode fiber. Every 125 microseconds, the slot generator at the node from which a bus start, transmits bursts of slots. The number of slots depends on the data rate (13 slots at T-3, 45 slots at STS-3). When permitted, nodes write data to the slots, and read and copy data from them.

(1) DQDB Access Protocol

The distributed-queue, dual-bus access protocol provides connectionless, connection-oriented and isochronous transfer services. Each node gains access to a bus by joining a queue of nodes waiting to transmit on the bus (there are two queues, one for each bus). For one direction (sending on bus A), the procedure is shown in **Figure 11.17**. When a node has a message to transmit on bus A, it sets a request bit on bus B. (Each node keeps a tally of downstream requests by monitoring the request field of slots that pass by on bus B.) It joins the queue for bus A by transferring the request count to the countdown counter and resetting the request counter to zero. The countdown count is reduced by one each time an empty slot on bus A passes the node. When the countdown counter is zero, all current requests to transmit from downstream stations have been satisfied, and the node transmits in the next empty slot. The countdown counter is not reduced below zero. The number of stations in the distributed queue on bus A is obtained by counting requests passing on bus B and counting free slots passing on bus A. The difference between the two counts is the number of outstanding requests for slots on bus A. Because MANs carry voice and video traffic (i.e., isochronous traffic), arrangements must be made to ensure that a station with voice (or video) to send obtains regular access. To do this, a number of *pre-arbitrated* slots are included in the slot stream generated at the beginning of each bus. Within these slots, octets are assigned to the stations requesting isochronous service.

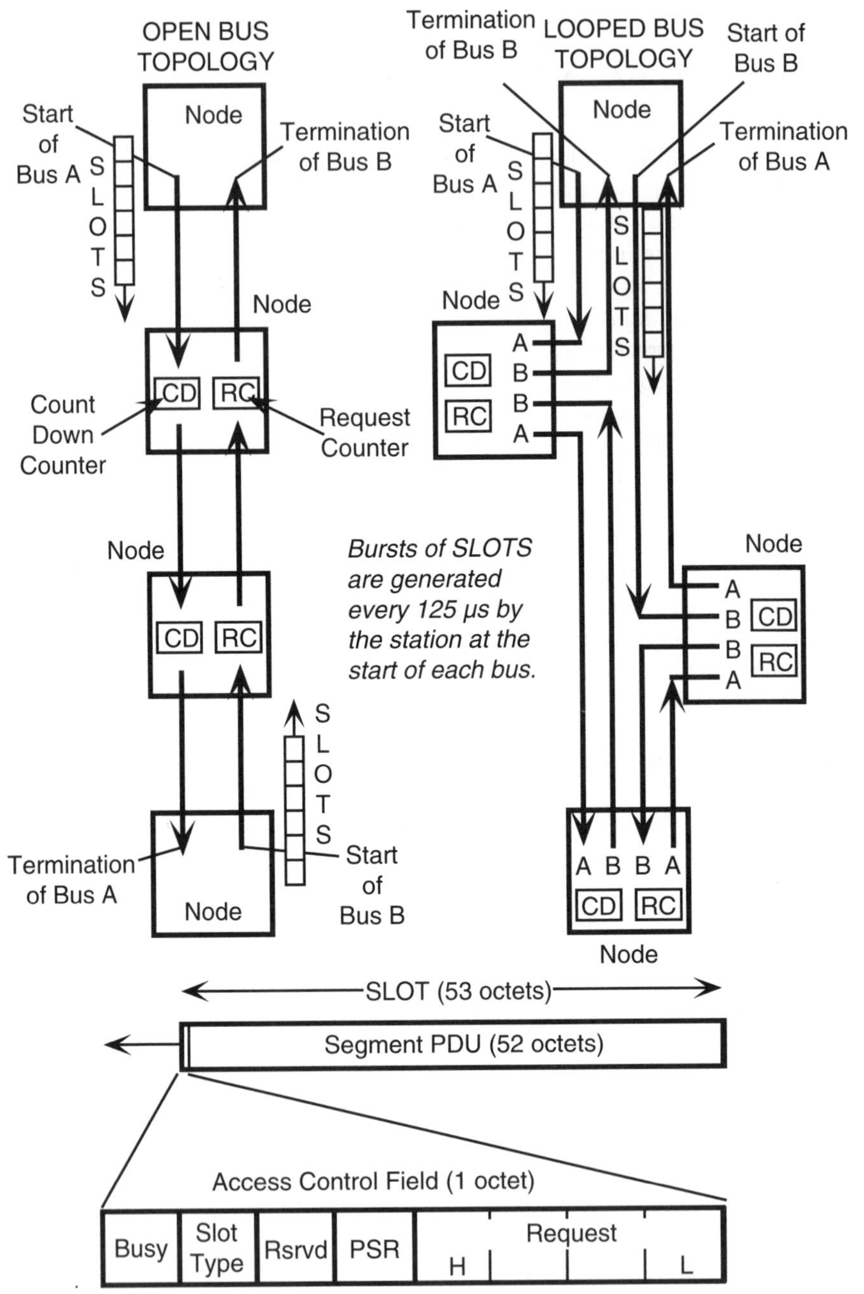

Figure 11.16 MAN Topologies
In either configuration, two busses (A and B, one for each direction) handle information transfer between stations. Stations gain access to one or the other through the use of the distributed-queue, dual-bus (DQDB) protocol.

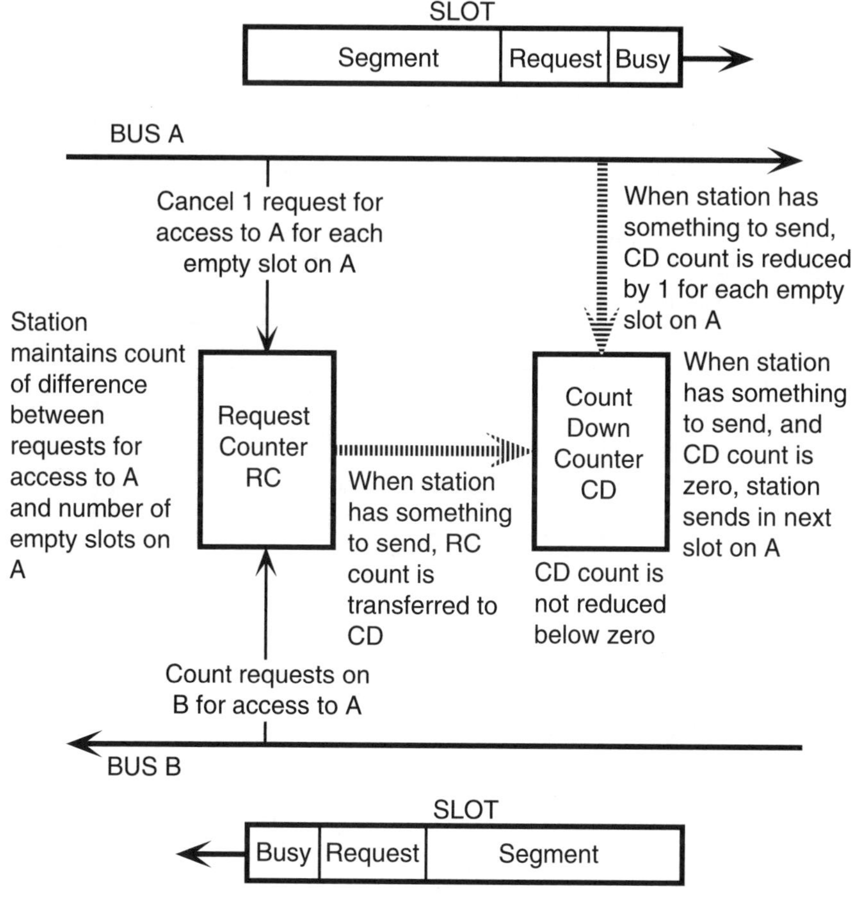

Figure 11.17 DQDB Access Procedure

(2) From LLC PDU to Slot PDU

Figure 11.18 shows the components of the DQDB medium access control sublayer. Divided into two blocks, its operation is directed by a layer management entity. The convergence block adds a 24 octet header (including destination address, source address, and class of service) and a 4 octet trailer to the LLC PDU to form the *initial* MAC PDU. If it is greater than 44 octets, the convergence block divides the PDU into 44 octet fragments, adds 2 octet headers (including segment type and message identifier) and 2 octet trailers (including error checking) to form 48 octet, *derived* MAC PDUs. In turn, by adding 4 octet headers (including virtual channel identifier), they are formed into 52 octet *segment* PDUs and then into 53 octet *slot* PDUs. Routing information within the MAN is provided by the virtual channel identifier in the segment header; physical (source and) destination information is

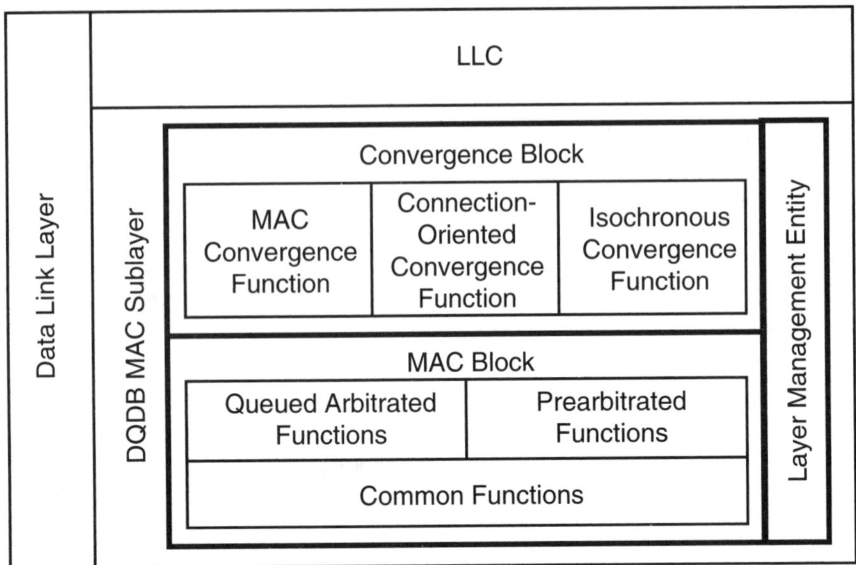

Figure 11.18 DQDB Medium Access Control Sublayer
The medium access control (MAC) sublayer for DQDB is divided in two. The convergence layer adds information to the PDU, divides it into 44 octet fragments, adds information to aid reassembly and creates 48 octet PDUs. The MAC block adds information concerning routing and control to create 53 octet slot PDUs.

contained within the initial MAC PDU; and (source and) destination service access point information is provided within the LLC PDU. The essential steps in this transformation are shown in **Figure 11.19**. For illustration, an initial MAC PDU of xx octets is assumed, where $88 < xx < 132$ octets. More details of the information contained in the headers and trailers are given below each diagram. When receiving, the sublayer functions reverse the procedure and assemble the LLC PDU from the incoming segments.

(3) Deployment

MANs may be deployed as individual open-bus structures, as single rings or interconnected rings. In **Figure 11.20**, we show the way in which an open bus can provide high-speed connections between the individual locations of an enterprise centered on an urban core. Traffic from many nodes passes over each bus raising questions of information security. With operations restricted to a single company, this may not be an important issue. For public networks, such is not the case, and separate access lines are provided between each customer and the MAN node that is located in a central office, or other protected location. **Figure 11.21** shows the use of a ring-connected MAN to service several independent (perhaps competing)

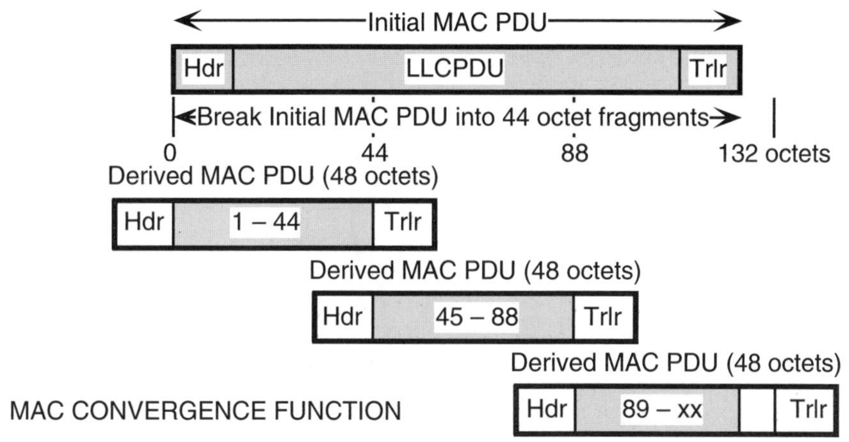

Initial MAC PDU -- Header: 24 octets, includes beginning-end tag, length, destination address, source address, protocol identifier, etc. Trailer: 4 octets, includes beginning-end tag and length.
Derived MAC PDU -- Header: 2 octets, segment type (2-bits), message identifier (14-bits). Trailer: 2 octets, length (6-bits), CRC (10-bits).

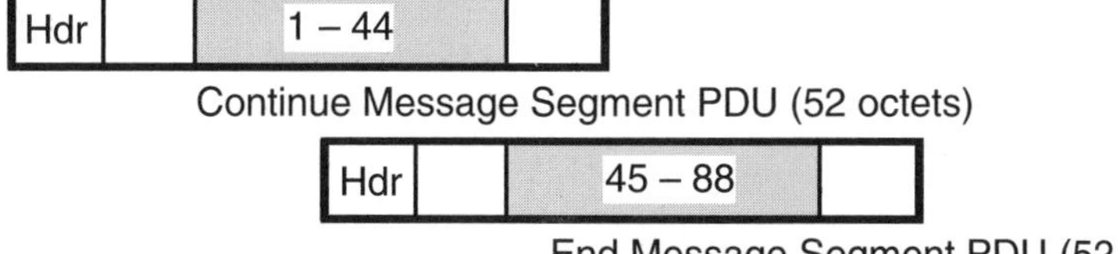

End Message Segment PDU (52 octets)

QUEUED ARBITRATED FUNCTIONS | Hdr | 89 – xx

Segment PDU
Header: 4 octets, virtual channel identifier (20-bits), payload type (2-bits), segment priority (2-bits), header check sequence (8-bits).

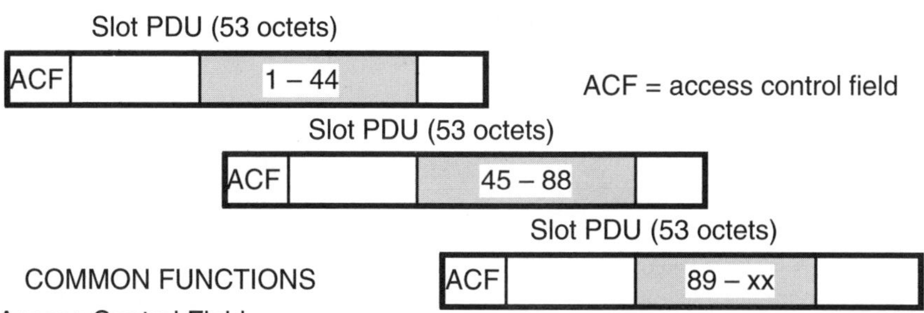

Access Control Field
1 octet, busy (1-bit), slot type (1-bit), reserved (1-bit), previous segment received (1-bit), request (4-bits).

Figure 11.19 Formation of Slot PDUs
Detailed look at the process of dividing the initial MAC PDU into 53 octet slot PDUs.

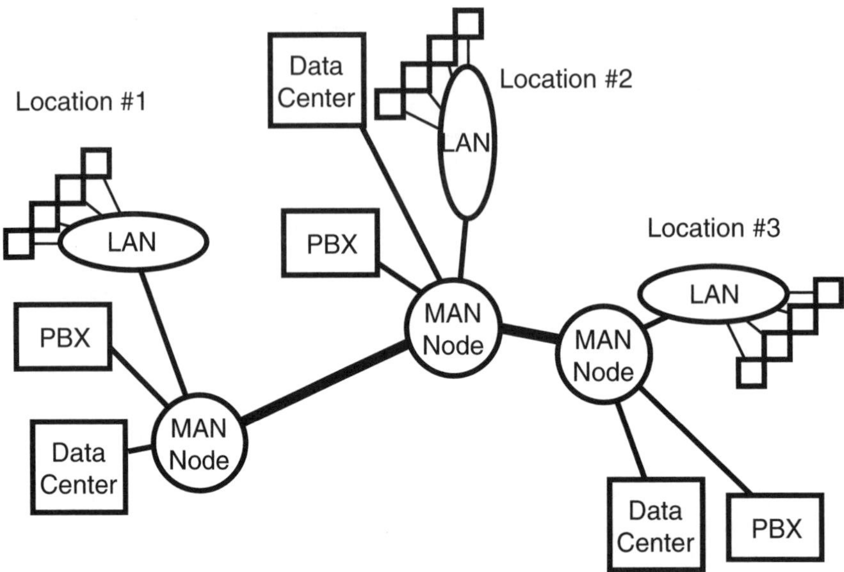

Figure 11.20 Single Enterprise MAN
A DQDB open-bus MAN that provides high-speed connections between the separated locations of an enterprise.

enterprises. Individual customer locations are connected through central office equipment to the high-speed node. In high-density urban areas, public MANs are likely to consist of orthogonal rings with common nodes. One of these structures is shown in **Figure 11.22**. Known as the Manhattan Street Network (MSN), it consists of orthogonal rings that pass information in alternating directions. A disadvantage of the topology is that routing decisions must be made at each node. The advantages are that connections between two nodes are likely to be shorter than in a conventional network, and alternate routes exist so that individual failures need not be catastrophic.

11.6 COMPETITION FROM ATM

Competing with FDDI–II and FFOL, and MANs, asynchronous transfer mode is a practical technology to interconnect multimedia terminals and servers that employ data, voice, and video at rates of 10 Mbits/s, and perhaps up to 100 Mbits/s. With the ability to handle multimedia environments, and to provide connection-oriented or connectionless service, a single ATM node can function as the central hub (ATM hub) of a star-connected arrangement. Individual equipment with ATM user-net-

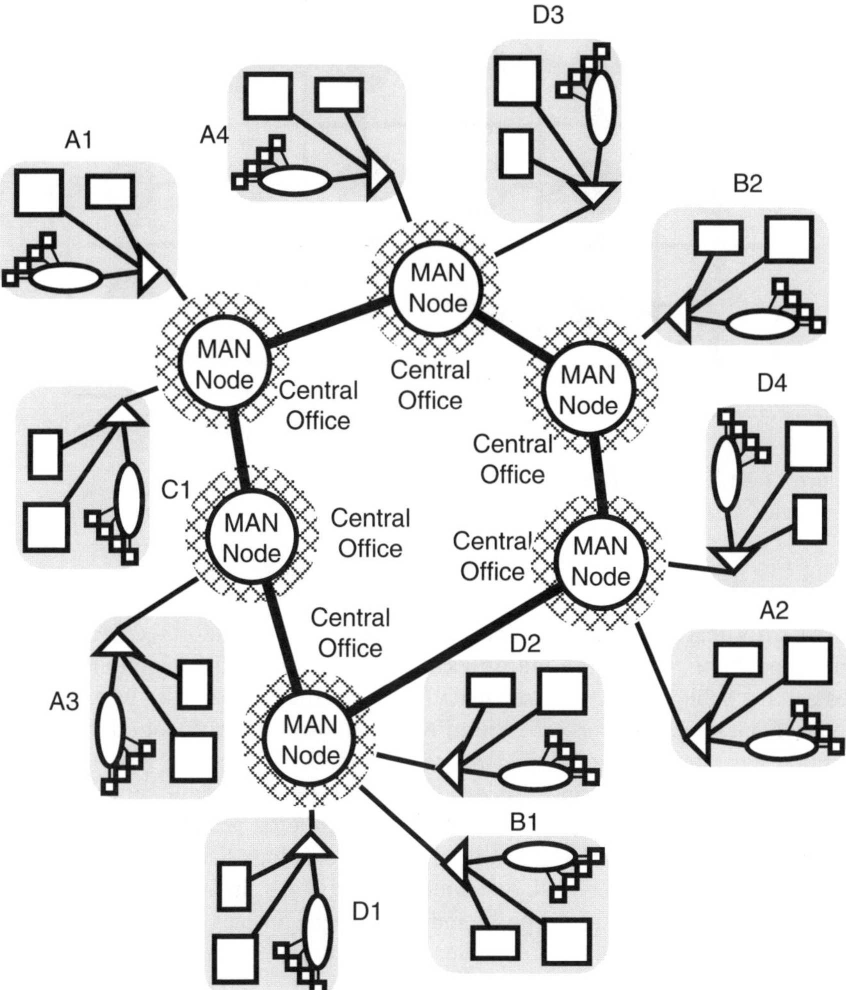

Figure 11.21 Public MAN
In a public MAN, provisions are made to safeguard each customers traffic. The MAN nodes are located in central offices and individual customers are connected through the distributing frame. The internode busses do not go through the customer's premises.

work interface (UNI) cards would be directly connected to it, and connections could be made between the hub and other public or private ATM networks. Regional private networks can be established with small ATM nodes (see § 9.2.4(2)) interconnected by SONET fibers. **Figure 11.23** shows the concept of an

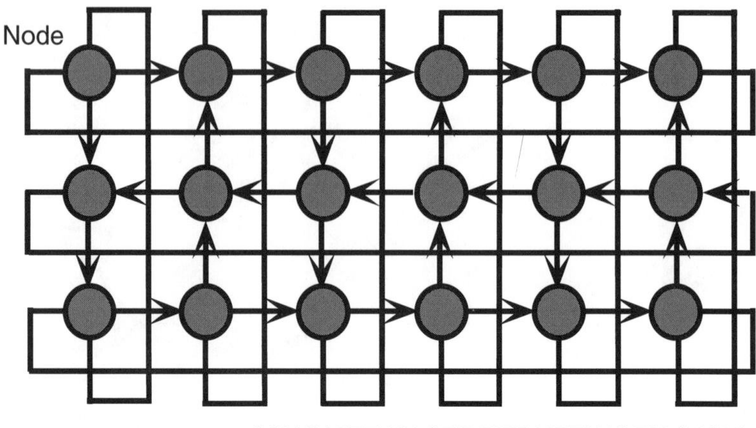

Figure 11.22 Urban Core MAN
The network consists of orthogonal rings that pass information in alternating directions.

ATM LAN that connects high-speed terminals, servers, and other LANs over a fiber backbone. Using the multiplexing capabilities of the segmentation and reassembly sublayer of AAL (see § 10.7.2(4)), traffic from LANs (which provide connectionless services) is carried from ATM gateways to servers or other LANs as multiplexed, connection-oriented streams. Across the UNIs, AAL type 1 or type 2 may be used for video and voice, and AAL type 5 for data services (see § 10.7.2(3)). In turn, the LAN is connected to public or private ATM networks. Without distance limitations, ATM is a routing technology that can be applied universally, i.e., to local, regional, and wide area networks. Its use would free multimedia communications from a proliferation of special devices designed to make the transition from one network to another and create a seamless network extending from the local to the worldwide.

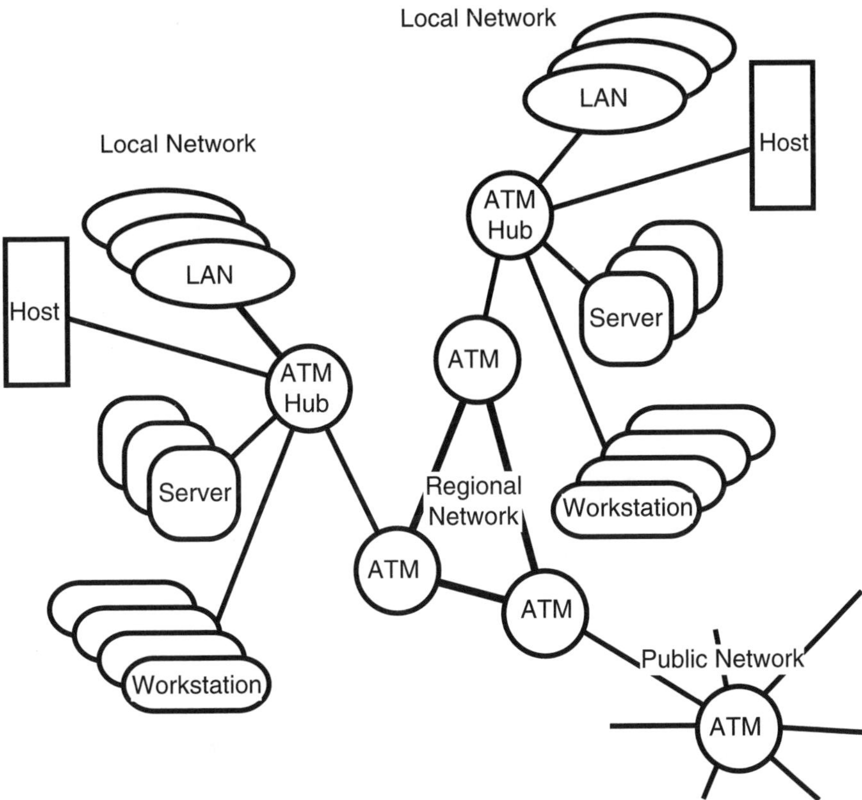

Figure 11.23 Application of ATM to LANs and Regional Networks
ATM hubs connect high-speed terminals, servers, and hosts in local networks that are joined into a regional network and connected to the public network. Use of ATM frees multimedia communications from a proliferation of special devices designed to make the transition from one network to another.

12

Wide Area Networks

In this Chapter we focus on data-only networks that serve a wider area than LANs and MANs. X.25 packet networks provide worldwide, fully-featured services to several million users. These networks assume that the transmission path is error prone, that the integrity of the frame must be checked (and corrective action taken if necessary) at each network node, and that flow control must be implemented across each link, as well as end-to-end. As a consequence, X.25 packet networks are quite reliable, but relatively slow. With the proliferation of terminals, personal computers, and workstations throughout the business community and the explosive growth of LANs, demand developed for faster data communication services. We describe two techniques that fill this need. One, called *frame relay*, provides permanent virtual circuit service between specially configured routers (or other devices) at access speeds up to DS-1. The other, called *switched multimegabit data service*, uses IEEE 802.6 concepts to provide connectionless data transport at speeds into the range of SONET facilities. Finally, we describe what may be the most used of all wide-area network techniques, IBM's System Network Architecture.

12.1 PACKET NETWORKS

In the late 1960s, the Advanced Research Projects Agency (ARPA) of the U.S. Department of Defense began to construct a packet-switching network. Called ARPANET, it was the precursor of many of today's commercial X.25 ventures and gave rise to a host of networks that linked military, defense industry, and academic research facilities around the world. Based on these developments, specialized common carriers offered data communication over packet-switched networks. Known as value-added networks, they provide connection-oriented and connectionless, point-to-point and point-to-multipoint, data services.

12.1.1 X.25 Networks

(1) What is an X.25 Network?

X.25 packet-switched networks multiplex packets of data across links between store-and-forward nodes so as to share the resources among all network users. Users employ X.25 protocols (see § 6.2.1) to connect to the network. Within the network, the nodes route packets in accordance with information carried in individual packets, information stored in the node, and the state of the network (see § 9.2.1). Between networks (a transport layer concern), ITU-T's Recommendation X.75 applies. The realms of application of the standards governing connections to, and connections between, packet-switched networks, are shown in **Figure 12.1**. At the top of the network is a FEP/Host combination that connects to the network using X.25 procedures. At the bottom of the network, a synchronous DTE employs a similar technique. Asynchronous DTEs are incapable of handling synchronous X.25 protocols. To provide them with the ability to use packet-switched networks, a special device, known as a packet assembler and disassembler (PAD), is required. As the name implies, it assembles packets from characters received from asynchronous DTEs and disassembles packets into characters for transmission to asynchronous DTEs. ITU-T Recommendation X.3 defines the PAD that is usually implemented in software resident at the boundary of the network. ITU-T Recommendation X.28 defines the procedures whereby the asynchronous DTE connects to the PAD, and X.29 defines the procedures that permit a packet-mode device to control the operation of the PAD. At top center is an asynchronous DTE that connects to a PAD through the use of X.28 procedures. At bottom center is a signaling terminal exchange (STE), a DCE that implements X.75 procedures to connect to another packet network.

(2) Packet Flow

Packets migrate across the network in different ways depending on the routing technique employed. In **Figure 12.2** we show three examples. DTE A sends groups of packets to terminals B, C, and D:

- Packets 1, 2, 3, and 4 are sent to DTE D using a *connectionless* service and *distributed* routing. Starting off in sequence, the packets are sent over different links at the direction of the individual nodes based on current knowledge of the traffic loads in the networks. They arrive at DTE D in the order 2, 3, 1, and 4. It will be up to the receiving DTE to reorder them if necessary.
- Packets 5, 6, and 7 are sent over a *permanent virtual circuit* to STE C for transmission to another network. They arrive at C in sequence.

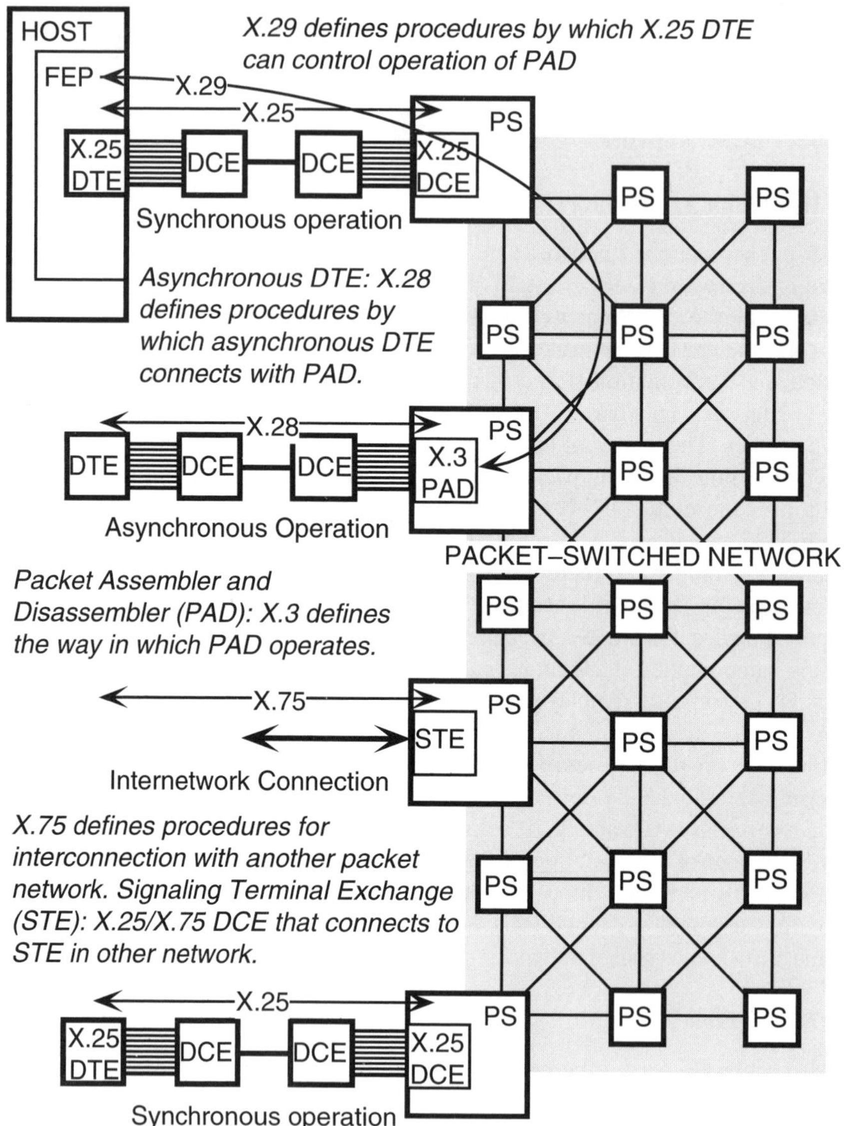

Figure 12.1 Packet Switching Network

- Packets 8, 9, and 10 are sent to DTE B using a *connection-oriented* service and *centralized* routing. A call request packet is used to establish a virtual circuit (primary and alternate route). When packets 8 and 9 have traversed the primary route, we assume a failure occurs that requires packet 10 to traverse the alternate route.

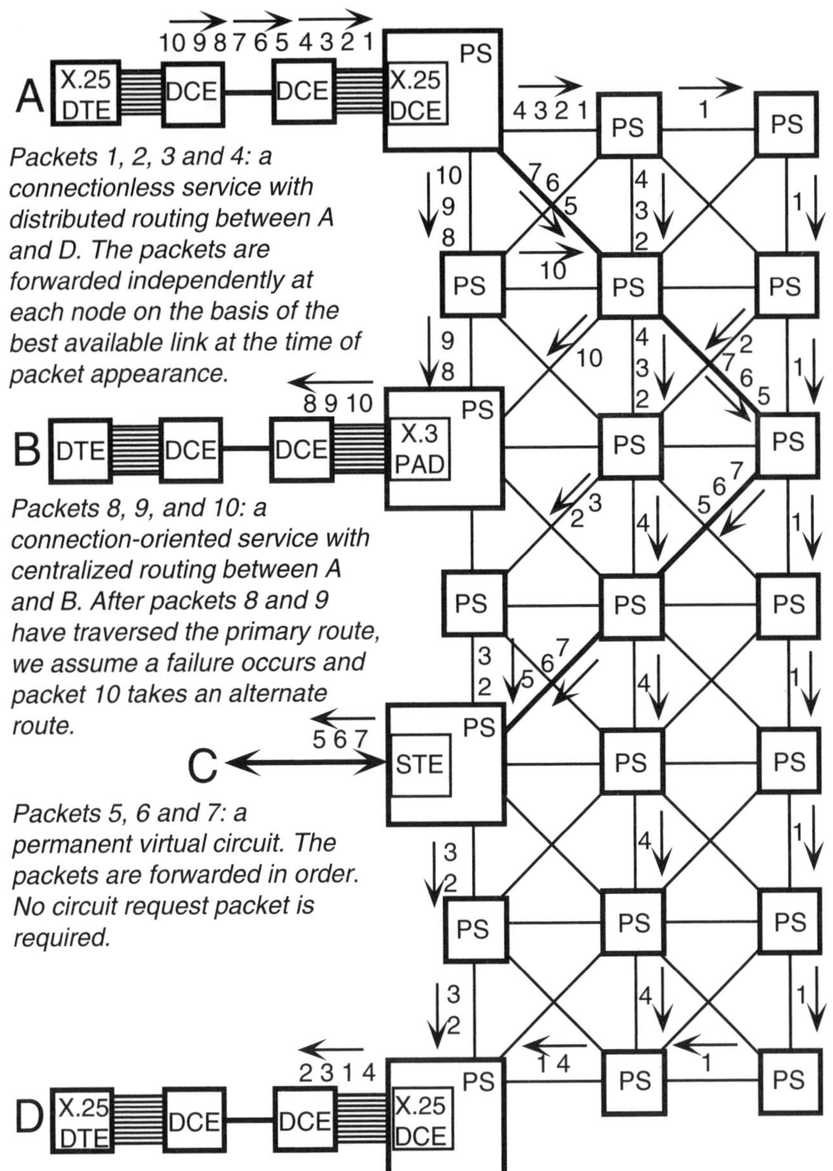

Figure 12.2 Effect of Routing Techniques

In a packet network, the integrity of each packet is checked at a node before the packet is passed on to the next node. This means that the entire packet must be received before it can be routed towards its destination, or destroyed. Thus, at each node, packets are delayed for at least one packet time. Further slowing the throughput is the fact that long messages are segmented into limited-length packets that contain additional overhead information to facilitate control, identification, sequencing, and delivery. In all, some 30 processing steps may be required before the node can begin to pass a packet to the next node on its path through the network. The presence of so many steps reflects the fact that conventional packet networks were designed to link dumb terminals to mainframes over relatively unreliable transmission facilities and that much checking is required to achieve acceptable network performance. Because these procedures are time consuming, X.25-based networks provide slow-speed transport service.

(3) Flow Control

A packet is a small segment of information. If 128 bytes is the limit of the information field, it can contain approximately 20 average words—and that may be less than two lines of text. Strings of packets, then, are common, and procedures are needed to ensure that they are not sent so rapidly as to block the receiving network links, DCE, or the receiving DCE/DTE combination. One technique employs *sliding window* flow control. In this process, the transmitter can transmit up to W packets and must then wait for an acknowledgment from the receiving DTE. The value of W is $1 \leq W < 8$ for data packets in which P(R) is a 3-bit number, and for packets that employ an extended format $1 \leq W < 128$, i.e., P(R) is a 7-bit number. If the sequence of W packets in the window is received correctly, the receiving DTE will acknowledge them by updating the value of P(R). This permits the sending DTE to transmit another W packets. The value of W is negotiated between sending and receiving DTEs. At any time, the receiving DCE may send a Receive Not Ready (RNR) indication to the sending DTE; also, the receiving DTE may send RNR or Reject (REJ) indications to the sending DCE. Thus, the terminating equipment has several options to exercise control over the flow of packets between the end DTEs. The value of the D-bit in the format identifier field of octet #1 in the header of an information packet (see Figure 6.6) distinguishes between local (i.e., DTEn ⇔ DCEn) and end-to-end (i.e., DTEn ⇔ DTE1) flow control.

EXAMPLE 12.1 Effect of Access Speed on Throughput

What is the effect of access line speed on relative throughput when sending blocks of data over a packet network? We will answer the question by calculating the throughput achieved when sending 4096 bits of user's data from DTE 1 to DTE 2 over three nodes of a packet network with internal speed of 1.544 Mbits/s. The network model and time diagram are shown in **Figure 12.3**.

Assumptions: (1) an X.25 UNI; (2) access speeds of 56 kbits/s or 1.544 Mbits/s; (3) transmission between DCE 1 and DCE 2 is by means of a permanent virtual circuit (PVC); (4) packets are delayed by one packet time at each node or device, and checking and routing are performed in no additional time so that the packet is sent

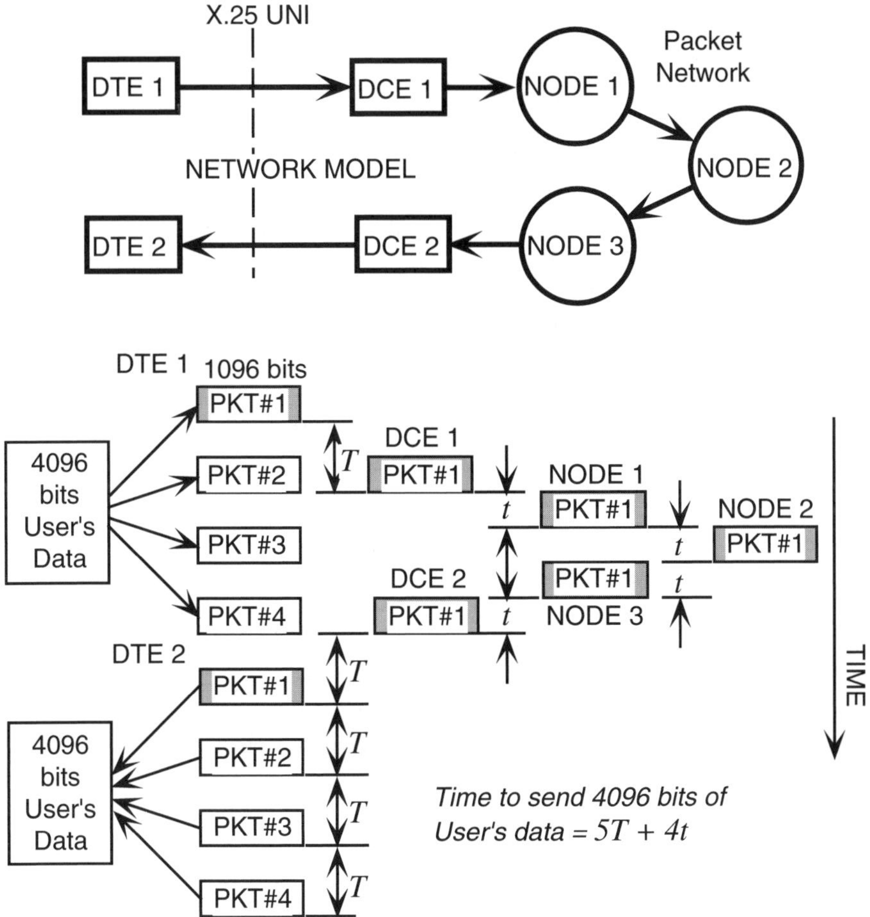

Figure 12.3 Network Model and Time Diagram for Example 12.1

on as soon as reception is completed (packet time = total number of bits in packet divided by the line speed in bits/s); (5) no errors occur; and (6) propagation delays may be neglected.

Derivation: the 4096 user's data bits are divided into exactly 4 ×1024 bits, a 3 octet packet header is added to each segment, and the whole is encapsulated in an HDLC frame with a 3 octet HDLC header and a 3 octet HDLC trailer. The total number of bits in each packet is 1096.

At 56 kbits/s, the packet time is:

$$\frac{1096 \text{ bits}}{56 \text{ kbits / s}} = 0.019571 \text{ seconds}.$$

At 1.544 Mbits/s, the packet time is

$$\frac{1096 \text{ bits}}{1.544 \text{ Mbits / s}} = 0.000710 \text{ seconds}.$$

A study of Figure 12.3 should convince you that, to send the *first* packet from DTE 1 to DTE 2, requires $2T + 4t$ seconds. (This is the time from when the first bit begins its journey from DTE 1 until the last bit arrives at DTE 2.) When the access speed is 56 kbits/s, $T = 0.019571$ seconds. If the network speed is 1.544 Mbits/s, $t = 0.000710$ seconds. Hence, the time required to send the first packet from DTE 1 to DTE 2 is 0.041982 seconds. If packets 2, 3, and 4 are sent immediately after packet 1 (this is known as *pipelining*), the total time to send 4092 user's data bits is $0.041982 + 3 \times 0.019571 = 0.100695$ seconds. Thus,

$$\text{throughput} = \frac{4096 \text{ user's data bits}}{0.100695 \text{ seconds}} = 40.677 \text{ kbits / s}$$

Expressed as a fraction of access line speed,

$$\text{relative throughput} = \frac{40.677 \text{ kbits / s}}{56 \text{ kbits / s}} = 0.7264 .$$

When the access speed is 1.544 Mbits/s, $T = t$, and the time to send 4092 user's data bits from DTE 1 to DTE 2 is $9t = 0.00639$ seconds, so that

throughput = 641.0 kbits/s, and…

relative throughput = 0.4151.

Comments: it is to be expected that as access line speed increases, the time required to transfer a given number of user's data bits decreases. Comparing time components:

- With an access speed of 56 kbits/s, the time spent crossing the UNI is $5T = 0.097855$ seconds and the time spent in the network is $4t = 0.00284$ seconds, a ratio of over 34:1.
- With an access speed of 1.544 Mbits/s, the time spent crossing the UNI is 0.00355 seconds and the time spent in the network is 0.00284 seconds, a ratio of 1.25:1.

Thus, as access speed increases, the time spent in the network becomes a greater fraction of the total time, and throughput becomes more dependent on the packet processes.

EXAMPLE 12.2 Effect of Packet Network Delays on Throughput

What is the effect on relative throughput of delays encountered at network nodes when sending blocks of data over a packet network? We will answer this question by generalizing Example 12.1.

Assumptions: (1) an X.25 UNI; (2) access speed of 1.544 Mbits/s; (3) packets are delayed by one packet time at each node or device for checking and routing and they may be delayed further by congestion, or for other reasons; (4) no errors occur; and (5) propagation delays may be neglected.

Derivation: building on the time diagram in Figure 12.3, letting $T = t = 0.00071$ seconds, and assuming that the average nodal delay is a packet times (in addition to the 1 packet time delay for checking), the time required to send the first packet from DTE 1 to DTE 2 is

$$t + (n + 1)t + nat + t, \text{ i.e.,. } t[3 + n(1 + a)].$$

If packets 2, 3, and 4 are sent immediately after packet 1, the total time to send the user's data bits is:

$$t[6 + n(1 + a)].$$

Thus,

$$\text{relative throughput} = \frac{4096 \text{ bits}}{t[6 + n(1 + a)]1.544 \text{ Mbits / s}},$$

$$\text{i.e., } \frac{4096 \text{ bits}}{[6 + n(1 + a)]1096 \text{ bits}} = \frac{3.73723}{[6 + n(1 + a)]}.$$

When $a = 0$, we have the situation described in Example 12.1—the packets are sent over a permanent virtual circuit. **Figure 12.4** shows values of relative throughput for a range of values of n and a.

Comments: Network delays significantly degrade the relative throughput of packet networks with high speed access lines. Based on Figure 12.4, performance can be no better than that achieved over a PVC and is likely to be a great deal worse.

(4) Making a Packet Call

Virtual calls are identified and managed on the basis of logical channel numbers that are assigned as required. An example of the procedure is shown in **Figure 12.5**. To initiate a virtual call to DTE2, DTE1 assigns an outgoing logical channel number that defines the virtual connection from DTE1 to DCE1 and sends a Call Request packet to DCE1. The packet identifies DTE1 and DTE2, and contains the

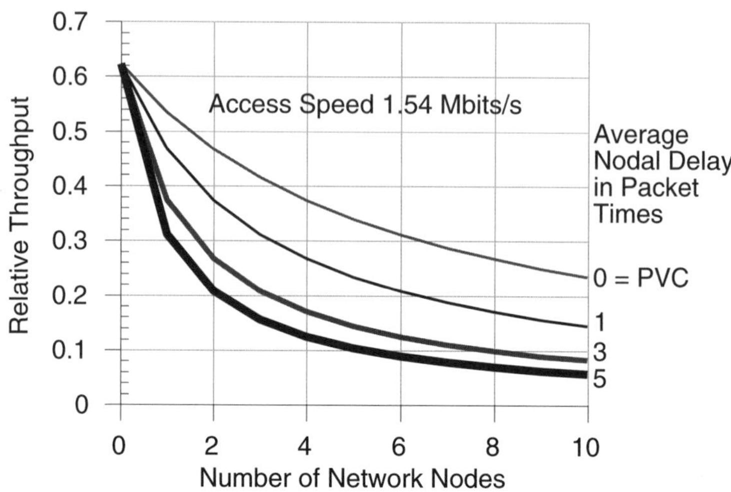

Figure 12.4 Effect of Number of Nodes and Average Delay on Throughput
In a packet network, as the average delay at each node increases, the throughput decreases significantly. Permanent virtual circuits perform the best.

outgoing logical channel number. At DCE1, the packet enters the network. On the basis of its destination, the Call Request packet is routed in a distributed fashion to DCE2. DCE2 assigns an incoming logical channel number that defines the virtual connection between DCE2 and DTE2 and sends an Incoming Call packet to DTE2. Assuming DTE2 is authorized and ready to communicate with DTE1, DTE2 responds to DCE2 with a Call Accepted packet. It contains the same incoming logical channel number as DCE2 assigned to the Incoming Call packet so that DCE2 shall know to which incoming call DTE2 is responding. At DCE2, the Call Accepted packet enters the network and is routed to DCE1. DCE1 assigns the original outgoing logical channel number to a Call Connected packet and sends it in an HDLC frame to DTE1. DTE1 may now proceed to send packets to DCE1 for DTE2. They are routed through the packet-switched network on the basis that they arrive at DCE1 with the outgoing logical channel number already assigned by DTE1, and exit from DCE2 with the incoming logical channel number already assigned by DCE2. All of the packets in the call follow the same series of links through the network so that they arrive at their destination in sequence. The virtual call is cleared by DTE1 sending a Clear Request packet to DCE1. DCE1 replies with a Clear Confirmation packet and sends the Clear Request to DCE2. DCE2 sends a Clear Indication packet to DTE2; it answers with a Clear Confirmation packet. The logical channel numbers are now available for assignment to other calls.

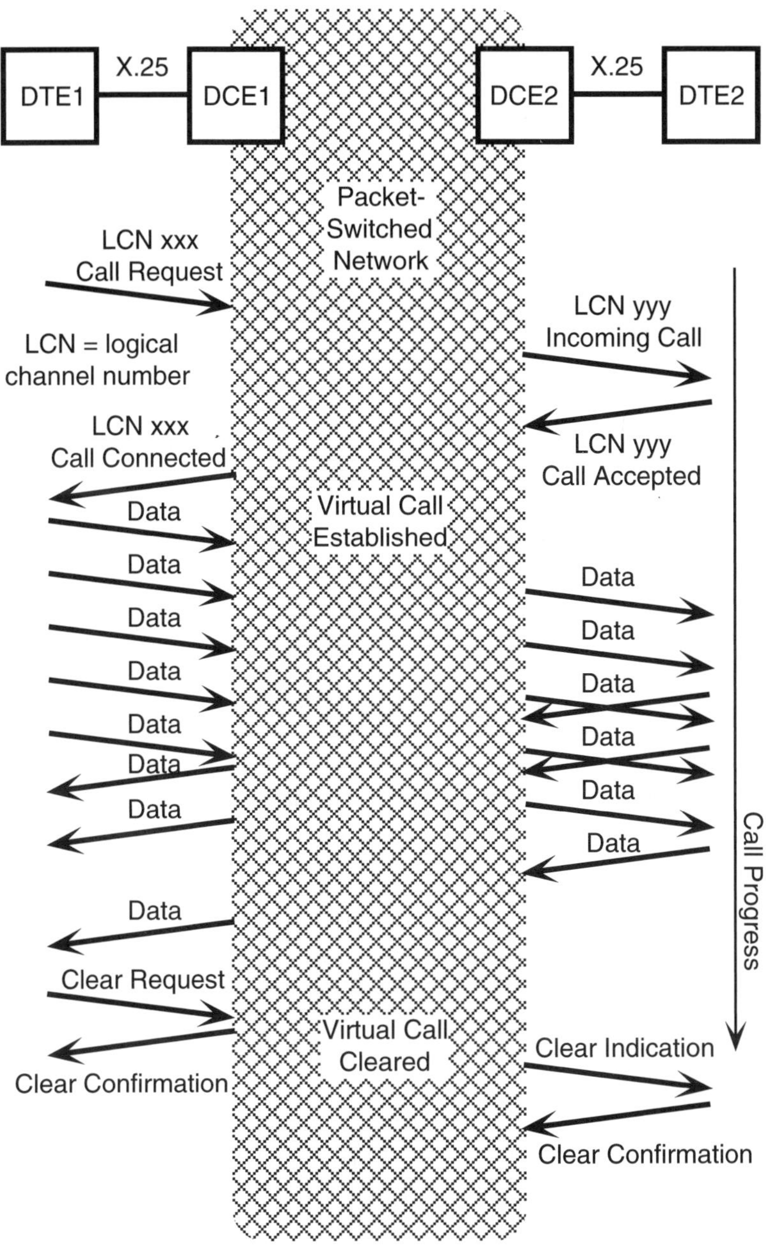

Figure 12.5 Packet Sequence in Virtual Call

(5) Fast Select Service

For short messages such as are encountered in automatic teller machine transactions and similar applications, the call connect and disconnect procedures described above can be circumvented by using a Fast Select Call Request packet in which up to 128 bytes of user's data can be included. The network delivers the packet in the usual way as an incoming call request. The receiver responds to the data with a Fast Select Clear Indication packet that contains up to 128 bytes of information in reply. It is delivered to the original sender as a Clear Indication packet.

12.1.2 Internet

(1) Internet Community

ARPANET has evolved into an international *network of networks* called *Internet* that serves data users around the world. Changes in organization and mission in the early 1990s, made the network resources available for commercial uses and transferred operations to a government-university-industry consortium. At the same time, advanced network services (ANS) were added to form ANSnet, a national, data network utility that provides data transport at DS-3 and SONET speeds. Recent additions include CommerceNet, that provides services designed to allow billing and paying bills, and World Wide Web, that permits documents to be searched and viewed by any connected computer running suitable software. Estimates of Internet's size are hard to give because of the rate at which it is expanding. At the beginning of 1995 it was likely that almost 40,000 networks were linked by Internet, and that approximately half of them were outside the United States. At the same time, it is likely that there were some three million computers connected. The number is rising exponentially.

(2) TCP/IP Suite

Members of the Internet community share a family of protocols. Called TCP/IP, the name is shorthand for a suite of protocols whose major members are TCP (transmission control protocol), a transmission layer procedure, and IP (internet protocol), a network layer procedure. Shown in **Figure 12.6**, the major members of the TCP/IP protocol suite are:

- **Application layer**: interfaces user processes with lower level protocols. Some specific capabilities include:
  - *TELNET*: virtual terminal service that provides access to remote hosts for terminals attached to a local host
  - *file transfer protocol* (FTP): transfers files between remote hosts
  - *simple mail transfer protocol* (SMTF): transfers mail between existing host mail systems

TCP/IP PROTOCOL SUITE

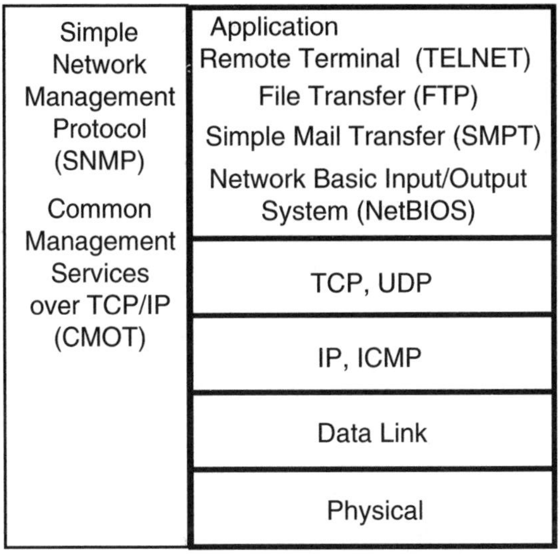

Simple Network Management Protocol (SNMP) Common Management Services over TCP/IP (CMOT)	Application Remote Terminal (TELNET) File Transfer (FTP) Simple Mail Transfer (SMPT) Network Basic Input/Output System (NetBIOS)
	TCP, UDP
	IP, ICMP
	Data Link
	Physical

TCP = Transmission Control Protocol
UDP = User Datagram Protocol
IP = Internet Protocol
ICMP = Internet Control Message Protocol

OSI PROTOCOL SUITE

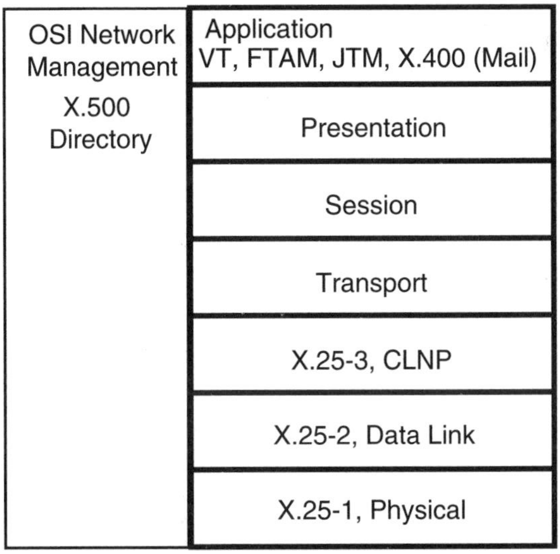

OSI Network Management X.500 Directory	Application VT, FTAM, JTM, X.400 (Mail)
	Presentation
	Session
	Transport
	X.25-3, CLNP
	X.25-2, Data Link
	X.25-1, Physical

VT = Virtual Terminal
FTAM = File Transfer and Management
JTM = Job Transfer and Management
CLNP = Connectionless Mode Network Protocol

Figure 12.6 TCP/IP And OSI Protocol Suites

  - *domain name service* (DNS): provides Internet addresses and other information concerning remote host
  - *NetBIOS*: supplements IBM's NetBIOS with internetworking capabilities of TCP/IP.
- **Transport layer**: contains protocols that establish, control and terminate network connections between data structure endpoints on source and destination hosts. Specific capabilities include:
  - *transmission control protocol* (TCP): provides reliable, connection-oriented, network communications between peer processes; maintains and controls the state of the connection and ensures that all data has been received by the destination host
  - *user datagram protocol* (UDP): provides connectionless network communications between peer processes.
- **Internet layer**: implements addressing that supports communications between network devices and provides routing for transmitting data between networks. Specific capabilities include:
  - *internet protocol* (IP): provides the Internet address to identify network devices and physical networks; fragments or reconstructs data segments as needed to suit diverse network requirements
  - *internetwork control message protocol* (ICMP): returns error messages to source when appropriate
  - *address resolution protocol*: translates 32-bit IP addresses and 48-bit Ethernet addresses.
- **Datalink layer**: employs HDLC, LAP-B, LAP-D, and IEEE 802.2 (LLC) protocols.
- **Physical layer**: incorporates a wide range of physical interfaces such as EIA232/449 (see § 5.1.2), ISDN (see § 10.5.1), IEEE 802 (see § 11.1.2) and ITU-T V.35.

(3) Comparison with OSI Model

In Figure 12.6, the TCP/IP suite is compared with the protocols contained in the OSI model of a host working into a packet network. In the physical layer, both suites accommodate a range of connections that includes EIA232/449, ISDN, and IEEE 802 standards. In the data link layer, both suites can employ versions of HDLC (such as LAP-B in X.25 networks and LAP-D in ISDNs) and IEEE 802.2 (logical link control). It is in the OSI network layer that the two begin to separate. X.25 is a connection-oriented network interface procedure, while IP is a connectionless procedure. IP shares some features with CLNP (connectionless network protocol), but the latter is significantly more sophisticated and the protocol data unit formats are quite different. The internet control message protocol (ICMP) is an essential adjunct to IP. It provides information to the hosts about communications problems that exist at the gateways between networks. When sending messages, it constructs a data unit and passes it to IP for encapsulation in an IP PDU. In the transport layer, the transport protocols share the same objective of assuring end-to-end transport, but differ in the way it is achieved and the functions that are performed. TCP/IP does not have segregated procedures that map into the session and presentation layers of OSI. Because of this, TCP/IP applications are limited in scope and flexibility; they cannot invoke many of the services that their counterparts in OSI use. A similar comment can be made to describe the network management standards that support the two systems. TCP/IP has a loyal, worldwide following, and is well established. However, the United States government has declared that OSI will be the network protocol of choice and the capability must be included in network products and services purchased by federal agencies. Only time will tell what the *advantage of occupancy* that TCP/IP enjoys will bring. Recent activities suggest that a real effort is being made to bring the two systems closer together.

(4) Administration

The maintenance of existing Internet standards (such as TCP and IP) is performed by the Network Information Center (NIC) at SRI International (Menlo Park, CA). Technical reports, called requests for comments (RFCs), are distributed by NIC for review by users. Eventually, when an adequate consensus develops, they are declared standards.

12.1.3 Frame Relay

(1) Taking Advantage of Better Transmission

Today, the bulk of contemporary data processing power is distributed throughout the workplace, and desktop computing has led to the installation of a rapidly growing number of local-area networks and the need to interconnect them. Because LAN traffic is bursty and each message may have a different destination, X.25-based packet switching networks would appear to be well suited for the transport of such messages. However, as we have seen, X.25 networks are matched to dumb terminals and poor quality transmission facilities. When intelligent end-user systems are linked over reliable, high-speed, digital transmission facilities, most of the checks performed by X.25 networks are unnecessary. A technique that significantly speeds delivery is called *frame-relay*. However, a frame-relay network does not guarantee faultless delivery of data. It detects, but does not correct transmission, format, and operational errors, may discard frames for a variety of reasons, and does not acknowledge or retransmit frames. Frame-relay relies on the ability of intelligent end-user systems connected by quality transmission facilities to cope with the exigencies of communication. Despite these caveats, frame-relay is rapidly becoming a technique of choice for private data networks that must interconnect LANs separated by substantial distances.

(2) Frame Format

Frame-relay transfers data in variable length frames over permanent virtual circuits (PVCs). Frame relay is implemented through an interface that supports data link and physical layers:

- **Frame-relay user-network interface** (FRI): a broadband network access arrangement that transfers information in variable length frames up to 8189 octets. Error detection (but not correction) is performed and corrupted frames are discarded (without requesting retransmission). Normally, the interface is limited to DS-1 speed.

The user-network interface employs a set of *core* functions included in a data link layer protocol that is based on LAP-D (ISDN data link protocol). Called LAP-F core, it includes frame delimiting and alignment (using beginning and ending flags), inspection to ensure that the frame consists of an integer number of octets that is less than the maximum permitted, transparency (using zero-bit stuffing), frame multiplexing and demultiplexing (using the address field), and detection of transmission errors (using CRC). Flow control (using windowing) and error correction (using ARQ) are included in LAP-F, but not in the core. **Figure 12.7** shows the LAP-F core frame format; the fields are:

- **Flag**: 01111110.
- **Address field**: a 2–octet routing label that may be expanded to 3 or 4 octets. It contains:
  - *data link connection identifier* (DLCI): 10 bits that identify the logical connection between end-point and nodal processor. While 1024 separate logical channels are possible, the actual number employed on any one connection will be much smaller. The DLCI may be expanded to 16 or 24 bits
  - *C/R* (control field): this bit notifies the receiver that the frame is a command or a response
  - *forward explicit congestion notification* (FECN): this bit notifies the receiver that the path through the network is congested, and that following frames may be delayed
  - *backward explicit congestion notification* (BECN): this bit notifies the sender that the path through the network is congested, and the sender should institute congestion avoidance procedures. It is set in a frame passed backward from node to sender. If congestion continues to increase, the network will begin to discard frames (see DE bit, below)
  - *EA* (extended address): bits that indicate the length of the address field (01 = 2 octets and DLCI = 10 bits; 001 = 3 octets, and DLCI = 16 bits; 0001 = 4 octets and DLCI = 24 bits)
  - *DE* (discard eligibility): this bit is set by the sender to indicate that the frame can be dropped before others (not so marked) in case of congestion.

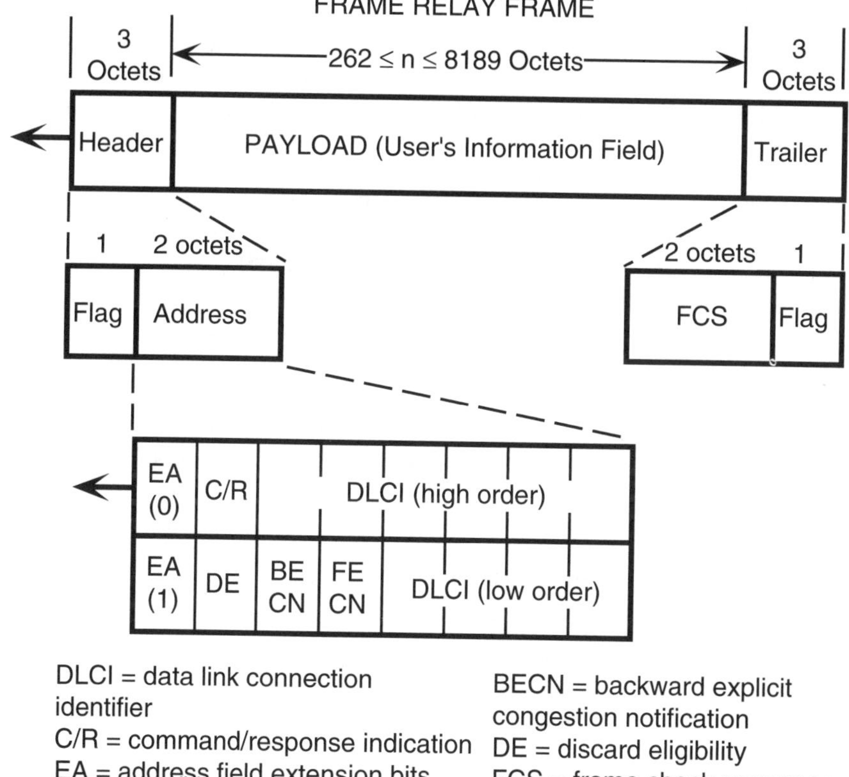

Figure 12.7 Frame-Relay Frame Format

- **Information field**: an integral number of octets between 262 and 8189. Contains user's information at the data link level (i.e., LLC + MAC, SDLC, or X.25-2). When the number of octets exceeds 4096, the 16-bit FCS is ineffective.
- **FCS** (frame check sequence): 16-bit remainder from cyclic redundancy check (CRC) performed by entity generating frame-relay frame.

As soon as the header has been received and the DLCI validated, the frame is started on the way to the next node. When an invalid frame is detected (for any reason), the node terminates transmission and discards the frame. It is left to the receiving end-user system to invoke an end-to-end, error control strategy to restore the frame.

(3) Two Networks

Figure 12.8 shows a group of LANs fully-connected by routers and private lines. When they are separated by any distance, such an arrangement can be expensive. **Figure 12.9** shows the same group of LANs connected in a hub arrangement using frame-relay. Through the use of frame-relay configured routers and a frame-relay nodal processor, the number of transmission lines is reduced, and the speed at which the frames are transported, is increased. This is the situation shown in the top diagram. The routers encapsulate the LLC/MAC frames in frame-relay frames (LAPF) that include data link connection identifiers (DLCIs). The frame-relay nodal processor (FRNP) interprets this information and forwards the frames over the circuits they define to the routers serving the destinations. There, the routers strip off the frame-relay headers and trailers and restore the frames to formats compatible with the receiving LANs. At the bottom of Figure 12.9, the protocol stacks for this operation are shown. Error control is performed by the transport layers of the users' stacks, and routing is performed by the network layers of the routers' stacks.

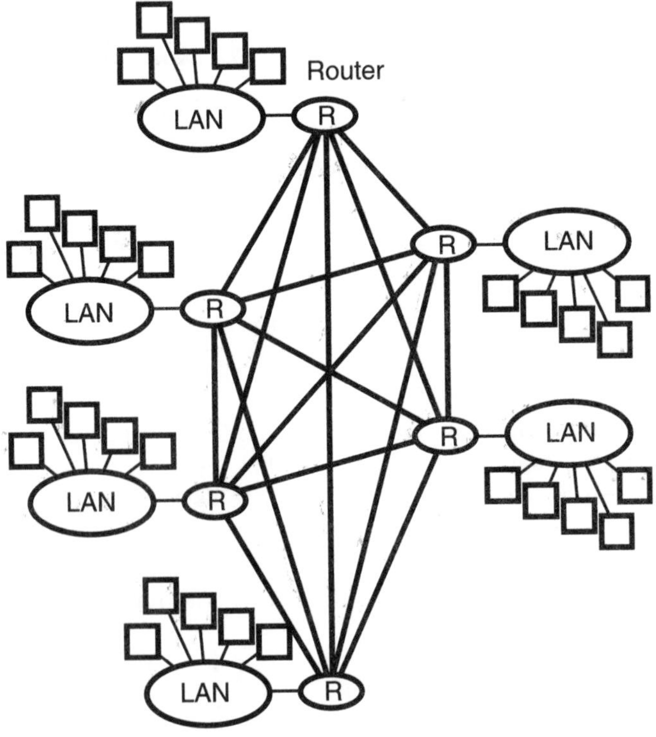

Figure 12.8 LANs Interconnected by Routers

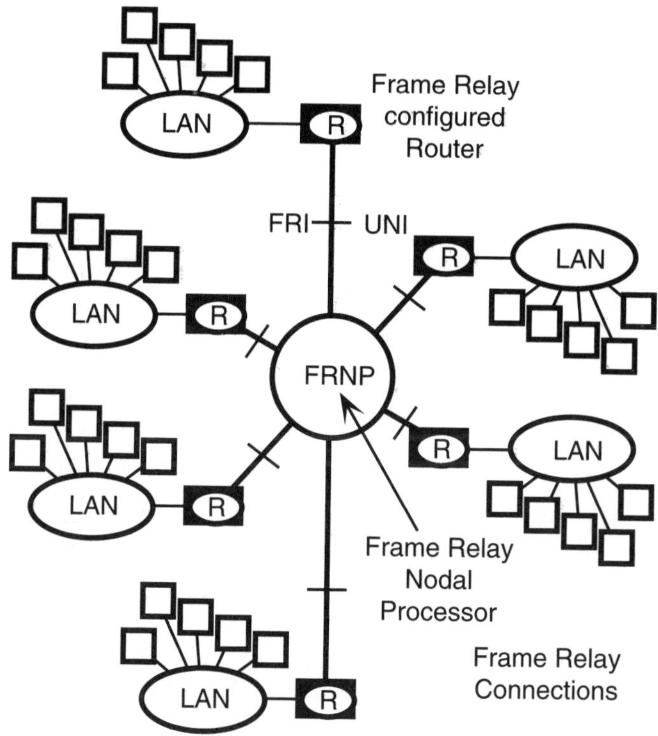

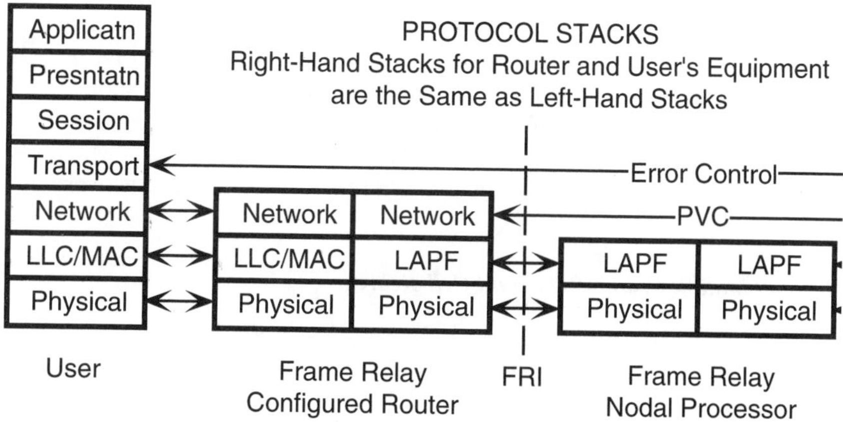

Figure 12.9 LANs Interconnected by Frame-Relay Nodal Processor

EXAMPLE 12.3 Costs of Router Network and Frame–Relay Network

Compare the monthly charges for the networks shown in Figures 12.8 and 12.9.

Assumptions: All transmission occurs at DS-1. In the mesh-connected router network, the total number of miles of T-1 is 18000. The monthly charge is $800 per 100 miles. In the FRPN network, the node is co-located with one router and T-1 facilities connect to the other five routers. Their total length is 4500 miles. The monthly charge is $900 per 100 miles. Routers are connected to IXC POPs by T-1 circuits at an average monthly charge of $250, and the monthly charge for POP access is $300. Converting a standard router to a frame-relay configuration costs $1000. A frame-relay processor costs $50000.

Derivation: The mesh connected router network consists of 30 loop circuits, 30 POP access connections, and 18000 miles of T-1 circuits. Hence the monthly charges are

$$\$7500 + \$9000 + \$144000 = \$160500.$$

The FRNP network consists of 5 loop circuits, 10 POP access connections, and 4500 miles of T-1 circuits. Hence the monthly charges for transmission circuits are

$$\$1250 + \$3000 + \$40500 = \$44750.$$

In addition there are capital costs of

$$\$6000 + \$50000 = \$56000.$$

Amortizing this amount over 60 months, the monthly equipment expense is approximately $1000. Thus, the total monthly charge for the FRPN network is $45750.

Comments: For the numbers chosen, the monthly charges for the FRPN network are approximately 30% of the mesh-connected router network. Obviously, real situations require real tariffs and real distances. Nevertheless, this simple example shows the kind of dollar savings that are possible with some frame-relay installations. This is particularly true when the transmission distances are large, an FRNP can be used as a hub, and the alternative is a fully-connected mesh network of T-1 lines.

(4) Combination Frame-Relay and Cell-Relay

The arrangement shown in Figure 12.9 is characteristic of private networks in which there is a limited number of destinations. The arrangement in **Figure 12.10** can handle many more data sources and destinations. It is typical of public networks that employ frame-relay nodal processors to feed cell-relay networks. The cell-relay node processors (CRNPs) are ATM-like nodes that operate on 53 octet cells at high speed as described in § 10.7.2. They may be linked by SONET transport facilities. In the FRNP protocol stacks, the UNI is the same as in Figure 12.9.

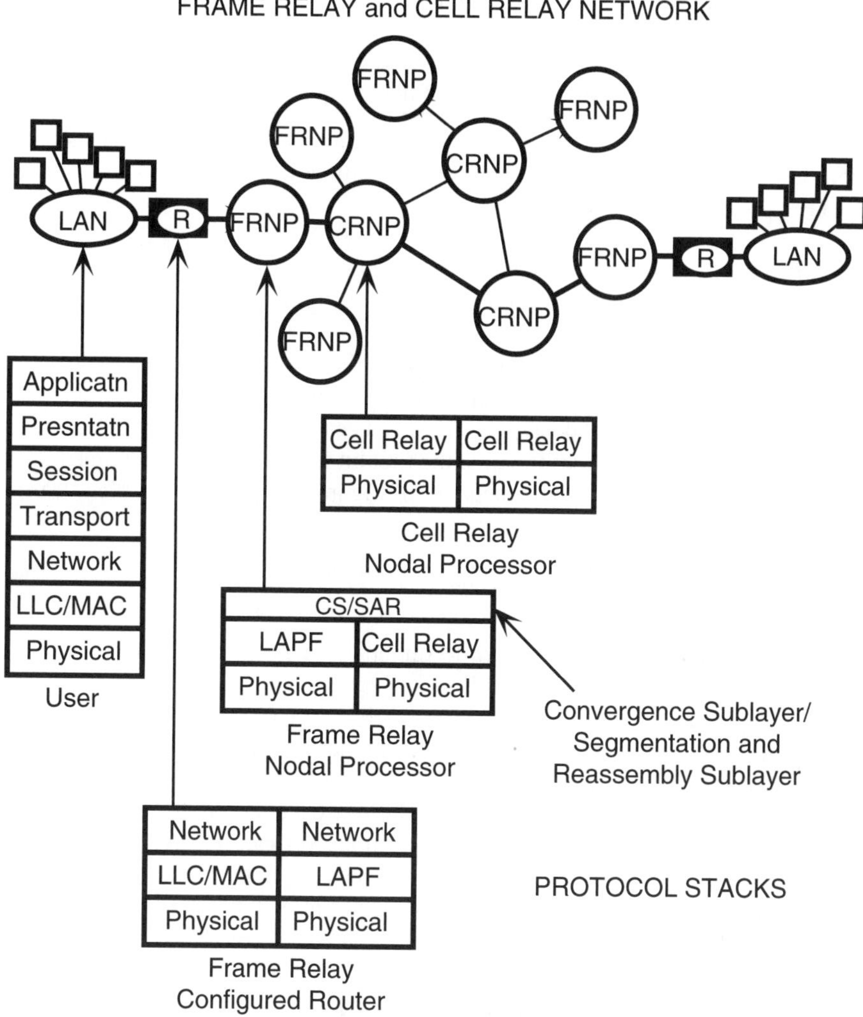

Figure 12.10 LANs Interconnected by Frame-Relay and Cell-Relay Processors

However, at the NNI, the stacks include the segmentation and reassembly, and the convergence sublayers (SAR and CS), of the ATM adaptation layer (AAL) to accomplish the *frame* ⇔ *cell* transformations. Cells are created and transmitted as soon as they can be formed from the incoming frame. In the cell-relay portion of the transmission path, frames are routed using the virtual channel identifiers (VCIs) assigned to the logical path defined by the DLCI. At the FRNP that serves the receiver, the segments are reassembled and the frame is passed to the receiver. End-to-end congestion control is implemented by the FECN and BECN bits in the address field of the LAPF frame.

EXAMPLE 12.4 Frame-Relay *versus* Packet Operation

Compare the relative throughput of packet and frame-relay networks for the situation described in Examples 12.1 and 12.2. The time diagram is shown in **Figure 12.11**.

Assumptions: (1) an FRI UNI; (2) access speed of 1.544 Mbits/s; (3) frame delay at a node is negligible; (4) propagation delays are negligible.

Derivation: for the 4096 bits (512 octets) of user's data, an FR frame contains 4144 bits (518 octets). Consider the path of packet 1 in Figure 12.3. Because frame-relay significantly reduces nodal delay, $t \to 0$. Also, because there is only one frame, two frame times are required from the time the first bit leaves DTE 1 to the time the last bit arrives at DTE 2 (in the nomenclature of Figure 12.3, this is $8T$). Hence, using the same reasoning as in Example 12.2,

$$\text{relative throughput} = \frac{4096 \text{ bits}}{2 \times 4144 \text{ bits}} = 0.49421 .$$

Figure 12.12 compares the relative throughputs of packet and frame-relay networks for an access speed of 1.544 Mbits/s.

Comments: For 2 or more nodes, a frame-relay network outperforms a packet network operating in PVC mode with high speed access lines. For one or more nodes, frame-relay outperforms packet when there are delays at the packet nodes. Comparative testing shows that frame-relay transport can be several times faster than X.25 packet transport. For 1000 octet frames in public frame-relay networks, carriers have stated the objective of achieving end-to-end delays of ≤ 250 milliseconds for DS-0 access and ≤ 20 milliseconds for DS-1.

12.1.4 Switched Multi-Megabit Data Service

Switched Multi-Megabit Data Service (SMDS) is a high-speed, public, packet switched, connectionless data service. Built on IEEE 802.6 MAN standards (see § 11.5.1) and extensions, SMDS supports higher access speeds and greater throughputs than frame relay networks. Without the DQDB requirement of 802.6 MANs, SMDS has no distance limitations—and is being implemented as a public service across the United States.

(1) SMDS UNI

Based on IEEE 802.6 standards, SMDS is implemented through an interface that supports data link and physical layers:

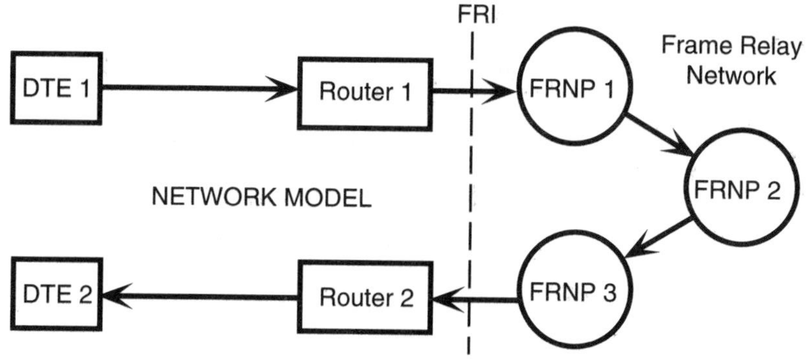

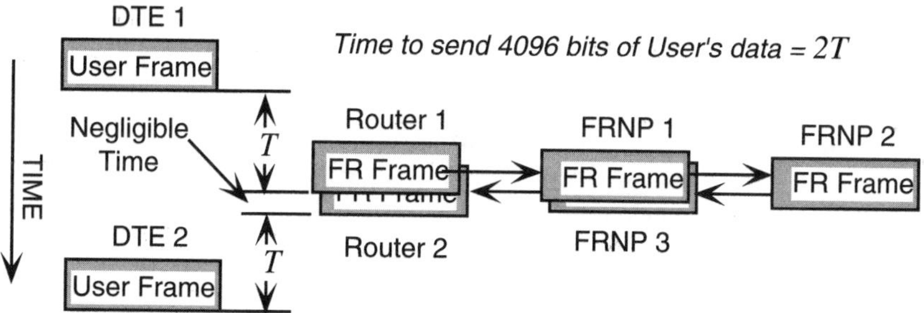

Figure 12.11 Network Model and Time Diagram for Example 12.4

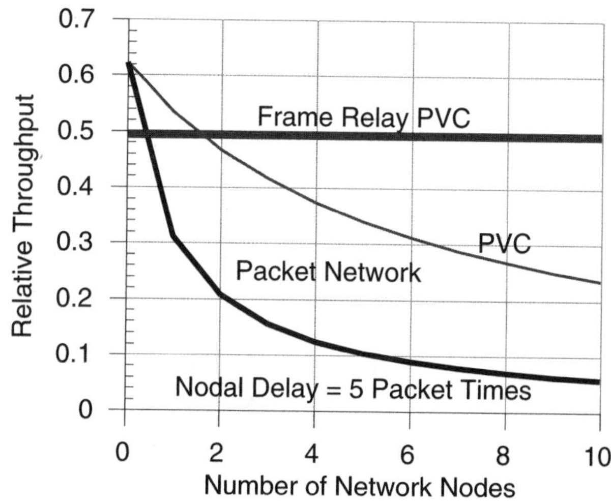

Figure 12.12 Throughputs of Packet Network and Frame-Relay Networks
On the basis of throughput, for more than one node, a circuit in a frame–relay network performs better than a permanent virtual circuit in an X.25 packet network, and substantially better when delays occur at the packet nodes.

- **SMDS network interface** (SNI): a broadband network access arrangement that supports switched, high-speed, connectionless data services. Through the use of a 3-level, *SMDS interface protocol* (SIP), SMDS transfers information in variable length frames up to 9188 octets over connections that employ cell-relay techniques. Each PDU is routed independently to its destination using a ten digit address based on the North American Numbering Plan. Error detection (but not correction) is performed and corrupted frames are discarded (without requesting retransmission). Connectionless service shifts the onus for communication integrity to a higher layer. SNI supports DS-1, DS-3, and SONET speeds.

SIP is a three level protocol that addresses, frames, and transmits PDUs, using cell-relay techniques. **Figure 12.13** shows the protocol stacks, PDUs, and frame formats associated with the use of SIP. Up to 9188 octets of user's data is collected in an SDU (called a Service Data Unit or an SMDS Data Unit) and passed from the upper layers of the router stack to Level 3 of SIP. There, it is combined with a 36-octet header and a 4-octet trailer to become the Level 3 PDU. At Level 2 of SIP, the PDU is segmented in 44-octet payloads and joined to other headers and trailers to become Level 2 PDUs. This action is accomplished *on-the-fly*—as soon as enough octets are present, the first PDU is formed and sent on its way, to be followed immediately by the remaining PDUs. They form a stream that passes to Level 1 of SIP where it is converted into signals suited to cell-relay. Except for the sizes of the headers and trailers, this is the same sequence of operations executed in a DQDB MAN (see Figure 11.19). SIP Level 1 corresponds to the IEEE 802 Physical layer. SIP Levels 2 and 3 correspond to the MAC sublayer of the IEEE 802 Data Link Layer. The SDU is formed in the LLC sublayer of the IEEE 802 Data Link Layer, or in some other compatible fashion.

(2) SIP Level 3 PDU

The fields in the header and trailer of the SIP Level 3 PDU are:

- **Reserved** (Rsvd): carried as all zeros. Reserved for possible use later.
- **Beginning tag** (Btag): binary number between 0 and 255. In conjunction with end tag, used to define first and last segments of Level 3 PDU.
- **BA size**: buffer allocation size. Gives length in octets from next field (DA) to end of SDU.
- **Destination address** (DA): contains two subfields:
  - *address type*: four bits set to 1100 when the address is a 60-bit individual address and to 1110 when the address is a 60-bit group address
  - *address*: ten, binary-coded decimal (BCD) digits that are the SMDS address of the destination equipment (similar to ten digit telephone number). The 40 bits are framed by 4 bits set to 0001 and 16 bits set to zero. It

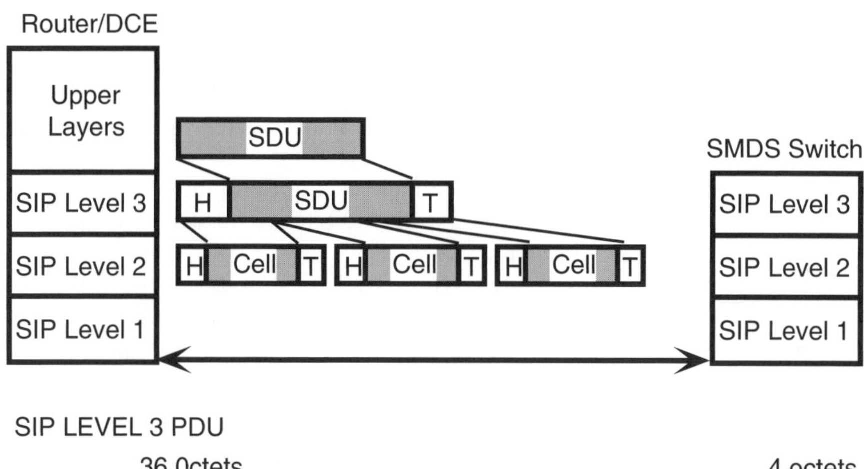

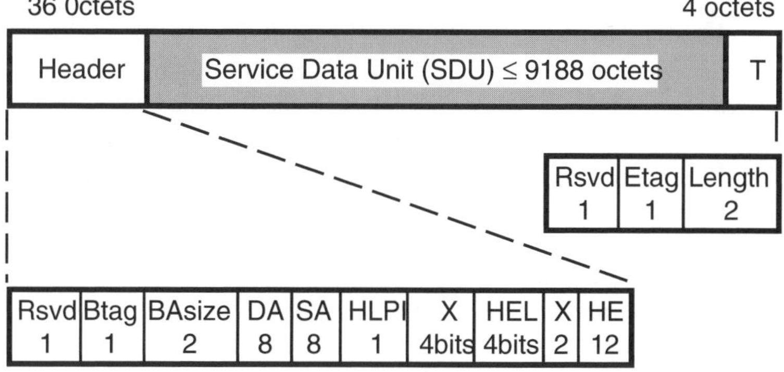

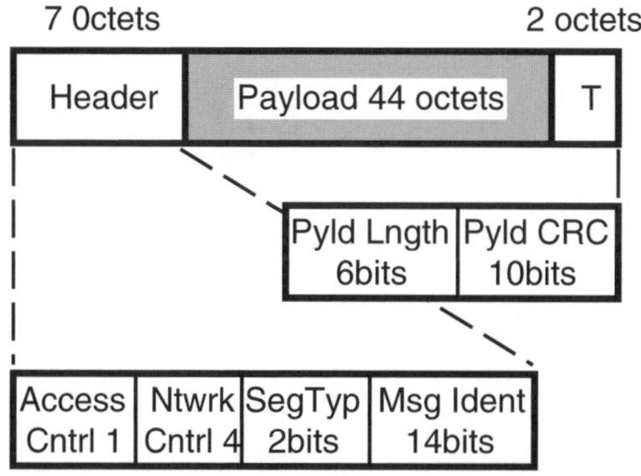

Figure 12.13 Protocol Stacks, PDUs, and Frame Formats In SMDS

is this address that guides the stream of Level 2 PDUs through the network.

- **Source address** (SA): contains two subfields:
  - *address type*: four bits set to 1100. (The source address is always an individual address.)
  - *address*: ten, binary-coded decimal (BCD) digits that are the SMDS address of the source equipment. The 40 bits are framed by 4 bits set to 0001 and 16 bits set to zero.
- **Higher-level protocol identifier** (HLPI): contains information on SIP/DQDB protocol formats for use by higher layers at destination.
- **X**: these fields are for management purposes. They are carried across the network unchanged.
- **Header extension length** (HEL): indicates the number of 4-octet groups in the HE field.
- **Header extension** (HE): repeats key header values.
- **Service data unit** (SDU): also called SMDS data unit. Variable length field that contains user's data up to 9188 octets. If needed, the field is padded by 1, 2, or 3 octets of 0s to bring the number of octets to a value divisible by four.
- **End tag** (Etag): binary number between 0 and 255. In conjunction with beginning tag, used to define first and last segments of Level 3 PDU.
- **Length**: length indication field. Used to check that reassembly has been completed successfully.

(3) SIP Level 2 PDU

The fields in the header and trailer of the SIP Level 2 PDU are as follows:

- **Access control**: bit 1 indicates whether the Level 2 PDU contains information (1) or is empty (0). The remaining 7 bits are used by the SMDS switch and other devices.
- **Network control**: distinguishes between an empty PDU and one with user's data.
- **Segment type**: 2 bits set to:
  - *10*. Segment begins message
  - *01*. Segment ends message
  - *00*. Segment is continuation of message
  - *11*. Single-segment message.
- **Message identifier**: identifies Level 3 PDU to which segment belongs.
- **Payload**: 44-octet field that contains segment of Level 3 PDU.
- **Payload length**: indicates how many octets in payload contain data (may not be 44 if the segment type is 01 or 11).
- **Payload CRC**: contains 10-bit FCS for error detection.

(4) SIP Level 1

The physical layer supports transmission over the SNI at DS-1, DS-3, and SONET (STS-3) speeds. Each access path is dedicated to a single user. Several devices (belonging to the same user) can access SMDS over this path.

(5) Example of Protocol Stacks

Figure 12.14 shows the arrangement of protocol stacks employed to transfer an IP datagram (see Figure 6.8) over an 802 LAN through an SMDS-configured router to

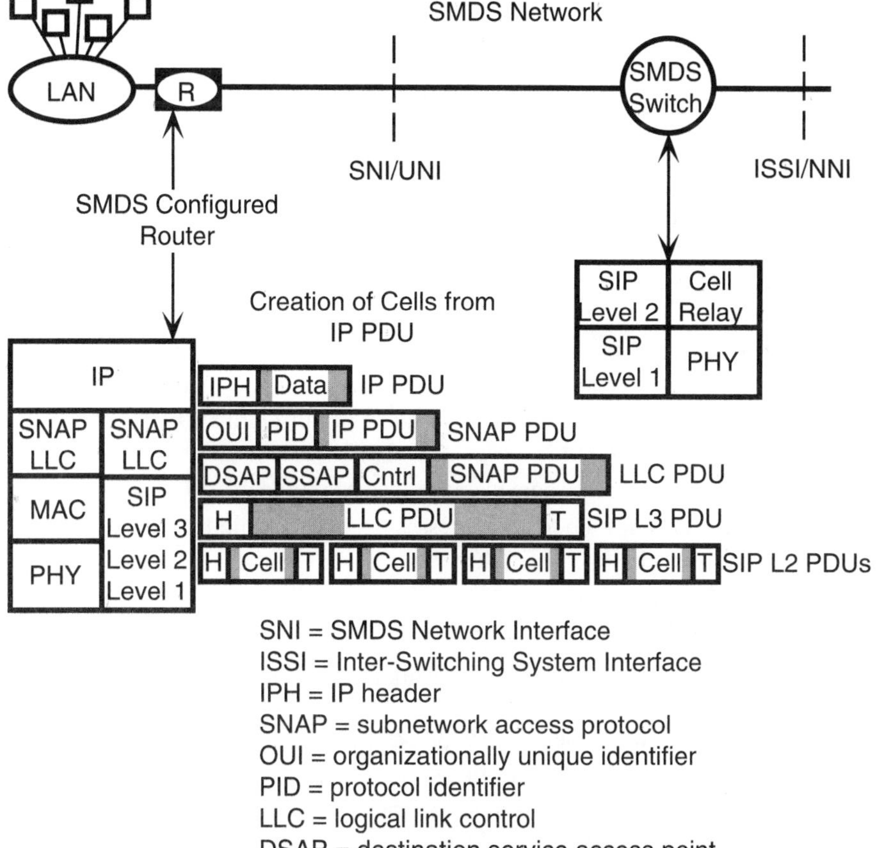

SNI = SMDS Network Interface
ISSI = Inter-Switching System Interface
IPH = IP header
SNAP = subnetwork access protocol
OUI = organizationally unique identifier
PID = protocol identifier
LLC = logical link control
DSAP = destination service access point
SSAP = source service access point
SIP = SMDS Interface Protocol

Figure 12.14 Creation of Cells in SMDS

an SMDS network. Data originates in the upper layers of the DTE stack. It is encapsulated in an IP datagram that includes the IP destination address. The subnetwork access protocol (SNAP) adds a unique identifier (OUI) and a protocol identifier (PID). Among other things, the LLC level adds destination and source service access points, and SIP Level 3 adds ten digit destination and source network addresses for the sending and receiving routers. The entire PDU is segmented, or reassembled, at SIP Level 2 and pipelined through the network by SIP Level 1.

As well as SNI, two other interfaces are defined:

- **Interswitching system interface** (ISI): facilitates the interconnection of switches from different manufacturers in the same network (shown in Figure 12.14).
- **Intercarrier interface** (ICI): facilitates the interconnection of intra-LATA and inter-LATA networks (not shown in Figure 12.14).

They permit the implementation of a country-wide, public SMDS network.

(6) Addresses

SMDS users can have significant flexibility in the way in which they use SMDS services. They have a number of options to choose:

- **Logical private network**: restricts communication to a selected set of addresses
- **Source address screening**: lists addresses of sources from which data will/will not be accepted
- **Destination address screening**: lists addresses of destinations to which data may/may not be sent
- **Group addressing**: the same message may be sent to up to 128 destinations
- **Multiple addresses**: up to 16 addresses (associated with the same customer) can employ one SNI.

(7) Network Performance

The carriers have established goals for network performance:

- **End-to-end network delay**: for 9228 octet PDUs:
  - *2 DS-3 access lines*. 95% of PDUs are delayed ≤ 20 milliseconds
  - *1 DS-3 and 1 DS-1 access line*. 95% of PDUs delayed ≤ 75 milliseconds
  - *2 DS-1 access lines*. 95% of PDUs delayed 130 milliseconds
- **Lost PDUs**: ≤ 1 in 10^4 SIP Level 3 PDUs lost
- **Duplicate PDUs**: ≤ 5 in 10^8 SIP Level 3 PDUs duplicated
- **Misdelivered PDUs**: ≤ 5 in 10^8 SIP Level 3 PDUs delivered to wrong address

- **Missequenced PDUs**: ≤1 in 10^9 SIP Level 3 PDUs delivered out-of-sequence
- **Errored PDUs**: ≤ 5 in 10^{13} SIP Level 3 PDUs delivered with errors.

12.2 SNA NETWORKS

Announced in 1974, SNA (Systems Network Architecture) describes a network in which the communication functions are tightly integrated with the computing resources. Linking the whole range of IBM (and compatible) data processing equipment, SNA has been carefully upgraded and documented to evolve into tens of thousands of private networks. Originally designed for a single-host, one-vendor environment, interconnected by leased lines, today's SNA accommodates a multiple-host architecture that accepts dial-up connections, and interconnects with local–area, packet-switched, and frame-relay networks.

12.2.1 Nodes, Access Methods, Monitors, and Addressable Units

(1) Node Types

An SNA network is divided into subareas and domains as shown in **Figure 12.15**. It consists of nodes interconnected by communication paths under the control of software resident in the nodes and devices incorporated in the network. The nodes are divided into peripheral nodes and subarea nodes. Peripheral nodes are of two types:

- **Terminals**: they provide the interfaces between the human users and the data processing systems.
- **Cluster controllers**: communications devices that act as interfaces between clusters of terminals and hosts or communications controllers.

Peripheral nodes form the lower levels of the network. They are assigned to a specific subarea. A subarea is a network address associated with a node (subarea node) and its attached peripheral nodes. Subarea nodes are of two types:

- **Communications controllers**: intelligent communications hardware that manages communication lines, cluster controllers, workstations, and the routing of data through a network. Communications controllers are further divided into the functional categories of front-end processors, concentrators, and intelligent switches. They may support a host subarea node in the subarea assigned to a host, or operate in a subarea of their own.
- **Hosts**: combinations of hardware and software that comprise the main data processor in a subarea. They are responsible for the application, and the communications access method software.

Some features of peripheral and subarea nodes are summarized in **Figure 12.16**.

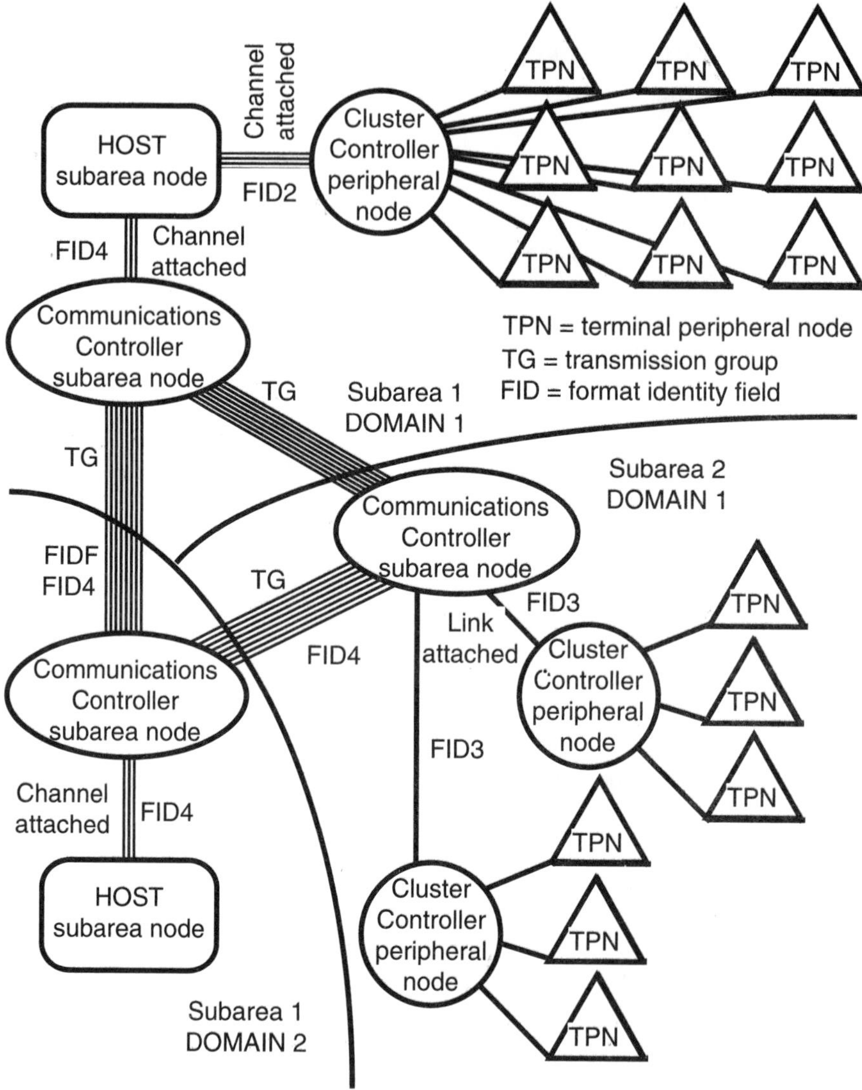

Figure 12.15 SNA Network

(2) Teleprocessing Access Methods

In SNA, the telecommunication software resides in the host(s) and the communications controller(s). The major components are the teleprocessing access method and the network control program. Teleprocessing (TP) access methods provide a software interface between the host main memory and its peripherals that assists in

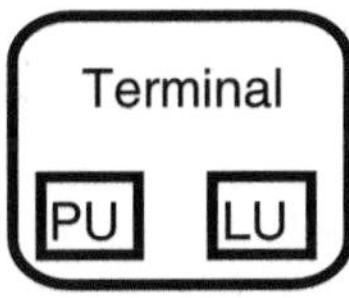

TERMINAL: provides interface between end users and the data processing system.
May be addressed as a Logical Unit (LU) or a Physical Unit (PU). Examples are IBM 3270 family of terminals.

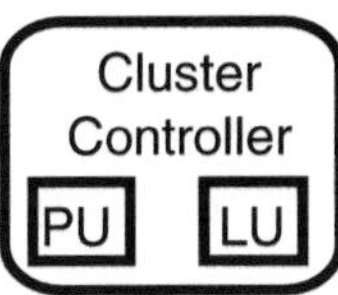

CLUSTER CONTROLLER: acts as interface between clusters of terminals and communications controllers or hosts.
May be attached to host channel or by a link to communications controller. Isolates terminals from data link protocol. May be addressed as a Logical Unit or a Physical Unit. Examples are IBM 3274 and 3174.

SUBAREA NODES

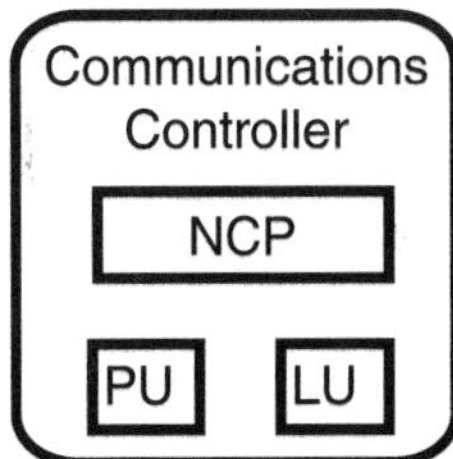

COMMUNICATIONS CONTROLLER: manages communication functions associated with lines, cluster controllers, workstations, and routing message units through network.
May be attached to host channel or by a link to other communications controllers. Sub-divided into front-end processors, concentrators and intelligent switches. May be addressed as a Logical Unit or a Physical Unit. Examples are IBM 3720, 3725 and 3745.

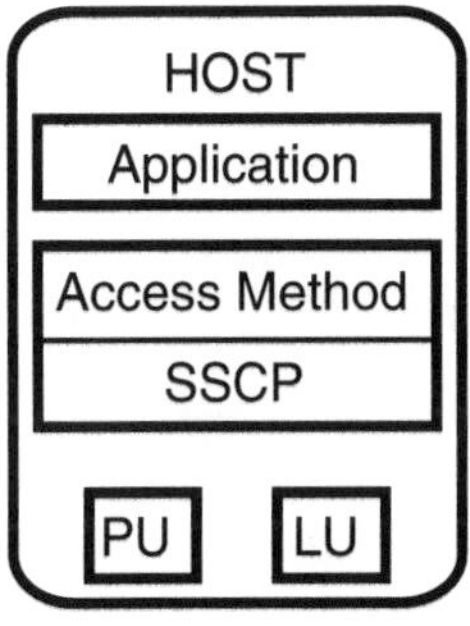

HOST: the main processor in a subarea. Executes applications and communication access method.
May be addressed as System Services Control Point (SSCP), Physical Unit or Logical Unit. Examples are processors in System 370/390 such as IBM 3090, 4381 and 9370

Figure 12.16 SNA Network Nodes

moving information between the network and the host over the I/O channel—a software-driven processor that executes commands and programs. Teleprocessing access methods reside in a host and run as independent jobs under the operating system. TP access methods that execute in an MVS environment include:

- **TCAM** (Telecommunications Access Method): a queued access method that transfers messages from one remote terminal to another, and between terminals and application programs. Because of its complexity, TCAM is being replaced by VTAM.
- **VTAM** (Virtual Telecommunications Access Method): an access method that directly controls transmission of data to and from channel-attached devices. It is the most common TP access method in use today.

Each TP access method manages a communications network for the exclusive use of the applications it supports. To send data to, and receive data from, remote devices over SNA links, TCAM and VTAM require the use of Network Control Program (NCP).

(3) Network Control Program

Resident in the front-end processor (communications controller), NCP performs as follows:

- Provides network information to the TP access method
- Controls cluster controllers and attached terminal nodes
- Routes data to communications controllers in other subareas
- Performs as a primary station for communication terminations
- Polls secondary stations, dials and answers switched lines, detects and corrects errors, and maintains operating records
- Allocates buffers for data coming from and going to the host, and to other stations.

NCP should not be confused with the generic description network control program that identifies any program that performs network control functions.

(4) Teleprocessing Monitors

In order to implement real-time, on-line, transaction processing, some applications are separated from the access method by an additional software module. In a manner that mimics the operating system, these modules supervise multiple transaction programs that run in a single partition. Called TP monitors, to the programs they appear as the operating system. For programmers, TP monitors make it possible to develop applications in a high-level language without detailed knowledge of the terminal characteristics. Two important TP monitors are listed:

- **Customer Information Control System** (CICS): supports multiple, on-line applications in a common network. Provides interfaces to files and database systems. Handles errors, security, and priority transactions.
- **Information Management System/Data Communications** (IMS/DC): originally a database management system, IMS has been augmented by commu-

nications capabilities to provide a comprehensive, database management and data communications capability.

In **Figure 12.17**, we conceptualize the major partitions of the host and front-end processor software.

(5) Network Addressable Units

As shown in **Figure 12.18**, Network Addressable Units (NAUs) are entities that terminate the SNA network. Resident in peripheral and subarea nodes, and inter-

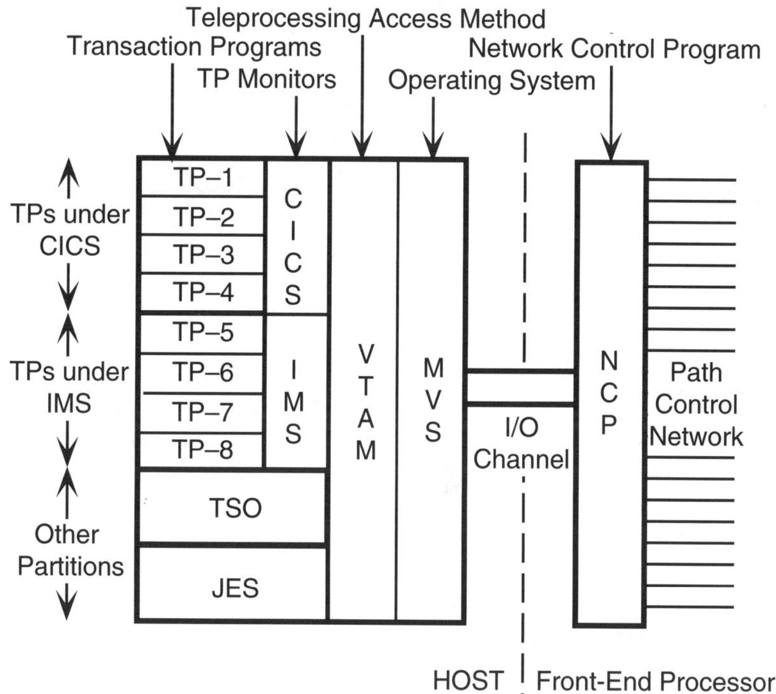

Figure 12.17 Organization of Host Software

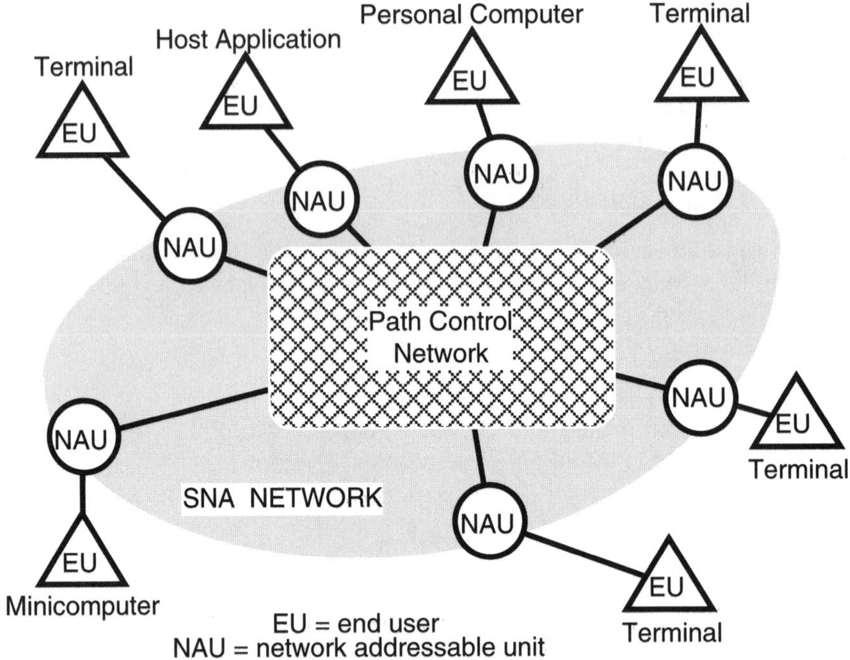

Figure 12.18 End-Users, Network Addressable Units, and Path Control Network

connected by the path control network, they provide services that permit end users to send and receive data, and help perform network control and management functions. NAUs are divided into three types:

- **Logical unit** (LU): the entity through which an end-user (human or application) accesses an SNA network to communicate with another logical unit
- **Physical unit** (PU): an entity that manages attached logical units and acts as a routing node for communication
- **System services control point** (SSCP): a part of the TP access method, SSCP runs only on a host node and instructs PUs to activate, control, and deactivate resources that have been defined as being in its domain.

Each NAU is assigned a unique address within the SNA network. It consists of two parts, subarea number (8 or 16 bits), and element address (16 bits). With the maximum number of bits, and discarding the all-0s address, this scheme identifies a total of 65,535 subareas, and 65,535 elements in each subarea. **Figure 12.19** shows the arrangement of LUs, PUs, and SSCP in a single-host environment with transaction programs running under CICS and VTAM. End-users at the terminals can access one of the applications (the other end user) running under CICS. The SSCP

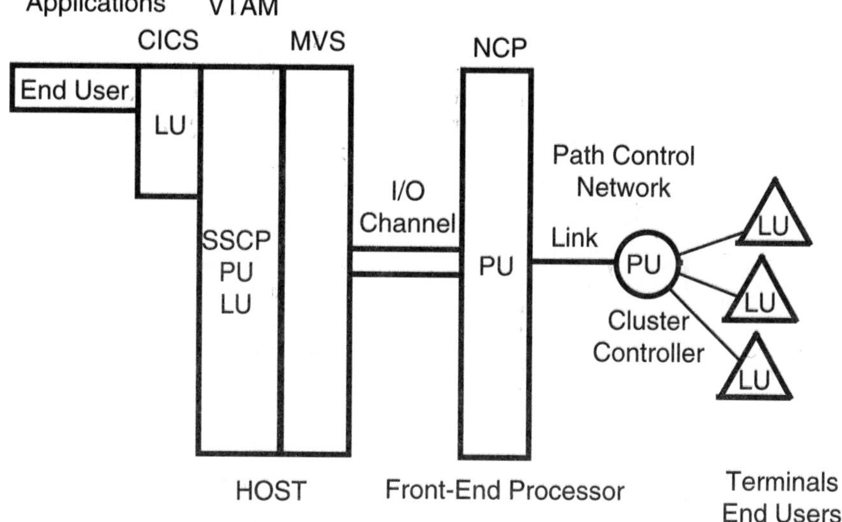

Figure 12.19 Distribution of Types of NAUs
Arrangement of LUs, PUs, and SSCP in a single-host environment with transaction programs running under CICS and VTAM.

in VTAM employs PUs in NCP and the cluster controller to establish, maintain, and terminate communication between the host and the terminal cluster over a link in the path control network. Also, the SSCP employs the PU resident in VTAM to activate, control, and deactivate CICS and the end-user application.

12.2.2 SNA Model

Like the OSI model, the SNA model consists of seven layers. However, their functions reflect a single-vendor environment, not an environment of machines from any vendor. Each of the layers performs a set of functions designed to achieve reliable data transfer between the logical units in a network composed of IBM products. **Figure 12.20** shows the layers arranged in model form.

(1) End-User or Transaction Services Layer

The end-user or transaction services layer provides:

- Application services to the end-user, including end-user access to the network, and requests services from the network addressable unit or presentation services layer.
- Network management services, including:

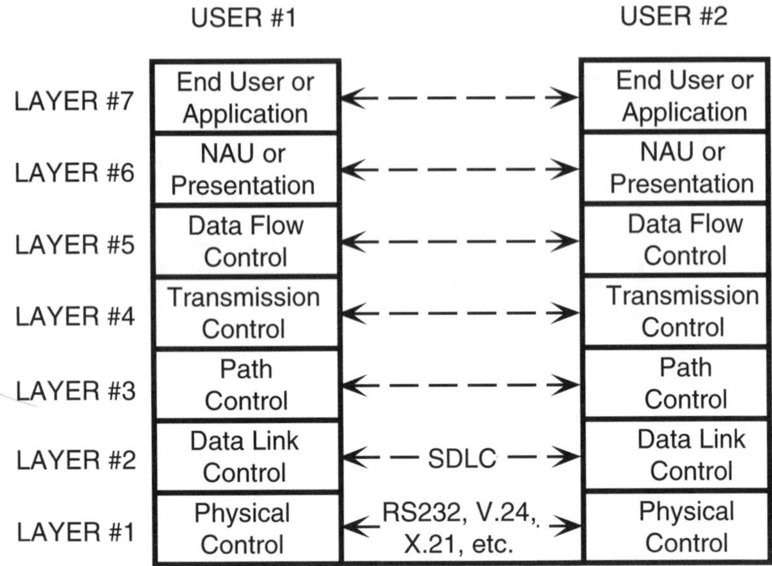

Figure 12.20 SNA Model for Communications

- configuration control associated with the physical configuration during a SSCP - PU session.
- session services associated with establishing LU - LU sessions including verification of user passwords and access authority, confirmation of session parameters, and translation of network names to network addresses.
- monitoring, testing, tracing and recording of statistics for resources used in SSCP - LU and SSCP - PU sessions.

(2) Network Addressable Unit or Presentation Services Layer

The NAU, or presentation services layer, is responsible for defining and maintaining communication procedures for transaction services programs; also, it controls the interaction between these programs by implementing functions for loading and invoking other transaction services programs, by enforcing correct syntax use and sequencing restrictions, and by processing transaction services program commands.

(3) Data Flow Control Layer

Orderly data flow in a LU - LU session is the responsibility of the data flow control layer. Units of data being transmitted are given sequence numbers and an end-user request is correlated with its response. Related request units are grouped into chains and related chains are grouped into brackets of data. This layer enforces communication procedures between LUs and coordinates which LU can send and which LU can receive at any given time.

(4) Transmission Control Layer

The transmission control layer is responsible for the synchronization and pacing of session-level data traffic. It does this by checking the session sequence numbers that were assigned to the request units from the data flow control layer. Another function it performs is the enciphering and deciphering of data.

(5) Path Control Layer

The path control layer is responsible for routing data units through a network to the desired destination. The function is performed in subarea nodes and consists of several steps:

- Selecting the transmission group to be used between subarea nodes for a specific session
- Determining which of the forward and reverse links available in the transmission group are to be used to make the explicit connection between the subarea nodes
- Controlling the flow of data over the virtual route formed from all of the explicit links assigned to the session.

(6) Data Link Control Layer

The data link control layer is responsible for scheduling and error recovery for the transfer of data across a link connecting two nodes. The data link control layer supports the link level flow of data for both SDLC and System 370/390 data channel protocols.

(7) Physical Control Layer

The physical layer defines the electrical and transmission signal characteristics necessary to establish, maintain, and terminate all physical connections of a link. SNA makes no statement concerning preferred procedures. RS232, V.24, and X.21 or other standards may be used.

(8) Path Control Network Functions

By routing and transmitting message units between end-users across peripheral nodes, and subarea nodes, Layers 1, 2, and 3 of the SNA model implement the path control network. Normally, connections between subarea nodes are implemented on parallel transmission links known as transmission groups (TGs, shown in Figure 12.15). To the rest of the network, the group appears to be a single, logical link. To the path control layer, however, each transmission group consists of a set of individual links, some in the forward direction, and some in the reverse direction. An individual session is allocated an explicit forward link, and an explicit reverse link on a non-exclusive basis. Together, they are known as a *virtual route*. High, medium, or low priority is assigned to each session message unit and higher priority messages are handled ahead of lower priority messages in the subarea nodes serving the links. All messages carry a sequence number that identifies their position in the particular data exchange to which they belong. Using this, the subarea node can handle them in sequence, and in order of priority. Because the connections may be used by a diversity of traffic, flow control must be employed to prevent message units arriving faster than they can be processed—an event that will lead to their destruction and subsequent requests for retransmission—and a reduction in the throughput of the network.

12.2.3 Formation of Message Unit

(1) Headers and Trailer

In the SNA model, peer-to-peer communication is effected by adding headers (and a trailer) at stages of the descent of the originating end-user's information (called a request unit or a response unit) through the layers of the model to the physical control layer, and removing them as the information unit ascends through the layers of the model to the terminating end-user. The formation of such a message unit is shown in **Figure 12.21**.

(2) Request/Response Unit

The request/response unit (RU) is variable in length and contains the information that is to be sent from the originating end-user to the terminating end-user. It may be enduser information or a command to control the operation of the network. For most logical units, the size of RU is ≤ 256 octets. The actual number of octets is carried in the transmission header.

(3) Request/Response Header

The request/response header (RH) is a 3-octet header that describes the information in RU. If the first bit of RH is 0, the header is a request header; if the first bit of RH is a 1, the header is a response header:

- **Request header**: provides information on the data format and the protocol that governs the session. Tells PU whether a response is required, if data brackets are being used, where in the chain of data the BIU (BIU = RH + RU) is positioned, and if a pacing request applies to this transmission.
- **Response header**: returns a positive or negative response to request from other LU.

(4) Transmission Header

The transmission header (TH) is concerned with path control. Its format and length depend on the types of traffic and the types of nodes employed. The first octet in TH, called the format identification field (FID), is used to identify them:

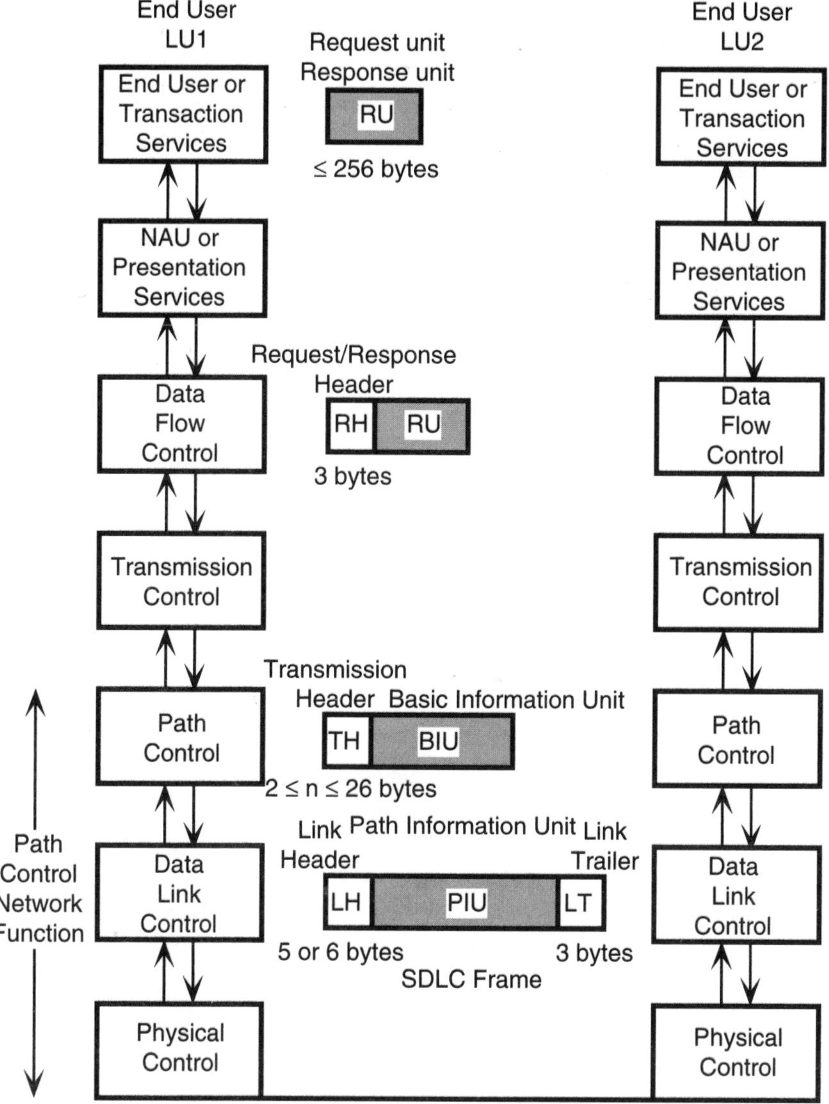

Figure 12.21 Message Unit Formation

- **FID Type 0** (TH = 10 octets): identifies non-SNA traffic between adjacent subarea nodes not supporting explicit and virtual route protocols
- **FID Type 1** (TH = 10 octets): identifies SNA traffic between adjacent subareas not supporting explicit and virtual route protocols
- **FID Type 2** (TH = 6 octets): identifies SNA traffic between a subarea node and an adjacent peripheral node

- **FID Type 3** (TH = 2 octets): identifies SNA traffic between an NCP subarea node and a type 1 peripheral node
- **FID Type 4** (TH = 26 octets): identifies SNA traffic between adjacent subarea nodes supporting explicit and virtual route protocols
- **FID Type F** (TH = 26 octets): identifies specific SNA commands between adjacent subarea nodes supporting explicit and virtual route protocols.

The use of some of these FID types is noted in Figure 12.15. The remainder of TH includes originating and terminating address fields; the sequence number of this PIU (PIU = TH + BIU) in relation to other PIUs for this request or response; the length of the RU; whether it is the only, first, middle, or last, in the transmission; explicit and virtual route information; and the addresses of the originating subarea and element and the terminating subarea and element.

(5) Link Header and Link Trailer

The link header and trailer (LH and LT) are standard SDLC frame items. The header contains a flag (1 octet), the station address (3 or 4 octets), and a control octet. The trailer contains a 2-octet frame check sequence and the terminating flag (1 octet).

12.2.4 Sessions

(1) Session Defined

Communication between network addressable units is called a session. Four specific sessions are required to implement the operation of SNA:

- Logical unit-to-logical unit (LU - LU) session
- System services control point-to-logical unit (SSCP - LU) session
- System services control point-to-system services control point (SSCP - SSCP) session
- System services control point-to-physical unit (SSCP - PU) session.

(2) LU - LU Session

End-users communicate with one another through LU - LU sessions. The LUs establish the communication procedures (roles, rules, and protocols). Data flows between them as SDLC frames called message units (MUs).

(3) SSCP - LU Session

In order to connect two end-users in the same domain, the request for communication from the originating LU (logon TSO, for instance) must be converted to the network address of the terminating LU (TSO). During the logon process, an existing SSCP - LU session is employed so that the SSCP can supply this information and direct the request to the terminating LU.

(4) SSCP - SSCP Session

In order to connect two end-users in different domains, the SSCP in the domain containing the originating LU must communicate with the SSCP serving the domain that contains the terminating LU. During the network generation process, this SSCP - SSCP session is established.

(5) SSCP - PU Session

These sessions are used for internal network purposes such as configuration and reconfiguration, gathering performance information, etc.

(6) Establishing a LU - LU Session

Some of the steps required to establish an LU - LU session across domains are illustrated in **Figure 12.22**:

- The originating end-user enters a logon message that requests connection to a specific service such as TSO (step 1).
- Over an already existing LU - SSCP session, LU1 sends the message to SSCP1 (in VTAM1) for logon name-to-network address translation (step 2).

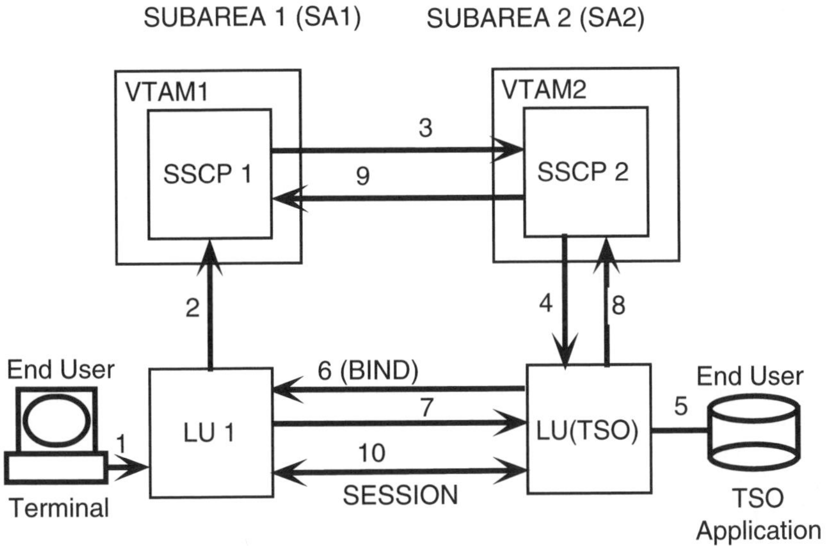

Figure 12.22 Steps Required to Establish a Session Across Domains

- Having determined that LU1 is permitted access to TSO, and that TSO is not within its domain, SSCP1 consults a listing of cross-domain resources and contacts the owning SSCP (SSCP2 in VTAM2) over an already existing SSCP - SSCP session (step 3). SSCP1 informs SSCP2 of the communication capabilities associated with LU1.
- Over an existing SSCP - LU session, SSCP2 delivers this information, together with the address of LU1, to LU(TSO) (step 4).
- LU(TSO) sends a session initiate request to the TSO application and receives a session initiate response (step 5).
- At the same time, SSCP2 causes the PU in NCP2 to establish explicit paths between LU(TSO) and the boundary unit in the subarea (SA2). Route selection is done by consulting tables or using other means.
- SSCP1 causes NCP1 to establish explicit paths between LU1 and the boundary unit in its subarea (SA1).
- When a virtual route has been established between LU(TSO) and LU1, LU(TSO) sends a command (called a bind) to describe the parameters proposed for the session. The bind customizes the session by specifying the roles, rules, and procedures that will be used (step 6).
- If LU1 is satisfied with the bind, it replies affirmatively (step 7).
- LU(TSO) sends a session established data unit to SSCP2 (step 8).
- SSCP2 sends a session established data unit to SSCP1 (step 9).
- As the primary LU, LU(TSO) issues a start data request to LU1, and the session proceeds over the virtual route established directly between the units (step 10).

Note that the initial communication steps are accomplished over pre-existing sessions. They are established when the system is generated, and form part of the communication structure required for operation. Unless an LU is connected to its owning SSCP in the original startup procedures, it cannot communicate. To connect an LU after system generation, the system operator must instruct the owning SSCP to activate a session between them using the intermediate PUs.

12.2.5 Gateways to Other Networks

(1) Packet Networks

The ready availability of packet-switched data networks, and the wide geographic distribution of terminals, creates a need to pass data through such networks to an SNA network. Special software products, NPSI (NCP Packet-Switched Interface), NIA (Network Interface Adapter), and PCNE (Protocol Converter for Non-SNA Equipment) have been developed for this purpose. SNA terminals employ NIA to convert the stream to X.25 standards. Terminals that have an X.25 port dispense with NIA. The strategy employed encapsulates the SNA information (transmission header, request/response header, and request/response unit) in a packet envelope (LAPB header, packet header, and LAPB trailer) for passage through the network. When it enters the SNA environment, NPSI strips the packet envelope and encapsulates the SNA information in an SDLC frame. Start-stop (asynchronous) terminals employ an X.3 PAD (packet assembler and disassembler) at the PSDN boundary and a PCNE (protocol converter for non-SNA equipment) at the SNA network boundary. Hardware and software upgrades to the 3745 FEP and NCP permit connection to, and passage, of SNA traffic over frame-relay networks. A frame-relay terminal adapter completes the path to a 3270 cluster controller.

(2) Other SNA Networks

LU - LU connections between SNA networks employ an SNA - SNA gateway. Called SNA Network Interconnect (SNI), the gateway function is provided by the combination of SSCP and NCP and can be implemented in one or the other network, or both. As described earlier, SSCP is required to establish cross-network sessions. Once the bind has been confirmed, data transfer occurs through the NCPs alone. They handle address interchanges as the PIUs cross the network boundaries.

(3) Local-Area Networks

Gateways are available that interconnect SNA resources and token-ring, Ethernet, TCP/IP, and MAP resources. Employing adapters, additional software and microcode, and VTAM(SSCP)/NCP, the gateways perform frame transformations, address translations, and whatever is required to represent the LAN as a fully implemented PU/LU resource to the target host. Terminal and host adapters are available to connect 3270 cluster controllers through a frame–relay network to an SNA host.

13

Enterprise Networks

In the United States, annual expenditures on private networks account for as much as one-third of the total spent on telecommunications facilities. The largest private network is maintained by the United States government. FTS (Federal Telephone System) serves the needs of federal government offices across the country, and provides citizens with toll-free calling to Social Security, Internal Revenue, and other departments. Individual corporations maintain smaller private networks.

13.1 FORCES THAT CREATE, GROW, AND SHAPE PRIVATE NETWORKS

13.1.1 Company Networks

(1) Do-It-Yourself Telecommunications?

Global competition has dramatically changed the way in which corporations operate. They have been forced to use their information resources wisely and to take advantage of as many new, technical developments as possible to increase productivity and boost profitability. Taking responsibility for telecommunications within a company may be justified by managers for the following reasons:

- *Telecommunications is a* **critical resource**: if telecommunications is recognized as a vital part of present and future operations, common sense dictates that management protect and strengthen this capability. Managers may say that communications is just too important a part of the business to be left to others—that management must control this resource just as they do engineering, manufacturing, marketing, and distribution.

- *Telecommunications is an* **expensive resource**: if telecommunications expenses are perceived to be high, management must attempt to contain and to reduce them.
- *Telecommunications is a* **competitive resource**: if telecommunications is seen as vital to maintaining and expanding market share, and to maintaining customer loyalty, management is likely to conclude that they must be the first to have the latest technology. Further, they will argue that creating a competitive edge through the use of telecommunications should be easier to achieve with their own facilities than by sharing the same public facilities as their competitors.

If better communications can make the difference between success and stalemate, many corporate managers feel the need for telecommunications facilities that they can control. Predictably, one of their first proposals will be to establish a private network.

(2) Demands on Existing Networks

For the vast majority of companies, the question of whether or not to invest in telecommunication facilities is no longer an issue. Already, they have PBXs, LANs, and corporate tie-lines that connect some of their business functions together. The question for the future is...*how much more to invest in our private network*? not...*should we have a private network*? Much more activity is devoted to improving, enhancing, and evolving existing networks, than creating new networks. The challenges faced by today's private networks are associated with various needs:

- **Corporate needs**:
  - serving more locations effectively (as corporations are acquired, merge, or expand)
  - accommodating equipment from different vendors that are of different technology vintages (particularly in the wake of acquisitions and mergers)
  - increasing traffic (as volume grows and business becomes more information intensive).
- **User needs**:
  - satisfying more diverse transmission and connection requirements (as product lines and services become more specialized)
  - improving responsiveness (faster turnarounds beget demands for even faster turnarounds)
  - improving network services (more sophisticated, more user-friendly, applications)
  - expanding applications services delivered by current capabilities (every improvement cannot be implemented with new hardware and software)

  - improving support services (as the network becomes more intelligent, users understand it less).
- **Operating staff needs**:
  - improving cost effectiveness (managers must show a continual increase in performance and a continuing decrease in cost/benefit ratios)
  - increasing network flexibility (keeping up with organizational changes requires the ability to reconfigure facilities quickly, and surely)
  - adding intelligence (diminishing the number of routine tasks performed by persons)
  - expanding network information and network control capabilities (knowing more about the utilization of current resources, and optimizing their use)
  - improving feature and service transparency (adding new features and services should not change the way existing functions are performed).

Fortunately, rapidly evolving, computer-based technologies provide ways to meet many of these requirements; however, they increase the sophistication and the complexity of the operating environments in which users, support staffs, and network managers must perform.

(3) Needs Assessment

The orderly evolution of private telecommunication facilities should be based on a continuing, comprehensive assessment that addresses the following factors:

- **Network requirements**: a reliable assessment must be based on a thorough review of the users' future requirements and the capabilities of existing facilities. With this knowledge, likely gaps in coverage can be discovered. However, defining future requirements is not easy, particularly if wide spectrums of users and uses are involved. When asked, each category of users may express future requirements in different terms, and individuals will vary greatly in their understanding of the scope of the inquiry and in their ability to describe and to estimate their future activities. Further, it is in the nature of work that individuals are likely to overestimate the importance of their activities to the corporation, and to overestimate the growth of their activities. When all of the information is gathered there remains a significant level of editing that must be done by the assessors.
- **Value of information**: all information is not equally valuable, and all communications are not equally important. An analysis of the relative values of messages with respect to system failures, or system compromises, can give insight into the relative importance of some capabilities vis-a-vis others. Will the firm really be unable to function if so-and-so does not have a working terminal for an hour or two? If response time is increased by x% does that really mean that everyone will be less productive by at least that amount, or does it mean only that they will have less waiting time between jobs? etc.

- **System architecture**: an architecture is required to provide the framework for future networking decisions. Without an architecture to guide its growth, even the best network is only a set of connections created to solve yesterday's problems. In large measure, the geographic distribution of users, and the communities of interest that form around certain applications, establish the architecture, and determine the set of alternatives to be considered in the evolution of a network. Properly conceived, the architecture will separate the network into independent parts that can be changed to meet the demands of new markets or to incorporate new technologies. It will permit change and growth in one network service segment without unduly affecting other segments.

(4) Owning the Switches and Leasing the Lines

For many corporations, having a private network means owning the switches (or other network node equipment) and leasing the lines from common-carriers. Some popular arrangements for specific classes of private network facilities include:

- Terrestrial transmission equipment—most corporate entities lease from carriers
- Satellite transmission equipment—many corporate entities own earth stations and lease satellite transponders from satellite carriers
- Network switching and other network node equipment—corporate entities own, or lease from carriers
- Customer premises equipment (CPE)—corporate entities own, or lease from manufacturers, third parties that provide maintenance, or carriers.

As carriers see opportunities to provide new public network options the amount of equipment owned directly by users is likely to decrease.

13.1.2 Network Applications

Telecommunication networks support a wide range of applications that are related to the sorts of connections they provide, and the types of information exchanges they facilitate. The sorts of connections networks provide can be described as:

- **Public—anyone-to-anyone**: pervasive, dial-up, interactive connections used for voice or low-speed data communication and provided by public switched telephone networks
- **Public—anyone-to-one**: pervasive, dial-up, interactive connections, provided by PSTN (e.g., 1-800… call, or low-speed data call to public database)
- **Interorganization—any vendor's equipment**: a network serving the special needs of several organizations that connects standard equipment made by any qualified vendor (e.g., FTS)

- **Interorganization—specific vendor's equipment**: a network serving the special needs of several organizations that connects equipment made by a specific vendor (e.g., a banking system clearing network)
- **Intra-organization—all departments**: a network that serves the communication needs of an entire company, or similar entity (e.g., corporate private network)
- **Intra-organization—one department**: a network that serves the communication needs of a single department (e.g., a local-area network).

This range of connections covers most person-centered and machine-centered needs.

The types of information exchanges supported by these networks can be described as:

- **Batch**: transmission of information to generate, or add to, files (e.g., update records of hours-worked by employees)
- **Simple transaction(s)**: question and answer exchanges probably using structured formats (e.g., request for person's telephone extension, address, or driver's license number); or the transmission of information that is processed and returned (e.g., conversion of hours worked and hourly wage rate into paycheck)
- **Related transactions**: a series of simple transactions that are related to one another (e.g., inquiry that relates payroll status, vacation entitlements, medical benefits, and retirement benefits, for a particular employee), including time-sharing sessions on remote hosts
- **Interactive sessions**: implementation of any style of transactions, at will (e.g., telephone conversation, computer conference, video seminar, etc.).

Examples of applications supported by networks exhibiting these styles of connections, and types of information exchanges, are shown in **Table 13.1**. Person-centered and machine-centered applications are distributed over the connection versus information exchange plane. It is divided into areas that are served mostly by public networks and mostly by private networks. By no means are the categories exclusive, nor are the listings exhaustive. The three left-hand columns contain applications that employ data, and the right-hand column contains applications that employ data or voice or video, or combinations thereof.

13.1.3 Communication Requirements

(1) Traffic Categories

While we do not pretend that Table 13.1 is a complete description of the activities supported by a private network, we can use it to define some of the categories of traffic that must be accommodated to support an enterprise. Defining the network

Table 13.1 Applications Arranged According to Styles of Connections and Types of Information Exchanges

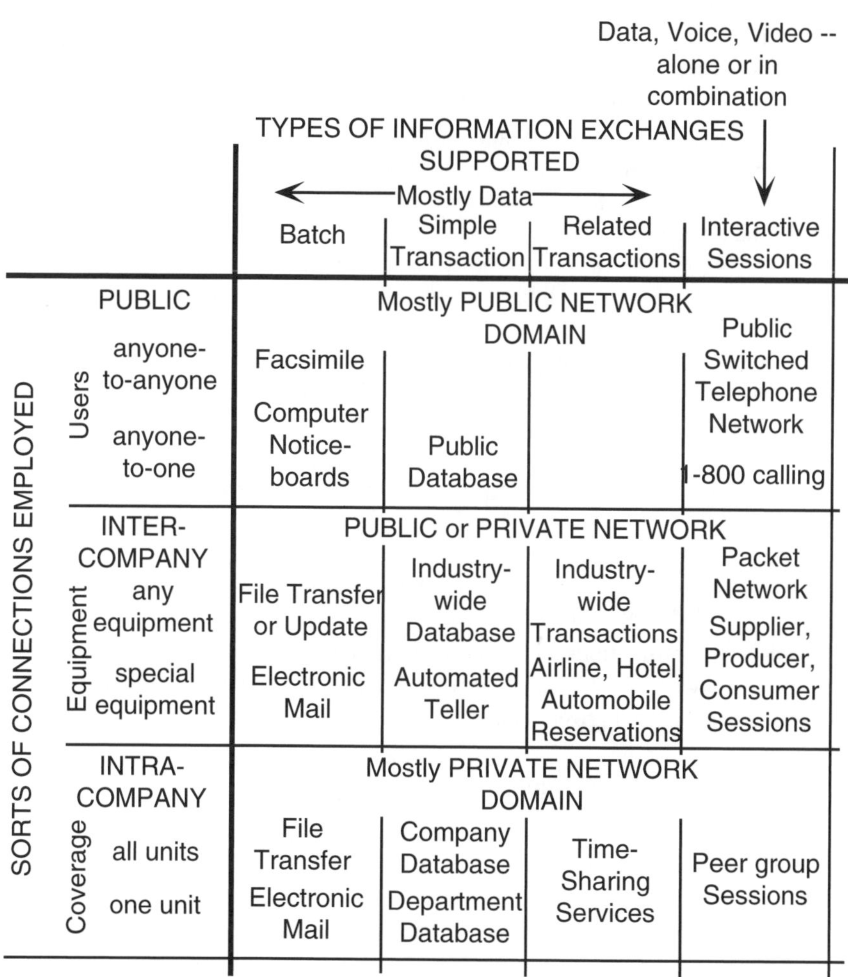

SORTS OF CONNECTIONS EMPLOYED	TYPES OF INFORMATION EXCHANGES SUPPORTED (Mostly Data): Batch	Simple Transaction	Related Transactions	Interactive Sessions (Data, Voice, Video -- alone or in combination)
PUBLIC — Users: anyone-to-anyone, anyone-to-one	Mostly PUBLIC NETWORK DOMAIN			
	Facsimile Computer Notice-boards	Public Database		Public Switched Telephone Network 1-800 calling
INTER-COMPANY — Equipment: any equipment, special equipment	PUBLIC or PRIVATE NETWORK			
	File Transfer or Update Electronic Mail	Industry-wide Database Automated Teller	Industry-wide Transactions Airline, Hotel, Automobile Reservations	Packet Network Supplier, Producer, Consumer Sessions
INTRA-COMPANY — Coverage: all units, one unit	Mostly PRIVATE NETWORK DOMAIN			
	File Transfer Electronic Mail	Company Database Department Database	Time-Sharing Services	Peer group Sessions

users as Hosts, LANs, and terminals (DTEs, telephones, or video/television terminals), we can divide the network traffic into two categories:

- **Asynchronous**:
  - *high-speed data.* 1) Point-to-point, possibly half-duplex, between hosts. The number of connections is small and terminations are likely to be within the enterprise. 2) Point-to-point, possibly half-duplex, between LANs. The number of connections is fairly small and terminations are quite likely to be within the enterprise.

  - *low-speed data*. Point-to-point, mostly duplex, between terminals, between terminal and LAN, or between terminal and host. The number of potential connections can be large, although only a small fraction may be in use at any time, and most terminations will be within the enterprise.
- **Synchronous**:
  - *video*. 1) Point-to-point, duplex, between principal locations. The number of connections is small, and terminations are likely to be within the enterprise. 2) Point-to-point, simplex, between major locations. The number of connections is fairly small and terminations are quite likely to be within the enterprise.
  - *voice*. 1) Point-to-point, duplex, between all locations and interconnecting with PSTN. The number of connections is large and terminations can be inside or outside the enterprise. 2) Many-to-one, duplex, between members of the public and an answering point. The number of connections can be large, and originations are likely to be from terminations in the PSTN to a central facility within the enterprise.

From a different perspective we can divide the network connections categorically:

- **Pervasive connections**: they carry voice and low-speed data and are likely to be provided over switched networks.
- **Discrete connections**: they carry high-speed data or video and are likely to be provided over dedicated or routed connections (i.e., connected through a router or a digital cross-connect).

(2) Off-Net and On-Net Calling

The calling requirements of an organization are not limited to calls that can be completed within their private network. For instance:

- Vendors and customers must be called—they are contacted through public networks
- Incoming calls originate from many facilities that are served by public networks
- At times of high usage, a private network may become congested—a restorative measure is to employ the public network for some of the traffic.

Thus, a private network must have access to public networks. Those calls that originate and terminate within the firm and are completed wholly over the private network are known as *on-net* calls; those calls that originate or terminate at points off the private network are called *off-net* calls.

(3) Topology, Hierarchy, and Architecture

Networks are characterized by topology, hierarchy, and architecture:

- **Topology** is concerned with the ways in which nodes and links are interconnected. Topological complexity can be expected to increase with the number of elements a network contains.
- **Hierarchy** is concerned with the ways in which nodes (switching centers, multiplexers, etc.) are arranged to achieve efficient division and routing of different classes of communications traffic.
- **Architecture** is concerned with the ways in which messages interact with topology and hierarchy so that communication can take place.

To understand the functioning of a network requires an appreciation of the unique properties present in all three features. Many factors must be weighed in establishing the topology, determining the hierarchy, and selecting an architecture. Usually, cost is given greatest importance; but, if the goal is a growing, thriving, competitive business, the immediate cost cannot be the only thing to be considered—future options and opportunities must be safeguarded. The possible consequences of each alternative must be evaluated to determine both short-term savings, and potential, long-term, performance liabilities.

13.2 SWITCHED NETWORKS

Recognizing that many corporations have been in the business of providing their own communication facilities for some time, we will describe a pervasive network of the sort that is maintained to transport voice and low-speed data in support of the business. It is based on specialized switches known as PBXs (private branch exchanges). Before describing the network, we discuss the characteristics of PBXs and other custom facilities.

13.2.1 PBXs and Related Equipment

(1) PBX Functions

A private branch exchange (PBX) is a circuit switch that serves a community of stations. Usually, the community is located in a limited area that includes the PBX. PBXs perform the following functions:

- By interconnecting stations (extensions), they provide communication paths among the community of users

- By connecting users to nodes in the public network, they assist in providing communications paths to remote stations, and *vice versa*; these calls are *off-net* calls
- By connecting users to other nodes in the private networks to which they belong, they assist in providing communications paths to remote stations, and *vice versa*; these calls are *on-net* calls
- Set up, maintain, terminate, and record details of on-net calls, and assist in these activities for off-net calls
- Provide special services according to the capabilities with which they are equipped. Many PBXs, particularly the larger ones, support networking—in conjunction with the public network, or as nodes in a private network.

Worldwide, PBXs are manufactured by many companies. In the United States, the leaders are AT&T (with approximately one-quarter of the domestic market) and Northern Telecom (with approximately one-fifth of the domestic market). PBXs range in size from those that serve 10s of lines, to those that serve 1000s of lines.

(2) PBX Organization

Figure 13.1 shows the major components of a generic, digital PBX. Voice, data, and trunk interface cards are attached to a non-blocking, digital switch. A digital processor responds to activity indications from the line scanning unit, allocates switching resources and/or service circuits as needed, maintains the connection for the duration of each call, and returns the resources to idle state to await other calls. The PBX is likely to offer voice messaging services, to contain an electronic mail center, and to have an automated attendant capability that permits the caller to select a department or service at the time of call reception, or to be referred to a human attendant for directory assistance, call routing instructions, and other information. At the same time, it records operational and management data for use later in billing or expense allocation activities, and in system testing. Also, it supports administration and maintenance terminals that are used to make changes to the system database, update station information, request statistics and status reports, initiate diagnostic routines, analyze test results, etc.

(3) PBX Features

Vendors seeking to differentiate their products from competing systems employ the full power of computer control to provide customized feature sets. Some common features include:

- Extension-to-extension calling, direct outward dialing, direct inward dialing
- Line selection on handset pick up, call transfer, call forwarding, camp on, call waiting, call restriction

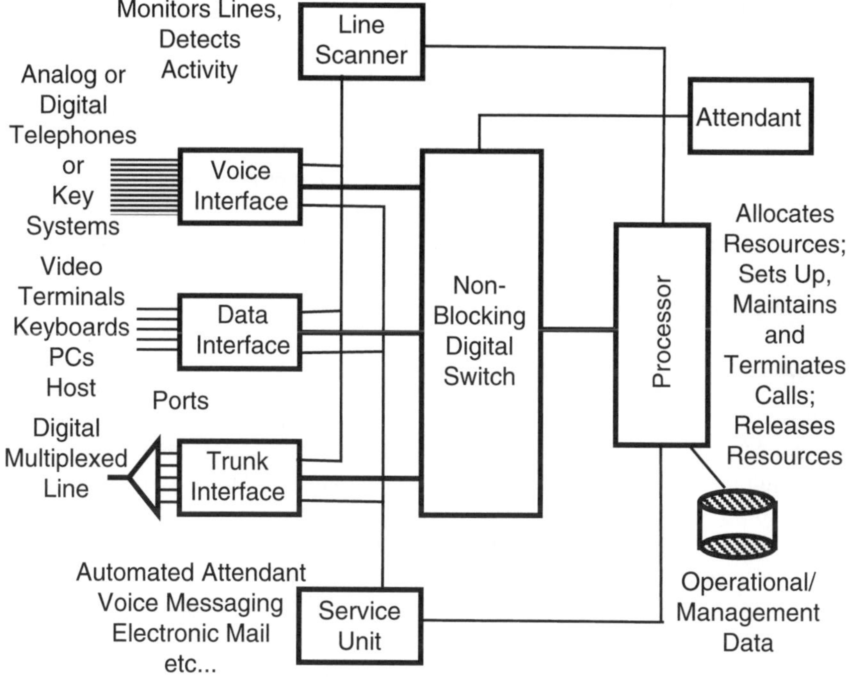

Figure 13.1 Major Components of a Digital PBX

- Last number dialed, saved number dialed, abbreviated dialing
- Conferencing, paging, answering, and message services
- Automatic route selection
- Automated moves and changes, message detail reporting
- Automatic call distribution
- Modem pooling, (i.e., modems are treated as common equipment and assigned to calls as needed), etc.

These features are selected from a generic capability that is likely to include several hundred voice and data options.

(4) Station Equipment

Advanced electronic telephones offer easy access to proprietary features. They include programmable keys, multiple line appearances, calling party identifiers, time of day, call duration, etc. Usually, because they are feature-rich, and most features are not standardized, it is not possible to interchange station equipment between PBX systems produced by different manufacturers.

(5) Distributed PBX Arrangements

In common with many electronic devices, PBXs need not be implemented as centralized systems. For instance, the geographic distribution of users may make it cost effective to place the line interfaces in the user communities and employ multiplexed facilities to connect to a central switching element. In this case, all switching activities occur in the central switch matrix. If the environment contains a few, large concentrations of users, another alternative places line interfaces and limited local switching capabilities with the communities, and employs the central switch to complete connections among the communities, and to the world. These options are shown in **Figure 13.2**.

(6) Connecting the Public to an Attendant

In consumer-oriented businesses, communications applications such as 800- and 900-number services, product service centers, and reservation centers, are supported by equipment that receives calls and directs them to one of a group of attendants waiting to provide service. Called *automatic call distribution* (ACD) systems, they may be stand-alone, integrated with a PBX, or they may be central office-based. A basic configuration for a moderate size ACD system that is integrated with a PBX is shown in **Figure 13.3**. Suppose customer calls are routed through the public network to an ACD system equipped with 14 access lines. When the 14 lines are occupied, further calls are blocked and the attempting traffic receives a busy signal; the caller may retry later, or abandon the endeavor. On the 14 access lines, the traffic is distributed to available attendants and to a holding device. With a pool of 12 attendants, it is possible that an average of 10 calls are being serviced and 4 calls are waiting for service.

(7) Serving Blocked Callers

If the ACD is one of several units that are employed for a common purpose, the blocked calls at one ACD may be redirected to other ACDs where service is available. Such a situation is likely to exist if units are located across the country in different time zones. Through the use of alternate routing instructions at the SCPs in the public networks, several attempts can be made to find an open line. All possibilities will be exhausted before returning a busy signal to the caller. Successful use of such time-of-day routing moves the concern of the system operator from the number of blocked calls to the number of callers on hold. If the wait is too long, these potential customers will hang up, and may dial a competitor.

(8) Centrex

To have the use of the sophisticated services that are provided by a modern PBX, it is unnecessary for an organization to install and maintain PBX equipment on its premises. Called *centrex* (from central exchange) services, the local exchange carrier can provide them from equipment located at a central office. Each centrex sys-

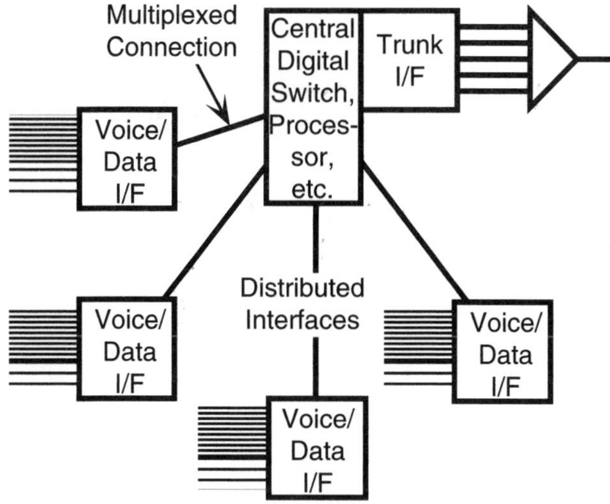

PARTIALLY DISTRIBUTED SYSTEM

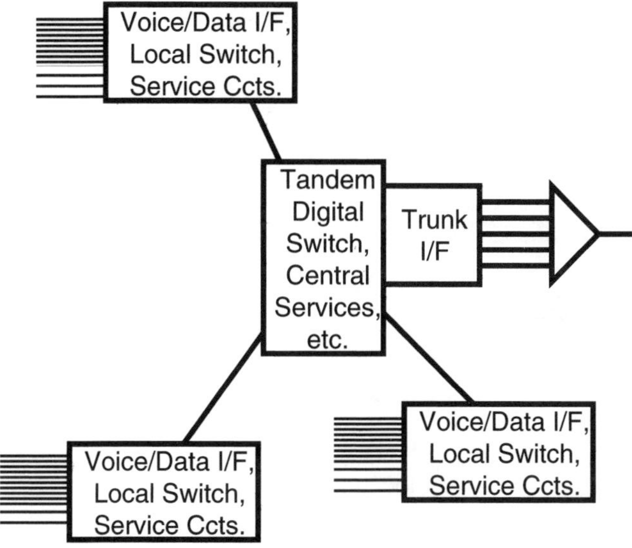

DISTRIBUTED SYSTEM

Figure 13.2 Optional PBX Architecture

tem is assigned a set of standard, seven digit numbers that are used to complete incoming and outgoing calls in the normal way. In essence, the customer's extensions are multiplexed to the serving office; there, switching and other functions that

PHYSICAL ARRANGEMENT

Automatic Call Distributor

Headsets Terminals ATTENDANT POSITIONS

14 Incoming Calls

A C D

Hold

10 Calls being Serviced

Terminals are connected to a DATA network

PBX

4 Calls Waiting

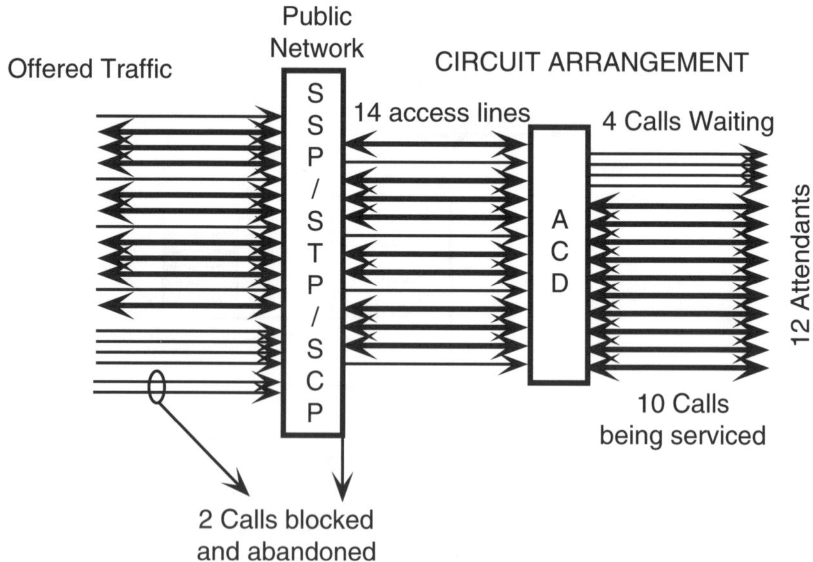

Figure 13.3 Automatic Call Distribution System

mimic the actions of a PBX are performed in the LEC's equipment. Centrex services include voice mail, electronic mail, message center services, modem pooling, protocol conversion, automatic route selection and packet switching, as well as connecting to standard trunking facilities such as WATS, FX, tie lines and IXC terminations. Centrex facilities can function as network end-points (in the manner of the PBXs in Figure 13.5), or as tandem switching points (in the manner of the main switches in Figure 13.5). Centrex users receive station message detail records and other cost-management information, and can have access to the appropriate facilities to effect station rearrangement, and to see to other matters. Even though the physical facilities are owned and operated by the LEC, centrex services provide firms with the same feelings of being in control of their local communications as are engendered by owning and operating their own PBXs. The provision of centrex services is illustrated at the bottom of **Figure 13.4**.

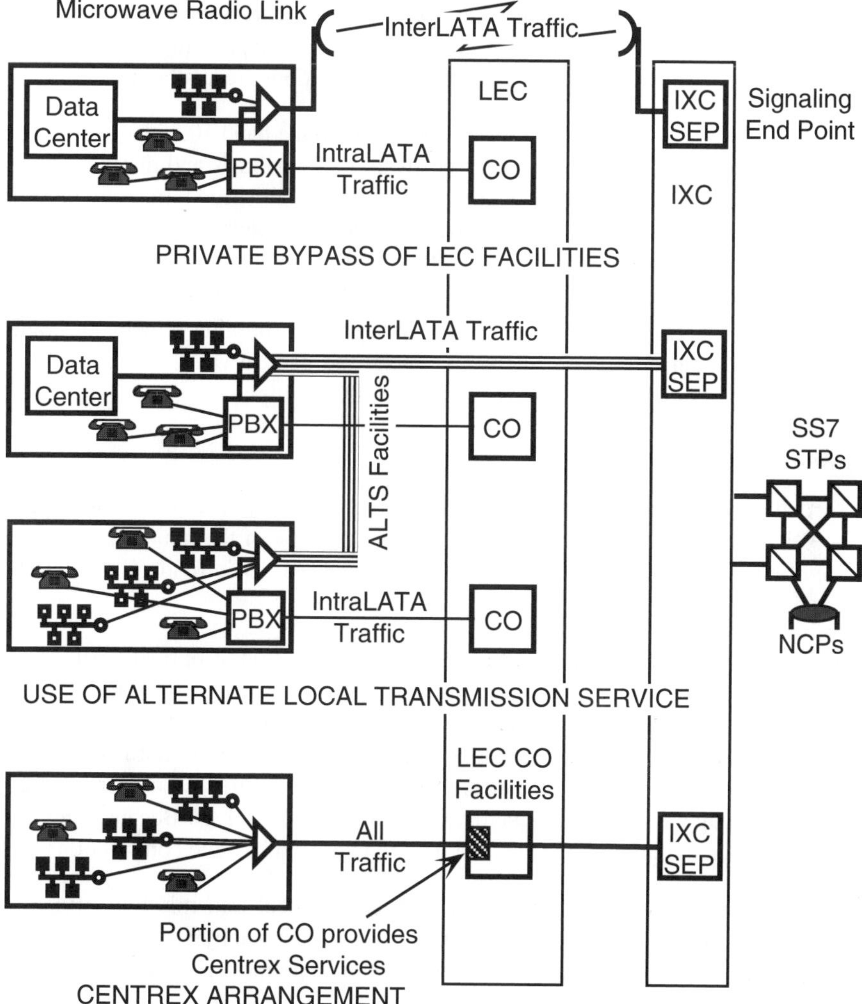

Figure 13.4 Custom Connections

Enterprises can employ bypass facilities to leapfrog carriers and use carrier facilities to substitute for customer provided facilities.

13.2.2 Custom Connections

(1) LEC Bypass

Businesses with large volumes of interLATA calls may wish to deliver them directly to IXCs over private connections that bypass the facilities of the local exchange carriers. In this manner they circumvent access fees, and may be able to take advantage of superior transmission performance or technical capabilities. Microwave radio systems are convenient bypass facilities. One end of the link can be located on the premises of the business seeking to bypass the local exchange carrier, and the other end of the link can be located at an IXC facility. A representation of the provision of microwave bypass is given at the top of Figure 13.4. Some other options are available:

- **Digital termination service** (DTS): operating in the band 10.55 to 10.68 GHz, DTSs provide digital channels at speeds up to 1.5 Mbits/s and distances up to six miles. At the DTS node, data is transferred to high-speed terrestrial links, or to other radio links for transport over DEMSNETs (digital electronic message service networks) to other cities, or other carriers. By dividing the area into sectors and reusing the assigned frequencies, several DTSs can operate in a city.
- **CATV service**: cable television system operators provide wideband, point-to-point connections for videoconferences and data services. Franchise limitations permitting, they connect directly to IXC equipment.

(2) Intra-City Bypass

A second problem facing corporations may have to do with providing high-speed connections between locations in the same city. In many of the major cities of the United States, *alternative local transmission service* (ALTS) providers have established networks:

- **Alternative local transmission companies**: employing optical fibers, they provide point-to-point digital circuits to organizations that wish to interconnect their locations within the metropolitan area, and/or to pass traffic directly to IXCs. Because they are independent entities, ALTS are not restricted by LATA boundaries—for instance, they carry traffic across entire urban areas, even though the areas may be divided into several LATAs. Because they are (specialized) common-carriers, ALTS are regulated by state commissions.

A representation of the provision of alternate local transmission service to interconnect separate locations of a company, and to provide bypass of the LEC, is shown in the middle of Figure 13.4. Another option is point-to-point digital radio:

- **Point-to-point digital radio**: operating in the 18 and 23 GHz portions of the spectrum, digital radio links can provide wideband, high-speed connections over several miles. Provided line-of-sight is available, they may be the simplest solution to linking different locations in the same city.

(3) Inter-City Bypass

Finally, corporations can implement communications independent of terrestrial carriers through the use of satellite services:

- **Satellite services**: satellite carriers provide point-to-point, point-to-multipoint, and broadcast, data, voice, and video services on dedicated or on-demand connections. Through the use of a geostationary satellite, points across town, or points separated by several thousand miles can communicate at the same cost. Satellite systems are described in § 14.1.2.

13.2.3 Model of Switched Network

(1) Wide-Area Enterprise Network

Figure 13.5 shows a wide-area enterprise network that supports business applications requiring pervasive voice and low-speed data connections. Network elements

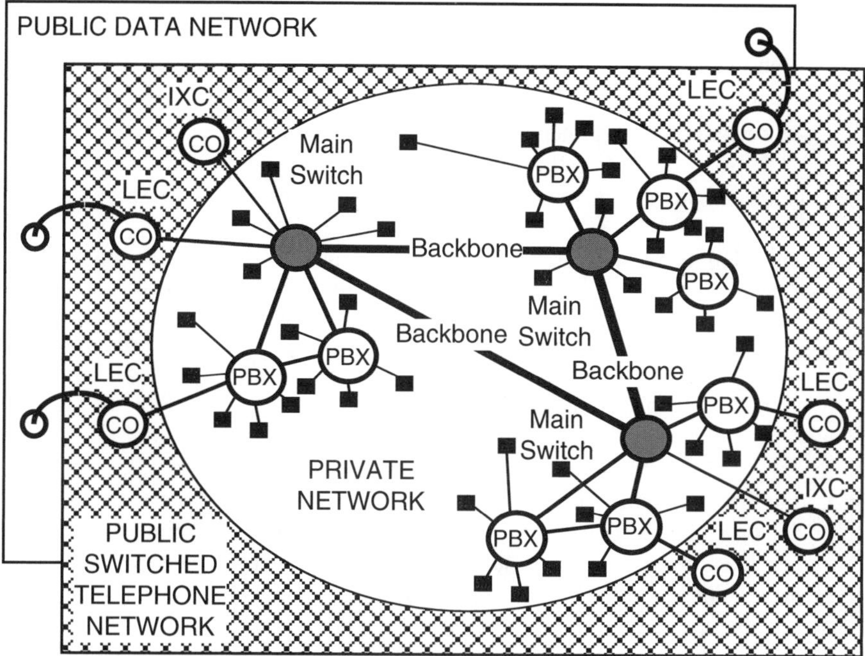

Figure 13.5 Wide-Area Enterprise Network

The private network transfers voice and data among the community it supports, and connects them to the public switched telephone network and one, or more, public data networks.

are clustered together in geographic areas served by a local network, and the clusters are interconnected by a long-distance network to facilitate calling from one community to another. The network can be described in the following terms:

- **Topology**: the resources provided to each area consist of star-connected links that terminate on local nodes (PBXs). The nodes are interconnected in star, tree or mesh fashion to form the local network. In turn, one node in each area (main switch) is the focal point for long-distance traffic, and is mesh-connected to nodes that serve the long-distance needs of the other areas. These nodes form the long-distance network (the backbone). At various points, access is provided to the public-switched telephone network, and, through PSTN end offices, to a public data network (PDN).
- **Hierarchy**: within the private network there is a two-level switching hierarchy:
  - PBXs handle local connections, and connections to other local switches
  - PBXs (main switches) handle local connections, connections to other local switches, and make long-distance connections.

  In addition, we have a two-level network hierarchy that is composed of:
  - local networks
  - long-distance backbone network.
- **Architecture**: calling from one telephone to another requires that plans have been made, and hardware and software are in place, to handle the following situations:
  - the reception and interpretation of the called number
  - the routing of the call to its destination
  - the signaling and supervision associated with the communication session.

This plan, or architecture, must include provision of facilities to condition the signal (analog to digital, for instance) and to provide alternate routes if needed.

(2) Transmission Paths

Network nodes are interconnected by transmission links that are likely to be provided by common carriers. They are divided into three categories:

- **Tie lines**: non-switched lines directly connecting two private network facilities (e.g., PBXs, data centers, etc.) for the exclusive, full-time use of those that lease them. Cost depends on mileage and performance required. They may be analog voice circuits, or digital voice or data circuits with several circuit speeds. Sometimes, tie lines are called tie trunks.
- **Foreign exchange lines**: tie lines that connect a private network facility (e.g., a PBX) to a public exchange in a distant city so that callers may have the benefit of local service in the distant calling area. Cost consists of the cost of the tie line between the cities, plus the cost of local service in the distant city, plus the cost of terminating the tie line in the distant city.

- **WATS lines**: connections that support wide-area telephone service—a service that offers bulk rates for station-to-station calls that are dialed directly. Calling categories are intrastate and interstate, and incoming or outgoing:
  - *outgoing WATS*. On a dedicated local access line, the subscriber may make calls to specific areas of the United States. The calling categories are intra-state, and inter-state. Interstate calling areas are divided into bands that span the continental United States, and serve Alaska, Hawaii, St. Thomas, St. John, and Canada. Costs depend on the originating area code, the band selected, and the number of calls made per month. Interstate WATS tariffs are offered exclusively by inter-exchange carriers. Rate structures vary somewhat by company.
  - *incoming WATS*. On a dedicated local access line, the subscriber may receive calls from specific areas of the United States and certain other countries. The calling categories are intrastate, interstate and international. Usually called *800 service*, the costs associated with all calls are borne by the subscriber.

Table 13.2 compares these classes of connections and includes direct-distance-dialing (DDD), the transmission path of last resort. A DDD circuit is seized by the calling party when placing a call to a distant station over the PSTN.

(3) Model of Private Network that Includes Public Network

The model presented in Figure 13.5 is a distortion of reality. In practice, the private network, the PSTN and the PDN extend over a common geographic area, and public facilities are intermingled with the facilities that serve the entity that created the private network. A way of drawing these public and private network relationships is shown in **Figure 13.6**. (For simplicity we have ignored the PDN.) To the right-hand side are the signaling and message networks of the interexchange carriers, and the message networks of the local exchange carriers. To the left-hand side are the PBXs and main switches that form the nodes of the private network. Each PBX is connected to one of the three main switches by tie lines (leased from the public network operators). Each main switch is connected to the other two by tie trunks (leased from public network operators) that form the network backbone. Access to the public-switched network is obtained through five connections to the local exchange carriers' networks. To make these arrangements clearer, **Figure 13.7** presents four partial views of Figure 13.6; they show the backbone network, PBX and Main Switch connections, connections to the PSTN, and station connections.

(4) Application of Signaling System #7

In the same way that it improves the operation of public networks by reducing the amount of call-carrying equipment that is occupied with call setup, and by providing the opportunity to access special services through information stored in a database (SCP), SS7 can be applied to private networks to improve equipment utiliza-

Table 13.2 Some Special Connections

TIE LINE	Line directly connecting two private telephone facilities (e.g. PBXs) for the exclusive, full-time use of those that lease them.	*May be between any two points selected by user. Cost depends on mileage.*
FOREIGN EXCHANGE (FX) LINE	Tie line that connects a private telephone facility (e.g. a PBX) to a public exchange in a distant city so that callers may have the benefit of local service in the distant calling area.	*May be between user's facility and any local exchange in US. Cost consists of the cost of the tie line between the cities, plus the cost of local service in the distant city, plus the cost of terminating the tie line in the distant city.*
WIDE AREA TELEPHONE SERVICE (WATS) LINE	Connections that support wide-area telephone service -- a service that offers bulk rates for station-to-station calls that are dialed directly. On a dedicated local access line, the subscriber may make calls to, or receive calls from, specific areas of the United States.	*Calling categories are intrastate, and interstate. The service may be outgoing, or incoming (in this case it is known as 800 service). Interstate calling areas are divided into bands. Cost depends on the originating area code, the band employed, and call volume.*
DIRECT-DISTANCE-DIALING (DDD) LINE	Line available on-demand that connects user to any other user.	*Calling categories are local, long-distance and international. Cost depends on distance, time of day, and holding time.*

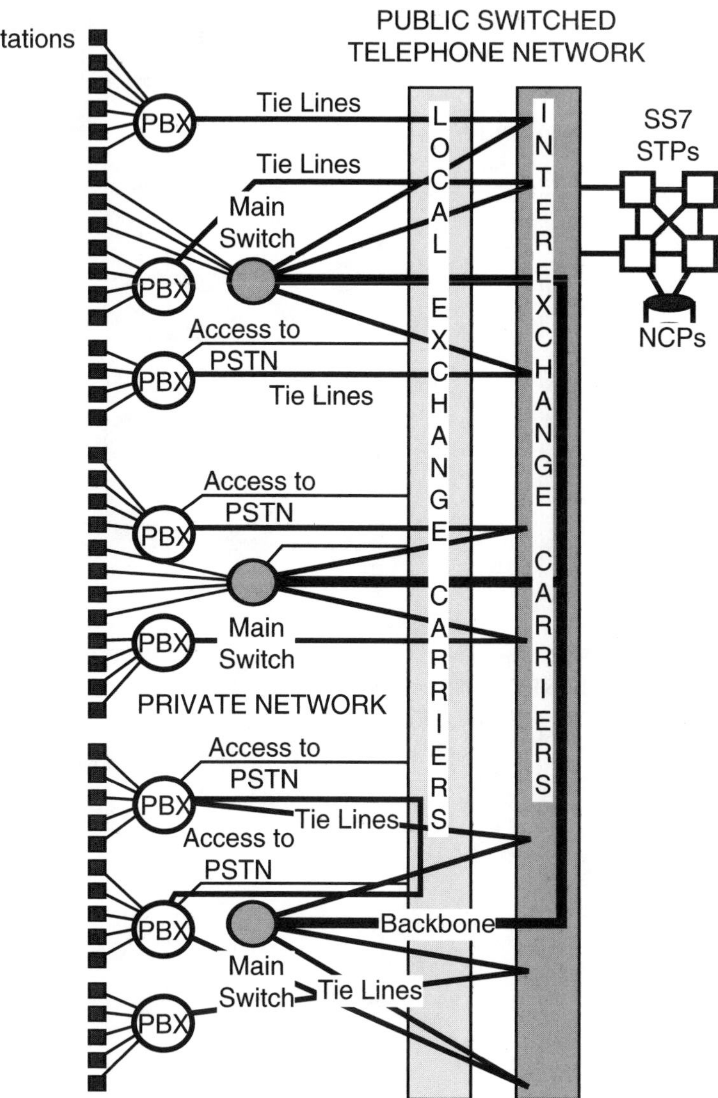

Figure 13.6 Enterprise Network Facilities Related To Public–Switched Telephone Network Facilities

For an enterprise that owns the switches and leases the lines, the network of Figure 13.5 is likely to be configured thus.

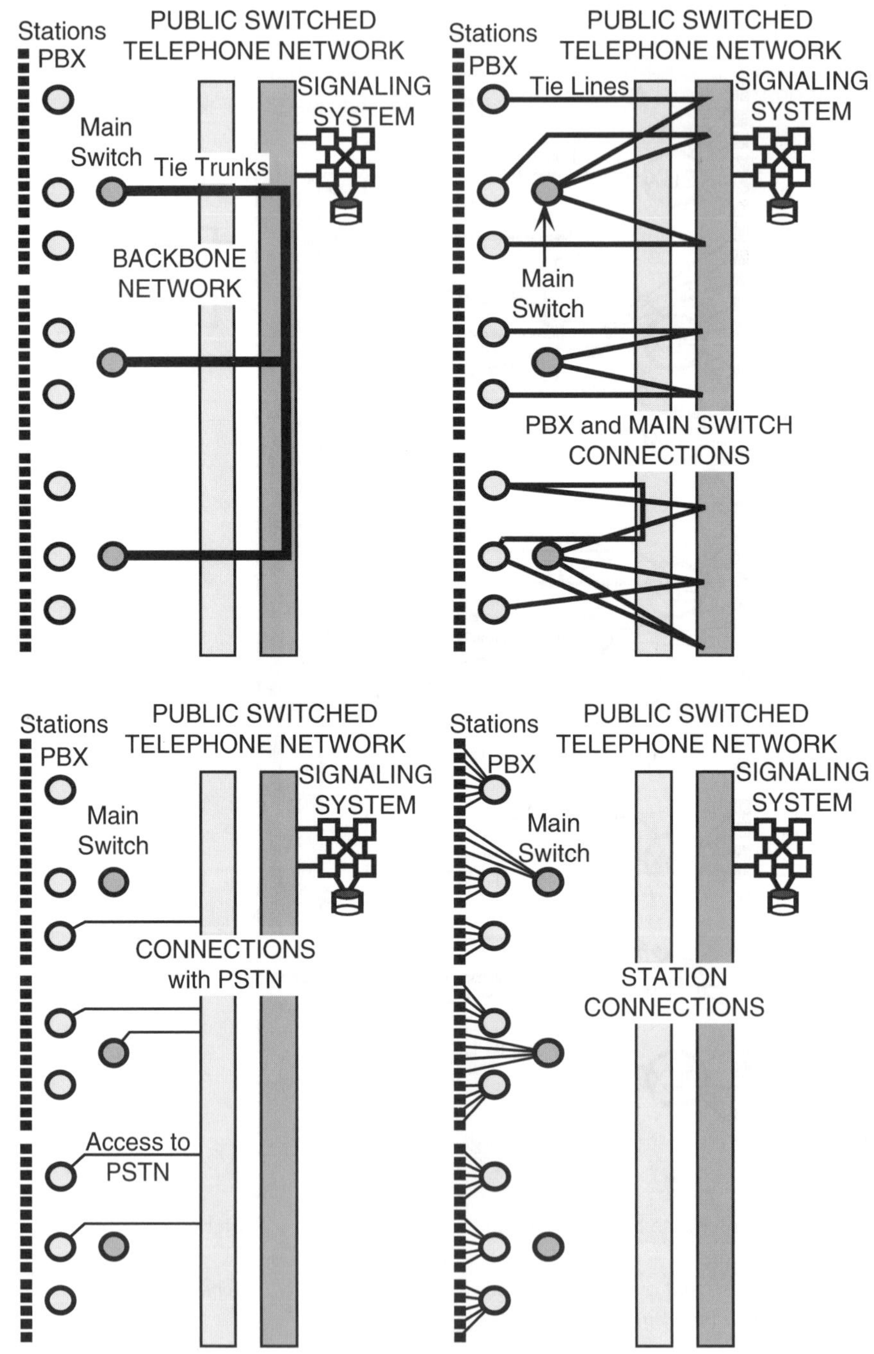

Figure 13.7 Four Components of Figure 13.6

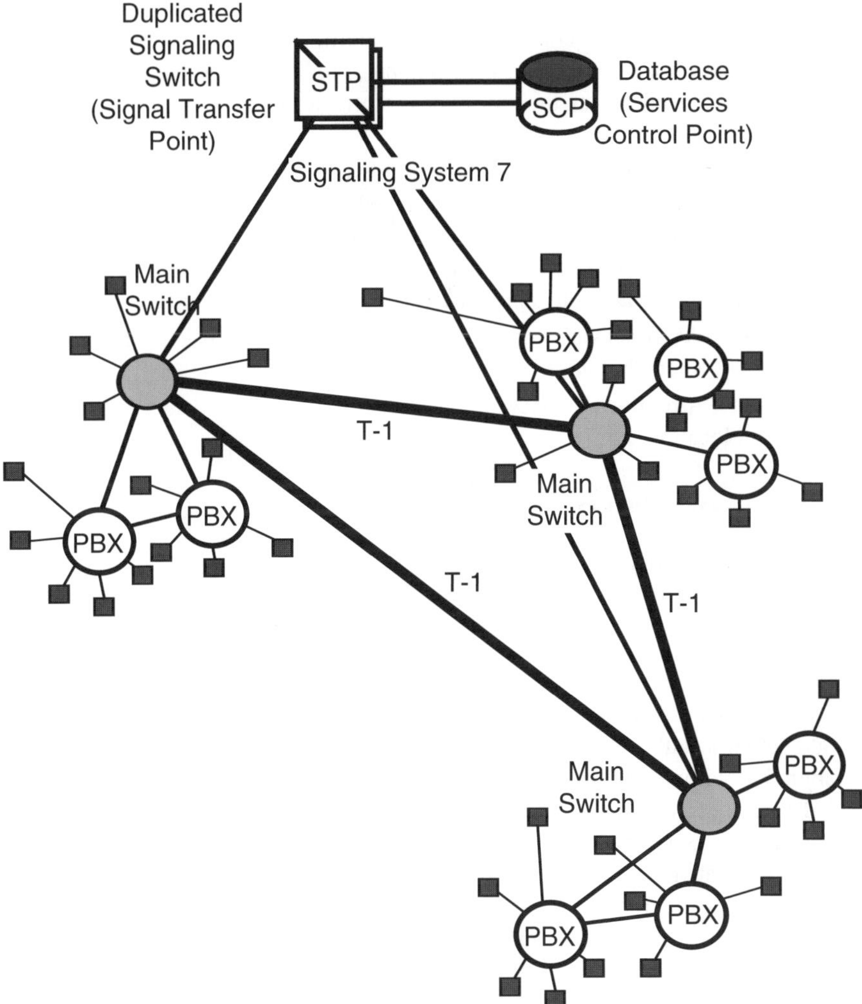

Figure 13.8 Use of SS7 in an Enterprise Network
The power of SS7 can be incorporated in an enterprise network to provide many of the same services as exist in the public network.

tion and provide enhanced services. **Figure 13.8** shows the addition of a signal transfer point and a services control point—elements that implement SS7—to the network drawn in Figure 13.5. They enable the use of special features across the network, and support database-related services such as:

- **Corporate credit card service**: completes off-net calls over the corporate network for corporate credit card holders. Upon receiving a request for connection to an 800 number, the LEC routes the caller to the corporate point-of-presence (POP). The corporate node supporting the POP prompts the caller for additional information (called number, credit card number, authorization code, etc.), and routes the call over the private network in accordance with established procedures.

- **Corporate 800 service**: completes traditional 800 Service calls over the corporate network. Upon receiving an assigned 800 number, the LEC routes the caller to the corporate point of presence. Depending on time of day, and any other criteria established by the corporation, the node supporting the POP translates the 800 number into a private network number, and completes the call.
- **Network ACD service**: routes calls intended for overloaded service centers to other centers that have idle agents.
- **Network attendant service**: routes calls requiring attendant assistance to a central location.

13.2.4 Virtual Networks

(1) Virtual Private Network

The network shown in Figure 13.8 can duplicate the enhanced signaling capabilities present in intelligent networks (see § 10.6.1). In fact, an option for implementing this arrangement is to lease STP/SCP capability from a carrier. Another alternative is to create a complete network out of carrier facilities—one that is operated as a private network with the special capabilities and performance required by the corporation, but without the burden of investment, maintenance, and disaster recovery. Called *virtual* networks, they are configured to provide users with reports on calls, equipment usage, configuration status, and other information that is required to manage corporate telecommunications activities. **Figure 13.9** redraws the dedicated private network of Figure 13.5 as a virtual network that is distributed throughout the carrier's network, yet under the control of the user.

The principal components of a virtual network include:

- **Access arrangements**: how does the customer connect to the carrier's network?
- **Switching**: what services will the customer receive from the serving switch?
- **Record creation and keeping**: how will call processing records be created and used, and what customer specific data will be kept?
- **Customer premises equipment**: what equipment will the customer expect the network to support?

(2) Network Access

Entry to a virtual network (VN) can be made over direct, switched, or remote connections. Egress is made over direct or switched connections. Direct access arrangements can include:

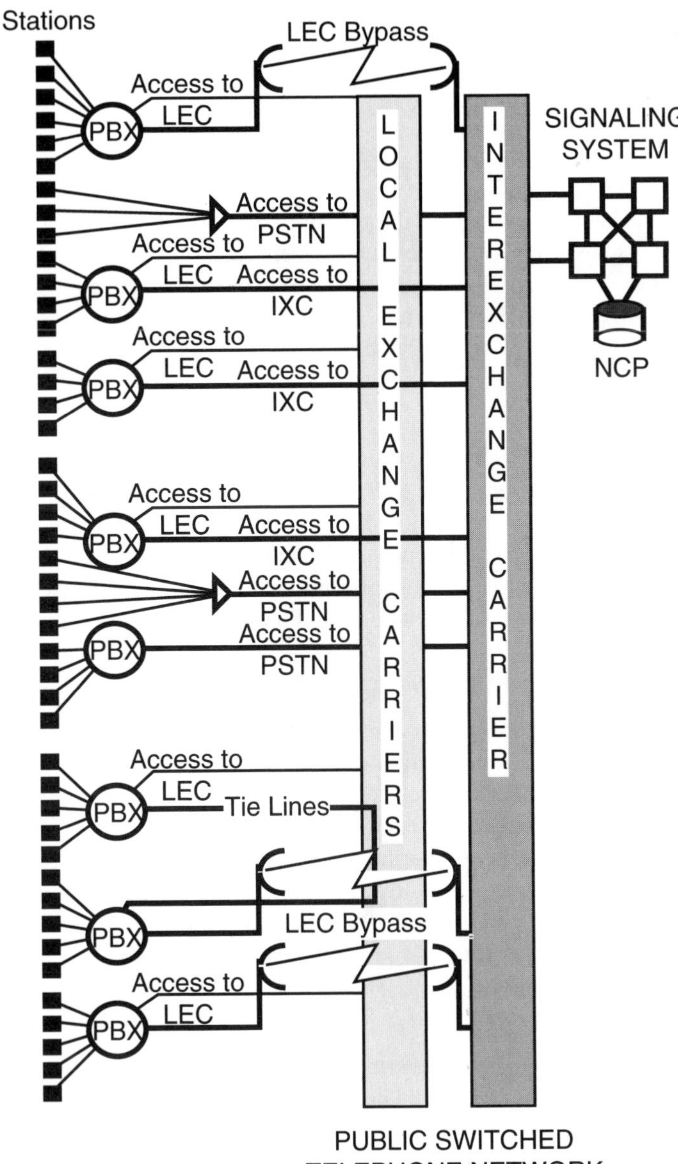

Figure 13.9 Enterprise Network as a Virtual Network
The network of Figure 13.5 converted to a virtual network provided by the IXC. Connections to IXC are made by bypass, leased lines (through the LEC SCO) and switched connections (through the SCO). The enterprise is supported by SS7 through IXC facilities.

- Voice grade private lines
- DS1 lines
- T-1 links equipped with bit-compression multiplexers (44 or 88 channels)
- 56 kbits/s data channels
- Other speed data channels.

Switched access is obtained through the LEC serving Class 5 office. It recognizes VN calls based on automatic number identification (ANI) capability and forwards them to the IXC, VN-equipped office. Remote access is obtained through the use of a VN access number from any remote terminal. The number is used by the VN switch to identify it as the precursor of a VN call.

(3) VN Switch

The VN-equipped, IXC switch receives VN traffic over direct, switched, or remote connections. For each event, over SS7, the switch communicates with the network element storing the customer specific data and call processing record. This unit verifies the call, may request additional information, and instructs the VN switch on how to process it. When it receives instructions to do so, the VN switch sets up a circuit to the terminating LATA where the call is routed to its destination.

(4) Call Processing Record/Customer Specific Data

The call processing record (CPR) contains the tests to be applied in processing a call from the specific access location (based on ANI). If an authorization code is required, the network element sends a message to prompt for and collect the extra digits. They are verified against customer specific data (CSD). Authorization codes are used to identify callers and define their privileges. If an authorization code is not required, privileges are defined by the calling number. After examination of the situation by the call processing record, instructions are sent to the switch to complete the call, or take other action.

(5) Customer Premises Equipment

Virtual networks operate with a variety of common types of customer premises equipment. This includes analog and digital PBXs, multiplexers, and routers.

(6) Features

Because they are a small part of a much larger network, VNs can offer sets of features that match, or surpass, those present in private networks. For instance, small sites can be included in the corporate virtual network over switched access and, using remote access arrangements, calls can be placed over the VN from anywhere. In addition, securing alternate facilities and coping with disasters becomes as much the carrier's responsibility as the user's. Also, through protected access to the

CPR/CSD database, the private network manager can make changes to stations, authorization codes, alternate routes, and other matters, as desired. Against these advantages must be weighed the vulnerabilities produced by the larger network—perhaps less privacy and susceptibility to the effects of other persons' disasters.

13.3 LIMITED CONNECTION NETWORKS

With the arrival of speedier PCs and workstations, their distribution throughout the workplace, and the proliferation of LANs, demands arose for higher-speed data connections than the network described in § 13.2 can provide. Among other things, their satisfaction produced frame-relay and SMDS (see §s 12.1.3 and 12.1.4). We begin our discussion of limited connection networks with networks based on T-carrier.

13.3.1 T-Carrier

(1) Voice-Grade Lines

Early data networks were composed of individual, point-to-point, *voice-grade lines* (VGLs) carrying multibit symbols that occupied no more than the voice passband. Some of the larger networks contained several thousand tie lines. With the tariffing of T-1 multiplexing for private line service, the number of individual VGLs providing data services has declined rapidly. In their place are networks of T-carrier facilities that carry multiplexed streams organized by digital cross-connect systems (DCSs) and intelligent multiplexers (see below).

(2) Digital Data Service

In 1973 the Bell System introduced *digital data service* (DDS), a digital transport service for private line applications. It employed existing technology to multiplex subrate streams (2.4 kbits/s and up) into 56 kbits/s channels as described in § 8.2.1(4) (see Figure 8.17). To achieve the published goals of 99.96% availability and 99.95% error-free seconds, the DDS network is implemented as a set of interconnected hubs that perform as routing, maintenance, and administration centers. In turn, the hubs are connected to digital lines that collect customer channels from serving offices. One hub serves many SOs. Customers are connected to the SOs over 4-wire loops in existing distribution cables. They pay for the entire circuit. For end-points that are relatively close together but far from the hub, the line to and from the hub is the major element of cost. Over the years, DDS service has been improved and reduced in cost. With newer equipment and better facilities, it is possible to arrange passage over the shortest practical route. For intra-LATA use, *hubless* DDS has been introduced that eliminates the back-haul to a hub. Nevertheless, the limitations of a 56 kbits/s channel remain, and DDS is being overtaken by more capable services.

(3) Fractional T-Carrier

Small users can lease a fraction of a T-1 facility from IXCs. Known as *fractional* T-1 (FT-1) service; it permits them to obtain *n* DS-0 channels, where *n* = 2, 4, 6, 8, or 12. (The remaining channels are leased to other customers.) In addition, because the individual DS-0 channels are synchronized together, multiple channels can be used to achieve speeds of *n* 64 kbits/s. Thus, speeds of 128 kbits/s, 256 kbits/s, 384 kbits/s, etc., can be obtained for compressed video or other purposes. Such channels are known as FT128, FT256, and FT384 respectively. For users that need several T-1s, *fractional* T-3 lines are available. Both of these facilities are priced so as to encourage leasing a complete system when occupancy is as low as one-third (i.e., for 8 DS-0 channels the lease cost is approximately equal to a complete T-1, and for 9 DS-1 channels the lease cost is approximately equal to a complete T-3). The exact crossover point depends on distance and variations in tariffs from different carriers. In the local exchange area, connections between the customers' premises and the LEC's serving offices are made by dedicated T-1 facilities, or may be provided as channels derived from digital loop carrier systems (see § 10.3.1(9)). Depending on the LEC, the connections may be tariffed as full T-1s or as FT-1s. **Figure 13.10** shows the way in which three locations of an enterprise that share the same LEC serving office might be connected to the corporation's private network. The backbone is assumed to consist of T-1 lines between IXC central offices where digital cross-connect capability is provided. Undoubtedly, the T-1s are part of in-place T-3 or SONET circuits maintained by the IXC. Between the IXC office and the POP, transport service is provided by 7 or 8 DS-0 channels (= FT-1) derived from other in-place IXC circuits. At the POP, they are connected to LEC circuits that provide transport to the serving office. At the SO, a DCS splits the channels and directs them to locations A, B or C. We assume that each location is served by a dedicated T-1 facility, and that advantage is taken of subrate multiplexing to reduce the number of individual channels to the serving office. Through the use of the DCSs at the LEC SO, POP, and IXC CO, individual, point-to-point connections can be changed over time.

(4) Intelligent Multiplexers

Today's T-1 multiplexer is a far cry from the equipment first given that name (see § 8.2.1(1)). Spurred on by the availability of low-cost, DS-1 access lines, and augmented with cross-connect and other capabilities, it has become a digital system capable of supporting sophisticated private network applications. Contemporary T-1 multiplexers, called *intelligent* (T-1) multiplexers (IMs), support large, complex, mesh networks, by providing the ability to add, drop, and redistribute circuits among multiple trunks, on command. By way of illustration, **Figure 13.11** shows a drop and insert application among three locations. In **Figure 13.12** the configuration is changed to permit the use of 12 DS-0 channels (FT768) for a videoconference between locations A and C, yet maintain 6 DS-0 channels between each location for other purposes. When the conference is completed, the configuration can be

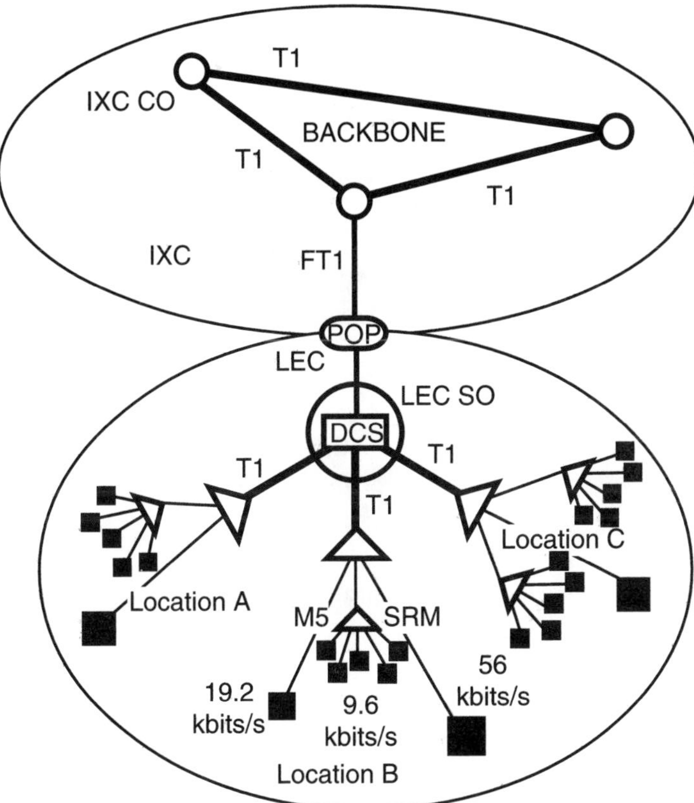

Figure 13.10 Enterprise Data Network Using T-1s and a Fractional T-1 (FT-1) Link
An FT-1 link can be used to carry enterprise traffic between the LEC SO and the backbone network. We assume T-1 links (probably derived from other equipment) between the enterprise locations and the SO, and between the backbone network nodes. They are likely to be implemented by DCS equipment in IXC COs.

changed back by electronic command. In addition, some IMs include the ability to interface with frame-relay and cell-relay equipment.

Intelligent multiplexers can be used to construct corporate data networks. An example is shown in **Figure 13.13**. With IM nodes located in San Francisco, Los Angeles, Denver, Dallas, Chicago, Atlanta, Miami, and New York City, point-to-point digital connections can be reorganized at will on commands from the network management center (NMC). The multiplexers connect to a variety of customer premise equipment including PBXs, computer hosts, terminal controllers, and routers. Figure 13.13 shows a network with significant traffic between New York City and Chicago, Chicago and Dallas, and New York City and Dallas. It employs

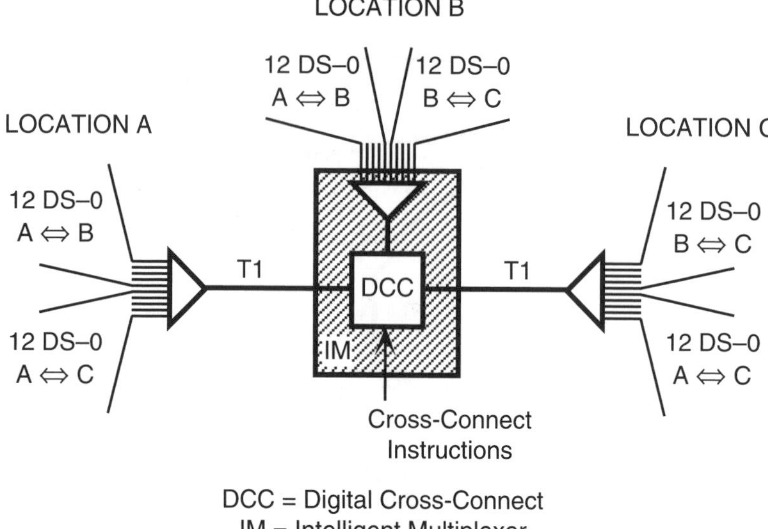

Figure 13.11 Intelligent Multiplexer Drop and Insert Application
The intelligent multiplexer splits the T-1 channels so that 12 are provided for traffic between A and B, 12 are provided for traffic between B and C, and 12 are provided for traffic between A and C.

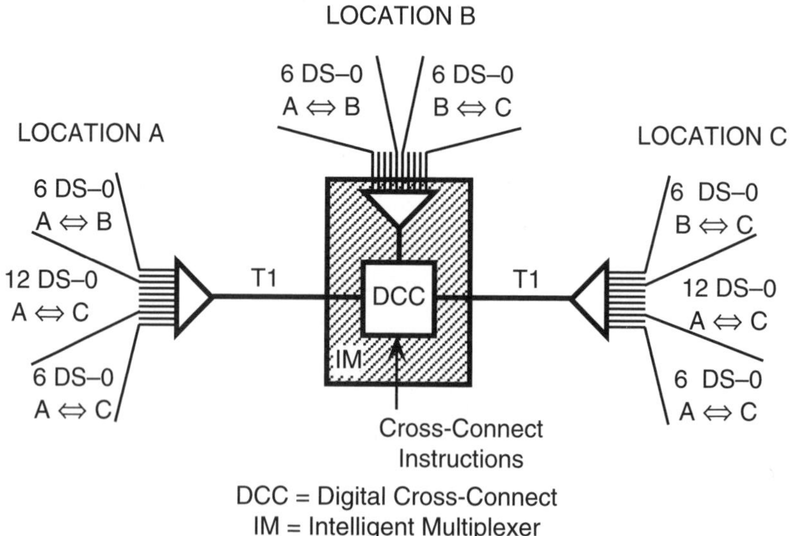

Figure 13.12 Reconfigured Drop and Insert Application
The arrangement in Figure 13.11 has been changed by instructions given to the DCC capability in the IM to become 6 channels between A and B, 6 channels between B and C, and 18 channels between A and C.

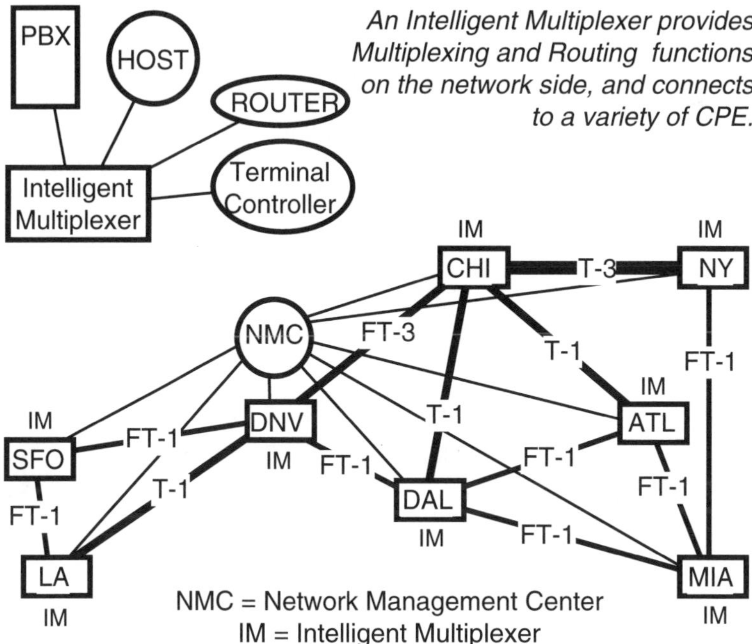

Figure 13.13 Enterprise Network Using Intelligent Multiplexers

a T-3 connection (NY—CHI), and a fractional T-3 connection (CHI—DAL). Other locations employ T-1 and FT-1 links. In a practical arrangement, IMs could be owned by the enterprise and interconnected by transport facilities leased from the carriers; the network management center (NMC) can be operated by employees of the enterprise.

(5) T-Carrier Networks

Private T-carrier networks exhibit a range of topologies. Different objectives, different locations, and different styles lead to many arrangements. Some points of general agreement are listed here:

- **Intra-LATA connections**: minimizing intra-LATA costs favors a star topology in which a hub connects to individual locations within the LATA.
- **Inter-LATA connections**: minimizing inter-LATA costs favors one connection between each pair of LATAs. This can be a hub-to-hub connection. These arrangements are shown in **Figure 13.14**.
- **Location of work sites**: the distribution of user sites will influence the topology. In smaller corporations it is likely that only one arrangement will fit the distribution, but, for many corporations, there will be several possibilities.

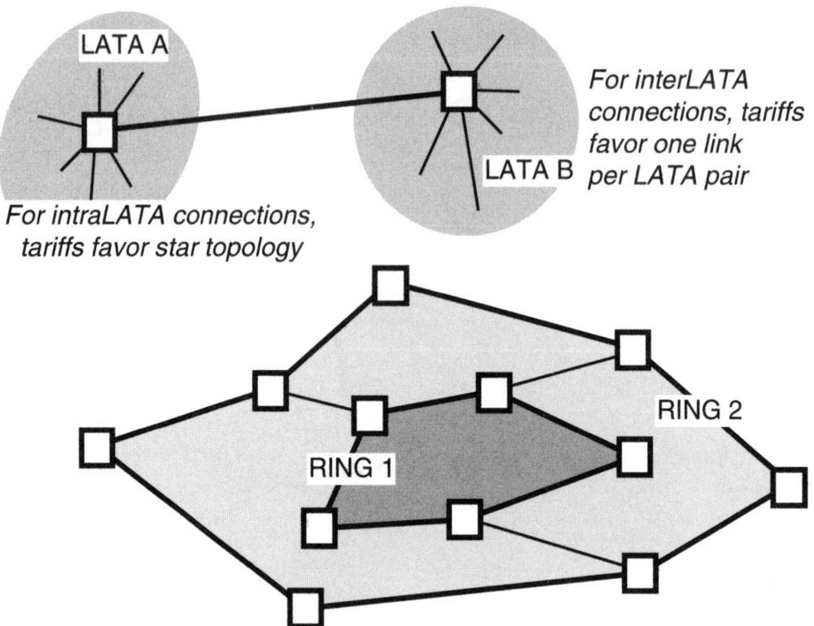

Figure 13.14 Points of General Agreement on T-Carrier Networks

- **Protection**: to protect against seriously degraded performance due to the loss of a line, traffic between nodes should have the opportunity to use as many alternate routes as possible.
- **Daisy chain connections**: several locations can be connected together in sequence by a single transmission system that includes T-1 multiplexers with drop and insert capability.
- **Multidrop connections**: multidrop connections between one primary and several secondary stations are possible up to speeds of 64 kbits/s.
- **Ring connections**: several locations can be connected by one, continuous transmission system so that, should the ring fail at any point, all of the stations remain connected in daisy chain fashion. By providing concentric rings, a network is formed that can have several options for recovery from the loss of one, or more, links. This arrangement is shown in Figure 13.14.
- **Mesh connections**: a fully connected mesh provides several options for recovery from the loss of one, or more, links. The expense of redundant connections must be balanced against the value of maintaining communication under adverse circumstances.

13.3.2 Frame-Relay and Cell-Relay

(1) Changing the Emphasis

Over the *public* network, it has long been held that voice traffic will always be substantially greater than data traffic. Indeed, a recent study confirms what was predicted several years ago—the number of bits associated with digital voice traffic is approximately four times the number of bits associated with data traffic. With these numbers in mind, many corporations have been satisfied to create the sort of network shown in Figure 13.5—a network that serves the needs of the voice community and does not do too bad a job for most data users. With the availability of virtual networks, voice traffic is migrating from private to public networks so that, over today's *private* networks, the ratio of voice to data is significantly less than four-to-one. Estimates place the current ratio of voice and data close to one, and project the volume of data traffic to grow to perhaps twice the volume of voice traffic. Under this circumstance the technology of private networks is changing from time division, synchronous techniques to statistical, asynchronous techniques that can also handle some synchronous traffic.

(2) From T-Carrier to Cell-Relay

In TDM systems, time-slots are allocated to sources on a permanent basis, whether they need them or not, and line speed is calculated by summing the maximum speeds of the circuits, whether they transmit continuously, or only once in a while. Thus, in the networks described in § 13.3.1, if we wish to increase the access speeds by n times, we must increase the speed (bandwidth) of all of the lines in the network by n. If the sources are continuous, this is the best we can do. If the sources are bursty, as data sources generally are, a better approach is to employ statistical multiplexing. For many, very short messages transmitted at random from a large number of sources with duty factors $1/n$, introducing statistical multiplexers allows us to increase the access speeds by n times without changing the speeds of the lines in the network. In a practical system, of course, the messages are of significant length and originate from a limited number of sources so that we can only increase the network bandwidth by some fraction of n. Its value will determine how many messages are delayed before they can gain entry to the network. Frame-relay and cell-relay are statistical multiplexing techniques that provide bandwidth-on-demand up to the limit of the facilities. Given a fixed capacity transmission facility, the inclusion of frame-relay and/or cell-relay processors allows the bandwidth to be fully utilized and the throughput to be increased.

(3) Frame-Relay Networks

Private frame-relay (FR) networks make use of FR configured routers and FR nodal processors to connect LANs and other data equipment. Two simple examples are given in **Figure 13.15**. The top diagram shows a direct connection between two LANs that is implemented over DDS, FT-1 or T-1 circuits. In the bottom diagram

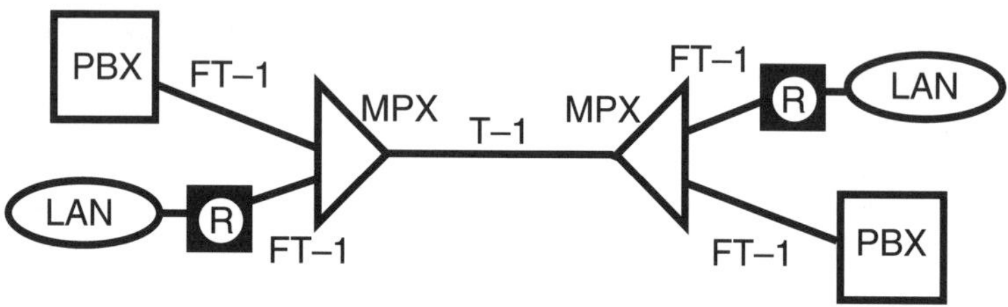

Figure 13.15 LAN-To-LAN Frame–Relay Connections

the same connection is made over a T-1 line in which some of the channels are employed by a voice PBX, and some of the channels are used to connect the two LANs using frame relay. In **Figure 13.16**, several LANs and other data processing equipment are connected together in a hubbed network through a single nodal processor. (This situation was discussed in Example 12.3.) At the hub, the nodal processor is connected to several FR routers that serve individual LANs, and local or remote FR PADs. The latter operate very much like PADs (packet assembler and disassembler devices) in X.25 packet networks. They create FR frames for equipment that is not able to do so and convert FR frames to formats that are compatible with the end-users. The local FR PAD (at the hub) connects to multiplexed terminals and a non-FR configured router that serves a LAN. The other FR PAD creates FR frames for three terminals before data are transmitted from a remote site to the hub. Connections across the network are implemented over FT-1 or T-1 circuits. **Figure 13.17** shows the elements of a larger network. Three FR nodal processors constitute the backbone network. LANs and other terminals are connected to these processors through appropriate devices. Within the backbone network, data are transported by cell-relay.

(4) ATM—the Ultimate Backbone

Figure 13.18 shows the elements of an enterprise network that serves the needs of the data community and carries high-volume voice and video traffic also. The backbone consists of mesh-connected ATM nodes that operate over SONET at 155 or 622 Mbits/s. They are fed from various data sources, from PBXs, and from videoconference rooms. The network transports PBX-to-PBX traffic within the enter-

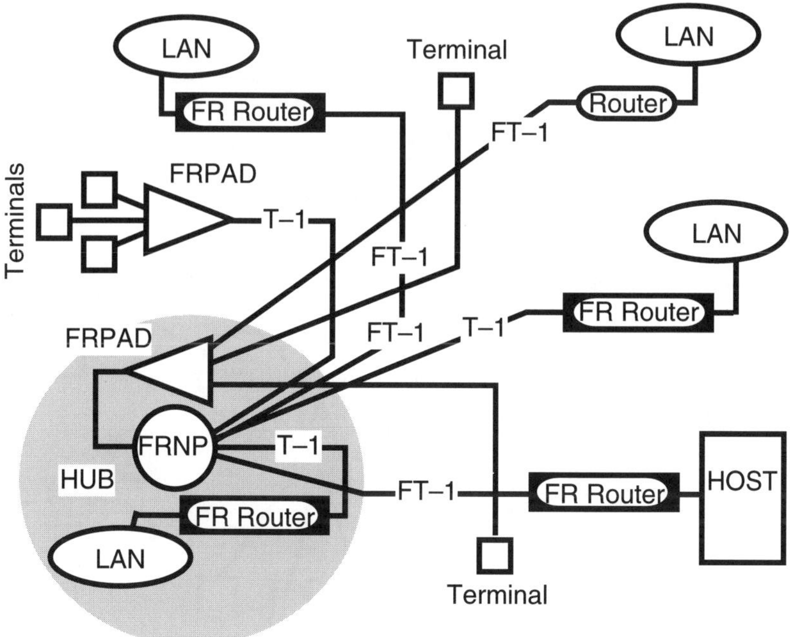

Figure 13.16 Connections Using Frame–Relay Hub
The Frame–Relay PAD assembles and disassembles FR frames for those terminals that are not served by FR routers.

prise but does not duplicate services that can be obtained from a virtual network created by a carrier. One cannot help but be concerned about the number of protocol changes that can occur in transferring messages from one end of the network to the other—yet all of the equipment shown is in use today, and is likely to stay in use for some time. Eventually, should ATM become the dominant technology, the environment can be simplified. Figure 11.23 shows how. Before this can happen, ATM nodes must be developed further and tested under all sorts of traffic loads.

13.4 COMPARISON OF TECHNIQUES

To end this discussion of enterprise networks, **Table 13.3** compares some of the parameters of T-carrier, X.25 Packet, frame-relay, SMDS and ATM networks. A number of items are considered in this comparison:

- **User's data unit**: what limits are imposed on the size of messages? For T-carrier, X.25 Packet, and ATM, there are no limits; for frame-relay, user's

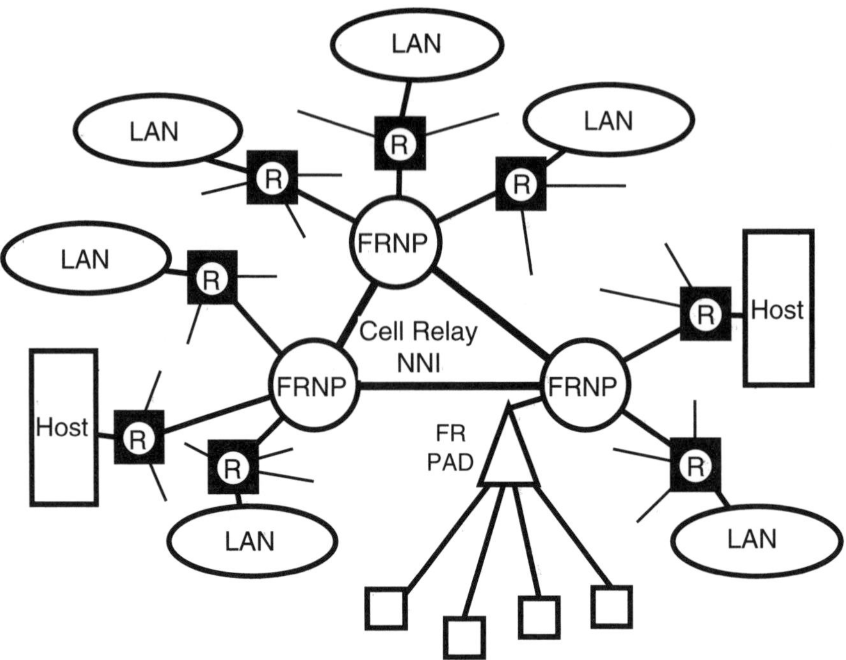

Figure 13.17 Network Employing Frame–Relay and Cell–Relay

data are limited to 8189 octets; and for SMDS, user's data are limited to 9188 octets.

- **Transport unit**: what size unit is used to transport the message within the network? The smaller it is, the less the network delay due to handling is likely to be. ATM, SMDS, and frame-relay employ 53-octet cells within the network. X.25 packet uses 136-octet packets, and T-Carrier (T-1) handles each stream octet by octet within the 193-bit frame.
- **Access line**: how many channels are supported by a single access line (i.e., the physical connection that crosses the UNI)? The greater the number, the fewer ports required at the sender and receiver. X.25 Packet, frame-relay, SMDS, and ATM support multiple channels. In X.25 Packet, there are 4095 possible logical channel numbers. While a practical number is significantly less than this, it is likely that X.25 Packet multiplexes the greatest number of channels on a single access line. Frame-relay, SMDS and ATM support fewer channels on each access line.
- **UNI**: how complex is the user-network interface? In the Table, we show UNI protocol stacks for each network. They are arranged in three levels that correspond approximately to IEEE 802 LLC, MAC, and PHY levels. T-Carrier is

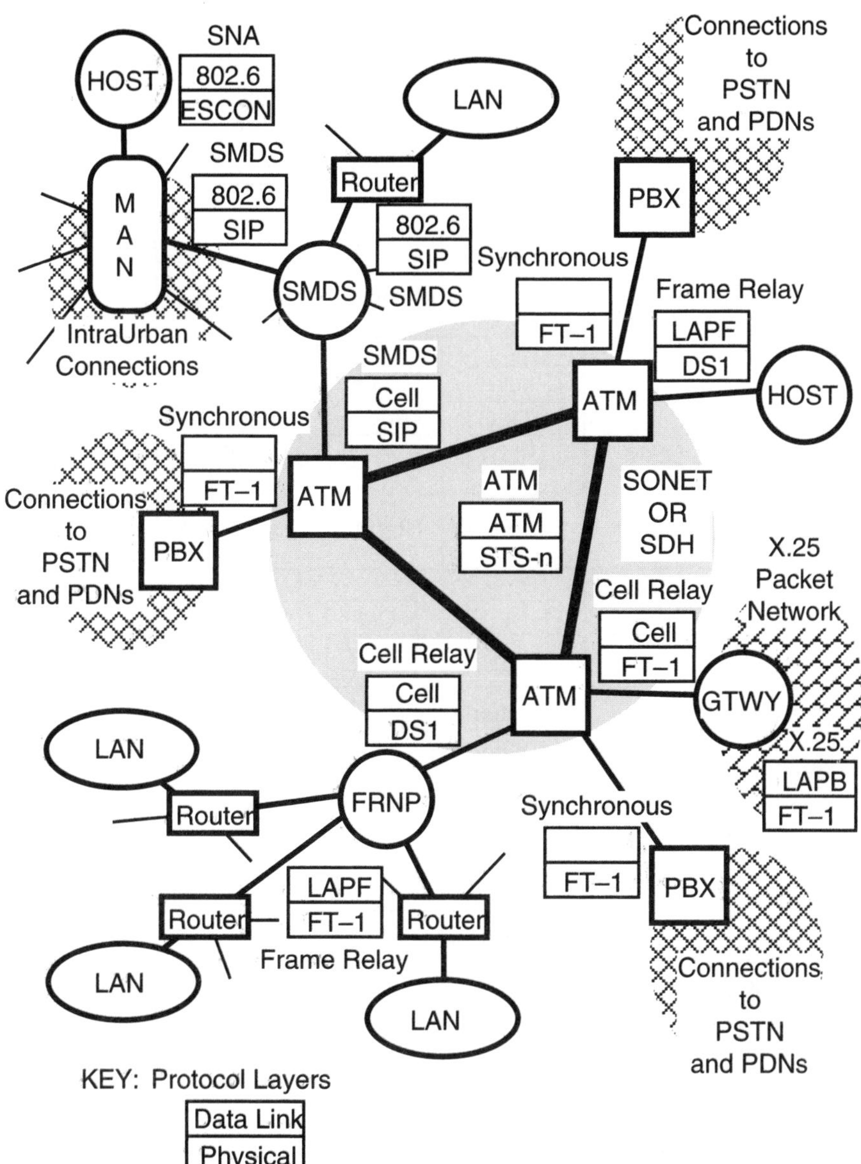

Figure 13.18 Modern Enterprise Network
Drawn to illustrate the range of equipment present and the number of different protocol stacks that are involved in a modern enterprise network.

Table 13.3 Comparison of Networks

	T–CARRIER	X.25 PACKET	FRAME RELAY	SMDS	ATM
User's Data Unit	No Limit	No Limit	≤ 8189 octets	≤ 9188 octets	No Limit
Transport Unit	Frame/ Octet	Packet	Frame/ Cell	Cell	Cell
Access Line	Single Channel	Multiplexed	Multiplexed	Multiplexed	Multiplexed
UNI	DS–1	Packet / HDLC / PHY	LLC / LAP-Fc / PHY	LLC / SIP3,2 / SIP1	AAL / ATM / PHY
Network Links	TDMPX	StatMPX	StatMPX	StatMPX	StatMPX
NNI	DS–1	HDLC / PHY	Cell / PHY	SIP2 / SIP1	ATM / PHY
Error Control	Detection, ESF	Detection, Correction, Link-by-Link	Detection Only	Detection Only	Detection Only
Routing	Permanent DCS	Distributed/ Centralized	PVC	PVC/SVC	PVC/SVC
Flow Control	Speeds Matched	Local and End-to-End	FECN and BECN	Congestion Managmnt	At UNI
Network Delay	Low	High	Low	Low	Low
Data Voice Video	Yes Yes Yes	Yes ? ?	Yes ? ?	Yes Yes Yes	Yes Yes Yes

? = technically possible; not likely in commercial service

only a physical connection. SMDS and ATM are probably the most complex since they include the segmentation, reassembly and convergence functions associated with cell usage.

- **Network links**: do the network nodes statistically multiplex data units on the network links so that bandwidth can be provided on demand? Only T-Carrier is without this capability.
- **NNI**: how complex is the node-network interface? In the Table, we show NNI protocol stacks for each network. They are arranged in two levels corresponding approximately to OSIRM Data Link and Physical layers. T-Carrier is only a physical connection. SMDS and ATM are equally matched, and frame-relay may be the most complex since it includes conversion from frame-relay to cell-relay (segmentation, reassembly and convergence functions).
- **Error control**: does the network provide error control (i.e., error detection and correction)? All networks provide some level of detection. Only X.25 Packet includes full correction capability.
- **Routing**: how are messages routed? ATM and SMDS provide permanent virtual circuit and switched virtual circuit capabilities. Frame-relay only operates over permanent virtual circuits. X.25 provides several routing procedures. T-Carrier provides permanent connections that can be changed from time to time using digital cross-connect systems.
- **Flow control**: does the network provide means for adjusting the rate at which data units are injected? Flow control is necessary if data units are not to be destroyed during periods of congestion. T-Carrier, in which transport speeds are matched so that there is no possibility of congestion, and X.25 Packet, in which comprehensive flow control capabilities are included, will not destroy data units to alleviate congestion. Frame-relay, SMDS, and ATM have less effective flow control capabilities and, if necessary, will destroy data units to control congestion.
- **Network delay**: does network operation create delays? Because X.25 Packet invokes link-by-link checks, it has the highest delay of all the networks. The others are considerably faster, and ATM probably has the lowest delay.
- **Data, voice, video**: what types of messages can the networks carry? All carry data; all except X.25 Packet and frame-relay carry voice routinely; and all except X.25 Packet and frame-relay can carry video. In addition, ATM provides one-to-many (multicast) and one-to-all (broadcast) connections.

But which network should I use?

There is no simple answer to this question. The choice of a network depends on the nature of the demand, the existing infrastructure, and the cost of different solutions. However, we can characterize the network options as follows:

- **T-Carrier network**: provides point-to-point, duplex, semi-permanent links (i.e., cross-connected not switched) that transport widely varying cross-sections (FT-1, T-1, FT-3, T-3). Carries synchronous traffic.

- **X.25 Packet network**: provides low-speed, switched, duplex, data-only transport. Best network solution when poor quality transmission links are present.
- **Frame-Relay network**: provides point-to-point, duplex, semi-permanent (i.e., permanent virtual circuits), higher-speed, data-only transport. Can be used as hubs in evolving, T-carrier, data-only network. Requires good quality transmission links.
- **SMDS**: provides point-to-point, duplex, switched, high-speed, transport of data, voice, and video. Requires good-quality transmission links.
- **ATM**: provides point-to-point, duplex, high-speed, switched data, voice, and video services, and multicast and broadcast services. Requires good quality transmission links. Can function in local, regional, and wide-area network.

14

Wireless Networks

In 1901, using the *wireless telegraph*, Guglielmo Marconi (an Italian physicist, 1874-1937) demonstrated radio communication across the Atlantic. Shortly thereafter, his invention was being employed for ship-to-shore communication. Using Morse code, it provided a slow-speed data channel that linked ocean liners to fixed installations on the mainland, or to other mobile platforms, and laid the foundation for today's wireless networks. Paced by improvements in portable telephones and personal computers, these networks provide flexible connections to a wide variety of *mobiles* (mobile stations). In this Chapter we describe the ways in which radio is used to remove geographic restrictions from those who create information *anywhere* and use it *everywhere*, without delay.

14.1 SATELLITE NETWORKS

Until the early 1960s, wireless networks were terrestrial networks characterized by fixed installations that had to be carefully sited so as to minimize the effects of terrain and other environmental constraints. Then, communication satellites appeared. Performing as *repeaters-in-the-sky* they made it possible for persons to communicate over thousands of miles without regard to geography, and stimulated discussions of the world as a *global village*. Today, satellite carriers provide point-to-point, point-to-multipoint, and broadcast, voice, video, and data services to anyone who has access to an earth station.

14.1.1 Communication Satellites

(1) Satellite Orbits

Most communications satellites are placed in orbit some 22,300 miles above the equator. At that distance, the satellites rotate about the earth's axis at the same angular velocity as the earth rotates about its axis (i.e., 360° or 2π radians in 24 hours). Called the *geostationary* orbit (GEO), the satellite travels at approximately 7000 miles/hr, appears stationary from earth, and can illuminate around one-third of the earth's surface. Other satellites are launched in *polar* orbit, that is to say they travel around the earth passing over the poles. Situated a few hundred miles above the earth's surface in *low earth orbit* (LEO), they require considerably less radio power. However, they are not stationary relative to an observer on earth, and tracking antennas must be used to achieve high performance. Furthermore, because they are closer to the surface, their breadth of view is less than GEO satellites. Other LEO satellites can be placed in inclined orbits between the polar and equatorial planes to provide additional coverage. **Figure 14.1** illustrates some of the properties of communication satellites and the radio beams they employ.

(2) Transmission Delay

In the simplest case, the spacecraft electronics captures the signals from earth, converts them to lower frequencies, and transmits them back to earth so that a message originating from a single point in the field of view of the satellite can be received at a second point on the earth's surface in the field of view of the satellite. The relayed message can be broadcast to the entire area, in which case it can be received by everyone, or dispatched to a specific area by a spot-beam antenna on the spacecraft. In this case, only those within the area illuminated can receive it. In either case, there will be a delay between transmission and reception back on earth due to the distance the energy must travel. For geostationary satellites it is approximately 0.25 seconds; for low-earth orbit satellites it is of the order of tens of milliseconds.

(3) Interference among Satellites

To avoid signal interference between satellites, both space and frequency diversity are employed:

- **Space diversity**: satellites are physically separated from each other. They occupy different orbit positions.
- **Frequency diversity**: satellites assigned to the same orbit positions operate in different frequency bands.

In the geostationary orbit, satellites positioned between approximately 90° and 130° West longitude can cover the 50 states of the United States, and satellites

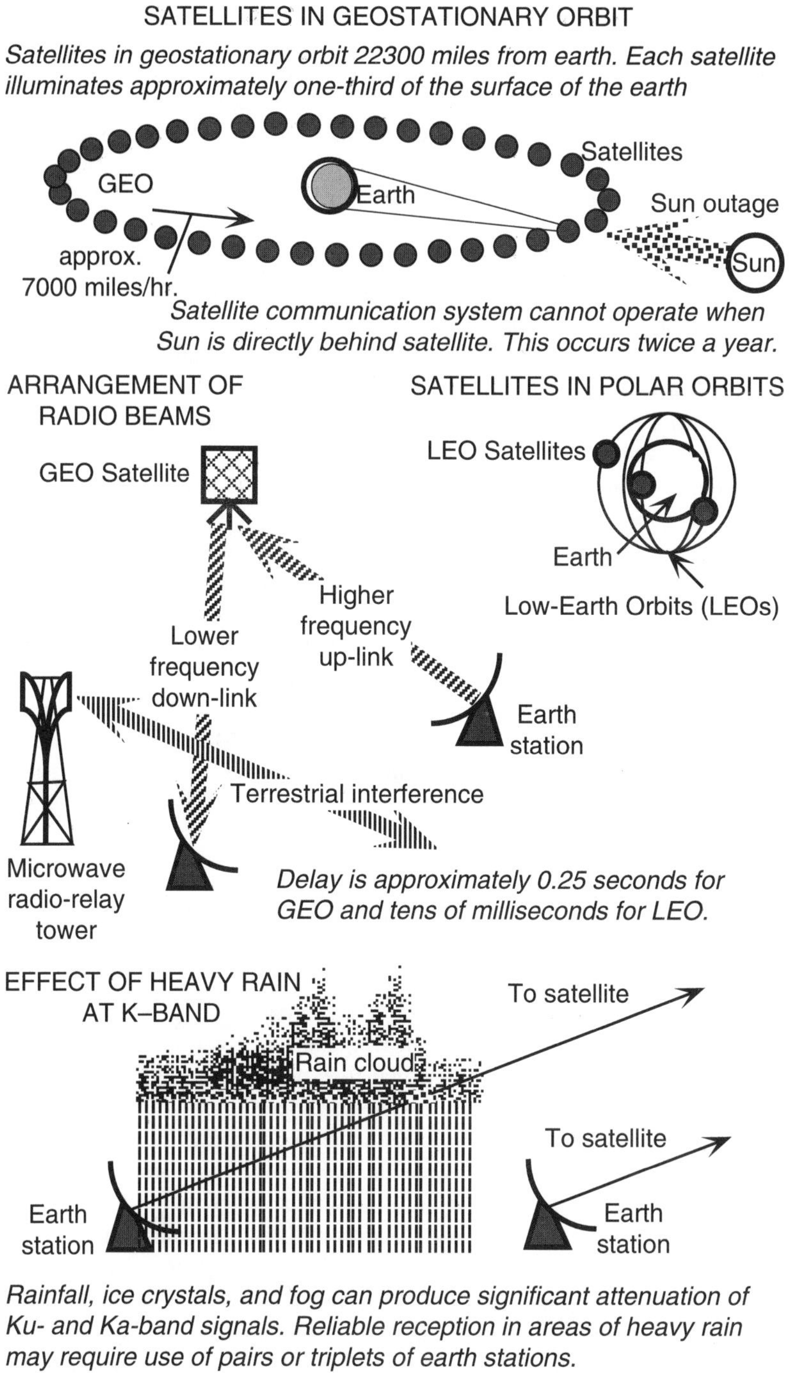

SATELLITES IN GEOSTATIONARY ORBIT

Satellites in geostationary orbit 22300 miles from earth. Each satellite illuminates approximately one-third of the surface of the earth

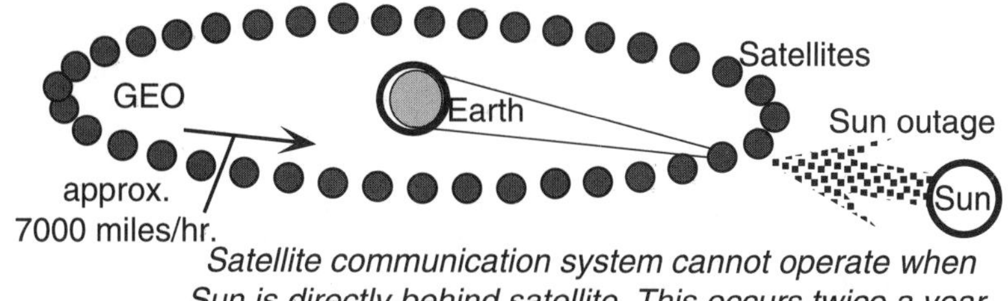

Satellite communication system cannot operate when Sun is directly behind satellite. This occurs twice a year.

ARRANGEMENT OF RADIO BEAMS

GEO Satellite

Higher frequency up-link

Lower frequency down-link

Earth station

Terrestrial interference

Microwave radio-relay tower

Delay is approximately 0.25 seconds for GEO and tens of milliseconds for LEO.

SATELLITES IN POLAR ORBITS

LEO Satellites

Earth

Low-Earth Orbits (LEOs)

EFFECT OF HEAVY RAIN AT K–BAND

Rain cloud

To satellite

Earth station

To satellite

Earth station

Rainfall, ice crystals, and fog can produce significant attenuation of Ku- and Ka-band signals. Reliable reception in areas of heavy rain may require use of pairs or triplets of earth stations.

Figure 14.1 Some Properties of Communication Satellites

positioned between approximately 60° and 90° West longitude can cover the 48 continental states. They are spaced at intervals of arc of ≥ 2°. GEO satellites employ frequencies in C band (around 4 GHz, space to earth, and 6 GHz, earth to space) and Ku band (around 11 and 12 GHz, space to earth, and 14 GHz, earth to space). Additional frequencies are allocated in Ka band (around 18 GHz, space to earth, and 30 GHz earth to space). At 4 and 6 GHz, interference with terrestrial microwave networks that share the same frequency bands may restrict opportunities for siting earth stations. At the higher frequencies (Ku and Ka bands), the absence of terrestrial facilities competing for the same frequencies makes the selection of sites for earth stations easier. However, atmospheric attenuation due to rainfall, ice crystals, and fog can be significant. During heavy rainstorms in the eastern and southern regions of the United States, pairs or triplets of stations may be required to assure reliable signal reception. On the basis of statistics describing the extent of local storms, they are sited so that at least one station is likely to be in the clear at all times.

At altitudes of several hundred miles, LEO satellites *see* a strip of the earth's surface whose width depends on their height, the land environment, and the frequencies employed. Allocations exist from as low as 150 MHz (149.9 to 150.05 MHz) to as high as 30 GHz for slant-path mobile/personal applications. Important bands for satellite provided personal communication services exist at 312 to 315 MHz (down-link) and 387 to 390 MHz (up-link), 1610 to 1625.5 MHz (down-link) and 2483.5 to 2520 MHz (up-link), and 1980 to 2010 MHz (down-link) and 2170 to 2200 MHz (up-link). For LEO satellites, lifetime is estimated to be five years; this is one-third of the lifetime of a GEO satellite.

(4) Power, Frequencies, and Footprints

Most spacecrafts employ power supplies that are charged from solar cells. In the design of the communications system, care is taken to minimize the power load presented by the electronics. Because atmospheric attenuation is greater at the higher frequencies, the frequency that requires the greater power is used by the earth station transmitter; thus, the down-link frequency is always less than the up-link frequency. Another way of saving down-link power is to employ shaped-beam antennas that concentrate the radiated power on specific areas of the global surface. For instance, geostationary satellites intended to carry traffic between continents separated by an ocean employ antennas that concentrate the radiated energy on the landmasses and not the ocean. For domestic coverage of the entire United States, the satellite antennas illuminate the 48-state continental landmass and provide additional spot beams focused on Alaska, Hawaii, and Puerto Rico. In the 48 states, spot beams may be employed to serve the major population centers. Restricting the radiated energy to areas surrounding terrestrial receivers allows better use of satellite power, provides some privacy by limiting the area in which the signals can be received, and permits the available radio frequency spectrum to be used over and over again. In each spot beam, the entire assigned bandwidth can be available for use provided adequate levels of isolation are maintained between the beams. This is accomplished by using an antenna with a radiation footprint that is a maximum in its intended service area and a minimum in areas served by other spot beams. To use these techniques with low-earth orbit satellites requires that the spacecraft antennas be able to track locations on the earth.

14.1.2 Satellite Communications

Commonly, satellite communication is used to provide the radio-relay of data, voice and video messages between widely separated points. Also, it is used to broadcast data, voice and video messages to particular areas of the world. The broad geographic coverage and multiple connectivity that a satellite provides require that access be carefully managed so that a large number of earth stations can be interconnected through the same satellite simultaneously. Most satellites contain multiple *transponders*, i.e., amplifying and frequency changing devices that operate over a portion of the frequency band assigned to the satellite. Channels may be pre-assigned (i.e., fixed allocation) for routes with heavy traffic (or for private usage), may be assigned on-demand (demand assignment multiple access, known as DAMA), or may be seized by earth stations when the need arises (random access).

(1) Pre-Assigned and Demand Assigned Access Techniques

Both frequency division and time division techniques are used to create channels that are assigned for communication over a satellite:

- **FDMA** (Frequency Division Multiple Access): a technique that assigns specific frequencies and bandwidths to create individual message channels in the transponder. All channels share the power of the transponder. Used with early satellites, FDMA is being replaced by digital techniques (TDMA and CDMA).
- **TDMA** (Time Division Multiple Access): a technique that assigns fixed time-slots to stations in the network; during its time-slot the full-capacity of the transponder is available to a station. A master station defines a superframe with specific transmit times for each station on the network; each of the transmitters is synchronized by a special pulse at the start of the frame, and transmits only during its assigned period. With TDMA, each station has a fixed allocation of time whether it has anything to send, or not. When most stations have a full, steady stream of traffic, TDMA is the best access technique to use.
- **CDMA** (Code Division Multiple Access): a spread spectrum technique (see § 2.3.3(2)) in which all stations in the network transmit at the same carrier frequency with approximately equal power and the same chip rate. Each station employs a code that is orthogonal to the codes used by the other stations. Each receiver sees the sum of the individual spread spectrum signals as uncorrelated noise. It can demodulate an individual signal if it has knowledge of the spreading code and the carrier frequency. With CDMA, collisions at

the transponder are of no consequence, and the full transponder bandwidth is available to each earth station at all times.

(2) FDMA System

In FDMA, a central control facility can assign operating frequencies at the request of the network stations. The calling earth station communicates to the control station requesting circuits for another station. Control selects a pair of frequencies and notifies both stations of the assignment. The principle is shown in **Figure 14.2**. It

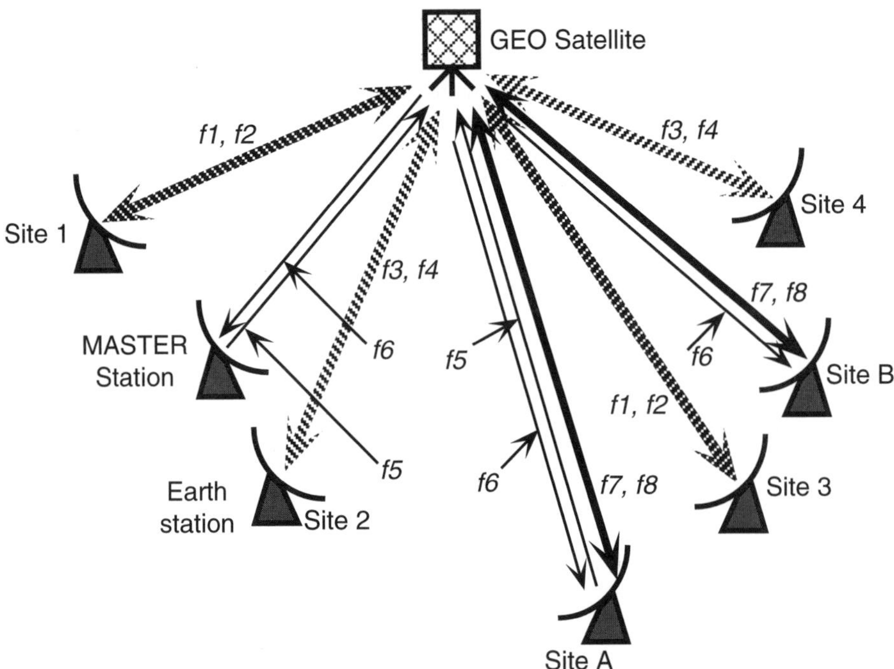

Master Station assigns frequency pairs to sites wishing one-time exchange of traffic. Also, manages power allocations among Sites to prevent transponder overload.

On f5 (up-link) and f6 (down-link), Site A requests Master to assign frequencies for exchange with Site B. On f5 (up-link) and f6 (down-link) Master assigns frequencies f7 and f8 to exchange.

On frequencies f3 and f4, Sites 2 and 4 exchange private traffic.

On frequencies f1 and f2, Sites 1 and 2 exchange public network traffic.

Figure 14.2 Satellite Communication Using FDMA

A control station assigns operating frequencies on request, or frequencies are assigned permanently.

includes examples of fixed-assignment and demand-assignment operations. Frequencies *f1* (up-link) and *f2* (down-link) are permanently assigned to Sites 1 and 3 for the exchange of public network traffic. Frequencies *f3* and *f4* are permanently assigned to Sites 2 and 4 for the exchange of private network traffic. Frequencies *f5* and *f6* are used to request channel assignments from the master station for a one-time exchange of traffic between Sites A and B. The master station assigns frequencies *f7* and *f8* to their exchange. Known as *single-channel per carrier* (SCPC) DAMA, this technique has been used with manual and computer controlled stations to provide on-demand, voice or data channel connections.

(3) TDMA System

In TDMA, time-slots are assigned to each earth station on a fixed allocation or demand basis. During each time-slot, the station has full and exclusive use of the entire transponder. The principle of a TDMA system is shown in **Figure 14.3**. Messages are buffered at the earth stations and sent in bursts timed so that they arrive at the satellite in a sequence that allows the transponder to serve each of the earth stations in turn. In the example, bursts are shown transiting the link in an organized way. Because of the long path, several discrete message blocks may exist between stations and satellite. A set of message blocks (one from each station) constitutes a frame. It starts with a synchronizing signal (synch burst) from the control station that includes a unique word for easy recognition. To determine when to send their message bursts, all other stations use this signal as a reference. Each message is opened by a preamble block that contains block and timing information, station identification, and other data, and may be closed by a block that performs error checking and other functions, and ends the sequence. A portion of the digital stream is reserved for channel requests and time-slot allocations. In the diagram, Site 1 functions as the master station that implements DAMA operation and provides network synchronization.

(4) Request Channel

Request and assignment channels are shared among all stations. The request channel may operate as a fixed allocation channel in which each earth station is assigned a portion of the channel, or it may operate as a contention channel in which each earth station may send its request when ready. With a fixed allocation strategy, the capacity of the request channel limits the absolute size of the network. With a random access strategy, the capacity of the request channel only limits the rate at which requests are made. The INMARSAT network that provides on-demand communications for ships at sea employs an ALOHA random access, request channel protocol.

(5) Random Access—Basic ALOHA

Invented for a microwave radio system linking the Hawaiian islands, ALOHA is a technique in which stations send data packets as they are ready. If the offered traf-

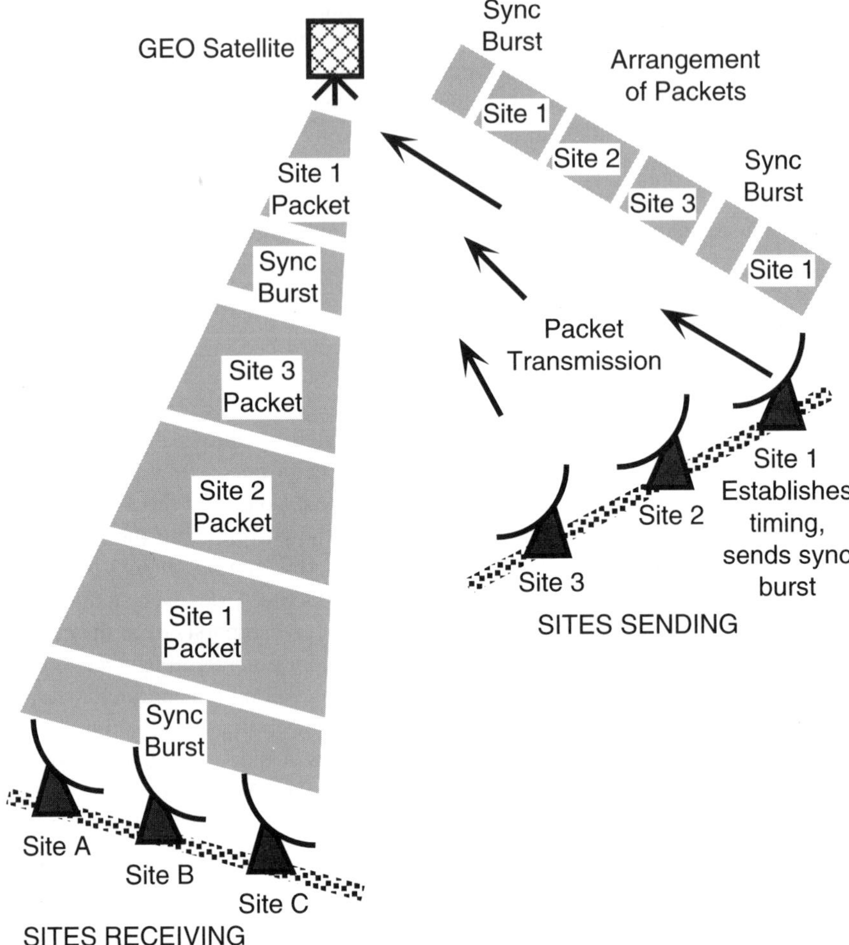

Figure 14.3 Satellite Communication Using TDMA
Time–slots are assigned to each earth station on a fixed allocation or demand basis. One station is responsible for establishing network timing.

fic from all the stations is 50% of full load (i.e., one-half of the traffic that will just fill the transponder), approximately 36% of the packets will get through on the first attempt. If the offered traffic is 10% of full load, approximately 80% of the packets will get through on the first attempt. Thus, so long as the total offered traffic from all stations is not a very large fraction of the capacity of the transponder, many packets have a reasonable chance of finding the transponder unoccupied upon arrival at the satellite and can be relayed without interference. Under this circumstance, the sender will receive an unchanged version of its message back from the transponder. However, if a second signal arrives at the transponder while the first

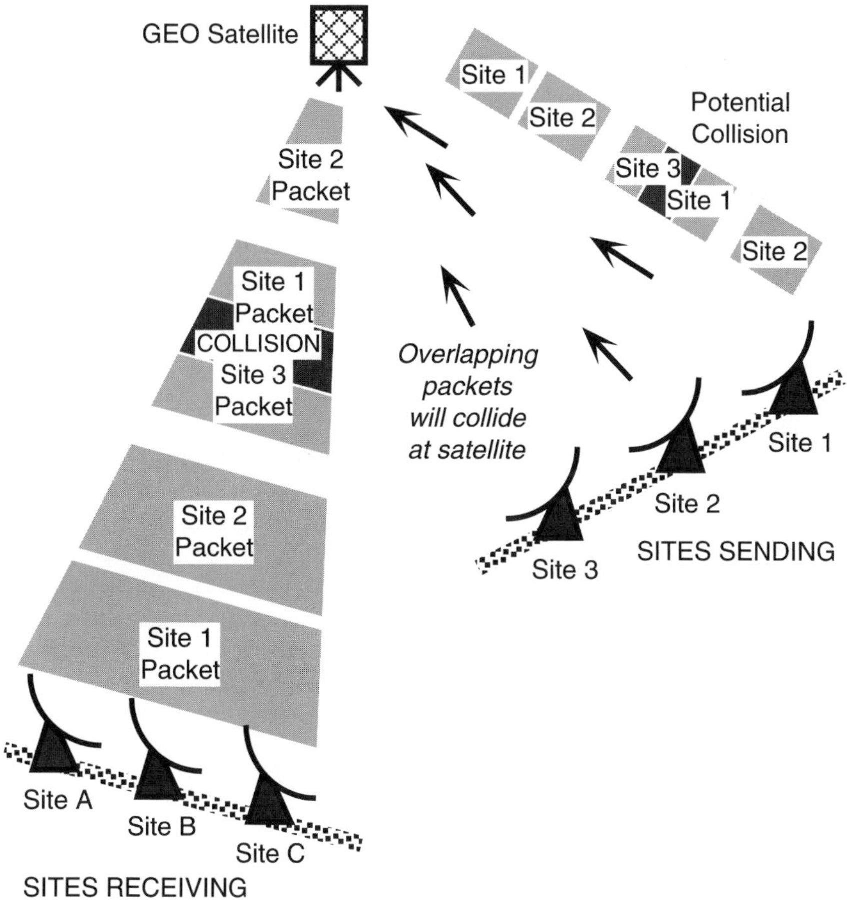

Figure 14.4 Satellite Communication Using ALOHA
ALOHA is a technique in which stations send packets when ready. If collisions occur, the stations wait a random time before sending again.

signal is still being received, a collision occurs and the sending stations receive garbled versions of their transmissions. When collisions occur, the sending stations wait (a random time) and then transmit again. The principle of data communication using ALOHA is shown in **Figure 14.4**.

(6) Slotted and Reservation ALOHA

By modifying the process so that all packets arrive at the transponder at the beginning of agreed upon time-slots, the throughput can be doubled. Called slotted-

ALOHA, the technique requires that all stations be synchronized in some fashion so that their packets only arrive at the transponder at the slot start times. An extension of this technique allows stations to reserve time-slots for their exclusive use. Called reservation-ALOHA, the technique guarantees that messages will arrive one-at-a-time at the transponder and will be relayed without collisions.

(7) Polling Over Satellite Links

The introduction of satellite links into a centralized computer network in which the FEP attached to the host is the primary station, and all units in the network are secondary stations, significantly reduces throughput. To speed things up, *satellite delay compensation units* (SDCUs) can be added at all earth stations. An exchange between the FEP and a remote DTE (DTE1), is shown in **Figure 14.5**. At the earth station serving the FEP, the SDCU acknowledges the polling command at the same time the polling packet is launched from the earth station towards DTE1. By doing this, the SDCU frees the FEP to poll other stations in the network. The SCDU continues to respond to polls until data arrives from DTE1. If the primary has something to send to DTE1, the SDCU at the earth station serving the FEP acknowledges receipt according to the ARQ procedure employed, while the packet stream is launched towards DTE1. At the earth station serving DTE1, the SDCU polls DTE1 and receives responses from it. When DTE1 has something to send, packets are launched in the direction of the FEP and the SDCU serving DTE1 provides acknowledgments that keep the stream going. If the FEP detects an error, an error message is sent to DTE1, and it implements the ARQ procedure in use.

14.1.3 Very-Small-Aperture Terminals

A VSAT network consists of a large number of geographically dispersed microstations that are star-connected to a hub station through a geostationary satellite. Private data networks that employ hundreds or even thousands of earth stations have been implemented. To reduce the cost of each earth station, a small antenna (3 to 6 feet in diameter) and a low-power transmitter (perhaps 1 watt) are employed, giving rise to the name *microstation*. Outbound transmissions (hub to satellite to VSATs) employ a TDM channel (time-divided on a packet basis) broadcast to all VSATs by the satellite. In this way, with a station unique addressing scheme, the hub can send messages to individual microstations. Inbound transmissions (VSAT to satellite to hub) from individual VSATs are accomplished using TDMA, ALOHA, or CDMA access procedures. Because of the small antennas and low powers employed at the microstations, the signals received by the hub from the individual VSATs are likely to be seriously distorted by noise. The hub employs sophisticated signal recovery techniques to regenerate them. Messages can be sent from VSAT to VSAT by going through the hub. The principle of satellite communication using VSATs is shown in **Figure 14.6**.

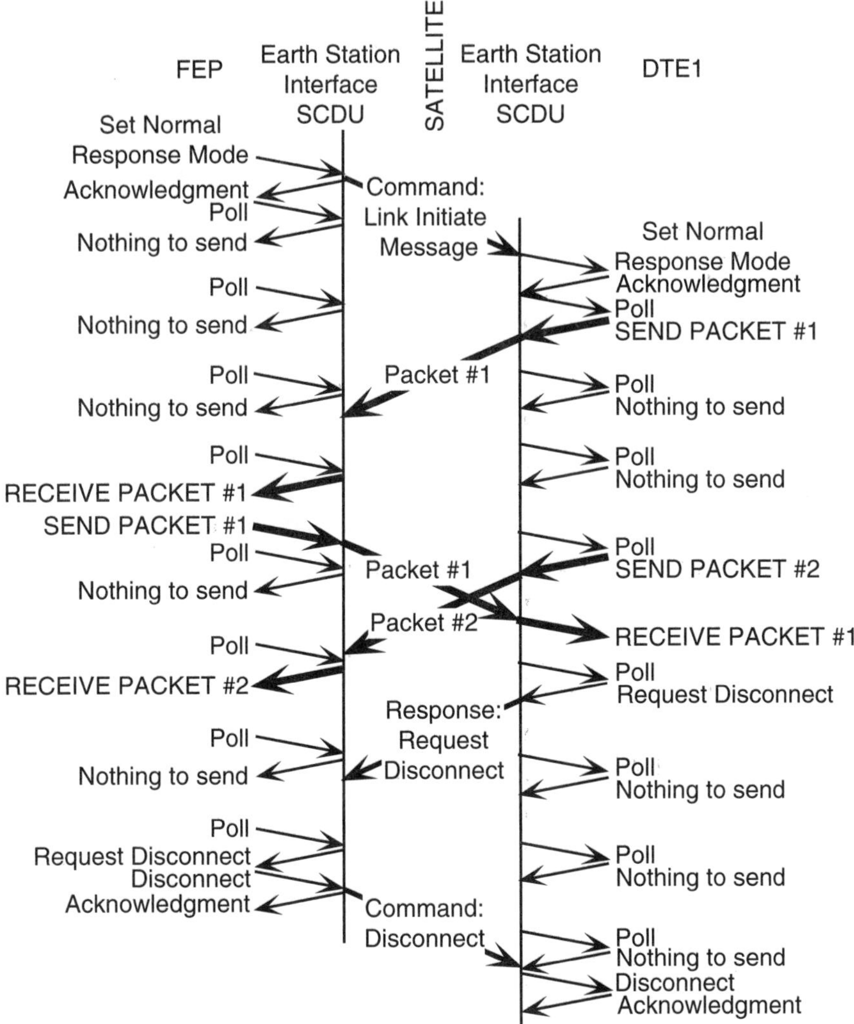

Figure 14.5 Polling over a Satellite Link
SDCUs at both ends provide positive responses so that satellite delay does not slow the polling process.

14.1.4 LEO Systems

Interest has redeveloped in the use of groups of low-earth orbit (LEO) satellites to provide communication networks with continuous global coverage. Using relatively low-cost VHF technology (< 1 GHz), several consortiums will offer electronic

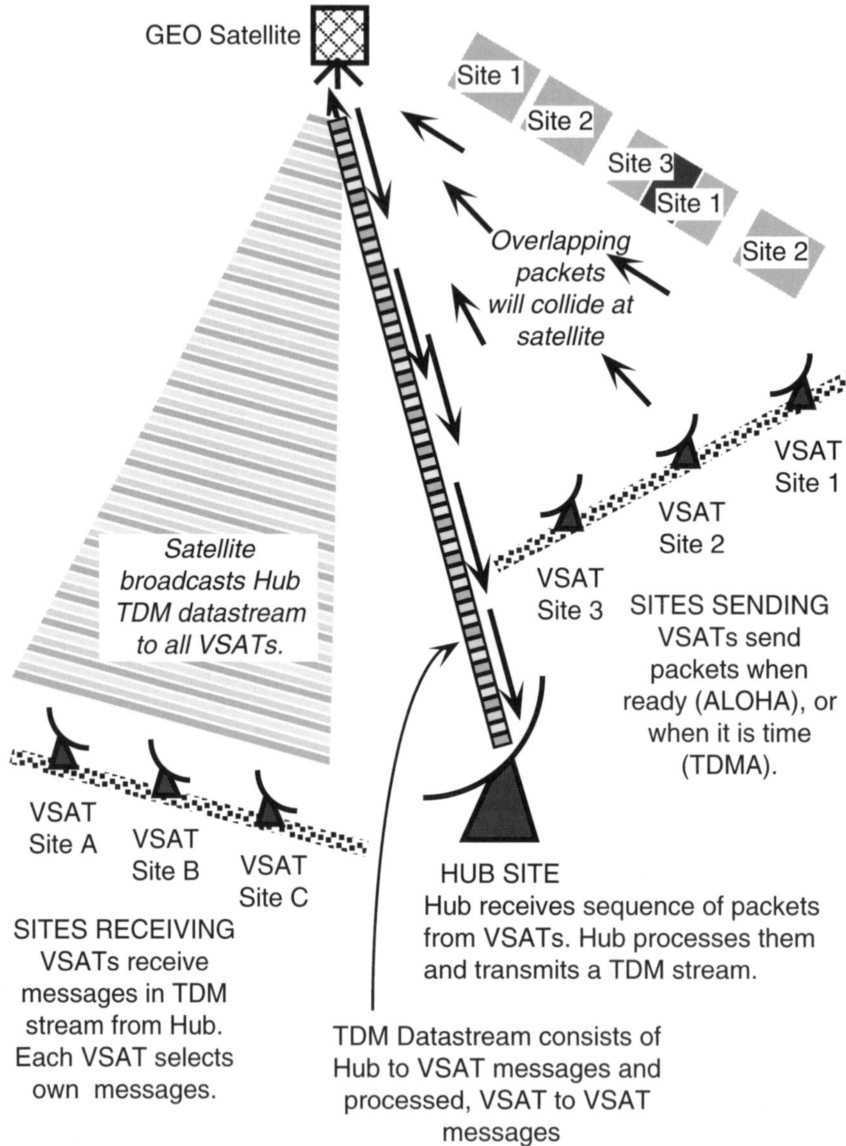

Figure 14.6 Satellite Communication Using VSATs
Hub to satellite to VSAT transmissions employ a TDM channel. VSAT to satellite to hub transmissions employ TDMA, ALOHA, or CDMA access procedures.

mail, paging, and other services that use brief digital messages. Using more expensive techniques, other groups will use frequencies above 1 GHz to provide services that will compete with mobile telephone providers.

14.2 CELLULAR MOBILE RADIO TELEPHONE SERVICE

Cellular mobile radio telephone service (CMRTS) has shown how to uncouple terminals from fixed service points. No longer is it necessary to be attached to a cable to send and receive information. Originally conceived as voice-only services, CMRTS now carry voice and data; CMRTS facilities are experiencing rapidly growing demand.

14.2.1 Cellular Radio

(1) Urban Radio Environment

In many applications, the propagation of radio waves can be modeled as if they were in free space. Under this circumstance, the energy drops off as the square of distance. In engineering terms, the energy decreases by 20 db for every tenfold increase in distance traveled. Thus, if a wave at one mile from the antenna in free space has a power level of 0 db, at ten miles from the antenna it has a power level of -20 db. In an urban environment, measurements in the range 30 MHz to 3 GHz show that the average path loss is proportional to the fourth power of distance. Thus, if a wave at one mile from the antenna in an urban environment has a power level of 0 db, at ten miles from the antenna it has a power level of -40 db. The presence of buildings, structures, and other impediments, produces a hundredfold increase in attenuation over the free space value. The increase in path loss is shown in the graph at the top of **Figure 14.7**. This effect is put to use in cellular radio systems.

In Figure 14.7, we show another effect of the urban environment. As the mobile unit moves, it receives radiation directly from the base station antenna. In addition it receives radiation that is reflected from buildings, mountains, and other features of the environment. Because the path they travel is longer than the direct path, these rays arrive out of phase with the main signal. Moreover, as the unit moves through the environment, the relationship between the reflecting objects and the mobile change, and the phase differences between the signals vary. The result is the strength of the received signal changes in the fashion simulated in the lower graph. Significant *fades* can occur at approximately half-wavelength intervals along the unit's forward progress. For this reason, frequency modulation, not amplitude-modulation is used in mobile systems.

(2) Frequency Reuse

For 50 years, the principle of *frequency reuse* has been employed in terrestrial microwave relay systems. In these systems, focused beams of microwave energy clustered about a specific carrier frequency are sent from one station to another. The same carrier frequency can be used in other beams so long as individual receivers only *see* their associated transmitter. The key to frequency reuse is providing adequate separation between entities that use the same frequencies so as to prevent one entity from receiving detectable signals from another entity that is not

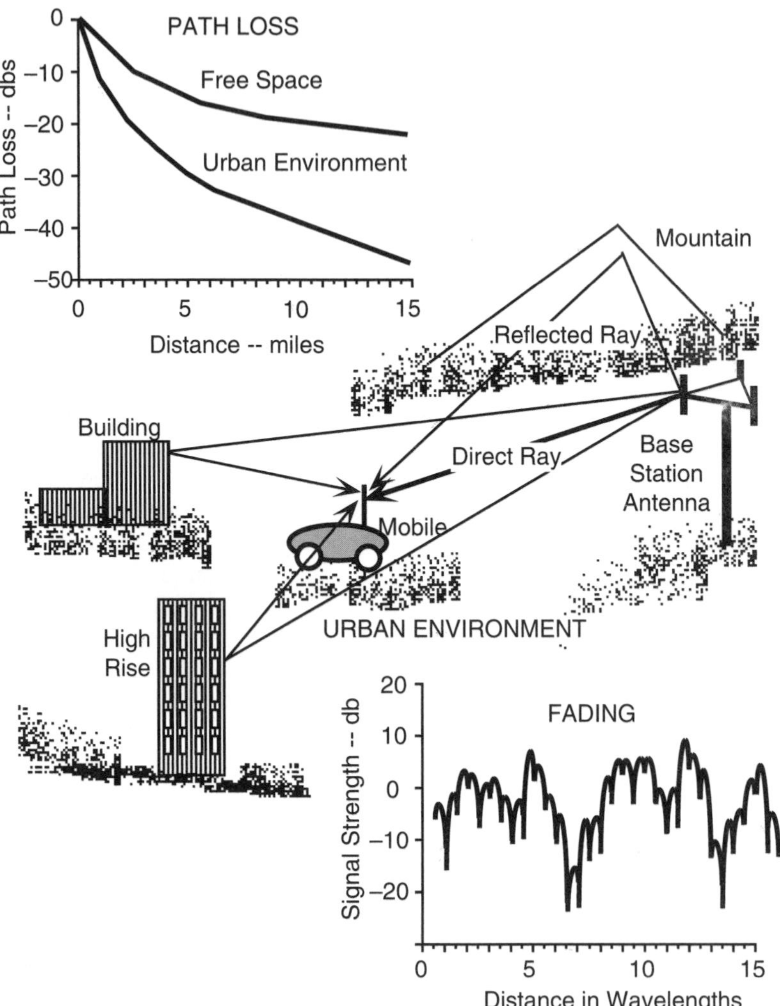

Figure 14.7 Urban Radio Environment
The strength of the received signal varies as the mobile moves in an urban setting. Significant fades can occur at approximately half-wavelength intervals.

assigned to it. The technique works well with fixed installations. With mobiles, we use a band of frequencies in a limited area and make arrangements such that other areas in which the same frequencies may be in use are separated by enough distance so that the energy is properly attenuated. These areas are known as *cells*.

(3) Cells and Microcells

Theoretically, cells can assume several geometric shapes (triangles, squares, pentagons, etc.). One that approximates the circular coverage of an omnidirectional antenna is the hexagon. In **Figure 14.8** we show a set of hexagonal cells packed together so as to cover an area in which *base* stations (also known as *cell sites*) are situated at the center of the cells in a regular pattern. Each base station consists of a transmitter to which a band of frequencies is assigned, and an omni-directional antenna so that energy is radiated uniformly, in all directions. Because a limited range of frequencies is available for use in each cell, the number of active mobiles in a cell is limited. To increase the system capacity, the cell size must be reduced so that the same number of mobiles can be served in a smaller area. To prevent interference in adjacent (smaller) cells, transmitter power must be reduced. But the cell size cannot be reduced without limit. Cells must be matched to the anticipated speeds of the mobiles. The smaller the cell, and the faster the mobile, the more often the base stations will have to *hand-off* to the next cell as the mobile traverses the system. Hand-off is a procedure we discuss in § 14.2.2(4). Suffice it to say that it requires time and system resources, and produces a discontinuity in the message

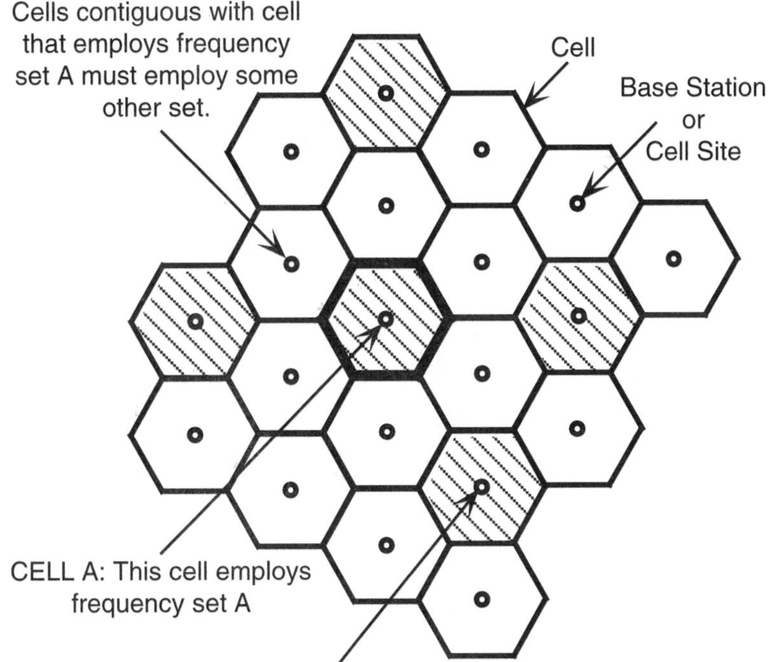

Figure 14.8 Cellular Radio Cell Structure

stream. The result is that cells for systems that serve high-speed vehicles (cars on an interstate highway, for example) may be several miles in diameter, while cells that serve pedestrians may be a thousand feet, or so, in diameter. For a given range of mobile speeds, there is a practical minimum cell size and thus a maximum limit to the system capacity that can be achieved by decreasing the size of the cells.

(4) Access Techniques

Connecting several active mobiles located in a single cell to the base station requires the adoption of an access technique. In existing and proposed systems, three methods are employed that are similar to those described in § 14.1.2 for satellite access:

- **FDMA**: using frequency division multiple access techniques, a pair of carrier frequencies is assigned to each mobile. One frequency is used for mobile-to-base station traffic and the other is used for base station-to-mobile traffic. In the center of the cell cluster in Figure 14.8, Cell A is assigned to operate with a set of frequencies we designate Frequency Set A. If cells surrounding Cell A are assigned the same set of frequencies, a mobile on the boundary between Cell A and its abutters will not have an unambiguous communication path to a base station. Therefore, at least one ring of cells around Cell A cannot be allocated Frequency Set A. Further, the abutting cells must be assigned different frequencies from each other so that mobiles on their borders have unique communication paths. In Figure 14.8, we have shaded four cells that have the potential to reuse Frequency Set A. Whether they do, or not, depends on the frequency sets assigned to their neighbors. Using only one ring of cells to separate those operating with the same frequency set is possible due to the blocking produced by buildings and other terrain-related attenuation in the urban environment. As described, this causes the received power to vary inversely with the fourth power of distance. Under this circumstance it is possible to locate transmitters closer together and to achieve sharper limits to the cells than might be expected. If the *one-way* bandwidth available to the *system* is B Hz, the reuse factor is $1/r$ (where r is the total number of cells in the system divided by the number of cells that use Frequency Set A), a fraction a of the circuits are used for overhead purposes, and each channel has a bandwidth of b Hz, the total number of mobiles that can be served cannot exceed aB/rb per cell. Reuse factors of 1/4, 1/7, and 1/12 are used.
- **TDMA**: using time division multiple access techniques, a pair of carrier frequencies are assigned to n mobiles and each mobile is assigned time-slots on the carriers. In this way the traffic from n mobiles is carried as a time-division multiplexed signal on the carrier pair. Each mobile transmitter sends bursts of information timed and synchronized so that they arrive at the cell site receiver at exactly the right time to follow the burst from the mobile

ahead in the sequence and lead the burst from the mobile that follows. Accurate slot timing and carrier synchronization is essential. Fortunately, all participants in the mobile-to-base station stream receive signals from the base station. They are used to pace the action. The total number of mobiles that can be served is *anB/rb* per cell.

- **CDMA**: CDMA produces a wideband, noiselike signal, and the entire range of frequencies allocated to the system can be used in each cell. Discrimination between mobiles in the same cell is achieved through the assignment of independent *spreading* codes (as described in § 2.3.3). Discrimination between mobiles in one cell and mobiles in an abutting cell is achieved through the assignment of a different set of spreading codes in the abutting cell. Under these circumstances, the number of mobiles that can be served is *aB/b* per cell. This is significantly greater than the number served with FDMA or FDMA/TDMA. But CDMA is not without its complications—so that users close to the base station do not overpower users farther away, each mobile must regulate its transmitted power in inverse proportion to some power of the distance it is from the cell site.

(5) Number of Mobiles Served by a Cellular System

By way of illustration, if $B = 10$ MHz, $1/r = 1/7$, $a = 0.95$, $b = 30$ kHz, $n = 3$, and the number of cells in the system is 28, FDMA systems can serve 45 mobiles per cell, and 1265 mobiles per system; TDMA systems can serve 135 mobiles per cell, and 3780 per system; and CDMA systems can serve 316 mobiles per cell, and 8857 mobiles per system. The number of mobiles served is the number of mobiles actively using the system at any one time. The number of subscribers to the system may be ten or twenty times this number.

14.2.2 Advanced Mobile Phone Service

(1) Existing System

In the United States, since the early 1980s, frequencies in the 800 MHz region have been allocated to cellular mobile radio telephone service:

- **Mobiles to cell sites**: 825 to 845 MHz
- **Cell sites to mobiles**: 870 to 890 MHz.

Known as *advanced mobile phone service* (AMPS), the system adopted by the operating entities has a channel bandwidth of 30 kHz, uses FDMA as the access method, and employs frequency modulation. The 666 radio circuits allocated by the FCC (i.e., 20 MHz ÷ 30 kHz) consist of pairs of one-way channels separated by 45 MHz. To encourage competition, in each *cellular geographic serving area* (CGSA), the FCC has assigned half the channels to an operating telephone company (wireline company), and half the channels to an independent company (non-wireline company) so that each provider operates 333 circuits in two 10 MHz bands. A few of these circuits are used for control purposes; the vast majority carry analog voice or low-speed data (up to 2400 bits/s).

(2) Cells and Cell Sites

Figure 14.9 shows the principle of AMPS. In this figure, cells are shown as overlapping circles, not neatly-fitted hexagons, to make the point that a cell is defined by the mobiles that are communicating with the base station. In a practical system, the geographical arrangement of cell sites, the topology of the area served, and the decision thresholds built into the electronics dictate when a mobile communicates with one base station, and when it communicates with another. The operation of AMPS depends on a central switch that serves the CGSA. This Mobile Telephone Switching Office (MTSO) is an electronic switching system programmed to provide call-processing and system fault detection and diagnostics. The MTSO is connected directly by wireline, or microwave radio, to a set of base stations, called *cell sites*, that communicate with the mobiles in their cells by radio. Small programmable controllers at each cell site perform call setup, call supervision, mobile location, hand-off, and call terminating. A microprocessor within each mobile implements signaling, radio control, and customer alerting functions. Two types of channels are provided:

- **Setup channels**: transmit data used to set up all calls
- **Voice channels**: provide message paths and carry signals needed for call supervision, i.e., to signal changes in state and to maintain the connection. The supervisory signals are:
  - *supervisory audio tone* (SAT). An out-of-band, continuous tone (either 5700, 6000 or 6300 Hz), the SAT is present whenever a call is in progress. Its absence indicates that one of the parties has gone on-hook, and the call is terminated.
  - *signaling tone* (ST). An out-of-band tone burst (10 kHz). Among other things, the signaling tone is used to initiate ringing, to signal hand-off to another cell site (tune the mobile to another set of frequencies), and to disconnect.

(3) Setting Up Calls

To set up a call to a mobile, the called number is broadcast on the outgoing set up channels of the cell sites. When the called mobile detects its number in the incoming stream, it sends its identification over the return set up channel. The nearest cell site (selected because it has the strongest signal) responds by assigning the mobile a pair of operating frequencies. The mobile tunes to the assigned receive frequency and prepares to transmit on the assigned sending frequency, and the mobile telephone is rung. When a mobile places a call, the nearest cell site designates a pair of

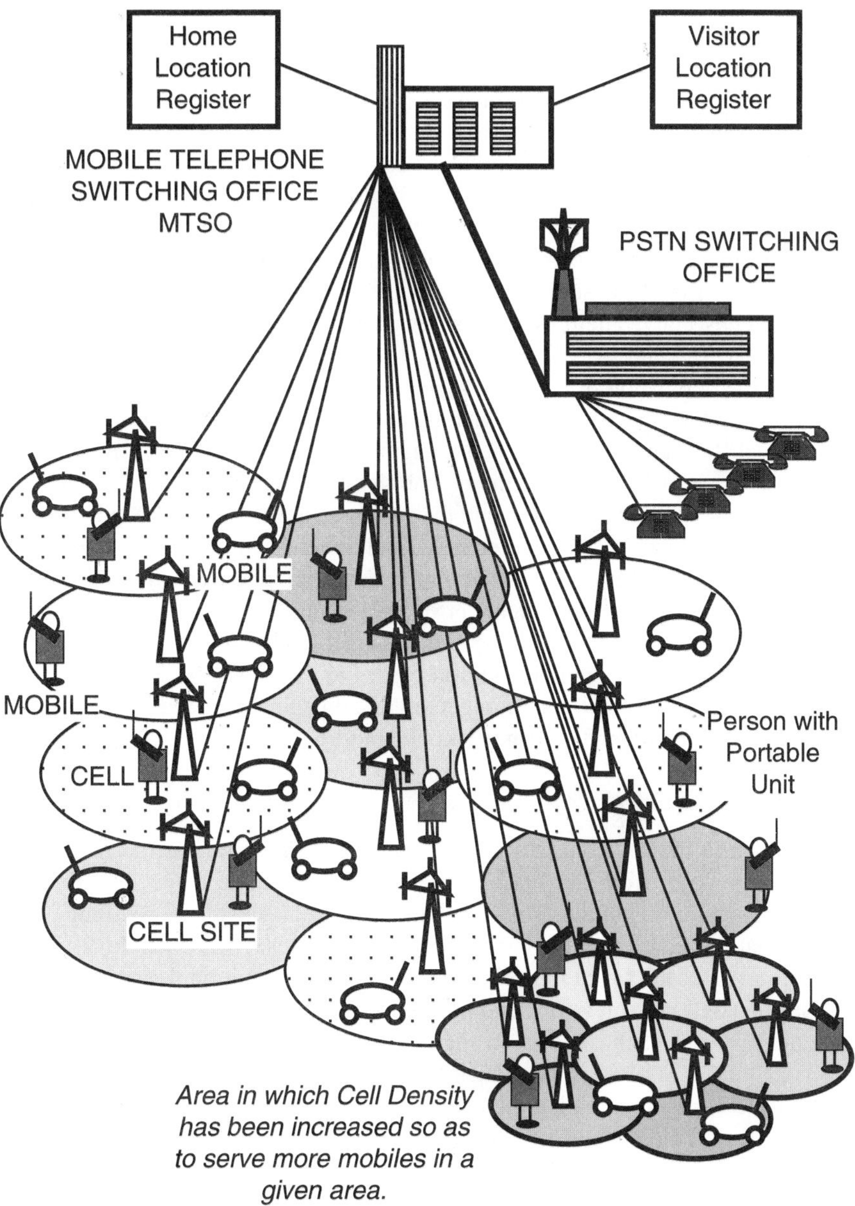

Transmission from mobiles to cell sites uses frequencies between 825 and 845 MHz. From cell sites to mobiles, the frequencies used are between 870 and 890 MHz.

Figure 14.9 Mobile Communication Using AMPS

frequencies for the call, notifies the mobile of the frequencies assigned, and prepares to receive the called number.

Using the setup channels is not as simple as described above. The mobiles are in contention for the attention of the base station. When it is processing one mobile it places a busy signal (i.e., sets every 11th bit of the paging stream to one) on the outgoing setup channel. This notifies other mobiles that they must wait to send their request for service. To prevent several waiting mobiles from sending immediately after the base station changes from busy to idle, each mobile waits a random time before attempting seizure of the base station. If the initial attempt at gaining the attention of the base station is not successful, the mobile retries automatically. If several consecutive attempts are unsuccessful, the mobile invokes a timeout period before trying again.

As the population of a mobile system grows, it becomes increasingly important to ensure the veracity of the signaling data on the setup channels, and its error-free receipt. In a large system, when a paging message is broadcast, it is received by thousands of idle mobiles. If the probability of a false, received message is one message in one million messages, and the system population is 10,000 idle mobiles, on average, one page in one hundred may result in a false response. If the probability of a false received message is one in ten thousand, one page in ten is likely to result in a false response, and higher error rates are likely to result in false responses to most every page. Over a fading radio channel, such error rates are quite likely. To prevent the system from choking on replies to false messages, the setup channels employ forward error correction coding, and paging calls are repeated several times. Before responding to a page, a mobile must repeatedly receive its number.

(4) Moving from Cell to Cell

As the mobile moves from cell to cell, the MTSO arranges for the talking path channels to switch from the site serving the cell the mobile is leaving, to the site serving the cell it is entering. This is done when the cell site that is handling the call reports that the signal strength has dropped below a preset threshold. In response to this information, the MTSO instructs the mobile to transmit on the set up channel, and instructs all cell sites surrounding the reporting cell site to monitor the setup channel and report the signal strength of the transmission. From the reports received, the MTSO determines which cell site is closest to the mobile and instructs the cell site currently handling the call, the cell site that will assume responsibility for the call, and the mobile, to prepare for hand-off. At a given instant, the connection between the MTSO and the first cell site is broken, a connection is made between the MTSO and the second cell site, and the mobile retunes to the new frequencies to complete the talking channels through the new cell site. The entire process takes place in milliseconds and the mobile voice user perceives no interruption in the call. If the mobile is transmitting data, the stream will be interrupted for at least the period between ceasing to transmit on one channel and beginning to transmit on the other.

(5) Data Transmission Over AMPS

The unreliable radio link (due to fading and/or hand-off) is a difficult connection to use for data. However, cellular modems are widely used to provide 2.4 kbits/s data service. Based on V.22bis standards, they employ error correction, as in V.42 and MNP classes 2-4 (see § 5.1.3), and data compression, and the terminals are likely to employ ARQ techniques (see § 4.2). Some carriers support an overlay data service. Recently standardized, and called *cellular digital packet data* (CDPD), it provides 19.2 kbits/s connectionless service on a 30 kHz cellular phone channel. Designed for broadcast information, electronic mail, dispatch, and similar services, packets are transmitted on idle cellular channels. When the channel is required for voice service, CDPD seizes another idle channel; if there is none, it waits for one to occur.

(6) Moving from System to System

Addressing is a major problem for mobiles. As they move from the domain of their home system to the domain of another system, roaming mobiles are assigned temporary addresses within the context of the new domain. For terminals or hosts seeking to engage them in communication, the customary addresses do not answer. To prevent loss of communication, mobiles must report temporary changes of address to their home MTSOs as soon as they occur. With this information, messages can be forwarded, or temporary address changes can be sent to the originating stations.

14.2.3 System Evolution

(1) Digital AMPS

In many urban locations, AMPS is congested and close to saturation. In addition, many subscribers would like to send higher-speed data. Consequently, an evolutionary strategy has been developed that will allow AMPS providers to serve more users and transport higher-speed data, yet use the 30 kHz channel spacing already established. Called D-AMPS (i.e., digital AMPS), and defined by the *North American digital cellular standard* (IS-54), it provides three 16.2 kbits/s digital TDMA channels on a 30 kHz cellular phone channel. Each channel contains 9.6 kbits/s coded voice, 3.4 kbits/s error control, and 3.2 kbits/s signaling. Under present frequency allocations, systems could operate 999 digital circuits over 333 radio circuits. Seen as a transitional system, carriers can gradually substitute 3-slot, TDMA channels for existing single-user analog channels so as to achieve graceful evolution from AMPS to D-AMPS. A 6-slot TDMA system is also possible. With 8.1 kbits/s assigned to each user, it is expected to employ 4.8 kbits/s vocoders.

(2) Global System for Mobile Telecommunication

At the time AMPS was installed in North America, national networks using cellular concepts were being installed in most developed countries. The result was a number of similar, but not identical systems, that cannot work together. During the 1980s, the European Community began to plan a fully-compatible, Pan European mobile telephone system. They produced the *Global System for Mobile Telecommunications* (GSM), a second generation digital, cellular, land mobile system. Phase 1, of what is now known as GSM 900, is being implemented. Some of its features include:

- **Radio channels**: GSM 900 operates in two frequency bands (890 to 915 MHz, mobile-to-base station and 935 to 960 MHz, base station-to-mobile). Individual circuits consist of 200 kHz channels separated by 45 MHz. Each channel is time-divided into time-slots (1733 slots/s) that are allocated in sequence to eight subchannels. A sequence of eight slots is known as a frame.
- **Organization of slots**: the slots carry bursts of user's data and signaling or control information as needed. The datastream employs FEC coding to alleviate some of the effects of fast fades (which produce bursts of errors). The slots are organized in two multiframe structures of 26 frames and 51 frames that can contain 13 kbits/s speech and 2.4, 4.8, and 9.6 kbits/s data. **Figure 14.10** gives an overview of the 26 frame structure. Frames 0 through 11 and 13 through 24 support eight *traffic channels* (TCHs) that carry 13 kbits/s speech, or 2.4, 4.8, or 9.6 kbits/s data. As required, slots are robbed in each TCH for *fast associated control channel* (FACCH) purposes—such as hand-off arrangements. Frame 12 contains eight *slow associated control channels* (SACCHs), one for each TCH, that perform call supervision and other functions. Frame 25 is reserved for later use.
- **Cells and cell sites**: cells range from one to several kilometers. Individual cell sites are allocated from one to fifteen circuits, i.e., 8 to 120 mobile circuits.
- **Hand-off**: the base station monitors the quality of channels in use, and will initiate a change to another circuit under its control, if warranted. Also, the mobile user may request a change in circuit if the circuit in use is perceived as poor quality. After the fact, the base station notifies the mobile switching center of these autonomous hand-offs. The MSC coordinates hand-offs that involve two base stations.
- **Network databases**: GSM employs four databases to manage mobiles and control calls:
  - *home location register* (HLR): maintains and updates subscriber's location and service profile.
  - *visiting location register* (VLR): maintains and updates visitor's location and service profile for local use. Allocates a mobile station roaming number (MSRN) to roamers and informs mobile's HLR.
  - *equipment identity register* (EIR): lists subscriber's authorized equipment. Denies service if attempts are made to use un-registered equipment.
  - *authentication center* (AC): supports encryption of radio channel signals.

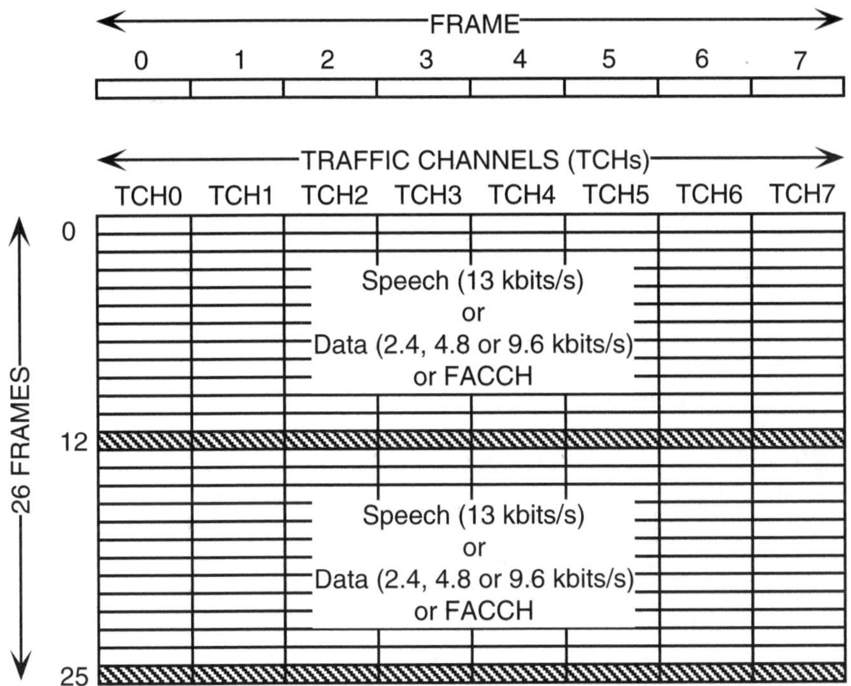

Figure 14.10 GSM Frame Structure
Global System for Mobile *Telecommunications (GSM) employs 200 kHz channels that are time-divided into 8 subchannels. Each subchannel occupies a slot in an 8 slot frame. Superframes consist of 26 and 51 frames.*

- **Interface with terrestrial network**: SS7 capabilities are employed between base stations and mobile switching center, and MSC and PSTN/ISDN.

ITU-R is developing standards for the next generation of FPLMTS. Called IMT-2000 (International Mobile Communications - 2000), it will operate in frequency bands allocated by WARC'92 (see § 14.3.2).

(3) Use of Intelligent Network Capabilities

The increased network functionality provided by IN capabilities (see § 10.6) such as *selective call forwarding*, *calling number exclusion* and *call waiting* can be used to advantage by busy mobile users. Made possible by SS7, they promise mobile subscribers the opportunity to manage mobile calls just as calls to fixed sites are managed. Also, IN facilities can be used to store the information required to operate the mobile network. Thus, **Figure 14.11** shows a way in which the home location register, the visiting location register, the equipment identity register, and the authentication center databases can be located in an intelligent network. The major problem in ensuring the availability of IN capabilities is the mobile's ability to update location records as it roams from area to area.

14.3 WIRELESS INFORMATION SYSTEMS

14.3.1 PCS, PCN, and UPT

CMRTS has stimulated demand for *wireless information systems* that operate in the high mobility environment of cars and planes, and in the low mobility environment of the workplace, at home and downtown where peripatetic users can stay in touch. The need most often expressed is for a communication system that can follow the user anywhere and provide a range of sophisticated services on demand. The user's entrance to this world of ubiquitous communications is a portable terminal that employs wireless or wired facilities transparently, and is recognized by network control no matter where it happens to be. This concept has been encapsulated in three acronyms—PCS, PCN, and UPT:

- **Personal communication services** (PCSs): a term employed by the FCC to mean a family of services provided over mobile or portable radio communications facilities for individuals and businesses on the move. These services will be provided in an integrated manner over a variety of competing networks. They are divided into three categories:
  - *narrowband PCS*: applications include two-way digital paging and data services for interactive television.
  - *broadband PCS*: applications include voice and data services that will compete with terrestrial mobile telephone systems.
  - *unlicensed PCS*: applications include cordless telephones and wireless LANs.
- **Personal communication networks** (PCNs): first used in the United Kingdom, and then adopted in Europe, a term employed to identify forms of metropolitan area, portable radio telephony networks.
- **Universal personal telecommunication** (UPT): ubiquitous telecommunication in which the users have single identifiers and are unaware of the means employed to consummate communications across town or around the world.

Because these concepts are quite new, there is no consensus that distinguishes one from the other. Indeed, they appear to overlap significantly. For our purposes we assume PCN is included in PCS, and PCS is included in UPT (i.e., PCN $\subset$ PCS $\subset$ UPT), and wireless information networks are a critical element of each.

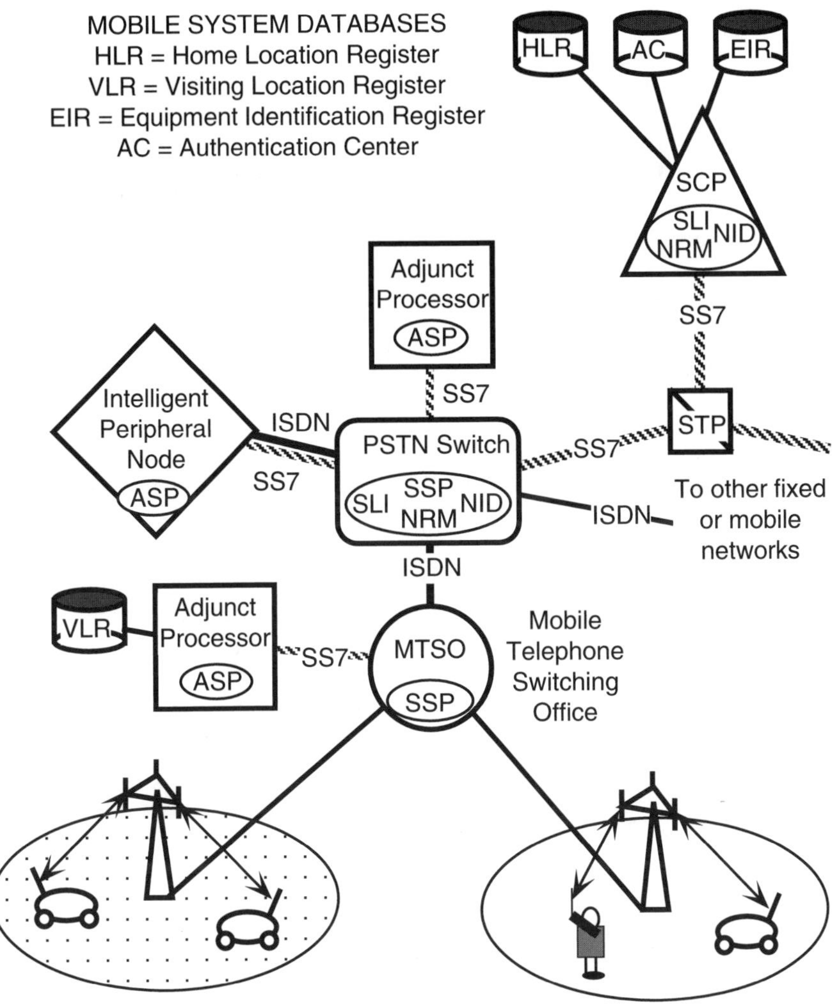

IN FUNCTIONAL RESOURCES
ASP = adjunct services point
NAP = network access point
NID = network information database
NRM = network resource manager
SLI = service logic interpreter
SSP = service switching point

Figure 14.11 Intelligent Network Capabilities Applied to Mobile Communication
Mobile users can benefit from the call management (and other) services provided by Intelligent Networks. IN facilities can be used to store the information needed to operate the mobile network.

14.3.2 Electromagnetic Spectrum

With the approval of WARC'92, ITU-R allocated 230 MHz around 2 GHz for use by wireless network services. Specifically, the allocation is in two bands—1880 to 2025 MHz and 2110 to 2200 MHz. In addition, two subbands—1980 to 2010 MHz and 2170 to 2200 MHz—were designated for mobile satellite links. National administrations and agencies are allocating frequencies in this band to wireless network applications. Within the United States, the FCC has allocated frequencies in the 1850 to 1970 MHz and 2130 to 2200 MHz bands for personal communication systems.

14.3.3 Wireless Network

(1) Network Model

Figure 14.12 shows an abstract model of a wireless network that supports UPT. It is divided into three planes:

- **User plane**: in which the network is described in terms of user services
- **Access plane**: contains those facilities and procedures that contribute to user mobility and provide access to the transport levels of the model

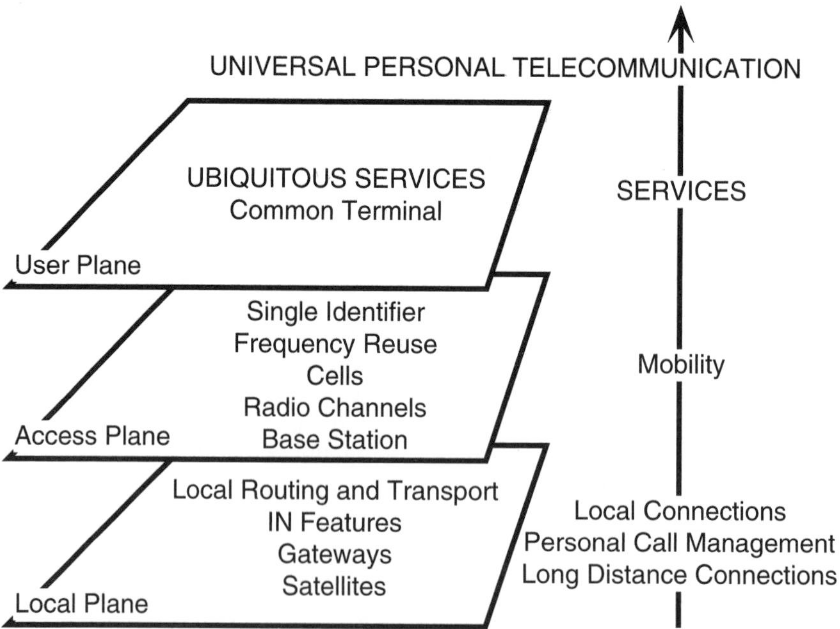

Figure 14.12 Model of Wireless Network for UPT

- **Local plane**: contains facilities that complete local connections, provide IN features for personal call management, and connect to long-distance facilities to provide a global, UPT network.

  We use this division to organize discussion of UPT networks and services.

(2) User's Perspective

In § 7.2.1(2), we introduced an empirical measure called *richness* to characterize personal messages that cater to an extensive range of human interests. It is reasonable to consider what level of richness UPT users expect; will it be the same in both directions; and will multimedia documents be included? The answers to these questions will have a profound effect on the implementation of UPT, and on the terminals that the subscribers will employ. According to ITU, the *future public land mobile telephone system* (FPLMTS) will provide global coverage for speech and low to medium bit-rate data (≤ 64 kbits/s), with high bit-rate data (≤ 2.048 Mbits/s) available in limited areas. FPLMTS will be compatible with ISDN and SS7. FPLMTS services are identified:

- **Dialog services**: anticipated to include voice, facsimile, and low-resolution videotelephone
- **Message services**: anticipated to include paging, voice mail, and text mail
- **Information services**: anticipated to include voice, music, and data.

These are not the requirements of virtual reality—but they are a significant jump from AMPS. We use them to guide discussion.

(3) UPT User's Terminal

UPT requires a truly portable, user-friendly, voice and data terminal. Better flat screens, higher density (i.e., more logic functions per chip) integrated circuits, and higher capacity, lower weight, faster recharging batteries, are but a few of the technical improvements required to implement such a terminal. To respond to an FPLMTS environment, in addition to a transmitter and a receiver, it must contain audio transducers for speech, voice mail, and music; a (color) display for data related functions and videotelephone; a videotelephone camera; a facsimile scanner and reproducer; a hard copy printer; and a keyboard or equivalent device for sending messages.

(4) UPT Service Set 1

Global and regional standards organizations are working to define a set of service features for UPT. Service Set 1 defines basic capabilities needed to support personal mobility. Some of them include:

- **UPT user identity authentication**: intended to protect the user and service provider, this feature verifies the identity of the user.

- **Incall registration**: to enable personal mobility, this feature allows the user to register for incoming calls at any terminal address.
- **Outgoing UPT call**: to enable personal mobility, this feature allows the user to make outgoing calls from any terminal address (with UPT user identity authentication)
- **Outcall follow-on**: allows the user to continue to make outgoing calls from the terminal address (without further UPT user identity authentication)
- **Incall delivery**: calls to the user's UPT number are delivered to the terminal address registered in incall registration
- **Variable default incall registration**: permits the user to establish a default routing matrix that forwards calls on the basis of caller ID, time of day, etc.
- **UPT service profile**: allows the user to change passwords, default services, etc.

14.3.4 Access Arrangements

The way in which users gain access to the network, and the way the network locates and gains access to the users, will determine how successful it is in providing UPT. Easy access depends on many things:

- **Single identifiers**: users of telecommunication systems are likely to have telephones, facsimile machines, video terminals, etc., at home and in the office. Some may have mobile phones, and some may also have pagers. Most likely, each of these devices has a separate number. If they work in two or three different places on different days of the week there are more numbers. For originating users, accessing the network is simply a matter of picking up the phone or switching on the modem. However, finding terminating users who may be in any number of locations is a different matter. As a goal, UPT requires the use of unique personal identifiers with which the network can track them down. Today, two approaches come part way to this goal:
  - *1-500...numbers*: assigned to specific persons, the callers' calls are transferred over SS7 to network control points (NCPs) that maintain listings of the numbers at which the called persons can be reached. They have been entered by the called parties who register their new numbers each time they change location. Of course, if the called persons forget to enter their new locations, they are as inaccessible as before. Safeguards, such as only authorizing the use of specific network numbers at the NCP, and the inclusion of default numbers if there is no answer, can be built in to the system.
  - *subscriber identity module* (SIM): in GSM, the user can have a subscriber identity module that is inserted at the mobile station. SIM carries the personal number assigned to the user. GSM associates it with the registered number of the mobile station and thus can route calls to the subscriber.

- **Frequency reuse**: we assume that a large number of users must be accommodated and that the frequency bands assigned for operations are limited to something much less than would be required to assign each user a unique channel. Frequency reuse, then, is an essential ingredient of UPT and, if users are to be able to move about freely, an area system is required. The frequency reuse factor depends on the potential number of active users, the way in which the radio channels are organized, and the total amount of spectrum available. Frequency reuse is provided by the terrestrial cellular systems incorporated in today's systems, and studies have shown that it could be provided by employing spot-beam antennas on geostationary, Ka-band satellites. We shall focus on the terrestrial approach.
- **Cells**: cell sizes must be matched to the traffic density, traffic speed, and the potential number of active users. If the traffic is mixed, with high-speed mobiles (cars) and low-speed mobiles (pedestrians) sharing the same space (as in an interstate highway traversing a downtown commercial area) the density of pedestrian users may be sufficient to require microcells within the macrocell that serves the speeding vehicles. The practicability of microcells for pedestrian use has spawned the notion of a telepoint:
  - *telepoint*: a single base station providing one-way wireless, pay phone access into the telephone network. It provides cordless pay phone service to customer-owned handsets within limited range of public base stations in malls and downtown areas. Cordless phones using FDMA and operating on frequencies between 864 and 868 MHz have been standardized as CT-2 in Europe, and telepoints for them are in limited use in several countries. For the future, ETSI has designated 1880 to 1900 MHz for digital European cordless telecommunications (DECT).
- **Radio channels**: at the heart of providing mobility, radio channels form a flexible link between the user and the base station located at the center of each cell. As described earlier, the channels may be based on discrete carriers that are assigned to mobiles on demand (FDMA; also, it may be described as SCPC-DAMA), or on discrete carriers that are divided up among several mobiles on the basis of time (TDMA), or on a discrete carrier that is changed into a collection of wideband, noiselike signals (CDMA). Because of its essential simplicity, FDMA was the technique chosen for early mobile systems. Because of its inefficiency (in terms of mobiles served per Hertz) it is being replaced by digital techniques. For a future UPT system, TDMA or CDMA will be employed. Two CDMA proposals merit mention. Because these approaches represent departures from the strategy of evolving analog cellular technology to digital cellular technology, their adoption depends on the support they receive from established system operators, not the excellence of their technical approaches:
  - *CDMA* (IS-95): IS-95 is a digital cellular standard endorsed by USTIA (U. S. Telecommunications Industry Association) and EIA that uses code-

division multiple access with a sophisticated power control algorithm to regulate the power of the mobile terminals so that users close to base do not overpower users far from base. Systems have been proposed that can handle up to 100 users with 16 kbits/s voice and data in a bandwidth of 1.5 MHz.

  - *digital microcellular system*: an all-digital system that switches and routes short, ATM-style packets and uses spread spectrum modulation. The base stations cooperate with one another to control power and arrange hand-off of mobile users. Cells have a small radius (approximately 300 feet) so that the power required is low and the antennas can be small. They are designed to transport data up to 1 Mbits/s.
- **Base stations**: at the interface between flexible access paths and the fixed, terrestrial network, base stations establish the topology of the wireless segment of the network. Further, the number of base stations sets a limit to the number of mobiles that can be served simultaneously in a given frequency band.

14.3.5 Connecting Arrangements

(1) Local Connections

The local network provides connections between mobiles in the same CGSA, between mobiles and local fixed facilities, and transfers mobile traffic intended for other areas to long-distance facilities. Within the local network, operations conform to PSTN/ISDN procedures and employ SS7. Through the network infrastructure, IN features can be invoked by the mobile user. The major elements associated with UPT include:

- **Mobile switching center**: the MSC is the end-office of the UPT network. Located between the radio loop and the higher levels of the network, it serves as the interface between the facilities and procedures of wireless technology and ISDN-based, SS7-directed facilities.
- **IN features**: when equipped with services switching point (SSP) capability, the MSC can invoke IN features from anywhere in the network to assist users to manage their calls. Combinations of SCP/NCP, AP/NCP, and IP/SN provide support that implements 1-500...number calling, corporate credit card service, virtual networks, voice mail, announcements, calling number identification, call-forwarding, call-blocking, and call waiting. While some of these features may not be available to users at present, they can be implemented without great difficulty if demand warrants them. Skillful use of IN capabilities should go a long way towards providing network functionality that permits UPT users to receive calls addressed to their personal numbers and manage calls according to their personal preferences.

(2) Satellite Applications

ITUs FPLMTS incorporates services delivered by terrestrial and satellite facilities. Several entrepreneurial organizations see the quest for UPT as an opportunity to build new LEO satellite networks that cover the globe and bypass existing facilities. Employing 66 satellites in polar orbits 500 miles above the earth's surface, one company proposes to provide worldwide telephone service. Using a special handset, the service will include facsimile and paging. Another proposes to employ LEO satellites to bounce signals between earth stations and portable telephones or computers. The modulation technique chosen is spread spectrum and the earth stations will serve as gateways to the PSTN.

(3) Indoor Wireless Networks

If users can be free to roam outside, and communicate with the world, why should they not be free to roam inside buildings and communicate with their colleagues, servers, hosts, and sundry other processors of information? Such things are happening. Based on the success of digital cordless phones, wireless PBXs and wireless LANs have been developed. Employing TDMA on multiple carriers or CDMA and ALOHA access techniques, they operate around 900 or 1800 MHz in portions of the spectrum set aside for unlicensed, low-power operation. Using the same cellular principles as the larger CMRTS, the base stations are distributed throughout the facility. For multibuilding facilities, traffic can be run in cables between the buildings to connect from one wireless subarea to another.

Indoor wireless networks can be implemented with omnidirectional, infrared transmitters and receivers. Because IR radiation is absorbed or reflected by solid structures, cells are limited to the confines of rooms and other closed areas. Because of its short wavelength, IR has the advantage that it is not subject to fading. However, it has the disadvantage that sunlight, and the radiation from incandescent and fluorescent lights contain IR so that there are many interfering sources. IR works best when it can be used at relatively high power over relatively short distances.

14.3.6 Future of Wireless Information Networks

Universal personal telecommunication has caught the imagination of many in the technical community. Given the necessary investment, technologists probably have the tools to finish the job of providing convenient, flexible, pervasive communication links between portable telephones or portable computers and fixed or mobile facilities anywhere in the world—and they will continue to push towards this goal. We are in a period known to business majors as *technology push*. It is the first major stage in bringing a product to market. Continuation requires demonstrations that arouse latent demand and build a constituency of potential subscribers that translates into *market pull*. Completion requires agreement on standards so that manufacturing efficiencies can be brought to bear to produce final products that offer services that people want at prices they are willing to pay. It will be several years before the answers are in—and during that period the competitive environment will change constantly as established providers seek to retain or increase market share despite entrepreneurial challenges.

15

Management and Administration

In the same way that other resources are managed to optimize their contributions to the enterprise, enterprise networks must be managed. As they become more complex and sophisticated, so the management task becomes more difficult, and the penalty for poor performance becomes greater. In this Chapter, we describe a framework for network management that has been developed by ITU. It provides a common set of concepts for managers, users and owners. Part of the management task is to ensure message integrity, and information privacy. These topics are discussed in later sections of the Chapter.

15.1 NETWORK MANAGEMENT

15.1.1 What is Network Management?

- **Network management**: the planning, organizing, monitoring, and controlling of activities and resources within a network to provide a level of service that is acceptable to the users, at a cost that is acceptable to the owner.

The activities that are managed are those associated with the use of the network to facilitate running the owner's business. The resources that are managed are the combination of persons, hardware, and software that has been approved by the owner for the support of telecommunications within the business. As in other management assignments, network management includes the selection, training, motivation, direction, and evaluation of persons. In addition, network managers must be willing to market network resources to potential users, and to assist them in solving their telecommunication problems. A manager of shared resources cannot neglect to treat users as special persons—*customers*. Except to note that the quality of the persons brought to bear is critical to the success of any operation, we will not discuss the human resources aspect of network management. Neither will we discuss the marketing of private networks. The principles involved in both of these activities are not markedly different from those that govern them in other areas of commerce. What we will address are the specialized, network-related aspects of network management; they involve tasks such as:

- **Monitoring** communications activities so as to detect and correct congestion, errors, or other problems
- **Planning and coordinating** equipment additions or changes to add new services or to improve the level of existing services, including the need for training, or the acquisition of additional persons
- **Administering** user-oriented information (station numbers, calling restrictions, security procedures, etc.)
- **Imposing** a communications architecture and maintaining accurate records describing nodes, links, and protocols
- **Operating** a network simulator, or other tools, to test opportunities for expansion or upgrade of equipment, and to examine the cost-saving potential of new tariffs or other changes in the environment
- **Analyzing** costs and performance to determine the cost-effectiveness of continuing to operate a private network.

It is possible to build a software description of the physical network that can be used to do the following tasks:

- Track the status of all network components
- Predict the effects of changes
- Provide a decision support tool that will handle the complexities of traffic analysis, alternate routing, and other network activities, automatically.

In large networks, the status of links and nodes can be reported to displays in the management center; in smaller networks, real-time reporting may be too expensive, and performance records must be entered off-line.

15.1.2 Network Management System

(1) Open Systems

In the 1980s, continuing growth in the applications of telecommunications-based information systems, the need to communicate with other companies, and the prevalence of takeovers and mergers that forced dissimilar networks together, focused attention on the need to implement a truly open system architecture—an architecture that would permit cooperating equipment to work together, and to be managed as an integrated entity. The reader will recognize this statement as containing the objective of the open systems interconnection reference model (OSI model) described earlier in § 5.3.1. With the support of many organizations and governments in developed countries, the International Standards Organization has extended the concept of OSI to include integrated management of heterogeneous networks. By defining the environment in an abstract way and concentrating on the interfaces between equipment (hardware and software), the ISO standards have extended and expanded the OSI model to cope with the problems of network management.

(2) Managed Objects

Using the concepts of object-oriented programming, ISO defines a number of building blocks to describe the management process. Chief among them is the idea of a managed object:

- **Managed object**: a network entity that is included in the management activity. A managed object may be a switch, a PBX, a router, a multiplexer, a workstation, or other hardware device, or a buffer management routine, a routing algorithm, an access scheme, or other software package.

The *management process* is concerned with the state of these objects as they exist in the network being managed. A managed object is described by four parameters:

- **Attributes**: they describe the current state and condition of operation of a managed object. Attributes consist of a type and one or more value(s). Thus, *status* is an integer type attribute that may take one of the values *disabled, enabled, active*, or *busy*.
- **Operations**: actions that are performed on the managed object, such as:
  - *create* a managed object
  - *delete* a managed object
  - *perform an action* on a managed object.

    Actions are also performed on specific attributes of the managed object
  - *get* value of an attribute
  - *add* value to an attribute
  - *remove* value from an attribute
  - *replace* value of an attribute
  - *set* value of an attribute.
- **Notifications**: reports provided by the managed object about events that occur within its domain.

- **Behavior**: actions exhibited by the managed object in response to operations performed on it, or to internal stimuli (e.g., notifications). Also, includes lack of response due to constraints placed on the behavior of the managed object (e.g., report on conditions only when critical).

  Managed objects are classified by how they fit into the OSI model:
- **(N)-layer managed object**: object specifically contained within an individual layer (e.g., data link protocol)
- **System managed object:** object pertaining to more than a single layer (e.g., router or gateway)

  and can be identified as a class:
- **Managed object class**: objects with similar characteristics (i.e., attributes, operations and notifications).

(3) Other Building Blocks

Per se, network management is performed with the aid of a network management program—an application layer process. In the ISO environment, it is divided into two parts:

- **Managing process**: that part of the application process that is concerned with overall management tasks
- **Agent process**: that part of the application process that is concerned with performing management functions on managed objects at the request of the managing process.

In addition, agent processes and managing processes share a common database through which they communicate with one another:

- **Management information base**: contains both long-term and short-term information concerning managed objects.

The top part of **Figure 15.1** shows the organization of these entities in a network management system. The network consists of assorted hardware and software entities that are designated managed objects. Managing agents communicate with them by means of operations, and they respond by changing behavior and/or providing notifications. The managing process attends to assimilating and analyzing information, initiating state changes as required, maintaining logs and records, and providing information in a comprehensible form to persons in the control center. The management information base serves as the record repository for the network. It contains descriptions of all network elements and their status over the last period. Using this concept, individual equipment manufacturers can produce products that will respond to operations expressed in standard form with notifications that can be understood by the managing process.

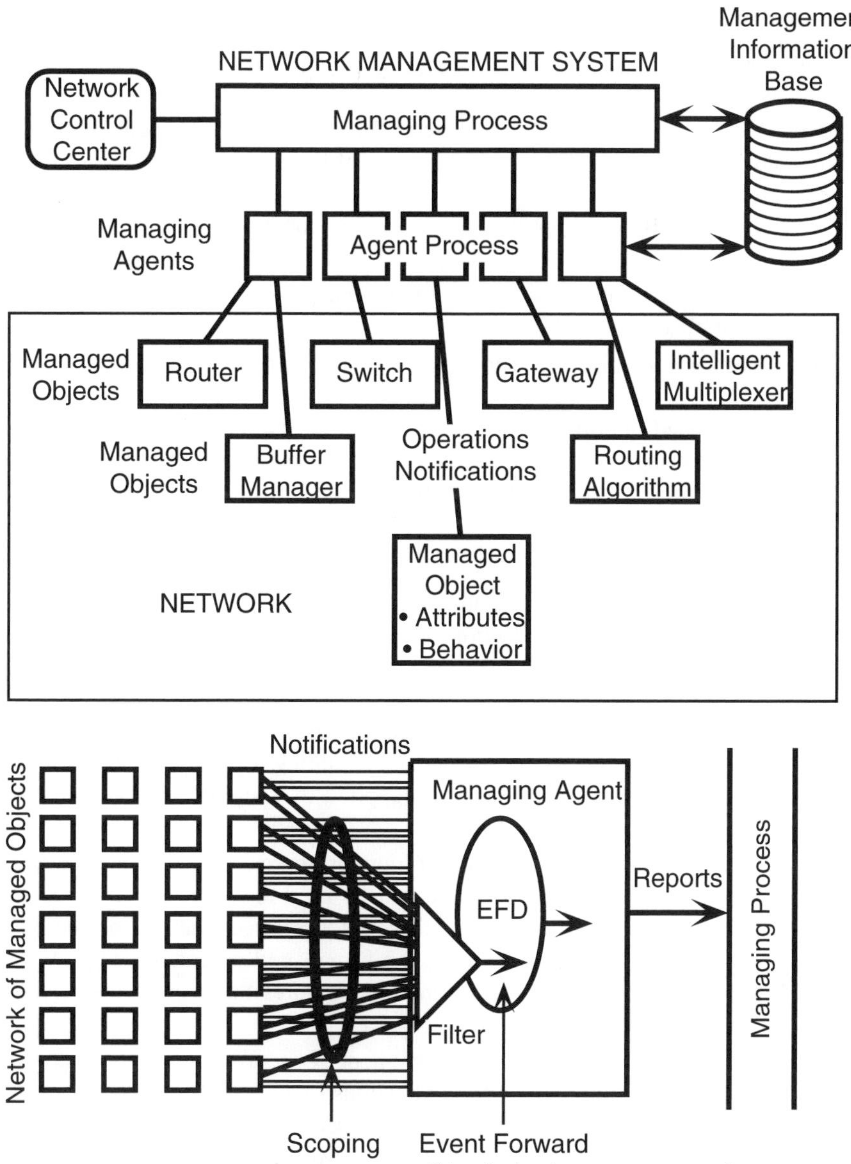

Scoping, filtering and discriminating are software functions.

Figure 15.1 Network Management System
Network management is divided into managing process (concerned with overall management), agent process (concerned with performing functions on managed objects at the request of the managing process) and management information base (contains short-term and long-term information about managed objects.

(4) Managing Agents

An important function of the agent process is to delete trivial notifications so that the managing process can focus on important events such as faults, congestion, and the like. This is achieved through the use of filters, scoping, and event forward discriminators. They are shown in the bottom part of Figure 15.1:

- **Filter**: uses Boolean operators to chain test conditions that determine which events in the network are to be reported
- **Scoping**: identifies the sub-tree of managed objects to which a particular filter is applied
- **Event forward discriminator**: directs reports to the appropriate destination.

Implemented in software at various points in the network equipment, these tools allow the managing process to manage by exception, and the network manager to investigate the utilization of any portion of the network.

(5) Architecture of ISO Network Management System

The system management application entity resides in the application layer of the OSI model. It consists of a collection of cooperating application service elements (ASEs). **Figure 15.2** shows the arrangement:

- **SMASE** (system management application service element): creates and uses the protocol data units that are exchanged during the management process. It employs communications services supplied by specific applications service elements (ASEs) or by CMISE.
- **CMISE** (common management information service element): provides management notification services that can be used to report any event about a managed object, and management operation services that define the operations required to perform actions on a managed object. Also, provides a framework for common management procedures that can be invoked from remote locations.
- **ROSE** (remote operations service element): general software support system employed in a number of OSI applications (e.g., X.400—Message Handling System, and FTAM—File Transfer and Access Method) that supports distributed, interactive applications processing between remote objects. In synchronous mode, ROSE requires the receiver to reply before the originator can invoke another operation. In asynchronous mode, ROSE permits the originator to continue invoking operations without waiting for a response.
- **ACSE** (association control service element): general software support system that works between application processes that are independent of application-specific needs. Employed in a number of OSI applications (e.g., X.400—Message Handling System, and FTAM—File Transfer and Access Method) it

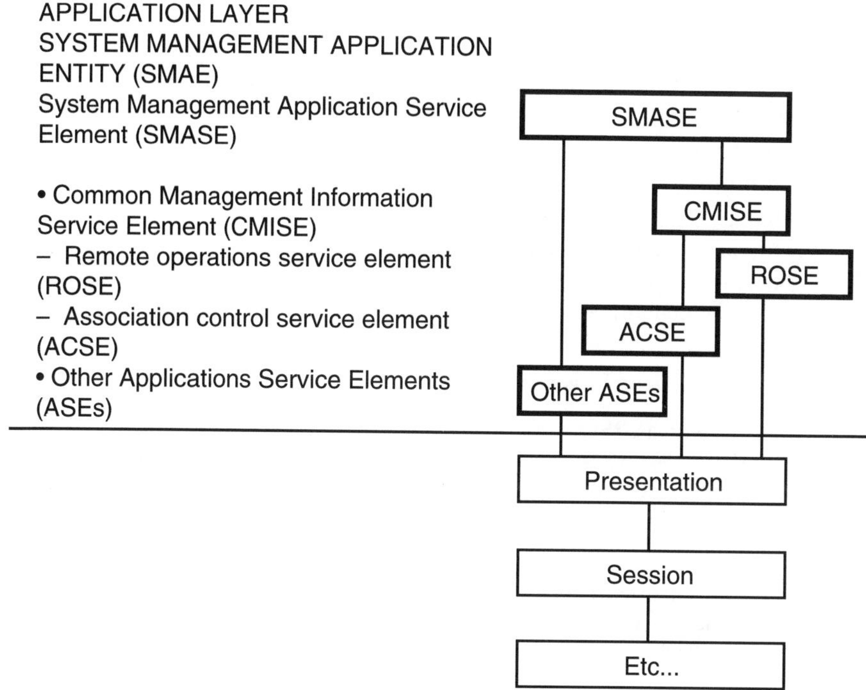

Figure 15.2 Network Management System Architecture
The system management application entity resides in the application layer. It creates and uses the PDUs that are exchanged during the management process.

supports the establishment, maintenance and termination of a cooperative relationship between application entities.

15.1.3 Management Activities

(1) Areas of Activity

ISO has defined five areas of specialized, technical responsibility; known as specific management function areas (SMFAs):

- **Configuration management**: maintaining, adding and updating records of hardware and software components, defining their functions and usability, initializing the network, and ensuring that the network is described correctly and completely at all times. Also, assign, maintain, and update records of users, including billing responsibilities and communication privileges. A critical feature of configuration management is the ability to cope with, and control, change.

- **Fault management**: provision, monitoring, and response to facilities that detect and isolate abnormal conditions in the network, so as to restore the network to normal functioning and correct the aberrant condition(s).
- **Security management**: control access to, and use of, the network according to rules established by the network owner.
- **Accounting management:** identify costs, collect and process billing information, prepare budgets, etc.
- **Performance management**: evaluate the performance of components, and the network as a whole, with regard to availability, response time, accuracy, throughput, and utilization.

(2) System Management Functions

For each of these areas, ISO working groups have identified a specific set of functions, and the procedures and services needed to implement them. Known as system management functions (SMFs), there is overlap between them, and many SMFs serve more than one SMFA. Briefly, some of them are listed:

- **Object management function**: describes the generation of reports on object creation, deletion, name changes, and changes in attribute values
- **State management function**: describes the generation of reports on the current state, and changes in state, of managed objects
- **Relationship management function**: defines relationships between managed objects
- **Alarm reporting management function**: facilitates reporting of problems by managed objects
- **Event reporting management function**: describes the selection of events that are to be reported to the managing process in a given period
- **Confidence and diagnostic testing**: supports means to determine if a managed object is able to perform its designated functions.
- **Log control function**: defines rules for construction of network operations management logs
- **Security alarm reporting function**: defines the rules for the generation of security alarms
- **Security audit trail function**: defines rules for construction of network audit trail management logs
- **Objects and attributes for access control**: defines levels of access controls
- **Accounting meter function**: defines rules for constructing usage and accounting data for managed objects
- **Workload monitoring function**: provides a model for monitoring the performance and measuring the workload of a managed object.

(3) Performance Management

On a day-to-day basis, the most important activity to be managed is measurement, analysis, and tuning of network performance. Constant monitoring should provide insight into the problems and faults that develop, and aid in selecting a strategy for recovery. Network performance information falls into two categories:

- **Service-oriented**: items such as availability, response time, and accuracy—parameters that are important to the users
- **Efficiency-oriented**: items such as throughput and utilization—parameters that are important to the network owner.

It is the network managers job to balance one with the other so as to maintain good relations with the users and the owner. Among other things, network performance is measured by the following:

- **Throughput**: effective rate of transfer of messages across the network.
- **Response time**: time between generating and satisfying a request. Associated with this measure are those times that have to do with setting up, using, and taking down a circuit, namely, connection establishment delay, transit delay, and connection release delay.
- **Waiting time**: time spent in queues waiting for service, time spent in processing and forwarding messages, and time between arrival events (at a queuing point, for instance).
- **Workload**: number of managed objects idle, number of managed objects active, number of managed objects busy, and number of managed objects rejecting service requests because of overload.
- **Statistical analysis**: analysis of event records to determine quantities such as utilization, availability, error rate, mean-time to failure, mean-time to repair, etc.

Network performance management requires continuous data gathering and analysis, the ability to make judicious allocations of resources to overcome potential problems, and the opportunity to hear from customers about their problems and concerns.

15.1.4 Fault Management

Once a fault occurs, its effect must be contained, and network performance restored as soon as possible. Fault management includes the ability to detect faults, to perform diagnostic tests on a managed object to define the fault, and to perform operations on the managed object to correct the fault, or to take the object out of service for physical maintenance. In addition, fault management includes the creation of a fault log that can be analyzed by the managing process and used as the basis for preventative maintenance activities. Among others, fault management makes use of the following SMFs: event reporting, confidence and diagnostic testing, log control, and alarm reporting.

15.1.5 Configuration Management

In today's dynamic network environment, resources can be enabled and disabled on command, and redistributed as required to follow changing demands for services. Under these circumstances, configuration management must report on hardware and software changes in a timely way and maintain an up-to-the-minute picture of the network. By operating on the management information base, configuration management programs maintain naming conventions, identify new managed objects, set initial values for attributes, supervise relationships between (among) managed objects, change operational characteristics of managed objects, report on changes in the state of managed objects, and delete managed objects. In describing the states of the objects in the network, configuration management distinguishes between administrative states and operational states. Administrative states employ the following descriptors:

- **Unlocked**: managed object can be used
- **Locked**: managed object cannot be used
- **Shutting down**: the managed object will honor current users, but will not accept new users.

Operational states employ the following descriptors:

- **Enabled**: managed object is operable, available, and not in use
- **Disabled**: managed object is not available
- **Active**: managed object is available for use and can accept new users
- **Busy**: resource is available but cannot accept new users.

Combinations of these states are permitted, and used. **Table 15.1** shows their meanings.

15.1.6 Security Management

Pervasive telecommunication facilities connected into networks are creating opportunities for ill-disposed persons to compromise transactions. In addition, there is always the possibility of a genuine mistake leading to an unexpected outcome. Security management is directed towards the detection of these occurrences. We discuss the subject in § 15.2.

Table 15.1 Meanings of Combinations of Administrative and Operational States

Configuration management must report on the states of managed objects in a dynamic environment. The Table shows the meaning of the terms used to describe possible states.

ADMINISTRATIVE STATE	OPERATIONAL STATE			
	Disabled	Enabled	Active	Busy
Unlocked	• *MO in service but disabled by some problem. State may change to locked.*	• *MO in service and can be used when required.*	• *MO in service, in use and has more capacity.*	• *MO in service, in use but has no more capacity.*
Shutting Down	• *MO with problem is being taken out of service.*	• *working MO is being taken out of service.*	• *MO in service, in use and has more capacity, is being taken out of service.*	• *MO handling traffic but can handle no more, is being taken out of service.*
Locked	• *MO out-of-service; disabled by some problem.*	• *MO out-of-service, can be used; probably in standby.*	*Not possible*	*Not possible*

MO = managed object

15.1.7 Accounting Management

Accounting management refers to the collection of data that can be used to charge for communications capabilities provided to the users on the basis that, in aggregate, it should cover the cost of doing business.

(1) Communications Charges

In a public network, a call-record is produced at the time of placing a call. It is used to charge for the service on the basis of the following:

- Time consumed and distance transmitted
- Bandwidth employed
- Number of PDUs (or octets, characters, or bits) transmitted
- Any combination of these and other parameters.

The carrier has adjusted the charge to cover the cost of capital equipment, running expenses, taxes, interest, and profit for the corporation. Because it relies on internal bookkeeping, in a private network, the cost of communications is what the accounting department says it is, or what the general manager stipulates. In truth, the network manager has only two controllable accounts—salary and outside expenses. Nevertheless, many organizations make a good-faith effort to determine the true cost of their communications.

(2) Annual Charge

The annual cost associated with purchasing, owning, maintaining, and using up plant is known as the *annual charge*. For items of plant that are owned by the private network operator, summing annual charges gives the total annual charge for them. Because they have been purchased on behalf of the network operator, these items must be managed throughout their economic and accounting lifetimes. They cannot be discarded without considering the point in their depreciation cycle and their salvage value. In the case of equipment that is retired early, a displacement charge (consisting of first cost, minus accumulated depreciation, minus salvage value) is incurred. Prudent management suggests that plant that is owned should be generally useful to the corporation for at least the depreciation period.

(3) Lease Cost

For items of plant that are leased by the private network operator, the annual cost includes the annual charges incurred by the lessor together with an additional charge for the lessor's profit. Within the terms agreed to in the lease, these items can be replaced at the initiative of the network operator. Often, early termination results in a penalty charge to compensate the lessor for loss of revenue that was intended to cover the initial investment. Prudent management suggests that plant that is leased should be generally useful to the corporation that leases it for at least the initial lease period.

(4) Rental Cost

For services that are used on demand over other companies' facilities, the network operator will be charged a fee by the service provider. For the use of public networks, the fees may be based on time and distance, and may be subject to the control of state and federal agencies. They will have been set to cover all the relevant costs of the provider. Available on-demand, the services are paid for as used. In addition, it is likely that there will be a basic monthly charge that represents a fee for connection. It is paid so long as the user wants to have the services available for immediate use. Prudent management suggests that services should be rented until demand is great enough to lease or buy plant that will provide them, or so long as rental costs are less than the cost of providing the service in other ways. It is the latter situation that is exploited by providers of virtual networks.

(5) Options

For private networks, some general options for allocating telecommunications costs exist:

- Usage records may provide the basis for charging the cost of the facilities to individual users or departments.
- The cost of communications may be included in general overhead and distributed to company departments on the basis of payroll, sales, or some other operational quantity.
- Communications may be provided by a separate cost center, with payments for services negotiated between the network manager and individual using departments.

All three arrangements (and several others) have been implemented in firms in the United States—and none is perfect.

(6) Equipment that is Owned

For general-use, equipment (PBXs, telephones, etc.) that is owned by the corporation and is generally useful for at least the depreciation period, a strong case can be made to treat the expense as a cost of doing business that is collected at the corporate level and allocated to each operating unit. Doing so simplifies accounting for each piece of equipment, and lends credibility to a corporate telecommunications unit. However, the procedure tends to mask inefficiencies in the telecommunications operation, and operating units can become frustrated by what they perceive as a lack of accountability of the telecommunications unit, particularly during periods of fiscal pressure.

(7) Equipment that is Leased

For general-use equipment that is leased, the cost can be bundled with the cost of other plant and allocated to each business unit. Leased equipment that is used by one, or more, units for a specific purpose, can be charged directly to the using units. In this way, there is the possibility of a critical review of its contribution to the business—at least every budget cycle.

(8) Services that are Rented

In providing these services, the supplier is likely to offer individual billing details that can be used to allocate exact costs to each user. Through their use, individual responsibility for the costs incurred is established, and there is no doubt who must pay. Extending the principle to virtual networks supplied by public carriers, it is possible to receive billing details that permit the allocation of charges to relatively small organizational units and have the charges match the use made of the facilities. However, because these records are an invitation to build an internal bureaucracy to check them, there is a point at which such detail becomes non-productive.

(9) Effect of the Number of Switching or Multiplexing Locations

In a network that serves several locations, switches, and/or multiplexers will be employed to minimize the total network cost. By connecting the backbone circuits to local networks, they make local connections available to anyone on the network. By concentrating traffic on tie lines between the switches, low-cost, long-distance connections make it possible for remote users to employ local area connections at a cost lower than the cost of individual WATS facilities. Generally, as the number of nodes increases the following effects occur:

- The investment in switches and multiplexers increases
- The cost of access lines decreases as the average connection to the serving switch or multiplexer gets shorter
- The cost per hour of calling may rise as the average loading on each link diminishes
- The inter-node trunk mileage (backbone network) costs increase.

The result is a complex tradeoff between switch and multiplexer investments and transmission expenses. Minimization of the total cost can lead to several, quasi-optimum, network topologies. The general nature of these relationships is shown in **Figure 15.3**. To discover what to do in any practical situation a computer-based study is required.

15.2 MESSAGE SECURITY

Telecommunications technology has made it possible to generate information anywhere, and to use it everywhere, without delay. We might add that this applies to unauthorized as well as authorized persons. Unless we take precautions, connecting an information system to a communication network makes it possible for anyone on the network to access the system—whether they are the intended users, or not, and whether they are connected with the full knowledge of the network managers, or surreptitiously connect themselves to the network for the express purpose of obtaining information to which they are not entitled.

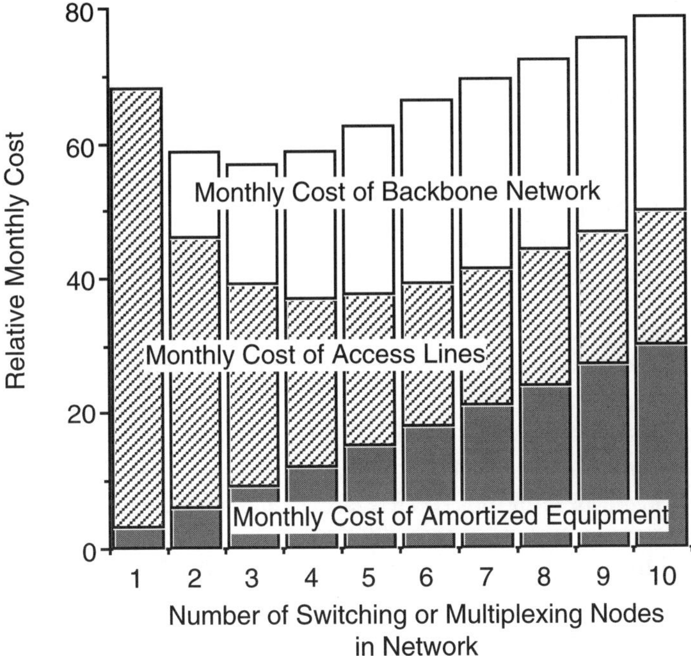

Figure 15.3 Major Components of Network Cost

15.2.1 Breaches of Security

(1) Objectives of Security Procedures

As they flow through the system, messages may be altered or destroyed by accidental events and deliberate acts. At the very least, procedures must be built in to the overall communication operations to detect these happenings. The ultimate objective of these procedures should be to guarantee that the system delivers authentic messages only to the persons authorized by the owners to receive them—no more, and no less—so that messages are not:

- **Stolen**:
  - by competitors, to gain an advantage
  - by blackmailers, to extort money or other valuables
  - by terrorists to disrupt the business
- **Distorted**:
  - by changing values or quantities to misrepresent facts
  - by changing addresses to steal money or goods from the organization
- **Damaged** by deleting information fields in order to disrupt an activity or organization

- **Disclosed** (leaked) to the public to cause embarrassment, or to other persons to create suspicion or undermine plans.

(2) Authentic, Corrupt, and Illicit PDUs

Message integrity can be destroyed by accidental events and deliberate acts. Because messages may be made up of several frames, we shall use the precise term, protocol data unit (PDU)—the entire sequence of bits transmitted at one time—to discuss the results of these happenings. In addition, we will define three states for the PDU:

- **Authentic protocol data unit** (APDU): the true PDU that the source wishes to send to one, or more, receivers for a purpose approved by the network owner
- **Corrupted protocol data unit** (CPDU): an authentic PDU that has been altered by accidental means
- **Illicit protocol data unit** (IPDU): a PDU introduced into the system for an illicit purpose. It may be an APDU that has been altered deliberately, an APDU that is received by an illicit source, or a PDU generated by an illicit source.

(3) Authentic and Illicit Sources and Receivers

Also, we shall distinguish between:

- **Authentic sources** (ASs): units that are connected to the network with the knowledge and approval of the network owner:
  - normally, they send authentic PDUs (APDUs)
  - sometimes, they send corrupted PDUs (CPDUs)
  - sometimes, they send illicit PDUs (IPDUs).
- **Authentic receivers** (ARs): units that are connected to the network with the knowledge and approval of the network owner:
  - normally, they receive authentic PDUs
  - sometimes, they receive corrupted PDUs
  - sometimes, they receive illicit PDUs.
- **Illicit sources** (ISs): units that are connected to the network without the knowledge and approval of the network owner:
  - always, they send illicit PDUs.
- **Illicit receivers** (IRs): units that are connected to the network without the knowledge and approval of the network owner:
  - they receive authentic PDUs, corrupted PDUs, and illicit PDUs.

  These definitions and concepts are diagrammed in **Figure 15.4**.

APDU = authentic PDU, i.e., the true PDU that the source wishes to send to one, or more, receivers for a purpose approved by the network owner
CPDU = corrupted PDU, i.e., an authentic PDU that has been altered accidentally.
IPDU = illicit PDU, i.e., a PDU introduced into the system for an illicit purpose.

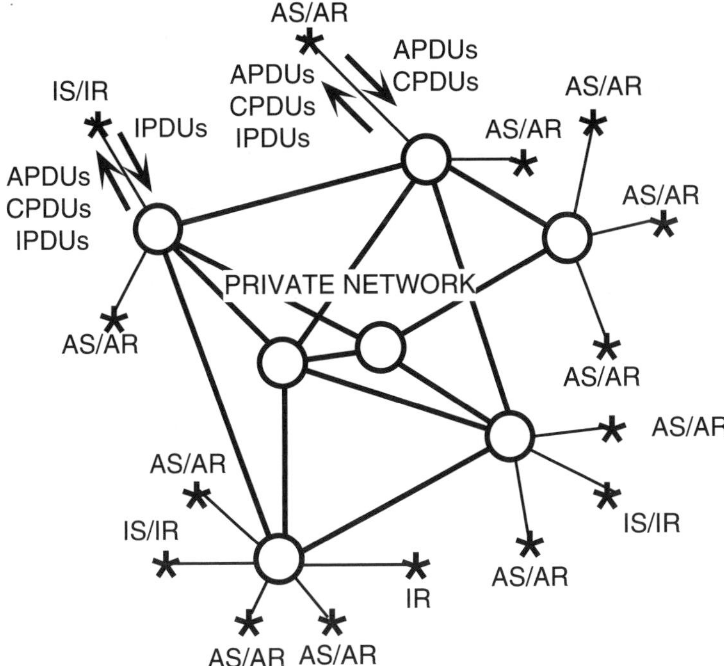

AS = authentic source, i.e., source connected to network with approval of network owner
AR = authentic receiver, i.e., receiver connected to network with approval of network owner
IS = illicit source, i.e., source connected to network without approval of network owner
IR = illicit receiver, i.e., receiver connected to network without approval of network owner.

Figure 15.4 Concepts Associated With Message Security

(4) Events that Affect the Proper Functioning of Communication Systems

In a communication system, PDUs may be subjected to one, or more, of the following events:

- **Loss**: the APDU never reaches the AR due to:
  - *misrouting*. An authentic PDU is routed improperly. This may be accidental (due to a one-time fault at a switch or router, for instance) or deliberate (due to unauthorized alteration of the routing table)
  - *wrong address*. The APDU is corrupted by an address that is not the address of the intended AR. This may be accidental (due to a mistake at the source, or to an error picked up in transmission that modifies the authentic address) or deliberate (due to the unauthorized alteration of the authorized address at the source)
- **Denial of service**: PDU delivery may be prevented by the failure of equipment as a component comes to the end of its life (a natural occurrence in electronic devices), or by an accident, natural catastrophe, or sabotage, that affects the entire system.
- **Alteration**: the APDU is corrupted so as to change the message from that which the sender intended. Alteration may be:
  - *accidental*. When it is due to a mistake made when entering the message into the system, or it is due to errors picked up in transmission that modify the text of the PDU
  - *deliberate*. When the authentic message is changed without authority prior to sending
- **Insertion**: an illicit PDU is sent from an authentic source, or introduced by an illicit source, so as to provide information (may be true or false) to a receiver.
- **Duplication**: copies of an authentic PDU may be sent to receivers other than the receiver(s) intended by the sender. This may be accidental (due to improper addressing, perhaps) or deliberate (desire to leak information, perhaps).
- **Disclosure**: an authentic PDU is delivered to an unintended authentic receiver, or an illicit receiver.

In each case, the events may be readily apparent, or may go unnoticed. Prudent management requires that measures be introduced to detect and confound them.

15.2.2 Measures that Detect Corrupt or Illicit PDUs

At all levels of the communication process, measures can be introduced to detect corrupt or illicit messages:

- **Sequence number checking**: message/PDU sequence numbers can be inserted in the text portion of the PDU. Their presence makes it possible for the human receiver to discover that PDUs are missing, or that PDUs have been inserted.

- **Identification of sender**: information can be included in the APDU to authenticate the identity of the sender and ensure that only the intended receiver can decipher the message. This is one of the applications of an asymmetric cryptosystem (see § 4.4).
- **Identification of receiver**: the use of an error detecting device (BCC or CRC/FCS, for instance) on the PDU address will provide a measure of confidence that, if the address is corrupted in transmission, any receiver will destroy it. This may prevent accidental disclosure.
- **Positive acknowledgment**: the receiver can acknowledge each PDU by message/PDU sequence number, confirming that the message reached the intended receiver. Such information is distinct from any acknowledgment scheme employed by the protocol, and is included in the text portion of the PDU.
- **Transaction journal**: by maintaining a copy of the header, message sequence number, and trailer of each PDU, and noting every communication event (such as transmit, arrive/depart intermediate nodes, and receipt) encountered by the PDU, an audit trail can be created for the communication process that will facilitate the detection and investigation of unauthorized events.
- **Time and date stamp**: the audit trail can be enhanced by the addition of time and date information showing when the PDU was generated, when it passed through each node in the network, and when it was received. If such information is added to a PDU during the communication process, it can be passed back to the originating point when the receiver acknowledges receipt of the PDU. As an alternative to collecting all of the information at the originating point, each node can maintain a log of its own activities (including message sequence number) that can be consulted by network control as necessary.
- **Periodic message reconciliation**: periodic, automatic testing of the audit trail should reveal loss, misrouting, disclosure, insertion, or duplication, of a PDU, and provide the basis for an investigation of unusual events.
- **Redundant information**: names, numbers, dollar amounts, etc., can be repeated within the text of the PDU, or within the text of a following PDU; examination at the receiver should reveal whether these key items have been transmitted correctly.

15.2.3 Physical Controls

(1) Access Controls

To a large degree, the security of data in a communications and processing environment depends on the ways in which access to the environment is authorized and controlled. We can distinguish three forms of access:

- **Physical access**: access to the installation, system components, input-output terminals, communication lines, switching points, and other physical entities that comprise the environment.
- **Operational access**: access to processing, storage, and input-output resources through the use of terminals, personal computers, or other devices that are connected over communications facilities to the processing entities in the environment.
- **Unauthorized monitoring**: access to the data transmission path for the purpose of intercepting the messages thereon.

The security of a communications and data processing environment depends upon preventing these types of accesses.

(2) Controlling Physical Access to Major Facilities

For facilities that contain major communications and processing equipment, the most straightforward technique is to control all entrances to the facility. This can be done through the use of the following:

- Guards who permit access based on a badge or other identification
- Manual or electronic keys that are carried by those authorized to gain access
- Magnetic or other electronic card systems, with or without a personal identifier
- Biometric parameters (voiceprints, fingerprints, etc.) that identify the person seeking entry.

These techniques may be employed singly, or in combination. A basic assumption is that possession of an appropriate access method identifies an authorized person who will use the facilities for the purpose access is authorized. None of them can prevent a person who is authorized access from stealing or mis-using information. The integrity of controls that limit physical access is based on the understanding that those who have access share common objectives. Because circumstances change over time, it is important to review frequently the list of those granted access, and to investigate changes in patterns of access.

(3) Controlling Physical Access to Remote Facilities

For remote facilities, such as cable vaults and switch rooms that are not normally occupied, or storage facilities to which persons need access occasionally the following security options can be considered:

- Alarm systems and intrusion detection devices can be installed
- Video surveillance can be maintained
- Access can be restricted to authorized workers identified by keys, cards, or biometric parameters.

However, if some of the transmission and switching facilities in the communications and data processing environment are part of someone else's network, these techniques cannot be implemented easily. While the messages are outside of the jurisdiction of the owner, they are susceptible to intrusion. Under these circumstances, private messages can be protected by encryption.

(4) Controlling Operational Access

Operational access does not require physical access. The moment a communication connection is made to a data processing center, it is vulnerable to electronic intrusion from around the world. Using a personal computer and a dial-up connection, computer *hackers* have demonstrated that unauthorized operational access can be gained to any system. Preventing operational intrusion relies on the use of passwords and procedures designed to make it practically impossible to penetrate the system:

- **Passwords** are used to authorize communication, access to resources, copy or alter file entries, etc.
- **Procedures** are used to ensure that certain types of data are restricted to certain locations, terminals may be designated *read-only* for certain classes of data, terminals may be restricted to working with one file at a time, etc.

A common technique to make dial-up penetration more difficult employs a call-back routine. On receiving a dial-up request for connection to the data center, the communications processor hangs up and then dials the number listed for the person whose password(s) accompanied the dial-in request. For the connection request to be honored, the requester must be at the authorized telephone location. If this is in a protected facility, the likelihood of unauthorized use is diminished.

(5) Protection from Monitoring

Transmission facilities that employ radio can be easy targets for eavesdroppers. Protection against monitoring relies upon making the cost uneconomical compared to other ways of obtaining the same information (compromising authorized personnel, for instance). Intrusion can be made more difficult (and more expensive) by the following methods:

- Employing multiplexed facilities in which each call uses a different channel (so that the eavesdropper must search a number of channels for the information)
- Using the highest speed data circuits available (so that messages are short, and therefore more difficult to intercept)
- Employing alternative transmission paths (so that eavesdroppers must monitor several routes)

- Using packet network technology (so that the message is segmented into packets that are mixed with other traffic)
- Encrypting the data.

15.2.4 The Security Challenge

If the rewards are great enough, those who practice illicit activities will circumvent any procedures that may be set up to thwart them. What one determined person constructs, a more determined person can overcome. Maintaining the integrity of a telecommunications-based information system that transports private information is a task that requires eternal vigilance and inquisitiveness.

15.3 INFORMATION PRIVACY

15.3.1 Individual and Company Privacy

(1) Personal Information

Records pertaining to any one of us can be collected, processed, and manipulated in telecommunications-based information systems to satisfy enquiries that may have to do with our performance as an employee, as a consumer, as a debtor, or as a vital statistic. They are all roles that most of us will play in our lifetime. Because they involve considerable sorting, review and validation, without telecommunications-based information systems, it is unlikely that such collections would exist. Doing anything out-of-the-ordinary is likely to lead to bureaucratic enquiries to some of these databases.

(2) Individual Privacy

Because these records may be concentrated in a few data processors that are connected to the public telephone network, records on an individual can be collected relatively easily by those authorized to do so, and with somewhat more difficulty by those without authorization to do so. From such a collection, trained persons can construct an accurate and detailed profile of an individual, or a family. Developing and using such knowledge is an invasion of an individual's privacy:

- **Invasion of individual privacy**: theft, interception, misuse, or aggregation of information concerning an individual's thoughts, sentiments, emotions, lifestyle, or status.

Depending on the individuals involved, some, or all, of this information can be obtained from computer files through the use of social security numbers, credit card numbers, telephone numbers, and zip codes. Many countries have laws to protect the individual against the disclosure or misuse of information stored in such databanks.

(3) Company Privacy

Company records are stored in interconnected systems that are accessed from many units every working day. Thus, it may be easier to collect information on company operations and conditions than on private individuals. Developing and using such knowledge is an invasion of company privacy:

- **Invasion of company privacy**: theft, interception, misuse, or aggregation of information concerning the operation of the business, particularly that information which may give a competitive advantage to the receiver. In many companies, this information is afforded the status of company proprietary.

(4) Protecting Private Information

The principal protective measures are:

- **Encryption** to make the meaning obscure to unauthorized persons
- **Access controls** to limit the number of persons who can use, modify, or examine data
- **Psychological deterrents** produced by laws that impose fines or imprisonment for improper use of personal information, and by company policies that proscribe dismissal for unauthorized use of proprietary information.

Of these measures, we shall address encryption and access controls. A discussion of psychological deterrents is beyond the scope of this book.

15.3.2 Encryption

(1) Use of Encryption

In § 4.4.1, we noted that encryption is the action of disguising information so that it can be recovered relatively easily by persons who have the key, but is highly resistant to recovery by persons who do not have the key. The encryption key may be changed from time to time to make an intruder's task more difficult. Restoration of a cryptogram to clear text is achieved by the action of decryption using a decryption key. In a symmetric, or single-key cryptosystem (DES, for instance), the encryption and decryption keys are the same. In order for two entities to exchange cryptograms, then, they must use the same key. The management of this distribution process requires extreme caution. In an asymmetric, or two-key cryptosystem (PKS, for instance), the encryption and decryption keys are different, and one cannot be derived easily from the other. The public keys of entities wishing to exchange cryptograms are available to all.

(2) On-Line Encryption

If private data are transmitted frequently, encryption devices are likely to be connected in the transmission path close to each terminal equipment. Thus, a terminal may have a separate encryption device, and can use a separate cypher key if desired.

(3) Off-Line Encryption

If private data are transmitted infrequently, or in low volumes, off-line encryption equipment can be employed. Upon receipt of an encrypted message, the same equipment can be used to decrypt the message. In this way, the cost of the equipment can be shared among several users. One cypher key may be employed for all messages, or separate keys that are invoked as required may be assigned to different terminals or different classes of data.

(4) End-to-End and Link-by-Link Encryption

Messages may be encrypted end-to-end, or link-by-link. The procedures are illustrated in **Figure 15.5**.

- **End-to-end encryption**: the transport layer PDU is encrypted before being transferred to the network layer, where the network header is attached, and to the data link layer where the data link header and trailer are attached. Thus, the bits that are transported over the physical connection represent an encrypted message (TPDU) in cyphertext, and other information (address, sequence number, channel number, etc.) in the clear (plaintext). On arrival at its destination, the frame is guided through the data link and network procedures by information contained in the plaintext portions, and the cyphertext is decrypted at the transport layer to reveal the TPDU. End-to-end encryption is private to the originating and terminating protocol stacks, and encryption and decryption are performed in software. To an eavesdropper, the plaintext headers and trailer may provide important information concerning traffic patterns (e.g., increasing frequency of communication between specific parties).
- **Link-by-link encryption**: the data link PDU is encrypted as it enters the physical layer. As a result, all of the bits transported over the physical connection are part of an encrypted message. On reaching the first network node, the cyphertext is decrypted at the interface to the data link layer to reveal the data link PDU which contains the cleartext message. The routing information is consulted in the network layer, a new network PDU is created for the next link, and the NPDU is encapsulated to become a new DLPDU. On entering the physical layer it is encrypted with a new key prior to being dispatched to the next node. In this way, routing information is concealed at the expense of having the cleartext message available at each node it transits. As a consequence, the network must provide physical security at all nodes to protect the

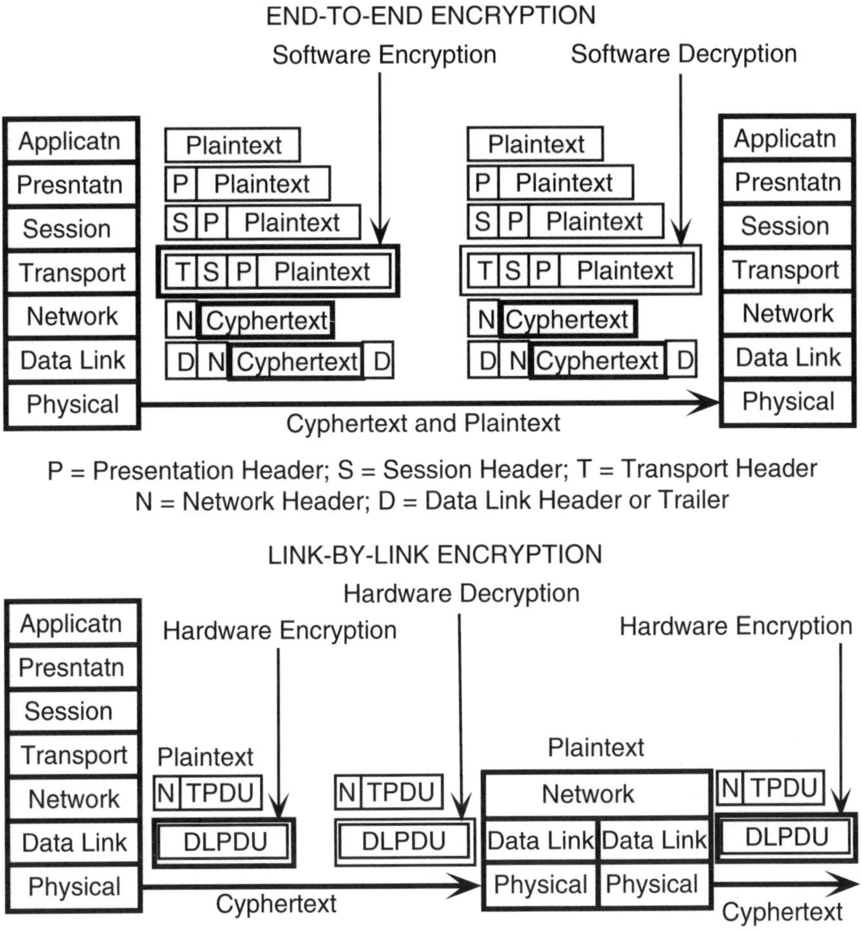

END-TO-END ENCRYPTION

Software Encryption Software Decryption

Applicatn, Presntatn, Session, Transport, Network, Data Link, Physical

Plaintext; P Plaintext; S P Plaintext; T S P Plaintext; N Cyphertext; D N Cyphertext D

Cyphertext and Plaintext

P = Presentation Header; S = Session Header; T = Transport Header
N = Network Header; D = Data Link Header or Trailer

LINK-BY-LINK ENCRYPTION

Hardware Encryption; Hardware Decryption; Hardware Encryption

Plaintext; N TPDU; DLPDU; Network; Data Link Data Link; Physical Physical

Cyphertext

Figure 15.5 End-To-End and Link-By-Link Encryption
Messages may be encrypted at the transport layer (in which case the address is carried in the clear) or encrypted at the data link layer (in which case the address is encrypted and the PDU must be decrypted at each intervening node).

cleartext messages available there. Link-by-link encryption is usually performed in hardware—thus it can be faster than software encryption.

(5) Encryption in Local-Area Networks

The greater number of users in a local-area network environment makes key management more difficult than in wide-area networks. Also, the limited processing power of many of the terminal units and devices may make encryption to a given security level a time-consuming process. Given that much of the information present on a local area network may be only marginally private, the sheer volume of data can be enough to discourage casual eavesdropping. To prevent more deleterious activities, end-to-end encryption with PKS can be employed. A point often overlooked—when the network supports common printers and other output devices, it is important to place them in a secure area and to control access to it.

15.3.3 Key Management

Key management is the greatest single challenge in administering a symmetrical cryptosystem. The keys must be protected to maintain security, yet they must be shared by at least two sites to permit mutual encryption and decryption.

(1) Key Management in Symmetrical Cryptosystems

Two basic methods of key management are employed in symmetrical systems:

- **Physical transportation**: if there is limited data traffic, and only a few communicating sites, keys may be carried to all sites by trusted persons who install them at pre-arranged times. However, if heavy data traffic, or many communicating sites are involved, and the keys are changed frequently enough to assure privacy, this option quickly becomes labor and travel intensive. A modification uses bonded delivery services, or other third-party means to distribute new keys. Obviously, the chance for compromise increases with the number of persons that knowingly handle them.
- **Automatic updating**: the encrypting device stores two keys. One is the key used for data encryption; the other is a key-encrypting key. The key-encrypting key is entered manually at some central site before the encryption device is shipped to a remote operational location. When the encryption key must be changed, a new encryption key is generated at the central site (or another location). Then it is encrypted with the key-encrypting key, and sent to the remote site(s) where it is decrypted and installed automatically. Because the key-encrypting key is used infrequently, it may be used over a longer time before being changed.

(2) Key Management in Asymmetrical Cryptosystems

The use of public key cryptosystems eliminates most of the concerns related to key management. An added burden is the need for a public directory that is updated as sites change and public keys are replaced. Another consideration is that each site must be able to generate its own secret key. However, by no means is the use of public key systems perfected, and users should be aware that the underlying strength of the technique is still being questioned.

(3) Escrowed Encryption Standard

The emergence of practically unbreakable, commercial encryption capabilities in integrated circuit chips that can be incorporated in digital equipment, ensures that persons can use communication facilities for private purposes with confidence. Neither an ill-disposed competitor, nor the government, can understand the messages exchanged. However, in 1994, so that encryption shall not protect both law-abiders and law-breakers, the United States Government proposed the voluntary use of a special encryption chip. Described in a standard known as *Escrowed Encryption Standard* (EES), a secret key (Device Unique Key, KU) is created and embedded in a tamper-resistant chip at the time of manufacture. Called the *Clipper* chip (officially designated MYK-78) it is given a device-unique identifier (UID) and an 80-bit common Family Key (FK). UK is split in two Key Components that are encrypted and given to two Key Escrow Agents for safekeeping. With court approval, for a specific UID, government officials can obtain these keys. The combination of KU, KF, and a 128-bit Law Enforcement Access Field (LEAF) permits them to decypher intercepted messages. EES specifies the use of *SKIPJACK*, a classified encryption algorithm. It transforms a 64-bit input to a 64-bit output using an 80-bit secret key. The same key is used for decryption. The *Key Escrow System* (KES) has created significant concern that it can be exploited or abused by government agencies. It remains to be seen if the Clipper chip (or a successor) will be incorporated in enough communications equipment to make the concept effective.

16

Traffic Patterns

If facilities are not sized correctly, the advantages of a private network can soon dissipate. If there is too much traffic handling capacity, the users will be paying for unused facilities; too little traffic handling capacity and users will not get through. Accordingly, it is important to understand the principles involved in traffic engineering.

16.1 TRAFFIC ENGINEERING

16.1.1 Telephone Traffic

Telephone traffic is measured by the number of calls and their duration. We can define three quantities:

- **Traffic density**: the number of calls in progress at a given moment
- **Traffic intensity**: the traffic density averaged over a one-hour period
- **Holding time**: the time between the moment when the circuit is first seized (the user goes off-hook) and the moment when the circuit is released (the user goes on-hook) and equipment becomes available for use by others.

(1) Traffic Units

A measure of the telephone traffic flow produced by a call is the length of time that the call occupies a telephone circuit (the holding time). Thus, if n calls are made, and, on average, each call occupies h seconds, the total traffic flow is nh call seconds. Two traffic units are in common use:

- **Erlang**: 1 erlang is the traffic due to a call that lasts for one hour; it is also the traffic due to 60 calls that last 1 minute each, or any combination of calls for which the sum of the holding times (in seconds) is 3600.
- **ccs** [hundred (centum) call seconds]: 1 ccs is the traffic due to a call that lasts 100 seconds, or any combination of calls for which the sum of the holding times (in seconds) is 100.

Because there are 3600 seconds in one hour, it follows that 36 ccs are equal to 1 erlang. Also, the traffic in erlangs is the traffic density averaged over one hour, or the traffic intensity.

(2) Variations in Traffic Density

In most businesses, the private network is used intensively during the working day. Usually, for firms whose operations are in a single time zone, the traffic density peaks during the morning and afternoon hours. For firms that span several time zones, the activity on some links will exhibit other peaks as the business activities in different locations across the country overlap. For instance, when it is 9 am in San Francisco, it is noon in New York; and, when it is 5 pm in New York, it is 2 pm in San Francisco. If lunch is from noon to 1 pm in both cities, New York-based employees and their San Francisco counterparts are likely to call each other between 1 pm and 3 pm, and 4 pm and 5 pm, New York time. The company must anticipate two traffic peaks in the afternoon on the facilities linking New York and San Francisco.

(3) Call-Blocking

The traffic in a private network will vary from minute-to-minute, hour-to-hour, day-to-day, and possibly from season-to-season. What is more, each message varies both with respect to the time it consumes and its destination. During periods of heavy traffic, users may have to compete for the network resources needed to service their calls. If elements in the network are unable to handle all of the call attempts immediately, some of them will be blocked. The state is defined as blocking:

- **Blocking**: the traffic condition when a call attempt cannot be completed immediately because network resources are fully occupied serving other calls.

Call-blocking events are likely to occur at the local switch (where common equipment is shared) or within the network (where transmission paths are shared).

(4) Classes of Traffic

For the calling process we can write,

Calls attempted = Calls in progress + Calls blocked

Multiplying each term by the average holding time of the calls in progress gives

Offered traffic = Carried traffic + Blocked traffic.

- **Carried traffic** is the number of calls in progress in the network, times the average holding time. It represents the traffic that is carried to its destination.
- **Offered traffic** is the number of calls attempted, times the average holding time of the carried traffic. If all attempted calls are completed when placed, this traffic is the same as carried traffic. If some attempted calls are blocked, this traffic is a measure of the demand for service.
- **Blocked traffic** is the number of calls blocked, times the average holding time of the carried traffic. It represents the amount of traffic that cannot be carried to its destination on the first attempt because the equipment is fully occupied with other calls.

What happens to blocked traffic depends on the configuration of the facilities employed. The possibilities are that the blocked traffic becomes:

- **Lost traffic**: some blocked calls are likely to be abandoned by the callers and some of the blocked calls may be transferred to the public network at the initiative of the caller. In either case, they are lost to the private network.
- **Retried traffic**: other blocked calls may be retried at a later time at the discretion of the callers.
- **Delayed traffic**: some (or all) blocked calls may be delayed in the network until facilities are available to handle them.

16.1.2 Traffic Model

Figure 16.1 shows a model of the calling process that includes carried, delayed, retried and lost traffic. Some points to be made include:

- **Number of sources**:
  - if the number is very large and the messages are very short, the input traffic will be random
  - if the number is finite, the input traffic will be non-random.
- **Number of service channels**:
  - *full-availability*: a source has access to all service channels

Offered traffic ≥ Carried traffic

Offered traffic = Carried traffic + Delayed traffic + Retried traffic + Lost traffic

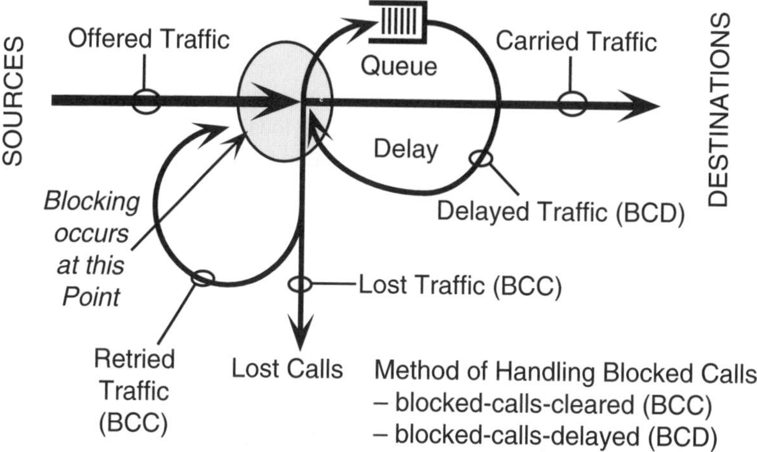

Figure 16.1 Traffic Model

- *limited-availability*: a source can connect with only a fraction of the service channels.

- **Method of handling blocked calls**:
  - *blocked-calls-cleared* (BCC). On receiving a busy signal that indicates all facilities are occupied and the call cannot be handled, the caller hangs up so that the call is lost to the system. The caller may re-try over the network at a later time. Alternatively, the caller may place the call over another network.
  - *blocked-calls-delayed* (BCD). The call is held in the system until it can be processed. In many systems, the caller may be alerted to the delay by an automated announcement, kept occupied with recorded music, and updated from time-to-time on the estimated waiting time remaining.

The degradation in service caused by competition for network elements can be measured by observing a large number of occurrences of several critical events that occur in placing and completing a call. Known as Grade of Service (GOS) criteria, the most important is the number of calls blocked. It is expressed in these terms:

- **Blocking probability**: the probability of the first call attempt being rejected because of the unavailability of facilities. A common value for blocking probability in a private network is P(blocking) ≤ 0.05, i.e., the probability of being blocked is less than or equal to 1 in 20 (5%) for each attempt; in public networks, P(blocking) is likely to be ≤ 0.01, i.e. the probability of being blocked is less than or equal to 1 in 100 (1%) for each attempt.

Other quantities can be measured to establish GOS; they include the times associated with the set-up and tear-down of calls:

- **Time to dial-tone**: the time between the moment the user goes off-hook, and the moment dial-tone is returned from the local switch. Common values of this parameter are μ (mean value) = 1 second and 95% of all callers receive dial tone within 3 seconds, i.e. P(0.95) = 3 seconds.
- **Connection time**: the time between the moment the last digit is entered and the calling circuit is completed.
- **Call set-up time**: the sum of the two times defined above, plus the time required to enter the called number.
- **Time to disconnect**: the time between the moment the user goes on-hook, and the moment the circuit is available to others.
- **Operating time**: the time to establish and to release a connection. It is the sum of call set-up time and time to disconnect.

These GOS parameters depend on the amount of common equipment at the originating and terminating switches. Performance is improved by adding more equipment so that each call attempt is less likely to be blocked, and more likely to receive electronic attention as soon as the caller goes off-hook.

16.1.3 Busy Hour

(1) Which Busy Hour?

The one hour period in any day in which most use is made of the network (greatest traffic intensity) is known as the *busy* hour. For obvious reasons, it is important that a good grade of service be maintained during this period, and networks are engineered to cope with busy hour traffic intensity. Consequently, at times other than the busy hour, the grade of service will be even better. However, because circumstances vary widely from day to day, there may be considerable variations in the daily busy hour traffic intensities. To overcome this difficulty, we may choose to work with an artificial busy hour that represents the average of the busy hours observed over an extended cycle of business activity. Alternatively, we may use that actual busy hour that registers the highest traffic intensity during the period of observation. Thus, busy hour may be defined in three ways:

- **Daily busy hour**: on any particular day, the 60 minute period in which the traffic intensity on a group of telephone facilities is the highest

- **Average busy hour**: a 60 minute period in which the traffic intensity is the average of the traffic intensities observed in a series of daily busy hours
- **Peak busy hour**: that daily busy hour selected from a series of daily busy hours that exhibits the highest traffic intensity.

Choosing the peak busy hour as the basis on which to engineer network performance to achieve a given grade of service, produces a GOS that is always equal to, or better than, the objective—but the network cost will be high. Choosing the average busy hour as the basis on which to engineer network performance to achieve a given grade of service, produces a GOS that, on average, is equal to the objective—at a lower network cost. Choosing the daily busy hour will produce a GOS that is equal to the GOS objective on the day the busy hour was observed, and on all other days with the same daily busy hour. On days on which the busy hour is different from the day observed, the GOS will be different from the GOS specified. Usually, as a practical matter, busy hour is taken to mean average busy hour.

(2) Example of Busy Hour Traffic

In **Figure 16.2**, the number of calls in progress (measured by sampling a group of lines at one minute intervals) during a daily busy hour, is plotted. The number ranges between two and eight calls, and the average number of calls in progress is five, i.e. the traffic intensity is five erlangs. We may envision this traffic intensity as equivalent to five circuits that are occupied throughout the busy hour.

While Figure 16.2 gives an impression of the traffic during busy hour, it is not easy to analyze. To do this we employ statistical techniques. The lower chart in Figure 16.2 shows the cumulative distribution function (CDF) and the probability density function (PDF) for this sample of traffic. The mean (or expectation) of these distributions is 5.12 erlangs, the variance is 2.23, and the variance-to-mean ratio (α is 0.435. The distribution of holding times encountered in the sample daily busy hour are shown in **Figure 16.3**. With an average holding time of 1.5 minutes, the total number of calls placed in the busy hour is 200.

EXAMPLE 16.1 Derivation of Mean and Variance of Traffic Density Distribution in Busy Hour

If, at one minute intervals, we count the number of simultaneous calls occurring in Figure 16.2, we obtain,

Number of Occurrences n_i	5	8	11	13	8	11	4
Number of Simultaneous Calls x_i	2	3	4	5	6	7	8

If,

- the total number of times the traffic was sampled during the busy hour is denoted by n (in our case $n = 60$),

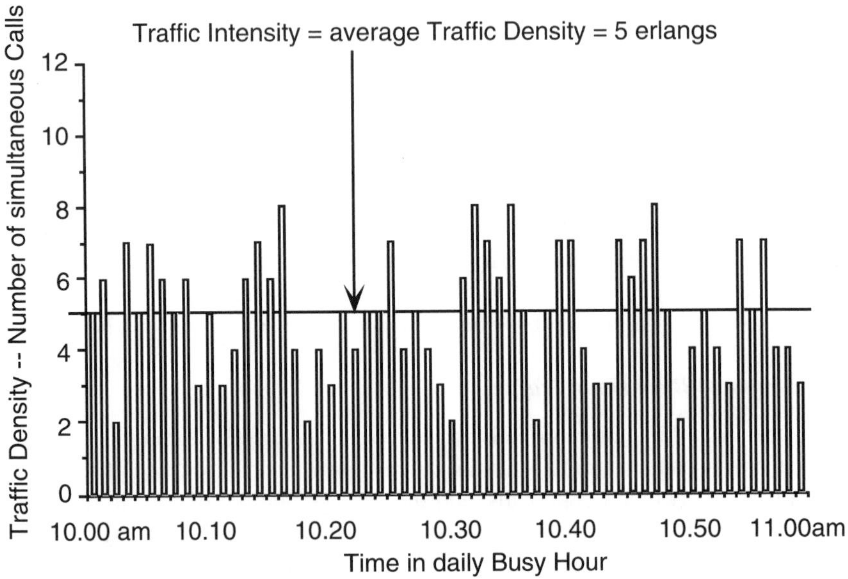

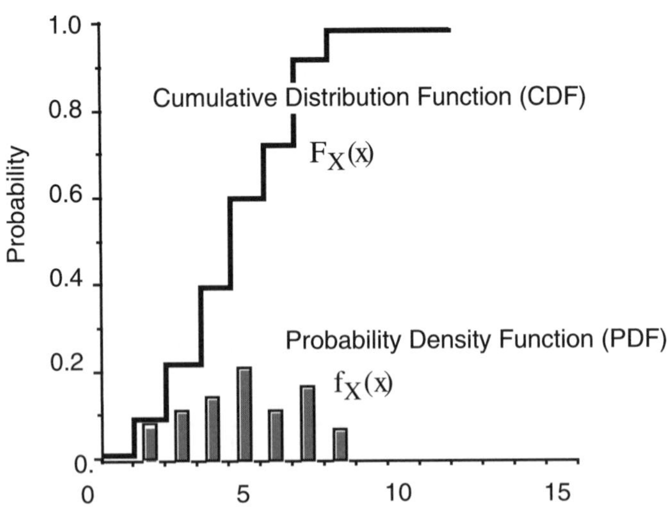

Figure 16.2 Daily Busy Hour Traffic Density

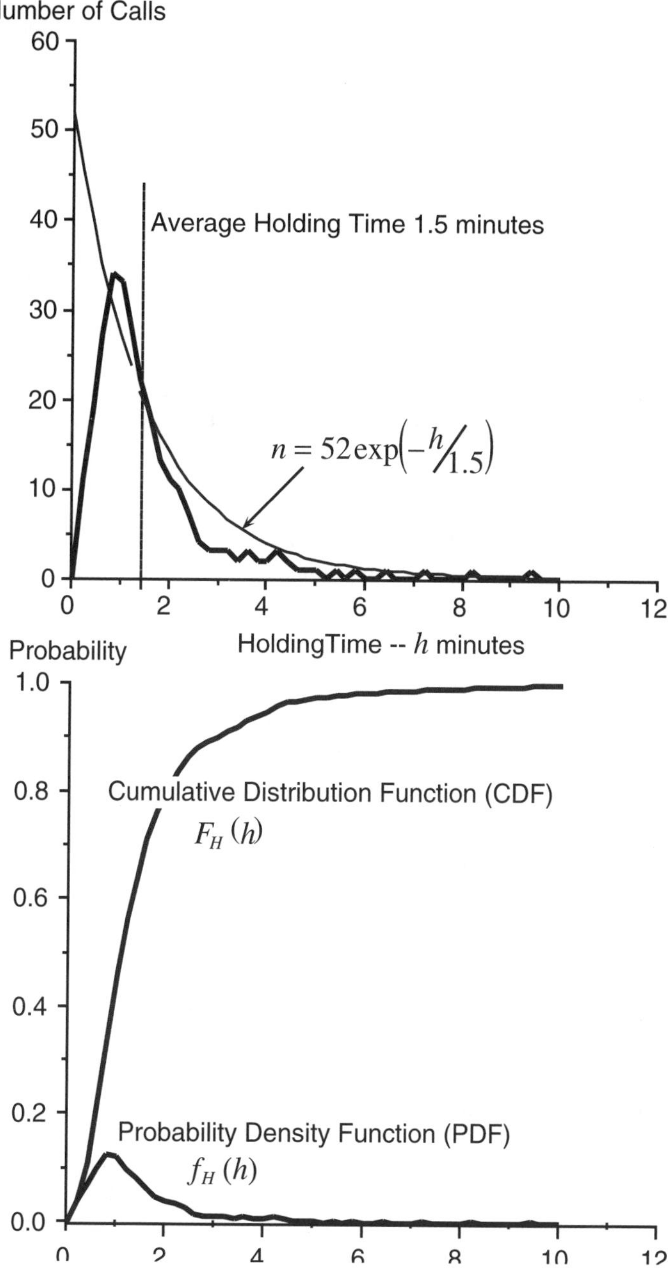

Figure 16.3 Distribution of Call Holding Times

- the number of simultaneous calls measured during a sample is x,
- the number of occurrences of a specific value (x_i) of simultaneous calls is denoted by n_i, and,
- the total number of discrete values of the number of simultaneous calls is t, then the average number of simultaneous calls, or the *average traffic density* during the busy hour is,

$$\overline{X} = \frac{1}{n}\sum_{i=1}^{t} x_i n_i = \frac{300}{60} = 5 \text{ erlangs}$$

This is the *traffic intensity*. If the average call-holding time is 1.5 minutes, the total number of calls during the busy hour is 200.

The probability of occurrence of x_i simultaneous calls is $\frac{n_i}{n}$, so that,

Probability of Occurrence $P(x_i)$	0.083	0.133	0.183	0.217	0.133	0.183	0.067
Number of Simultaneous Calls x_i	2	3	4	5	6	7	8

This is the probability density function of the random variable X that is shown in the lower diagram of Figure 16.2. Integrating the PDF gives the cumulative distribution function (CDF) that is included in the lower diagram.

The *mean* (μ), or *mathematical expectation* [E(X)] of X is,

$$E(X) = \sum_{i=1}^{t} x_i \, P(x_i) = 5.12 \text{ erlangs}.$$

The *variance* [σ^2 or Var (X)] of X is,

$$\sigma^2 = E(X^2) - [E(X)]^2 = \sum_{i=1}^{t} x_i^2 \, P(x_i) - \left[\sum_{i=1}^{t} x_i \, P(x_i)\right]^2$$

$$= 28.84 - 26.21 = 2.23.$$

Finally, the *variance-to-mean ratio* (α) is,

$$\alpha = \frac{2.23}{5.12} = 0.435.$$

16.1.4 Traffic Distribution

(1) Range of Distributions

Generally, traffic patterns can be characterized as smooth, peaked or random:

- **Smooth traffic**: consider a telemarketing operation in which five callers place a continuous stream of calls that average 1.5 minutes each. In a typical hour, they generate 200 calls, and the traffic intensity of their operation is 5 erlangs. Because of the particular application, calling is regular and steady, and five lines are required to serve them. The mean traffic is 5 erlangs, the variance is zero, and the variance-to-mean ratio is zero. The traffic they generate is said to be *smooth*.

- **Peaked traffic**: next, consider a brokerage house where 50 investors are watching the variation in the prices of stocks. Suppose the market begins to see-saw so that they all want to place buy or sell orders simultaneously. To satisfy their demand, 50 lines will be required. If each conversation results in a holding time of 1.5 minutes, and the event occurs four times in one hour, at the end of the hour, the traffic intensity will be 5 erlangs, the value in the previous example, but the variance will be 225, and $\alpha = 45$. The traffic generated in this example is said to be *peaked* (some persons describe it as *rough*).
- **Random traffic**: finally, consider an office of several hundred persons that place 200 calls with an average holding time of 1.5 minutes in the busy hour. They will require more than five lines, but less than 50 lines. If their calling pattern is random, the traffic intensity will be 5 erlangs, the value in the previous examples, the variance will be 5, and $\alpha = 1$. It can be handled with 11 lines with less than 1% blocking (see Figure 16.6).

In these examples, the dispersion of the calls is significantly different. To put a mathematical definition to them, we employ the value of the ratio of the variance to the mean of these distributions:

- **Smooth traffic**: $0 < \alpha < 1$
- **Random traffic**: $\alpha = 1$
- **Peaked traffic**: $1 < \alpha < \infty$.

Diagrams of traffic intensity, and probability of number of simultaneous calls, for the situations described above are shown in **Figure 16.4**.

(2) Effect of Traffic Distribution on Blocking

For a traffic intensity of 5 erlangs, **Figure 16.5** illustrates the increase in the number of serving channels that is required to maintain a given level of blocking as the offered traffic becomes more peaked. The random traffic case is calculated using the Erlang-B formula—something that is discussed below. For the cases of smooth and peaked traffic, we calculate the levels of blocking produced by less than the full number of channels from the scenarios given above. Blocking of 20% is produced by four channels for very smooth traffic ($\alpha = 0$), by six channels for random traffic ($\alpha = 1$), and by 40 channels for very peaked traffic ($\alpha = 45$). Obviously, for any calculation of the number of channels required to serve a given situation, an accurate knowledge of the traffic distribution is essential.

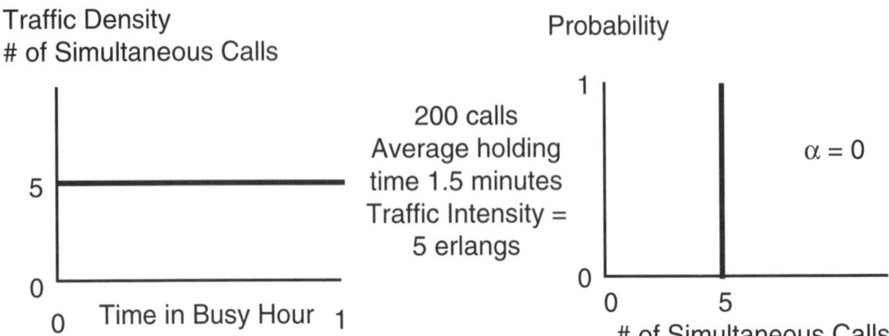

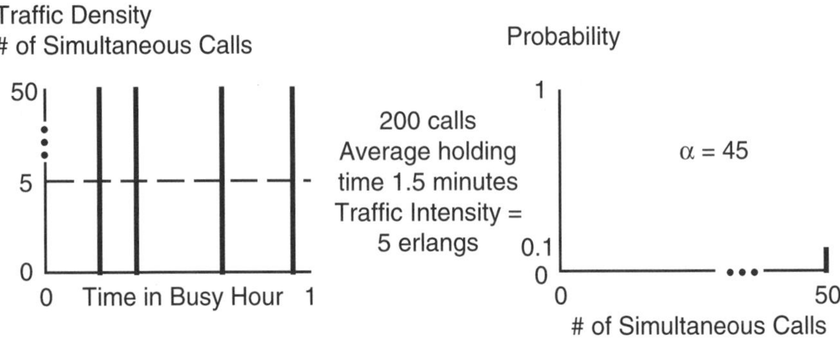

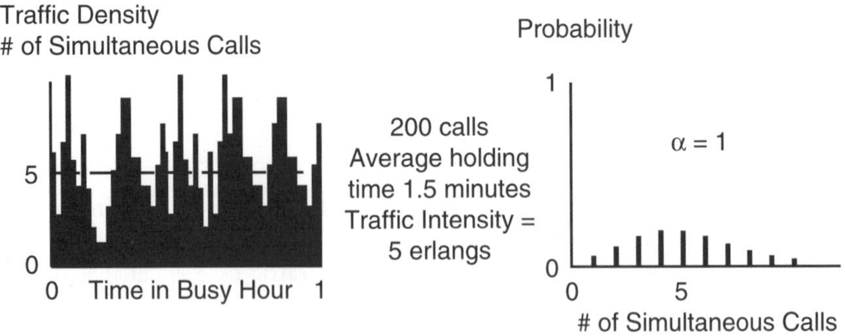

Figure 16.4 Smooth, Peaked, and Random Traffic Distributions
By itself, traffic intensity does not give enough information to define the calling pattern. In each of the cases shown, the traffic intensity is 5 erlangs. However, the variance ranges from zero to 45.

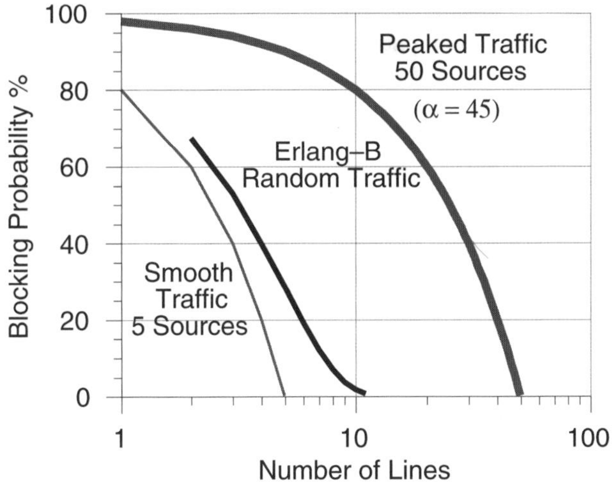

Figure 16.5 Blocking Probabilities
Blocking probability as a function of the number of lines available for the cases shown in Figure 16.4.

16.2 APPLICATIONS

Today's traffic engineering problems are solved by specialists using proprietary software who explore all possible alternatives and arrive at one, or more, near optimum solutions. They employ analysis and/or simulation techniques, and produce results that can guide business decisions. Generally, one of the strong points of their programs is that they can evaluate potential remedies, taking into account the spectrum of tariffs and arrangements that are available from an assortment of carriers. Faced with real-life telecommunication problems, the responsible manager will do well to take advantage of these services. Understanding the scope and validity of their solutions may be easier if we examine a few simple problems without the aid of automation.

16.2.1 Erlang's Formulas

Basic traffic analysis tools were developed in the early 20th century by the Danish mathematician A. K. Erlang. Known as *Erlang-B* (or Erlang Loss), and *Erlang-C* (or Erlang Delay), we shall use tabulated values derived from his formulas to predict facility blocking for two policies—blocked-calls-cleared (BCC), and blocked-calls-delayed (BCD). For situations in which a large number of sources generate relatively short messages at random, the results are generally in agreement with reality. For more usual situations in which a small number of sources generate messages that are related to business events, the results should be used as a guide. Under these circumstances, computer simulations must be employed to obtain better information.

(1) Outward WATS—BCC Mode

A common telecommunications arrangement employs WATS to provide telephone connections, and permits callers to overflow to the public network when all WATS lines are busy (i.e., BCC). The question arises—how many WATS lines are needed to minimize the total cost of telephone service? Among other things, the answer depends on the traffic intensity, the areas called, the carriers used, and the current tariff structure:

- **Random traffic**: suppose all calls are placed to an area located in a single WATS band, that there are many sources so that traffic may be considered to be random, and that the traffic intensity in the average busy hour is 5 erlangs. We can use the *Erlang-B* formula to draw the chart shown in **Figure 16.6**. It depicts the way in which random traffic will occupy up to 11 lines that are selected in rotary fashion, i.e., each call attempt tries line 1, then line 2, etc. until it finds an unoccupied line that can be seized. With a traffic intensity of

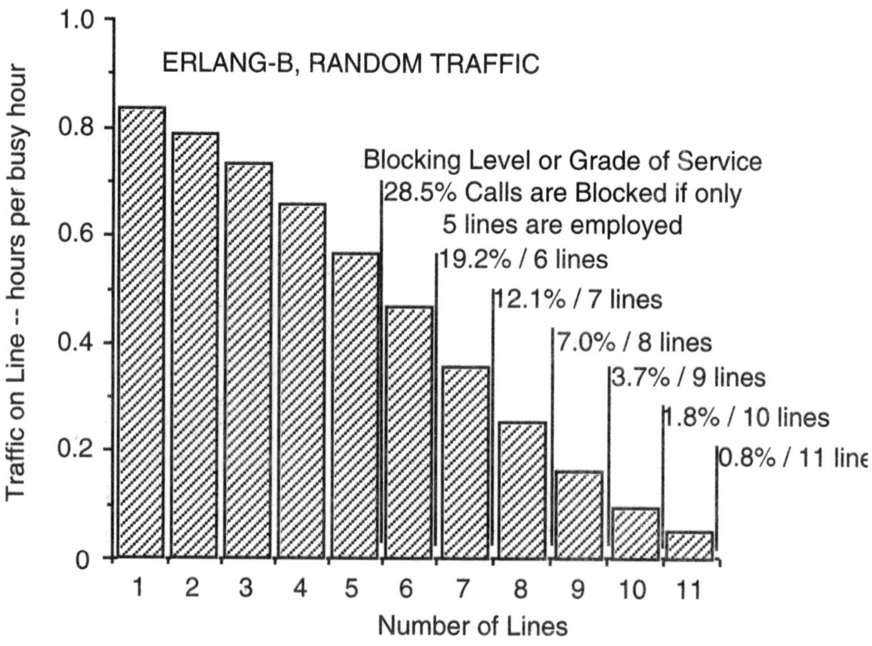

Figure 16.6 Blocking Levels for Random Traffic

5 erlangs from many sources, 28.5% of calls will overflow to the public network if 5 WATS lines are employed, 19.2% of calls will overflow to the public network if 6 WATS lines are employed, etc. If 10 WATS lines are employed, 1.8% of the calls will overflow to the public network. As an alternative to the presentation of Figure 16.6, we include **Figure 16.7**. It provides a more general view of blocking levels as a function of the number of lines available for random offered traffic. For almost certain transmission (i.e., probability of blocking = 0.001, or one-tenth of 1%), 14 lines are required when the offered traffic is 5 erlangs.

- **Peaked traffic**: but what if we have peaked traffic? **Figure 16.8** shows a diagram for the case of peaked traffic for which the variance-to-mean ratio is 4 and the traffic intensity is 5 erlangs. In contrast to the random environment, 21.1% of the calls will overflow to the public network if 10 lines are employed, and 19 lines must be employed to reduce blocking to 2%.

The message of this example is not the absolute number of lines that are likely to be required, but the fact that the nature of the traffic must be defined before the number of lines can be determined.

(2) Optimum Number of WATS Lines—BCC Policy

In a given BCC situation, the optimum number of WATS lines to use will be the number for which the cost of calls that overflow to the public network is less than the cost of an additional WATS line.

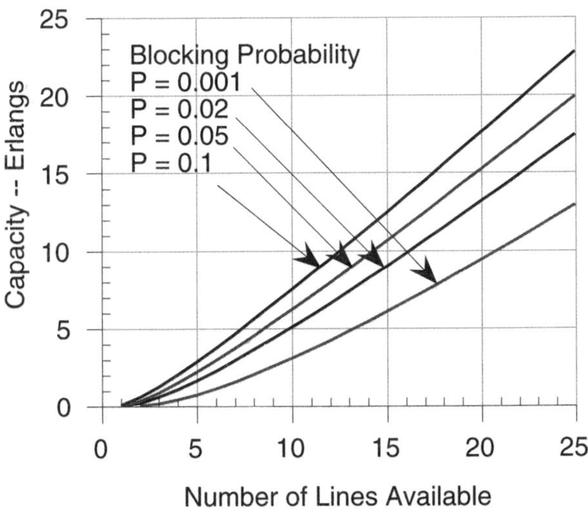

Figure 16.7 Traffic Capacity and Blocking Probability

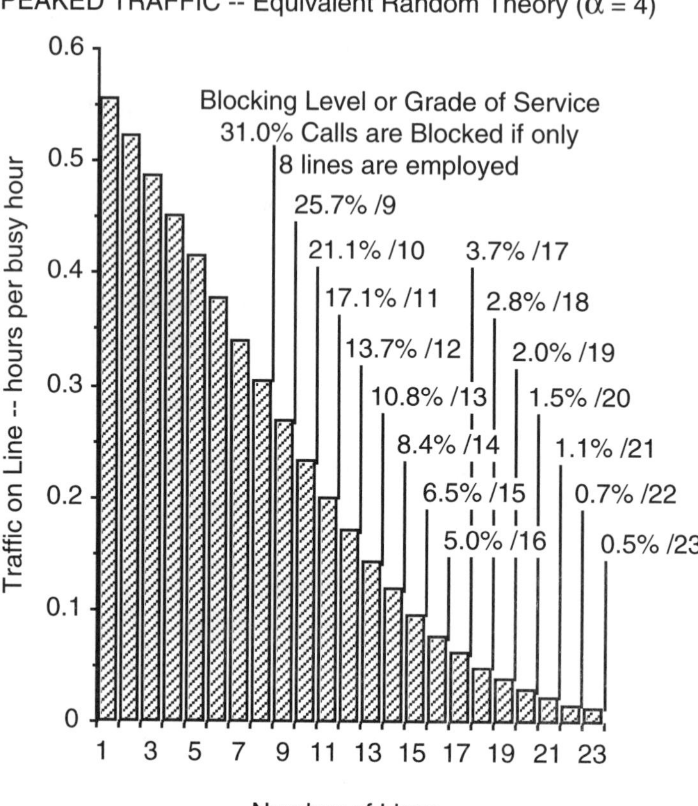

Figure 16.8 Blocking Levels for Peaked Traffic

EXAMPLE 16.2 Find the Optimum Number of WATS Lines Under a BCC Policy

Situation: a large cotton broker in Tennessee calls various destinations in California. Busy hour occurs twice per day. During the average business day the traffic intensity varies as follows:

Hour in Work Day	1st	2nd	3rd	4th	5th	6th	7th	8th	9th	10th
Traffic Intensity—E	1	3	5	3	1	1	3	5	3	1

Assignment: estimate the least cost of communications to California if WATS lines are employed and a *blocked-calls-cleared* policy is used. You should assume the following:

- Calls are charged at the same rate throughout the day.
- The monthly cost of calls to California using DDD circuits is $25,000.
- WATS lines to California cost $1000 per month.
- Erlang-B gives the following values for traffic carried on individual lines tried in sequence:

Offered Traffic — Erlangs	Traffic Carried on Each Line—Erlangs									
	Line Number									
	1	2	3	4	5	6	7	8	9	10
1	0.500	0.300	0.137	0.047	0.016*					
3	0.750	0.662	0.550	0.420	0.288	0.174	0.091	0.041	0.024*	
5	0.833	0.788	0.730	0.657	0.567	0.465	0.357	0.252	0.162	0.095

* These values have been adjusted to make the total traffic carried 100%.

Derivation:

1) Find the monthly cost of communication to California *per daily erlang*:
Total traffic per day is $4 \times 1 + 4 \times 3 + 2 \times 5 = 26$ erlangs.
Total cost of communication to California per month = $25,000
Therefore, the monthly cost per daily erlang is 25,000/26 = $961.54.

2) for numbers of WATS lines, calculate the offered traffic blocked each day, the cost of the number of WATS lines employed, the cost of sending overflow traffic on DDD, and the total cost:

Number of WATS lines	Daily Carried Traffic erlangs	Daily Overflow Traffic erlangs	Cost of WATS lines $/month	Cost of Overflow Traffic $/month	TOTAL Cost $/month
1	4×.500+4×.750+2×.833 = 6.666	26-6.666=19.334	1000	18590	19590
2	6.666+4×.300+4×.662+2×.788 = 12.09	26-12.09=13.91	2000	13375	15375
3	12.09+4×.137+4×.550+2×.730 = 16.298	26-16.298=9.702	3000	9329	12329
4	16.298+4×.047+4×.420+2×.657 = 19.48	26-19.48=6.52	4000	6269	10269
5	19.48+4×.016+4×.288+2×.567 = 21.83	26-21.83=4.17	5000	4010	9010
6	21.83+4×.174+2×.465 = 23.456	26-23.456=2.544	6000	2446	8446
7	23.456+4×.091+2×.357 = 24.537	26-24.537=1.463	7000	1407	8407
8	24.537+4×.041+2×.252 = 25.205	26-25.205=0.795	8000	764	8764
9	25.205+4×.024+2×.162 = 25.625	26-25.625=0.375	9000	361	9361
10	25.625+2×.095 = 25.815	26-25.815=0.185	10000	178	10178

Conclusion: under a BCC policy, the least cost is $8407 per month. It is achieved with seven WATS lines.

(3) Outward WATS—BCD Policy

In **Figure 16.9** we use Erlang-C to calculate the mean delay encountered by all traffic when blocked calls are delayed. Erlang-C assumes that:

- Blocked calls are delayed until they can be serviced
- There is an infinite queuing capacity.

Figure 16.9 shows that 5 erlangs can be completely served by six lines with a mean delay for all messages of one average holding time. If the number of lines is increased to ten, the mean delay will be reduced to 20% of one average holding time. Under a BCD policy, there can be no overflow to another network, so that the cost of telecommunication is limited to the cost of the WATS circuits employed.

EXAMPLE 16.3 Communication Cost Under a BCD Policy

Assignment: under the conditions described in Example 16.2, estimate the least cost of communications to California if WATS lines are employed, a *blocked-calls-delayed* policy is used, and the delay encountered during the busy hour cannot exceed one average holding time. Erlang-C gives the following values for the delay encountered by traffic that will wait as long as necessary for a circuit:

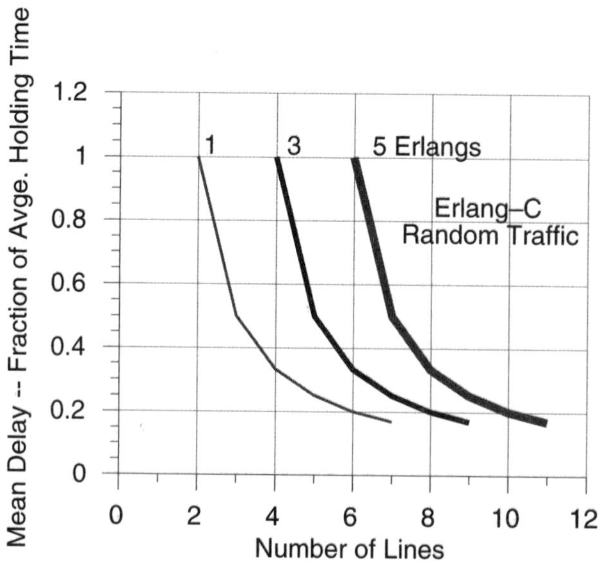

Figure 16.9 Delay Incurred with BCD Policy

Offered Traffic — Erlangs	Delay Encountered—Fraction of Average Holding Time Line Number									
	1	2	3	4	5	6	7	8	9	10
1		1.000	0.500	0.333	0.250	0.200				
3				1.000	0.500	0.333	0.250	0.200	0.167	
5						1.000	0.500	0.333	0.250	0.200

Derivation: for 5 erlangs offered traffic (busy hour intensity), find the number of WATS lines that will carry the busy hour traffic with a mean delay of one average holding time. From the Erlang-C table, 6 WATS lines are required. They will cost $6000 per month.

Extension: we can use the table to find the number of WATS lines that will carry the busy hour traffic with a mean delay of less than one average holding time. For seven WATS lines, the delay is half an average holding time, and for eight WATS lines, the delay is one-third an average holding time.

Comments: combining the results of Examples 16.2 and 16.3, we can state that the cost of communications to California can be reduced substantially:

- If we use six WATS lines and a BCD policy, the cost is $6000 per month, and the average waiting time for service during busy hour will be one average holding time
- If we use seven WATS lines and a BCD policy, the cost is $7000 per month, and the average waiting time for service during busy hour will be one-half an average holding time
- If we use eight WATS lines and a BCD policy, the cost is $8000 per month, and the average waiting time for service during busy hour will be one-third an average holding time
- If we use seven WATS lines and a BCC policy, the cost is $8407 per month, and there is no waiting time.

(4) Effect of Rate of Retries

Usually, callers to 800 (or 900) numbers have no other way to complete their calls. If all lines are busy, they must hang up and retry. The rate at which they do this will depend on the enthusiasm they have for the function they wish to perform. If a viewer is anxious to obtain a bargain offered on a television shopping program, the recall rate may be very high—less than one average holding time may elapse before retry; if a consumer is seeking weather, or sports information, several holding times may elapse between trials; and, if a well-disposed person is responding to a radio or television appeal for funds, the caller may lose interest and not attempt to call again. Retries cause blocking to increase significantly over the level that will prevail if their are no second attempts. Because most inward WATS arrangements are provided for customer convenience, and many are an important link in selling a product, it is important that the number of access lines be selected so as to achieve a blocking percentage commensurate with the likely rate of retries and the value of the service dispensed over them.

16.2.2 Implications for Private Networks

(1) Use of Tie Lines

Tie lines connect telephone facilities for the exclusive benefit of the lessor 24 hours a day. Spanning from point-to-point, they are used when the traffic pattern includes large volumes between two company locations, or between a company location and a specific local exchange area [when they are known as foreign exchange (FX) lines], or when immediate availability of a connection is important. Tie lines are not switched, and their cost depends only upon the mileage between the facilities they connect, the characteristics of the channels provided, and the types of termination arrangements. Because tie lines represent a fixed, usage-insensitive charge, they must be occupied for as long as possible so as to drive down the cost per call. If kept full, the cost per hour of calling on a tie line can be less than the same service over WATS, and both are cheaper than the same service over the DDD network.

(2) Automatic Route Advancement and Least Cost Routing

Given a set of circumstances that results in routes that are served by combinations of tie lines, WATS lines, and Direct-Distance-Dialing, attempting traffic should try tie lines first, WATS lines next, and then, if permitted, overflow to the public (DDD) network. The action of automatically changing from one category of service to another is known as *automatic route advancement* (ARA). Coupled with the requirement to start with the cheapest service and advance to the most expensive, the action is known as *least-cost routing* (LCR).

(3) Time-of-Day Routing for Incoming Traffic

Depending on the pattern of calls, it is possible that the busy hour in one section of a network occurs when another section is relatively quiet (because work has only just started, or is ending for the day). With a backbone of tie lines, incoming busy hour attempts that cannot be handled at the target location can be directed to other locations. Called *time-of-day routing*, this technique shares the calling load across the entire network and reduces the need for each location to be equipped to handle its own busy hour demand exclusively. Airline reservation systems, and similar networks, employ this technique to even the load presented to reservation centers. Time-of-day routing is a time-sensitive case of alternate routing.

(4) Data Traffic Patterns

Because the traffic patterns for voice are different from those for data, carrying voice and low-speed data through a PBX may cause difficulties. During busy hours, PBXs are susceptible to blocking so that call attempts fail entirely (BCC policy) or the connection setup process is delayed (BCD policy). However, once a connection is established, no amount of additional traffic can impair its performance. Data services such as electronic mail are characterized by a large number of sources; they generate traffic that is not greatly different from voice traffic to which we can apply the Erlang-B formula. Other data services such as scientific timesharing are characterized by few sources that generate long-holding time traffic. To this situation we may apply the Erlang-Engset formula. Yet other data services such as batch transfers between mainframes may represent smooth traffic that can be queued; to these situations, we can apply the Erlang-C formula.

Further Reading

For those who wish to dig deeper into the topics discussed, the following books and magazines will be helpful.

INTRODUCTION

Fortune Magazine, *Special Issue on Managing in a Wired World*, vol. 130, No. 1, July 11, 1994.

Peter G. W. Keen, *Competing In Time, Using Telecommunications for Competitive Advantage*, Updated and Expanded, Ballinger Publishing Company, 1988.

Thomas Valovic, *Corporate Networks: the Strategic Use of Telecommunications*, Artech House, 1993.

U. S. Congress, Office of Technology Assessment, *Technology and the American Economic Transition: Choices for the Future*, OTA-TET-283, U. S. Government Printing Office, May 1988.

SIGNALS

Ronald N. Bracewell, *The Fourier Transform and Its Applications*, 2nd Edition, Revised, McGraw-Hill, 1986.

A. Bruce Carlson, *Communication Systems—an introduction to Signals and Noise in Electrical Communication*, 3rd Edition, McGraw-Hill, 1986.

William A. Gardner, *Introduction to Random Processes—with Applications to Signals and Systems*, Macmillan, 1986.

Herbert Taub and Donald L. Schilling, *Principles of Communication Systems*, 2nd Edition, McGraw-Hill, 1986.

BUILDING BLOCKS

Ulysees Black, *TCP/IP and Related Protocols*, McGraw-Hill, 1992.

R. J. Deasington, *X.25 Explained—Protocols for Packet Switching Networks*, 2nd Edition, Ellis Horwood Limited, 1990.

Edsger W. Dijkstra, *A Discipline of Programming*, Prentice-Hall, 1976.

IEEE Communications Magazine, *Special Issue on High Speed Network Protocols*, vol. 27, no. 6, June 1989.

IEEE Communications Magazine, *Special Issue on Multimedia Communications*, vol. 30, no. 5, May 1992.

Leonard Gorney, *Queueing Theory: A Problem Solving Approach*, Petrocelli Books, Inc., 1986.

John Henshall and Sandy Shaw, OSI Explained—End-to-End Communication Standards, *2nd Edition, Ellis Horwood Ltd., 1990.*

M. T. Hills, *Telecommunication Switching Principles*, MIT Press, 1979.

M. T. Hills and S. Kano, *Programming Electronic Switching Machines*, Peter Peregrinus Ltd., 1979.

Gary C. Kessler and David A. Train, *Metropolitan Area Networks: Concepts, Services and Standards*, McGraw-Hill, 1992.

John C. McDonald (Editor), *Fundamentals of Digital Switching*, Plenum Press, 1983.

M. de Prycker, *Asynchronous Transfer Mode, Solution for Broadband ISDN*, 2nd Edition, Ellis Horwood, Ltd., 1993.

NETWORKS

Ulysees Black, *Network Management Standards—the OSI, SNMP and CMOL Protocols*, McGraw-Hill, 1992.

James Arlin Cooper, Computer and Communications Security—Strategies for the 1990s, *McGraw-Hill, 1989.*

IEEE Communications Magazine, *Special Issue on Signaling System,* vol. 28, *no.* 7, July 1990.

IEEE Communications Magazine, *Special Issue on BISDN: High Performance Networks*, vol. 29, no. 9, September 1991.

IEEE Communications Magazine, *Special Issue on Intelligent Networks*, vol. 30, no. 2, February 1992.

IEEE Communications Magazine, *Special Issue on Wide Area Private Networks*, vol. 30, no. 3, March 1992.

IEEE Communications Magazine, *Special Issue on Gigabit Networks*, vol. 30, no. 4, April 1992.

IEEE Communications Magazine, *Special Issue on Wireless Personal Communications*, vol. 33, no. 1, January 1995.

Atul Kapoor, *SNA: Architecture, Protocols, and Implementation*, McGraw-Hill, 1992.

Ramses R. Mina, *Introduction to Teletraffic Engineering*, Telephony Publishing Corporation, 1974.

Daniel Minoli, *Enterprise Networking: fractional T1 to SONET, frame relay to BISDN*, Artech House, Inc. 1993.

William Stallings, *ISDN: An Introduction*, Macmillan Publishing Co., 1989.

J. E. Trulove, *A Guide to Fractional T1*, Artech House, 1991.

A. S. Tanenbaum, *Computer Networks*, 2nd Edition, Prentice Hall, 1988.

List of Acronyms

2B1Q: code in which 2 bits are represented by 1 quaternary symbol.
2W: two-wire.
4B5B: code in which 4 bits are represented by a 5–bit symbol.
4W: four-wire.
7B8B: code in which 7 bits are represented by an 8–bit symbol.

AAL: ATM adaptation layer.
ABR: answer-to-bid ratio.
AC: authentication center.
ACD: automatic call distribution system.
ACSE: association control service element.
ADSL: asymmetric digital subscriber line.
ADTS: automated digital terminal equipment system.
AH: application header.
ALTS: alternative local transmission system.
AMI: alternate-mark inversion.
AMPS: advanced mobile phone service.
ANSI: American National Standards Institute.
AP: adjunct processor.
APDU: application protocol data unit; also, authentic protocol data unit.
AR: authentic receiver.
ARQ: automatic-repeat-request.
AS: authentic source.
ASCII: American Standard Code for Information Interchange.
ASE: application service element.
ASK: amplitude-shift keying.
ASP: adjunct service point.
ATM: asynchronous transfer mode.

B8ZS: bipolar with 8 zeros substitution.
BCC: block check character; also, blocked-calls-cleared.
BCD: blocked-calls-delayed.
Bellcore: Bell Communications Research.
BFSK: binary-FSK.
BISDN: broadband ISDN.
BISYNC: binary synchronous data link control.
bit: binary digit.
BORSCHT: battery feed, overvoltage protection, ringing, supervision, coding and filtering, hybrid and testing functions required by users' terminals.
BPSK: binary-PSK.
BRI: basic rate interface.
BS: bursty second.

CBMS: computer-based message system.
CCAF: call control agent function.
CCC: clear channel capability.
CCF: call control function.
CCITT: Comité Consultatif International Télégraphique et Téléphonique.
ccs: one-hundred call-seconds.
CCSN: common channel signaling network.
CDMA: code-division multiple access.
CDPD: cellular digital packet data.
CENTREX: central exchange service.
CGSA: cellular geographic serving area.
CICS: customer information control subsystem.
CIU: channel interface unit.
CLASS: custom local area signaling services.
CMISE: common management information service element.
CMRTS: cellular mobile radio telephone system.

CLNP: connectionless network layer protocol.
CLTP: connectionless transport layer protocol.
CO: central office.
CPDU: corrupted protocol data unit.
CPFSK: continuous phase FSK.
CPN: calling party's number.
CRC: cyclic redundancy check.
CS: convergence sublayer.
CS-1: capabilities set level-1.
CSMA/CD: carrier-sense multiple access, collision detection.
CSU: channel service unit.
CT-2: cordless telephone version 2.

DAMA: demand assignment multiple access.
DAML: digital added main line.
D-AMPS: digital AMPS.
DCE: data circuit terminating equipment.
DCS: digital cross-connect system.
DDCMP: digital data communications message protocol.
DDD: direct distance dialing.
DDS: digital data service.
DES: data encryption standard.
DH: data link header.
DLC: digital loop carrier; also data link control (protocol).
DNHR: dynamic nonhierarchical routing.
DTMF: dual-tone, multi-frequency.
DoJ: Department of Justice.
DPSK: differential-PSK.
DQDB: distributed queue, dual bus.
DPDU: data link PDU.
DS-0: digital signal level 0.
DS-0A: digital signal level 0 containing data from a single subrate station.
DS-0B: digital signal level 0 containing data from several subrate stations.
DS-1: digital signal level 1.
DS-1C: digital signal level 1C.
DS-2: digital signal level 2.
DS-3: digital signal level 3.
DS-4: digital signal level 4.
DS-4A: digital signal level 4A.
DSBSC: double sideband suppressed carrier.
DSU: digital service unit.
DT: data link trailer.
DTE: data terminal equipment.
DTMF: dual–tone multi-frequency.
DTS: digital termination service.
DX: duplex.

E-1: first level of voice multiplexing in Europe.
EBCDIC: Extended Binary Coded Decimal Interchange Code.
EDI: electronic data interchange.
EIA: Electronic Industries Association.
EIR: equipment identity register.
ES: errored second.
ESCON: enterprise systems connection.
ESF: extended superframe format.
ETSI: European Telecommunications Standards Institute.

FACCH: fast associated control channel.
FC: functional component.
FCC: Federal Communications Commission.
FCS: frame check sequence.
FDDI: fiber distributed data interface.
FDM: frequency division multiplexing.
FDMA: frequency division multiple access.
FDX: full duplex.
FE: functional entity.
FEC: forward error correction.
FEP: front-end processor.
FFOL: FDDI follow-on LAN.
FH: frequency hopping.
FISU: fill in signal unit.
FITL: fiber-in-the-loop.
FM: frequency modulation.
FPLMTS: future public land mobile telephone system.
FPS: fast-packet switching.
FR: frame–relay.
FRI: frame–relay interface.
FS: failed seconds state.
FSK: frequency–shift keying.
FT-1: fractional T-1.
FT-3: fractional T-3.
FTAM: file transfer and access method.
FTTC: fiber-to-the-curb.

FTTH: fiber-to-the-home.
FX: foreign exchange.

GEO: geostationary orbit.
GSM: global system for mobile telecommunication.

HDLC: High Level Data Link Control Protocol.
HDX: half–duplex.
HI: highway interchange(r).
HIPPI: high-speed serial interface.
HLR: home location register.
HTR: hard-to-reach.

ICC: Interstate Commerce Commission.
ICI: intercarrier interface.
IEC: International Electrotechnical Commission.
IEEE: Institute of Electrical and Electronic Engineers.
IM: intelligent multiplexer.
IMAC: isochronous medium access controller.
IMS/DC: information management system / data communications.
IN: intelligent network.
INSC: intelligent network switch capabilities.
IP: Internet protocol; also, intelligent peripheral.
IPDU: illicit protocol data unit.
IR: illicit receiver.
IS: illicit source.
ISDN: integrated services digital network.
ISI: interswitching system interface; also intersymbol interference.
ISO: International Standards Organization.
ITU: International Telecommunication Union.
ITU-D: Telecommunications Development Sector of ITU.
ITU-R: Radiocommunications Sector of ITU.
ITU-T: Telecommunications Standardization Sector of ITU.
IXC: interexchange carrier.

JTC1: Joint Technical Committee 1 (ISO/IEC).

LAN: local–area network.
LAP-B: link access procedure-balanced.
LAP-D: link access procedure-D channel.
LATA: local access and transport area.
LCAM: large-carrier amplitude–modulation.
LCN: logical channel number.
LEC: local exchange carrier.
LENO: originating line equipment number.
LENT: terminating line equipment number.
LEO: low-earth orbit.
LLC: logical link control.
LRC: longitudinal redundancy check.
LSSU: link status signaling unit.
LU: logical unit.

MAC: medium access control (sublayer).
MAN: metropolitan–area network.
MNP: microcom networking protocol.
MSC: mobile switching center.
MSK: minimum-shift keying.
MSN: manhattan street network.
MSRN: mobile station roaming number.
MSU: message signal unit.
MTP: message transfer part.
MTSO: mobile telephone switching office.

NANP: North American Numbering Plan.
NAP: network access point.
NAU: network addressable unit.
NCP: network control point; also, network control program.
NH: network header.
NID: network information database; also, network interface device.
NII: National Information Infrastructure.
NIST: National Institute for Standards and Testing.
NNI: node-network interface.
NPDU: network protocol data unit.
NRM: network resource manager.
NT1: network termination 1.
NT2: network termination 2.

OC-1: optical carrier level 1.
OCC: other common carrier.
OC-N: optical carrier level N.
OEM: original equipment manufacturer.

OMAP: operations, maintenance, and administration part.
OOF: out-of-frame event.
OQPSK: offset-QPSK.
OSA: open systems architecture.
OSI: open systems interconnection (model).
OSIRM: open systems interconnection reference model.
OSP: outside plant.
OSS: operator services system.

PAD: packet assembler and disassembler.
PAM: pulse amplitude–modulation.
PBX: private branch exchange.
PCM: pulse code modulation.
PCN: personal communication network.
PCS: personal communication services.
PDU: protocol data unit.
PE: physical entity.
PH: presentation header.
PKS: public key cryptosystem.
PMAC: packet medium access controller.
POP: point-of-presence (of IXC).
PPDU: presentation protocol data unit.
PRC: primary reference clock; also, preventive cyclic retransmission.
PRI: primary rate interface.
PRK: phase-reversal keying.
PSD: power spectral density.
PSK: phase–shift keying.
PSTN: public switched telephone network.
PTT: Post, Telegraph, and Telephone Administration.
PU: physical unit.
PVC: permanent virtual circuit.

QAM: quadrature-carrier amplitude–modulation.
QPSK: quadrature phase-shift keying; also, quaternary-PSK and quadriphase PSK.

RARC: regional administrative radio conference.
RBOC: Regional Bell Operating Company.
ROSE: remote operations service element.

SACCH: slow associated control channel.
SAI: serving area interface.
SAP: service access point.
SAR: segmentation and reassembly sublayer.
SAT: supervisory audio tone.
SCC: specialized common carrier.
SCCP: signaling connection control part.
SCDU: satellite delay compensation unit.
SCF: service control function.
SCO: serving central office.
SCP: service control point.
SCPC: single-channel per carrier.
SCPC DAMA: single channel per carrier, demand–assignment multiple access.
SDCU: satellite delay compensation unit.
SDF: service data function.
SDH: synchronous digital hierarchy.
SDLC: synchronous data link control protocol.
SDO: selective dynamic overload.
SEP: signaling end-point.
SES: severely-errored second.
SF: superframe format.
SH: session header.
SIM: subscriber identity module.
SIP: SMDS interface protocol.
SLC: subscriber loop carrier.
SLI: service logic interpreter.
SLP: service logic program.
SMASE: system management application service element.
SMDS: switched multimegabit data service.
SMF: system management function.
SMFA: specific management function area.
SN: service node.
SNA: systems network architecture.
SNAF: subnetwork access functions.
SNDCF: subnetwork dependent convergence functions.
SNI: SMDS network interface.
SNICF: subnetwork independent convergence functions.
SNR: signal-to-noise ratio.
SONET: synchronous optical network.
SP: switching point.
SPDU: session protocol data unit.
SPE: synchronous payload envelope.
SRF: specialized resource function.
SRM: subrate (data) multiplexer.

SS7: signaling system #7.
SSB: single–sideband.
SSCP: system services control point.
SSF: service switching function.
SSP: services switching point.
ST: signaling tone.
STE: switching terminal exchange.
STM-1: synchronous transport module level 1.
STM-N: synchronous transport module level N.
STP: signal transfer point.
STR: selective trunk reservation.
STS: space-time-space.
STS-1: synchronous transport signal level 1.
STS-N: synchronous transport signal level N.
SU: signal unit.
SUERM: signal unit error rate monitor.
SW56: 56 kbits/s switched digital service.
SW64: 64 kbits/s switched digital service.
SW384: 384 kbits/s switched digital service.
SW1536: 1536 kbits/s switched digital service.
SX: simplex.

T-1: T-carrier multiplexer, level 1.
T-1C: T-carrier multiplexer, level 1C.
T-2: T-carrier multiplexer, level 2.
T-3: T-carrier multiplexer, level 3.
T-4: T-carrier multiplexer, level 4.
T-4A: T-carrier multiplexer, level 4A.
TA: terminal adapter.
TCAM: telecommunications access method.
TCAP: transaction capabilities part.
TCH: traffic channel.
TCM: trellis code modulation.
TCP: transmission control protocol.
TDM: time division multiplexing.
TDMA: time division multiple access.
TE1: subscriber terminal 1.
TE2: subscriber terminal 2.
TH: transport header.
TOH: transport overhead.
TP: teleprocessing access method; also, teleprocessing (monitor) and transport layer protocol.
TP0, **TP1**, **TP2**, **TP3**, **TP4**: transport layer protocols (ISO).
TPDU: transport protocol data unit.
TSI: time-slot interchange(r).
TST: time-space-time.
TTC: Telecommunications Technology Committee.
TTY: teletypwriter.

UDC: universal digital channel.
UDP: user datagram protocol.
UNI: user-network interface.
UPT: universal personal telecommunication.
UP/TUP: (ISDN) user part/telephone user part.
USART: universal synchronous/ asynchronous receiver transmitter.
USTIA: United States Telecommunications Industry Association.

VCI: virtual circuit identifier.
VGL: voice–grade line.
VLR: visiting location register.
VPN: virtual private network.
VRC: vertical redundancy check.
VSAT: very-small–aperture terminal.
VSB: vestigial sideband.
VT: virtual tributary (signal).
VTAM: virtual telecommunications access method.

WARC: World Administrative Radio Conference.
WATS: wide-area telephone service.

ZBTSI: zero byte time–slot interchange.

Glossary

2B1Q digital subscriber carrier: digital loop carrier equipment that employs 2B1Q coding to provide 160 kbits/s, duplex connections over two-wire copper loops up to 18000 feet.

acknowledged connectionless service: messages are acknowledged but a logical connection is not established.

active: in network management, an operational state in which the managed object is available for use and can accept new users.

adaptive delta modulation: in digital voice, a coding technique that employs step size adaptation in regions that contain the faster signal changes.

adaptive equalizer: in digital transmission, a device that continuously corrects channel induced distortion.

adaptive DPCM: in digital voice, a coding technique that employs an adaptation algorithm to vary the quantizer levels in accordance with the rate of change of the input signal.

adaptive transform coding: in digital voice, the input signal is divided into short time blocks (frames) that are transformed into the frequency domain using techniques based on Fourier series. The result is a set of transform coefficients that describe each frame. These coefficients are quantized into numbers of levels based on the expected spectral levels of the input signal. The number of bits for any quantizer, and the sum of all the bits per block are limited to a predetermined maximum.

adjunct processor: in intelligent network, device that contains an adjunct service point that responds to requests for service processing. Directly connected to an IN switching system the adjunct processor supports services that require rapid response to users' actions.

adjunct service point: in intelligent network, a location that responds to requests for intelligent network services.

administrative unit: in synchronous digital hierarchy, the combination of a tributary group and pointer information.

advanced mobile phone service: a cellular mobile radio telephone service operating in the United States on assigned frequencies of 825 to 845 MHz (mobiles to cell sites) and 870 to 890 MHz (cell sites to mobiles). AMPS has a channel bandwidth of 30 kHz, uses FDMA as the access method, and employs frequency modulation.

agent process: in network management, that part of the application process that is concerned with performing management functions on managed objects at the request of the managing process.

aliasing: condition that exists when signal energy that is present over one range of frequencies in the original signal appears in another range of frequencies after sampling and reconstruction.

ALOHA: in satellite communication, an access technique in which stations send data packets as they are ready. So long as the total offered traffic from all stations is not a very large fraction of the capacity of the transponder, packets have a reasonable chance of finding the transponder unoccupied upon arrival at the satellite and can be relayed without interference.

alternate mark inversion signal: 1s are represented by return-to-zero pulses that alternate between positive and negative levels, and 0s are represented by the absence of pulses.

alternative local transmission system: a transmission system that bypasses facilities of the local exchange carrier. Operating in many cities, many ALTS provide point-to-point optical fiber transport services that are used to connect buildings in the same city, or connect customers directly to IXC equipment.

American National Standards Institute: manages the voluntary development of U.S. national standards.

American Standard Code for Information Interchange: employs a sequence of seven bits and contains 128, 7-bit patterns.

amplitude-shift keying: in digital modulation, the transmitted signal is a sinusoid whose amplitude is modulated in discrete steps. When two states are employed, the ASK-wave is known as a *binary*-ASK wave. To produce more complex symbols that represent groups of bits, several amplitude levels are employed.

analog function: single-valued, continuous state function that assumes positive, zero, or negative values. Changes in value occur smoothly and rates of change are finite.

analog line circuit: circuit that performs analog to digital conversion, 2-to-4 wire conversion, signaling detection, and battery feed.

angle modulation: result of varying the argument of a sinusoidal function in sympathy with the message signal.

annual charge: annual cost associated with purchasing, owning, maintaining, and using up plant.

answer-to-bid ratio: the ratio of the number of calls accepted by, and the total number of calls presented to, a specific switch, in a given time period.

anti-aliasing filter: in digital voice, a low-pass filter that passes frequencies up to W Hz and precedes a sampling device operating at $r \geq 2W$ samples/s.

application agent: in OSIRM, embedded in the local operating system, the application agent exists between the user and the operating system of the processing resource to provide access to specific processor capabilities.

application entity: in OSIRM, exists within the application layer. As required, provides services to the application agent by exchanging protocol data units with peer entities (application entities in other machines).

application header: in OSIRM, header added to user's data unit by application layer to create application PDU. It contains the unique identifier for the exchange agreed by the communicating hosts.

application layer: layer 7 of OSI reference model. When sending, the application layer organizes the user's data and negotiates a unique identifier with the other host. When receiving, the application layer strips off the application header and passes the information contained in the APDU to the user.

application protocol data unit: combination of application header and user's data.

area code: North America is divided into subareas that are assigned 3-digit area codes.

area-wide Centrex: based on an SCP that is connected to all switches within a LATA, an intelligent network service that provides PBX-like operations from designated stations across the LATA.

Argand diagram: consists of a phasor of amplitude A signal units that rotates anticlockwise about the origin. The phasor traces out sinusoidal functions on the real and imaginary axes.

associated signaling: signaling messages are sent over the same route as the information that is exchanged between the parties.

association control service element: in network management, general software support system that works between application processes that are independent of application-specific needs. ACSE supports the establishment, maintenance, and termination of a cooperative relationship between application entities.

asynchronous multiplexing: a digital multiplexing technique in which time–slots are assigned to signals as required. Arrival times at the multiplexer, not what input line they are on, determine the order in which octets are sequenced. As they are placed on the multiplexed line, the multiplexer adds an identifier so that the demultiplexer will know to which output line the octets are to go.

asynchronous operation: data are sent or actions occur when the sender is ready.

asynchronous transfer mode: a high-speed (155 and 622 Mbits/s), fast-packet switching technique employing short, fixed length cells that are statistically multiplexed over virtual connections. Each cell consists of a 5–octet header and a 48–octet payload of user information (data, voice, or video).

ATM adaptation layer: level 3 protocol employed in ATM bearer service. Receives different classes of information from higher layers and adapts it to occupy fixed length, ATM cells. Receives data from ATM layer and rearranges it to fit user's application requirements. AAL is divided into two sublayers...segmentation and reassembly (SAR) sublayer, and convergence sublayer (CS).

ATM bearer service: employs a three level protocol to provide fast–packet, cell–relay service.

ATM LAN: local ATM node that connects high-speed terminals, servers, and other LANs using the multiplexing capabilities of the segmentation and reassembly sublayer of AAL to provide connectionless services in connection-oriented streams.

ATM layer: level 2 protocol employed in ATM bearer service. Multiplexes and demultiplexes cells belonging to different connections identified by unique VCIs. Translates VCI at switches and cross-connection equipment. Passes cell payload to and from the ATM adaptation layer without error control or similar processing. Performs flow control at UNI. Implements connection-oriented and connectionless services.

audio conference: an interactive, real-time service {4,4} that may be augmented by text {5,5} or graphics {6,6}.

authentication center: in mobile telephone network, a database that verifies user's credentials and supports encryption of radio channel signals.

authentic protocol data unit: in communication security, the true PDU that the source wishes to send to one, or more, receivers for a purpose approved by the network owner.

authentic receiver: in communication security, receiver that is connected to the network with the knowledge and approval of the network owner.

authentic source: in communication security, sender that is connected to the network with the knowledge and approval of the network owner.

authorization access: in local–area networks, an operating mode in which each station sends when ready *and* authorized. Usually, authorization is granted through possession of a token—a data message that is passed from station to station.

automated digital terminal equipment system: a digital loop carrier system that multiplexes individual voice and data channels over DS-1, DS-2, or DS-3 facilities. Provides monitoring and maintenance functions, routing capability, clear-64 and ESF capability, fiber connections to the users' premises, and supports SONET connections.

automatic call distribution system: equipment that receives calls and directs them to attendants waiting to provide service. ACD systems may be stand-alone, integrated with a PBX, or central office-based.

automatic-repeat-request: an error correction procedure that requires the transmitter to re-send the portions of the exchange in which errors have been detected.

automatic route advancement: action of automatically changing from one category of transport service to another (i.e., tie line to WATS to DDD).

average busy hour: a 60 minute period in which the traffic intensity is the average of the traffic intensities observed in a series of daily busy hours.

band-limited signal: a signal that contains no energy at frequencies greater than W Hz, i.e., $0 < f \leq W$ Hz.

bandpass characteristic: performance such that energy is transmitted between mf Hz and nf Hz ($m < n$), but not at frequencies below mf Hz nor above nf Hz.

bandwidth: the range of frequencies that just encompasses all of the energy present in a given signal.

baseband: the bandwidth of the modulating (message) signal.

baseband LAN: digital signals are carried directly on the common bearer.

base station: in wireless networks, the fixed equipment at the center of a cell. The number of base stations determines the number of mobiles that can be served simultaneously in a given frequency band.

basic rate interface: ISDN channel consisting of 2 B-channels + 1 D-channel. In the United States the BRI is supplied over a connection that operates at 160 kbits/s. It contains 144 kbits/s of user and signaling data.

baud: unit of signaling rate, one symbol per second.

B-channel: ISDN information channel that supports digital transmission at 64 kbits/s, is circuit switched, and employs user-specified protocols. Subrates of 2.4, 4.8, 8, 16, and 32 kbits/s may be multiplexed on to the channel.

bearer: physical entity that supports an electronic medium.

bearer services: information services available at the user-network interface.

Bell Communications Research: central technical organization funded by the seven RBOCs.

binary-ASK: amplitude-shift keying that employs two amplitudes.

binary-FSK: frequency-shift keying that employs two frequencies. The presence of a signal at one frequency corresponds to a 1-state, and the presence of a signal at another frequency corresponds to a 0-state.

binary-PSK: phase-shift keying that employs two phase angles. The presence of the carrier signal in one phase condition corresponds to a 1-state, and the presence of the carrier signal 180° out of phase with it corresponds to a 0-state. Also known as phase-reversal keying.

binary synchronous data link control: a character-oriented data link control protocol.

biphase signal: a 1 is a positive current that changes to a negative current, and a 0 is a negative current that changes to a positive current.

bipolar signal: for 1s, positive current flows, and for 0s, negative current flows.

bipolar with 8 zeros substitution: in a T-1 system carrying data, a technique used to provide 64 kbits/s clear channels to the user yet maintain ones density requirements.

bit: binary digit.

bit-level multiplexing: in turn, bits from each input line are interleaved on the common bearer in a regular fashion. Timing is more critical than at the octet-level, but throughput delay is reduced.

bit-oriented DLC protocol: data communication procedure that makes use of a flag character to mark the beginning and ending of the frame. Between these markers, the header and the trailer fields are of predetermined lengths. Whatever data lie between them is the text field.

bit synchronization: state of transmission facility when receiver can identify bits within each time-slot and reconstruct the same bit sequence as the sender transmitted.

block check character: a character formed from bits created by the longitudinal redundancy check process.

blocked-calls-cleared: on receiving a busy signal that indicates all facilities are occupied and the call cannot be handled, the caller hangs up so that the call is lost to the system. The caller may re-try over the network at a later time. Alternatively, the caller may place the call over another network.

blocked-calls-delayed: a call that cannot be processed immediately is held in the system until it can be processed.

blocked traffic: the number of calls blocked times the average holding time of the carried traffic. It represents the amount of traffic that cannot be carried to its destination on the first attempt because the equipment is fully occupied with other calls.

blocking: the traffic condition when a call attempt cannot be completed immediately because network resources are fully occupied serving other calls.

blocking probability: the probability of the first call attempt being rejected because of the unavailability of facilities.

bridge: device that links local–area networks when the differences reach no higher than the implementation of the data link layers.

bridged tap: in the local loop, a new connection is bridged to an old one, leaving an active wire pair serving the new user attached to a portion of an inactive wire pair that served the former user.

broadband ISDN: digital network that distributes information at rates up to 600 Mbits/s, and supports interactive, switched services up to approximately 150 Mbits/s.

broadband LAN: signals are carried on parallel channels over a coaxial cable using frequency division multiplexing. Devices on the network are assigned to channels they share with other devices. A specific set of frequencies is used for sending, and another set of frequencies is used for receiving.

broadcast network: a non-switched network in which a transmitting node broadcasts information to a large number of receivers in a specific geographical area.

bursty second: in ESF error checking, any second in which from 2 to 319 CRC-6 error events are detected.

bus LAN: all stations are connected to the same bearer—the *bus*—which is open ended.

bus network: all nodes or user stations are connected to a common bearer.

busy: in network management, an operational state in which the managed object is in service but cannot accept new users.

byte-count-oriented DLC protocol: data communication procedure in which the number of bytes (or octets) in the text field is carried in a byte-count field in the header.

call processing software: in a switching machine, software that initiates, maintains, and terminates a message path between calling and called parties.

call set-up time: time to dial tone, plus the time to enter called number, plus time to make connection.

carrier-sense multiple access, collision detection: an access technique employed with IEEE 802.3 (Ethernet) LANs. When a period of no activity on the bus is detected, the station with something to send may send it. Once this station has begun transmission, the other stations should detect the activity and withhold their own messages. If two, or more, stations begin to transmit at the same time, they will interfere with each other, a collision will be declared, and both must try again at a later time.

carrier serving area: area served from remote terminal of digital loop carrier equipment.

carrier signal: see modulation.

carried traffic: the number of calls in progress in the network, times the average holding time. It represents the traffic that is carried to its destination.

C-bit parity DS-3: employs the format of traditional DS-3 frame. However, the 7 DS-2 signals created in the multiplexing process are generated at exactly 6.306272 Mbits/s and a stuffing bit is inserted in every subframe. The result is a DS-3 signal at exactly 44.736 Mbits/s. Because they are not needed to report the presence or absence of stuffing bits, the 21 C bits can be used for other purposes such as error detection, alarms, and status information.

ccs: 1 ccs is the traffic due to a call that lasts 100 seconds, or any combination of calls for which the sum of the holding times (in seconds) is 100.

cell: in cellular wireless networks, the area occupied by mobiles that communicate with a single base station.

cell site: in cellular wireless networks, the location of the base station.

cellular digital packet data: a cellular wireless information system that provides 19.2 kbits/s connectionless service on idle 30 kHz cellular phone channels.

cellular geographic serving area: in wireless networks, an area consisting of multiple cells served by a mobile telephone switching office.

cellular mobile radio telephone system: provides voice or low-speed data (up to 2400 bits/s) communication between mobile units and fixed sites.

central exchange service: services provided by a carrier that mimic the services the customer could obtain from a PBX. The customer's extensions are multiplexed to the serving office; there, switching and other functions are performed in the carrier's equipment.

centralized routing: in a packet network, a primary (and perhaps an alternate) path is (are) dedicated to a pair of stations at the time of need. For the sequence of packets that constitutes a block of user's information, the same primary (or alternate) *en-route* resources are used by each packet.

central office: an end–office.

channel bank: equipment that combines analog voice signals into an FDM or TDM signal. In the case of TDM: when transmitting, the channel bank performs quantization, coding, μ-law compression, and multiplexing; when receiving, the channel bank performs demultiplexing, decoding, μ-law expansion, and reconstruction of the voice signals.

channel interface unit: in FDDI, units that receive frames from their upstream neighbors, and send frames to their downstream neighbors over the primary fibers.

channel service unit: provides impedance matching, loopback (for testing), and limited diagnostic capabilities, so that the circuit provider can monitor and adjust the performance of a digital line. Also, generates bipolar, AMI signals, and provides span powering for repeaters, if needed. In addition, supports special framing requirements and may generate keep-alive signals when the connection is broken on the customer side.

character-framed data: each character is framed by a start bit (0) and one or two stop bits (1 or 11).

character-oriented compression: by substituting special characters, or small groups of characters, for larger strings of characters that appear in a message, the length of the datastream can be made shorter than the original.

character-oriented DLC protocol: procedure that employs special characters in the datastream to facilitate data communication.

checksum: an error detecting procedure. The result of adding the characters in a message together to create a total that can be checked after transmission.

cipher: in encryption, the action of transposing or substituting the bit patterns or characters in the original text.

circuit speed: the number of bits transmitted over a circuit in one second.

circuit switch: selectively establishes and releases connections between transmission facilities to provide dedicated paths for the exchange of messages between users. The paths are established before the information exchanges begin, and, until the session is terminated, they are maintained for the exclusive use of the calling and called parties.

circuit-switched network: one in which circuits are formed by interconnecting links at network nodes so that voice or data paths are established between pairs of users. Each connection is set-up using signaling information (dialed telephone number, for instance), is maintained for the duration of the call (holding time), and is taken-down upon completion of the call. The connections may be physical, i.e., actual segments of wire interconnected by contact devices to form metallic paths, or derived, i.e., produced by time- or frequency division of a transmission facility, or they may be part physical and part derived.

Class 4 Switch: toll center switch.

Class 4/5 Switch: switch that combines toll and end–office features.

Class 5 switch: switch that serves a community of business and residential users. Located in the network end–office.

clear 64: the customer has full use of a 64 kbits/s channel.

clear channel capability: the customer has full use of a 64 kbits/s channel.

Clipper chip: an asymmetric encryption chip proposed for use by the United States government so that the pervasive encryption of communication channels shall not protect both the law-abiding and law-breakers. Designated MYK-78, a secret key is created at the time of manufacture and placed in escrow. With court approval, government agencies can obtain this key and employ it to decipher encrypted messages obtained through wiretaps.

coaxial cable transmission system: consists of a central wire surrounded by a conducting, grounded tube in which electromagnetic energy is confined. As CATV, used to distribute television signals throughout communities. Also employed in megabit speed, data connections, including local–area networks.

code: in encryption, the action of transposing or substituting the words or phrases in the original text.

code division multiple access: a spread spectrum technique in which all stations in the network transmit at the same carrier frequency with approximately equal power and the same chip rate. Each station employs a code that is orthogonal to the codes used by the other stations. Each receiver sees the sum of the individual spread spectrum signals as uncorrelated noise. It can demodulate an individual signal if it has knowledge of the spreading code and the carrier frequency.

command: in SDLC, a frame that passes from the primary station to a secondary station.

Comité Consultatif International Télégraphique et Téléphonique: International Consultative Committee on Telegraphy and Telephony. In 1993, CCITT was absorbed in ITU-T.

Committee T1—Telecommunications: sponsored by the Exchange Carriers Standards Association (ECSA), accredited by ANSI and approved by FCC as an appropriate forum, an organization that addresses telecommunications networks standards issues in North America.

common-channel signaling: signaling messages relating to many circuits are transported over a single channel by addressed messages.

common channel signaling network: network comprising the combination of signaling data links between switching centers and signal transfer points, and the data links between STPs.

common information space: the communication mode(s) common to the downstream and upstream directions.

common management information service element: in network management, provides management notification services that can be used to report any event about a managed object, and management operation services that define the operations required to perform actions on a managed object. Also, provides a framework for common management procedures that can be invoked from remote locations.

communication: the activity associated with distributing or exchanging information.

communication channel: a one-way communication path.

communication circuit: a communication path consisting of two channels, one in each direction between communicating entities.

communication network: consists of a finite set of facilities (stations, links, and nodes) that can be interconnected to support communication among several subscribers. Resources may be allocated to users when they request the opportunity to communicate, and are sufficient to support many users at the same time.

communication protocol: rule or procedure to cope with normal or abnormal operating situations.

communications quality speech: contains a limited range of frequencies and may have discernible noise and distortion.

companding: action of reducing the dynamic range presented by a signal. At the transmitting end, a compressor decreases the amplitude of the stronger signal components with respect to the weaker signal components. At the receiving end, an expander, reverses the distortion introduced by the compressor and restores the voice components to their original relationship.

computer-based message system: an electronic mail system that is supported by one, or more, mainframe computers.

computer conference: an interactive, store-and-forward service [1,1] that allows participants to share a common data space and message community.

concentration and expansion processes: the number of ports on an end office switch is much less than the total number of loops installed in the exchange area, but larger than the number of active circuits (under normal circumstances). A concentration stage is employed to connect the active loops on the distributing frame to the input ports of the switching stages. In reverse, the concentration stage acts as an expansion stage to connect active lines from the switching stages to the loops that complete the calls.

configuration management: in network management, the actions of maintaining, adding, and updating records of hardware and software components, defining their functions and usability, initializing the network, and ensuring that the network is described correctly and completely at all times. Also, assigning, maintaining, and updating records of users, including billing responsibilities and communication privileges.

connectionless: communicating stations do not create an administrative dialog. The message is sent by the sending station without reference to the status of the receiver, and receipt goes unacknowledged.

connection-oriented: communicating stations create an administrative dialog that keeps track of the status of each station and the messages exchanged.

connection-oriented service: a logical connection is set up between originating and terminating stations. Acknowledgments, error and flow controls, and other features, are employed to ensure reliable data transfer between stations.

connection time: the time between the moment the last digit is entered and the calling circuit is completed.

container: in synchronous digital hierarchy, a part of the payload. Each container contains one or more tributary signals.

contention access: when using a contention technique, each station competes for transmission resources. They send when ready, send when ready *and* the bus is not occupied, or send when ready *and* the bus is not occupied *and* listen for a collision.

continuous phase FSK: digital modulation technique that avoids the phase discontinuities of FSK. Continuity of phase across the boundary between symbols is achieved by adjusting the phase of the frequency representing the next symbol to compensate for a phase shift introduced by the frequency keying process. In binary-CPFSK, the phase shift between symbols is 90°.

continuous-wave modulation: brings together a higher-frequency sinusoidal carrier signal and a lower–frequency message signal to create a modulated wave in which message signal energy is distributed at frequencies around the carrier signal frequency.

convergence sublayer: component of ATM adaptation layer. Performs functions required by specific services.

convolutional code: code generated by continuously performing logic operations on a moving, limited sequence of n bits contained in the message stream. For each bit in the message stream, the encoder produces a fixed number (two or more) of bits; their values depend on the value of the present message bit and the values of the preceding bits (i.e., the previous n - 1 bits).

cordless telephone: a telephone instrument in which the path between handset and base station is completed by radio or infrared radiation.

corporate 800 service: completes 1-800...calls over an enterprise network.

corporate credit card service: completes off-net calls over an enterprise network for corporate credit card holders.

corrupted protocol data unit: in communication security, an authentic PDU that has been altered by accidental means.

country code: a component of ITU's universal numbering plan that can have up to three digits.

CRC-6 error event: in ESF error checking, when the received FCS does not match the FCS calculated at the receiver.

customer information control system: in SNA, a teleprocessing monitor that supports multiple, on-line applications in a common network. Provides interfaces to files and database systems. Handles errors, security, and priority transactions.

customer loop: the pair of wires (twisted pair) that connects the customer's station to the distributing frame of the serving central office.

custom local area signaling services: in intelligent networks, based on the ability to know the calling party's number from messages on the signaling channel, a group of switch-controlled, common channel signaling features that use existing customer lines to provide end–users with call management capabilities.

cyclic redundancy check: an error detecting procedure. The sender divides the message by a prime number to produce a remainder known as the frame check sequence. It is placed in the trailer of the frame. The receiver divides the message by the same prime number. If agreement is reached as to FCS, the message has been received without error.

D4-channel bank: a digital loop carrier that multiplexes 48 or 96 individual voice or data channels over DS-1, DS-1C, or DS-2 facilities.

D5-channel bank: a digital loop carrier that multiplexes 96 individual voice or data channels over 4 , DS-1 facilities or a DS-2 facility. Includes a common control subsystem that manages up to 20 units, and supports fiber connections to the users' premises.

daily busy hour: on any particular day, the 60 minute period in which the traffic intensity on a group of telephone facilities is the highest.

data circuit terminating equipment: equipment that conditions data signals received from an associated DTE for transmission over communication connections, and restores received data signals to formats compatible with the DTE.

data compression: a technique that reduces the number of bits transmitted during a communication session so that more information is transmitted over a given facility.

datagram: data unit that is routed across network by nodal decisions (distributed routing) without establishing a connection or call record.

data link control protocol: a set of rules that governs the exchange of messages over a data link.

data link header: includes the flag, class of frame identifier, sequence number for use in the control byte, and other information.

data link layer: layer 2 in OSI model. Prepares data in a specified format for transfer over a single communication link without relays or intermediate nodes.

data link PDU: protocol data unit that consists of data link header, network PDU and data link trailer.

data link trailer: includes an FCS to check the frame and a flag.

data services: services that collect data for processing in equipment that executes applications (programs), and distribute the resulting information to persons and machines that must act on it.

data terminal equipment: data processing device that sends and receives data messages.

D-channel: ISDN channel that supports a packet-mode, ISDN protocol for call-control signaling, user-to-user signaling, and packet data. In addition, it is used for operations, administration, and maintenance messages. In the BRI, the speed of the D-channel is 16 kbits/s; in the PRI the speed is 64 kbits/s.

dc-term: in a Fourier series, the zero frequency component.

decoding noise: in digital voice, the difference between the reconstructed sampled signal at the output of the receiver's pulse generator, and the signal that would be produced if the sampled signal at the output of the sender's quantizer is transmitted without error.

delayed traffic: some (or all) blocked calls may be delayed in the network until facilities are available to handle them.

delta modulation: in digital voice, a coding technique in which the sampling rate is many times the Nyquist rate for the input signal so that the correlation between consecutive samples is very high, and the difference between them is very small.

demodulation: a signal processing technique in which the modulating signal is recovered from the modulated wave.

deterministic function: at every instant in time, a deterministic function exhibits a specific value (including zero) that is related to values at neighboring times; thus, it is possible to know the value of the signal at a future time, and to know the time in the future when the signal will have a specific value.

differential coding: in digital voice, coding that exploits the correlation between adjacent signal samples.

differential PCM: in digital voice, a coding technique in which the difference between successive samples is sent.

differential phase-shift keying: digital phase modulation technique that eliminates the need for the synchronous carrier in the demodulation process. At the transmitter, the datastream is processed to produce a modulated wave in which the phase changes by 180° when one symbol (0 or 1) is transmitted, and does not change when the other symbol (1 or 0) is transmitted.

digital added main line: 2B1Q digital loop carrier equipment that implements an additional voice circuit over a single pair of wires for circuit expansion purposes.

digital AMPS: a mobile phone service evolved from AMPS. Defined by the North American digital cellular standard (IS-54), it substitutes three 16.2 kbits/s TDMA channels for existing single-user analog channels so as to achieve graceful evolution from AMPS to D-AMPS.

digital channel bank: see digital loop carrier.

digital cross-connect system: a computer-based facility that reroutes channels contained in synchronous carrier transmission systems without demultiplexing.

digital data communications message protocol: data communications procedure employing a header that contains exactly ten octets, a text portion that contains a stated number of octets, and a trailer that consists of a 2-octet FCS.

digital data service: a 56 kbits/s transport service for private line applications. The DDS network is implemented as a set of interconnected hubs that perform as routing, maintenance and administration centers. In turn, the hubs are connected to digital lines that collect customer channels from serving offices.

digital function: single-valued, discrete state function that assumes a limited set of values. The most popular class of digital functions are two-valued (binary), i.e., they exist in two states only. Changes of state occur instantly, and the rate-of-change at that instant is infinite—otherwise it is zero.

digital loop carrier: a carrier system that time division multiplexes subscriber loops and operates between the user's premises and the serving central office. DLCs provide standard support functions for the user's terminal.

digital microcellular systems: all-digital, microprocessor-controlled, wireless information systems that switch and route short, ATM-style packets and use spread spectrum modulation.

digital service unit: converts unipolar digital signals received from DTE to bipolar signals and performs any special coding or other process required to make the signal compatible with the operation of the transport system—such as zeros-substitution codes.

digital termination service: a local loop bypass microwave radio system that operates in the band 10.55 to 10.68 GHz, and provides digital channels at speeds up to 1.5 Mbits/s and distances up to six miles.

direct-sequence spread spectrum signal: an information-bearing, modulated signal is used to modulate a high-speed, wideband, pseudo-random, digital sequence.

disabled: in network management, an operational state in which the managed object is not available.

disassociated signaling: signaling messages are sent over a route different from the route taken by the information that is exchanged between the parties.

distance learning: see education-at-a-distance.

distributed broadband service: broadband ISDN service in which information transfer is one-way, from distributor to customer.

distributed-queue, dual-bus: metropolitan–area network access arrangement that provides connectionless, connection-oriented, and isochronous transfer services. Each node gains access to a bus by joining a distributed queue of nodes waiting to transmit on the bus.

distributed routing: in a packet network, the node decides which path the packet shall take to its destination on the basis of information about traffic conditions and equipment status. Each packet is routed without regard to packets that may have preceded it.

distribution cables: in the local loop, they provide connections from serving area interfaces to service access points close to the users' premises.

double-sideband, suppressed-carrier AM: modulation produced by multiplying the carrier signal by the message signal. The modulated signal is double-sideband, and there is no separate carrier component.

downstream direction: the direction in which richer messages flow.

drop cables: connect from service access points (such as a pedestal, or other distribution point) to the customer's premises.

DS-0: digital signal level 0, 64 kbits/s.

DS-0A: a DS-0 signal containing data from a single subrate station. The bytes are repeated as necessary to match the sender's speed to the speed of the DS-0 line.

DS-0B: a DS-0 signal containing data from several subrate stations. The octets contain a subrate synchronizing bit (bit #1) and six data bits (bits 2 through 7). The eighth bit is set to 1 to ensure meeting the 1s density requirement.

DS-1: digital signal level 1, 1.544 Mbits/s.

DS-1C: digital signal level 1C, 3.152 Mbits/s.

DS-2: digital signal level 2, 6.312 Mbits/s.

DS-3: digital signal level 3, 44.736 Mbits/s.

DS-4: digital signal level 4, 274.176 Mbits/s.

DS-4A: digital signal level 4A, 139.264 Mbits/s.

dual–tone multi-frequency: signals used by telephone to signal called number, and other information.

dumb terminal: device that is dependent on the host for all but the most routine functions.

duobinary encoding: applied to the baseband signal, a coding technique that doubles the transmission capacity of the system. Duobinary signaling achieves a signaling rate of $2B_T$ symbols/s in a bandwidth of B_T Hz. This is the maximum rate predicted by Nyquist.

duplex: information can flow in two directions at the same time. A circuit is employed.

E-1: first level of voice multiplexing in Europe. Standardized by ETSI, combines 30 voice channels plus 2 common signaling channels at 2.048 Mbits/s.

education-at-a-distance: the use of telecommunication techniques to instruct students in several locations from a central location. The sound and sight of the instructor, and the images employed by the instructor, are transmitted to all receiving sites, and, when authorized to do so, students at the sites may ask questions of the instructor.

EIA232: standard that describes a multiwire cable that terminates in 25-pin connectors. At a speed of 20 000 bits/s, the cable is limited to 50 feet.

EIA449: standard that describes a cable that operates at 100 kbits/s, at distances up to 4000 feet.

electronic data interchange: common message formats and procedures established by ANSI for transactions between industrial organizations.

Electronic Industries Association: a nonprofit organization that represents the interests of manufacturers. The Association is well known for its work in standardizing connections between data terminals and other equipment.

electronic mail: a half-duplex, record service [1,1].

empty token: in an IEEE 802.5 LAN, a frame that consists of a start flag, an access control octet, and an end flag.

enabled: in network management, an operational state in which the managed object is operable, available, and not in use.

encryption: the action of disguising information so that it can be recovered relatively easily by persons who have the key, but is highly resistant to recovery by persons who do not have the key. Encryption is accomplished by scrambling the bits, characters, words, or phrases in the original message.

end–office: a facility that contains the lowest node in the hierarchy of switches that comprise the network.

end-to-end encryption: the transport layer PDU is encrypted before being transferred to the network and data link layers.

enterprise systems connection: an optical fiber connection operating up to 40 kilometers at 17 Mbits/s. ESCON was introduced by IBM to extend the mainframe, high speed I/O channel directly to major, remote peripherals.

entropy: the entropy of a source is the average self-information of the symbols it emits.

entropy encoding: technique that uses the statistical properties of the signal to implement variable length coding and reduce the entropy of the source.

equipment identity register: in mobile telephone network, a database that lists subscriber's authorized equipment and denies service if attempts are made to use un-registered equipment.

ergodic process: a random process in which time-averages are equal to corresponding ensemble averages.

erlang: 1 erlang is the traffic due to a call that lasts for one hour; it is also the traffic due to 60 calls that last 1 minute each, or any combination of calls for which the sum of the holding times (in seconds) is 3600.

errored second: in ESF error checking, any second in which one, or more, CRC-6 events are detected.

ESF error event: in ESF error checking, an OOF event, CRC-6 event, or both.

eternal function: function that exists for all time.

European Telecommunications Standards Institute: created to produce technical standards and achieve a unified, European telecommunications market.

even parity check: the parity bit makes (or leaves) the number of 1s even.

event forward discriminator: in network management, directs reports to the appropriate destination.

exchange area: the area that contains all the customers whose loops terminate at a given end office or wire center.

exchange code: identifies specific switching facilities located in an end–office or wire center. Many offices handle more than one exchange code.

extended binary coded decimal interchange code: code employing a sequence of eight bits that contains 256, 8-bit patterns.

extended superframe format: 24 DS-1 frames form an extended superframe that contains 4,632 bits and takes 3 milliseconds to transmit at the DS-1 rate. The 24 framing bits are used for synchronization and other purposes.

facsimile: an intercommunication, simplex, real-time service {2}. A communication service that captures images from paper, transmits them to a distant site, and reconstructs them on paper.

fade: a decrease in signal strength that occurs when a receiver moves in an environment in which it is illuminated by the direct ray and reflected rays (multipath). Fades occur at approximately half-wavelength intervals along the unit's forward progress.

failed seconds state: in ESF error checking, state achieved after ten consecutive severely errored seconds have occurred. The state remains in effect until the facility transmits ten consecutive seconds that are not severely errored.

fast associated control channel: in Global System for Mobile Communication, control channel used for hand-offs and other functions.

fast–packet switching: techniques for transporting data, voice, or video messages in a fixed length packet. FPS supports the transport of fixed length cells that can contain data, voice, or video information. Asynchronous transfer mode is the preferred FPS technique.

fast select service: in a packet network, the call connect and disconnect procedures are circumvented by using a Call Request packet in which up to 128 bytes of user's data can be included. The network delivers the packet as an incoming call request. The receiver responds to the data with a Clear Indication packet that contains up to 128 bytes of information in reply.

fault management: in network management, provision, monitoring and response to facilities that detect and isolate abnormal conditions in the network, so as to restore the network to normal functioning and correct the aberrant condition(s).

FDDI-II: a development of FDDI that handles asynchronous packet data and isochronous voice and video.

FDDI follow-on LAN: a gigabit version of FDDI that will be compatible with higher-speed SONET levels (STS-48, 2.488 Gbits/s, for instance) and BISDN.

feeder cables: the backbone of the local loop network, feeder cables connect the distributing frame in the serving office to serving area interfaces.

fiber-and-coax network: a hybrid network that provides duplex voice and data circuits and simplex broadband (television) channels to the public. Optical fibers are used for signal transport to local distribution points where the digital, TDM signal streams are converted to an FDM stream and distributed to individual homes over shared coaxial cable. In each home, an interface device separates television from voice and data. Two-way operation over cable is achieved by allocating a portion of the spectrum for upstream signals.

fiber channel standard (FCS): describes devices designed to operate up to 10 kilometers at speeds up to 800 Mbits/s on optical fiber.

fiber distributed data interface: FDDI is a dual-ring, optical fiber, data-only network that connects up to 500 stations in a ring of ≤ 100 kilometers. Data is passed at a speed of 100 Mbits/s using a token-passing protocol and 4B5B coding.

fiber-to-the-curb: extension of optical fiber loops to pedestals.

fiber-to-the-home: extension of optical fiber loops to residences.

filter: in network management, uses Boolean operators to chain test conditions that determine which events in the network are to be reported.

foreign exchange line: tie line that connects a private network facility (e.g., a PBX) to a public exchange in a distant city so that callers may have the benefit of local service in the distant calling area.

forward error correction: FEC techniques employ special codes that allow the receiver to detect and correct a limited number of errors without referring to the transmitter.

four-wire connection: a duplex circuit that employs two pairs of wires, one for each direction of transmission. Four-wire connections must be employed when the distance is great enough to require repeaters or when the existence of two signals on the same bearer will cause them to interfere.

fractional T-1: a fraction of a T-1 facility. FT-1 provides n DS-0 channels, where $n = 2, 4, 6, 8$, or 12, as required.

fractional T-3: a fraction of a T-3 facility, i.e., FT-3 = n T-1, where $1 < n < 28$.

frame check sequence: the remainder after dividing the bitstream between the start flag and the end of the text field by a binary prime number.

frame-level multiplexing: frames from different input lines are interleaved on a common bearer.

frame–relay: a fast–packet transport technique that transfers data in variable length frames over permanent virtual circuits (PVCs).

frame–relay interface: also known as frame–relay UNI. Consists of a data link layer and physical layer over which streams up to 8192 octets can be transmitted intact on permanent virtual circuits. Error detection (but not correction) is performed, and corrupted frames are discarded (without requesting retransmission). Congestion control is implemented by sending warning messages to the sender, and destroying frames if necessary.

frame–relay PAD: creates frame–relay frames for equipment that is not able to do so and converts frame–relay frames to formats that are compatible with the end–users.

frame synchronization: state of transmission facility when sender and receiver agree on beginnings of frames.

frequency allocation: the set of frequencies allocated to all stations offering a specific service.

frequency assignment: the specific frequency on which a particular station must operate to provide the service to which the frequency is allocated.

frequency diversity: in satellite communications, to avoid signal interference, satellites assigned to the same orbit positions operate in different frequency bands.

frequency division multiple access: a technique that assigns specific frequencies and bandwidths from the band of frequencies assigned to particular radio systems (communication satellites, mobile radios, etc.) to create individual message channels.

frequency division multiplexing: to each signal a fraction of the bandwidth of the bearer is assigned so that the multiplexed signals exist at the same time, at different frequencies, on the same bearer.

frequency domain coding: in digital voice, coding that divides the signal into a number of frequency bands and codes each separately. More bits may be employed in those bands in which activity is significant at the expense of those bands that are quiet.

frequency hopping spread spectrum signal: an information-bearing, modulated wave is used to frequency modulate a carrier signal.

frequency modulation: modulation produced by varying the frequency of the carrier in sympathy with the amplitude of the message signal.

frequency-shift keying: digital modulation technique in which the transmitted signal is a sinusoid of constant amplitude whose frequency is varied by the modulating datastream. When two states are employed, the presence of a signal at one frequency corresponds to a 1-state, and the presence of a signal at another frequency corresponds to a 0-state. This is known as a *binary*-FSK (BFSK).

frequency reuse: the principle of using a set of frequencies in one place and reusing them in another. The locations must be separated so that entities that use the same frequencies do not receive detectable signals from the wrong sources.

frequency reuse factor: in wireless networks, the ratio of the number of cells using the same set of frequencies to the total number of cells in the system.

frequency–shift keying: a technique in which a carrier at one frequency represents a mark and a carrier at another frequency represents a space.

front-end processor: equipment that performs all the communication-related functions at a host site. The FEP relieves the much faster host of activities associated with the relatively slow processes required for data communication.

full-duplex: information can flow in two directions at the same time. A circuit is employed.

full-load signal: in digital voice, a signal that just sweeps through all levels of a quantizer.

functional component: elementary call processing command that directs internal resources to implement specific intelligent network services.

fundamental component: in a Fourier series, the component with the lowest frequency (but not zero).

fundamental frequency: in a Fourier series, the frequency of the fundamental component.

future public land mobile telephone system: ITU system that will provide global coverage for speech and low to medium bit–rate data (≤ 64 kbits/s), with high bit–rate data (≤ 2.048 Mbits/s) available in limited areas. FPLMTS will be compatible with ISDN and SS7.

gateway: devices used to connect LANs that differ above the network layer.

generic program: in a switching machine, a general-purpose software package installed in all machines of the same class that includes sub-programs that perform call processing, system maintenance, and system administration.

geostationary orbit: orbit 22,300 miles above earth's surface in which satellite travels at approximately 7000 miles/hr, appears stationary from earth, and can illuminate around one-third of the earth's surface.

global functional plane: describes IN services from a service design perspective.

global system for mobile telecommunications: a second generation digital, cellular, land mobile system adopted by ETSI.

Now known as GSM 900, it operates in two frequency bands (890 to 915 MHz, mobile-to-base station and 935 to 960 MHz, base station-to-mobile). Individual circuits consist of 200 kHz channels separated by 45 MHz. Each channel is time-divided into time slots (1733 slots/s) that are allocated in sequence to eight subchannels.

go-back-n: an ARQ technique in which the sender sends frames in a sequence, and receives acknowledgments from the receiver. On detecting an error, the receiver discards the corrupted frame, and ignores any further frames. The receiver notifies the sender of the number of the frame it expects to receive (i.e., the number of the first frame in error). On receipt of this information, the sender begins re-sending the data sequence starting from that frame.

graphics mail: an intercommunication, half-duplex, record service [2,2]. Entered in facsimile machines, graphics messages are reconstructed on similar machines that serve the communicating parties.

Group 1 Facsimile: scans documents at a resolution of approximately 100 lines per inch. Uses frequency modulation to transmit a page in four to six minutes.

Group 2 Facsimile: essentially the same as Group 1, but includes signal compression techniques to speed up delivery to two to three minutes.

Group 3 Facsimile: scans black and white documents at 200 × 100, or 200 × 200 dpi. Scanning generates digital signals that are subjected to variable length coding to reduce redundant information. Transmits a page in less than one minute over modem-terminated, telephone line connections operating at speeds from 2400 to 9600 bits/s.

Group 4 Facsimile: a black and white, digital system that employs advanced, modified Read coding to reduce redundant information. Transmits a page in less than ten seconds over 64 kbits/s, digital connections. The group is divided into three classes according to combinations of horizontal and vertical resolutions that range from 200 dpi to 400 dpi.

half-duplex: information flows in two directions, but only in one direction at a time (go or return).

hand-off: in wireless networks, the action of changing the base station with which the mobile communicates.

hard-to-reach: a protective routing scheme. When the answer–to–bid ratio of a switch drops below a given threshold, it is classified as hard–to–reach. On subsequent unsuccessful bids for connection, the originating switch is informed over SS7 of the HTR status of the destination switch. Individual switches reduce the number of times they request connection through a congested switch.

harmonic component: in a Fourier Series, a component whose frequency is an integer multiple of the fundamental frequency.

harmonic frequency: in a Fourier series, the frequency of a harmonic component.

H-channel: ISDN information channel (H for higher) that operates at 384 kbits/s (H0), 1536 kbits/s (H11), or 1920 kbits/s (H12).

hierarchical network: sometimes called a tree structure, lower level nodes are connected to higher level nodes. The highest node is the root node.

high-pass characteristic: performance such that energy is transmitted at frequencies greater than *hf* Hz, but not at frequencies below *hf* Hz.

high-performance parallel interface: provides a connection operating up to 120 feet at 800 Mbits/s on a single cable, and 1600 Mbits/s on two cables. Over optical fiber, HIPPI operates up to 10 kilometers.

high-speed serial interface (HSSI): supports data transfer at speeds up to 52 Mbits/s.

highway interchange process: the action of a space switching stage that, on instructions from the stored program control, places octets from incoming frames in the same time–slots on different highways.

holding time: the time between the moment when the circuit is first seized (the user goes off-hook) and the moment when the circuit is released (the user goes on-hook) and equipment becomes available for use by others.

home location register: in mobile telephone network, a database that maintains and updates subscriber's location and service profile.

hookswitch: switch on telephone instrument that closes when the handset is removed from its cradle.

Huffman code: a variable length code whose members are assigned to the symbols in some way in proportion to their use.

hybrid: in a telephone, balance transformer that separates the user's talking signal from the signal to which the user is listening.

hybrid mode: in FDDI-II, asynchronous data and synchronous voice are carried in FDDI frames.

IEEE 802.2: LAN standard for logical link control frame.

IEEE 802.3: LAN standard for CSMA/CD operation.

IEEE 802.4: LAN standard for token-bus operation.

IEEE 802.5: LAN standard for token-ring operation.

IEEE 802.6: standard for metropolitan–area network.

illicit protocol data unit: in communication security, a PDU introduced into a communication system for an illicit purpose. It may be an APDU that has been altered deliberately or a PDU generated by an illicit source.

illicit receiver: in communication security, receiver that is connected to the network without the knowledge and approval of the network owner.

illicit source: in communication security, sender that is connected to the network without the knowledge and approval of the network owner.

in-band signaling: the associated signals are analog tones that exist within the bandwidth used for the information exchange.

information: that which is distributed or exchanged by communication. When associated with machines, it may be defined as organized or processed data. When associated with persons, it may be defined as the substance of messages—that which is known about, or may be inferred from, particular facts or circumstances.

Information Management System/Data Communications: in SNA, a teleprocessing monitor that provides a comprehensive, database management and data communications capability.

information system: software entity that organizes data to produce information for the benefit of an enterprise.

in-slot signaling: the associated digital signals use part, or all, of a bit position contained within a digital message octet (also known as bit-robbing).

Institute of Electrical and Electronic Engineers (IEEE): a professional society that develops technical performance standards.

integrated services digital network: a switched, digital network that provides a range of voice, data, and image services through standard, multi-purpose, user-network interfaces.

integrated STP: in an SS7 network, a signaling point that has an STP capability and also provides OSI model transport layer functions.

intellectual stage: with regard to using the telephone, managing the flow of information to create an understanding of the message.

intelligent multiplexer: a multiplexer that has the ability to add, drop, and redistribute circuits among multiple trunks, on command. Supports large, complex, mesh networks.

intelligent network: based on ISDN, a network that distributes call processing capabilities across multiple network modules, employs reusable network capabilities to create a new feature or service, and employs standard protocols across network interfaces.

intelligent network switch capabilities: identify calls associated with intelligent network services, formulate requests for call-processing instructions from service logic elements and implement instructions. Provide IN services to subtending users and lower level switches equipped as network access points.

intelligent peripheral: in intelligent network, contains an adjunct service point that responds to requests for service processing. Provides and manages resources such as speech recognition, voice synthesis, announcements, and digit collection equipment as required by IN service logic.

intelligent terminal: device programmed to perform some of the tasks usually assigned to a host machine—such as screen formatting and data editing, or more complex assignments—so as to reduce dependence on the central processor.

interactive broadband service: BISDN service in which there is an exchange of video and/or high-speed data information.

interactive network: a non-switched network in which several (secondary) users obtain services from a specialized (primary) node within the network (e.g., data processing services provided by a central host). Also, one in which many equal status users exchange information (e.g., a local-area network, or a VSAT network).

intercarrier interface: facilitates the interconnection of intra–LATA and inter–LATA SMDS networks.

intercommunication: telecommunication that takes place between two sites.

interexchange carrier: provides long-distance transport services between points-of-presence that are established for each LATA.

International Electrotechnical Commission: promotes safety, compatibility, interchangeability, and acceptability of international electrical standards.

International Standards Organization: the leader in standardizing communications between cooperating, data processing systems.

International Telecommunication Union: a specialized agency of the United Nations that regulates radio communications, standardizes international telecommunications, and develops global networks.

International Telegraph Union (ITU): forerunner of International Telecommunication Union.

Internet: an international network of packet networks operated by a government-university-industry consortium and serving data users around the world.

Internet protocol: a connectionless Internet network layer procedure.

inter-office signaling: the exchange of signaling messages between switches that serve the routes between calling and called parties.

interpersonal message service: with regard to X.400, a service in which messages are delivered to a mailbox as they are created, and stored until delivery is requested by the owner of the mailbox.

interswitching system interface: in SMDS, facilitates the interconnection of switches from different manufacturers.

invasion of company privacy: theft, interception, misuse, or aggregation of information concerning the operation of the business, particularly that information which may give a competitive advantage to the receiver.

invasion of individual privacy: theft, interception, misuse, or aggregation of information concerning an individual's thoughts, sentiments, emotions, lifestyle or status.

I/O channel: in an IBM mainframe, a software-driven processor that executes commands and programs and provides high-speed input and output channels for the host.

ISDN B channel: 64 kbits/s switched digital service.

ISDN H0: 384 kbits/s switched digital service.

ISDN H11: 1536 kbits/s switched digital service.

ISDN service node: in an SS7 network, a facility that supports provision of ISDN bearer services and supplementary services.

ISDN user part: an SS7 protocol that represents the location of model layers invoked when the network employs the transport capabilities of the MTP and SCCP to provide call-related services to the user.

isochronous medium access controller: in FDDI-II, supports voice and video applications.

ISO/IEC Joint Technical Committee 1: formed in 1987 by the merger of information technology committees of ISO and IEC, JTC1 is responsible for consensus standards in the area of information processing systems.

keying: the process of modulating a continuous wave carrier with a binary, digital message signal. See ASK, FSK, and PSK.

LAP-B: Link Access Procedure-Balanced—a component of ITU-T's X.25 standard.

LAP-D: Link Access Procedure-D Channel—for ISDN D-channel data transport.

LAP-F core: a link access procedure that includes most of the capabilities of LAP-D including frame delimiting and alignment (using beginning and ending flags), inspection to ensure that the frame consists of an integer number of octets that is less than the maximum permitted, transparency (using zero-bit stuffing), frame multiplexing and demultiplexing (using the address field), and detection of transmission errors (using CRC). Flow control (using windowing) and error correction (using ARQ) are not included.

large-carrier AM: a modulation technique in which the carrier signal is multiplied by the factor $[1 + \mu m(t)]$. The modulated wave is double-sideband, and contains a large carrier signal component.

least-cost routing: automatic route advancement that is coupled with the requirement to start with the cheapest service and advance to the most expensive.

linear block code: a block of k information bits is presented to the encoder; the encoder responds with a unique code block of m bits ($m > k$). The set of codewords is selected so that the logical distance between each codeword and its neighbors is approximately the same.

linear prediction model: in digital voice, speech analysis and synthesis are performed using models in which successive samples of speech are compared with previous samples to obtain a set of linear prediction coefficients (LPCs) that are used to reproduce the sound.

line overhead: in SONET, part of the transport overhead that locates the first octet of the synchronous payload envelope (pointer function), and monitors line errors and communications between line terminating equipment.

link-by-link encryption: the data link PDU is encrypted as it enters the physical layer.

local access and transport area: a grouping of exchange areas served by a single LEC. Traffic between LATAs must be carried by an interexchange carrier.

local–area network: a bus or ring connected, limited distance network that serves the data communication needs of users within a building or several buildings in proximity to each other.

local exchange carrier: provides public, switched telephone services within limited areas called LATAs. The LEC provides connections between users located within the same exchange area, between users located in different exchange areas that are in the same LATA, and delivers traffic directed to an exchange area outside the LATA to inter-exchange carriers (IXCs).

local network: a star or hierarchically connected network provided by a local exchange carrier company to transport voice and data traffic within a defined area and to connect with networks of inter-exchange carrier companies.

locked: in network management, an administrative state in which the managed object cannot be used.

logical channel number: in a packet network, defines the virtual connection from DTE1 to DCE1.

logical link control sublayer: defines the format and functions of the protocol data unit that is passed between service access points in the source and destination stations over an IEEE 802 LAN.

logical unit: in SNA, a network addressable unit through which an end–user accesses the network to communicate with another logical unit.

long-distance network: a mesh or hierarchically connected network that spans states, countries, continents, or the world.

longitudinal redundancy check: parity bits are assigned to bit sequences formed from bits in specific positions in a block of characters. They form a block check character.

lost traffic: some blocked calls are likely to be abandoned by the callers and some may be transferred to the public network at the initiative of the caller. In either case, they are lost to the private network.

low-earth orbit: orbit a few hundred miles above the earth's surface.

low-pass characteristic: performance such that energy is transmitted from 0 Hz (dc) to *gf* Hz, but not at frequencies above *gf* Hz

M1: T-carrier multiplexer that combines 24, DS-0s into one DS-1, also known as M24, M24/56, and M24/64.

M1C: T-carrier multiplexer that combines 2, DS-1s into one DS-1C.

M5: a subrate multiplexer with customer access = 5, 9.6 kbits/s data channels. Output = DS-0B.

M10: a subrate multiplexer with customer access = 10, 4.8 kbits/s data channels. Output = DS-0B.

M12: T-carrier multiplexer that combines 4, DS-1s into one DS-2.

M13: T-carrier multiplexer that combines 28 DS-1s into one DS-3.

M20: a subrate multiplexer with customer access = 20, 2.4 kbits/s data channels. Output = DS-0B.

M23: T-carrier multiplexer that combines 7, DS-2s into one DS-3.

M24: original T-1 multiplexer. Customer access = 24, DS-0 voice channels. Per channel signaling uses 8th bit robbing in every 6th frame. Framing uses 8000 bits/s. Output = DS-1.

M24/56: multiplexer with customer access = 23, 56 kbits/s data channels. One channel is used for common channel signaling. Output = DS-1.

M24/64: multiplexer with customer access = 23, 64 kbits/s data channels. Employs B8ZS, and 1, 64 kbits/s channel for data synchronization, alarms and signaling. Output = DS-1.

M34: T-carrier multiplexer that combines 6 DS-3s into one DS-4.

M34NA: T-carrier multiplexer that combines 3 DS-3s into one DS-4NA.

M44: multiplexer with customer access = 44, DS-0 voice channels that are converted to 44, 32 kbits/s voice channels. 2, DS-0 channels are used for common channel signaling. Output = DS-1.

M48: multiplexer with customer access = 48, DS-0 voice channels that are converted to 48, 32 kbits/s channels. Per channel signaling uses 8th bit robbing in every 6th frame. Output = DS-1.

M88: multiplexer with customer access = 88, DS-0 voice channels that are converted to 88, 16 kbits/s voice channels. 2, DS-0 channels are used for common channel signaling. Output = DS-1.

M96: multiplexer with customer access = 96, DS-0 voice channels that are converted to 96, 16 kbits/s channels. Per channel signaling uses 8th bit robbing in every 6th frame. Output = DS-1.

managed object: in network management, a network entity that is included in the management activity. A managed object may be a switch, a PBX, a router, a multiplexer, a workstation, or other hardware device, or a buffer management routine, a routing algorithm, an access scheme, or other software package.

managed object class: in network management, objects with similar characteristics (i.e., attributes, operations and notifications).

managing process: in network management, that part of the application process that is concerned with overall management tasks.

management information base: in network management, contains both long-term and short-term information concerning managed objects.

manchester signal: a 1 is a positive current that changes to a negative current, and a 0 is a negative current that changes to a positive current.

manhattan street network: metropolitan–area network that consists of orthogonal rings that pass information in alternating directions.

***m*-ary signaling**: digital modulation technique in which *M* distinct symbols are employed; each represents $\log_2 M$ bits. The symbols are generated by changing the amplitude, phase, or frequency (alone, or in combination) of a carrier (either a single carrier, or a coherent quadrature-carrier arrangement) in *M* discrete ways. *M*-ary signaling schemes are used to transmit datastreams over bandpass channels when the desired data rate is more than can be achieved using binary schemes.

mass communication: telecommunication in which information flows simultaneously from a single (transmitting) site to a large number of (receiving) sites.

medium: an entity that produces action-at-a-distance.

medium access control sublayer: in local–area networks, defines the format and functions of headers and trailers that are added to PDUs so that entire frames can travel between source and destination addresses over IEEE 802 LANs.

mesh network: in a fully interconnected mesh, each node is connected to every other node.

message-framed data: each message is framed by data devoted to the transmission process, i.e., a header and a trailer.

message-switched network: one in which complete data messages are received, stored, and, when facilities are available and the receiving station is ready, transmitted to their destinations.

message transfer agents: in X.400, agents that interact directly with user agents and other message transfer agents. They route messages, relay messages, and store messages as required.

message transfer part: an SS7 protocol that provides reliable transfer of signaling information across the signaling network. This includes the capability to respond to network and system failures so as to ensure that reliable transfer is maintained.

message transfer service: in X.400, a service (store, forward and store) that delivers a message as soon as the receiver is ready to receive it. At the receiver, the message is stored until called for.

metropolitan–area network: described by IEEE 802.6, a facility that spans an urban core, metropolitan area, or major industrial campus. Connecting the premises of many organizations, MANs provide the opportunity to employ public, switched, high-speed facilities to transport data, voice, and video.

microcell: in wireless networks, a small cell.

microcom networking protocol: by assembling asynchronous data in blocks and performing cyclic redundancy checks, MNP permits the receiving modem to verify the integrity of the received stream. Versions of MNP include data compression, automatic negotiation to achieve the highest data speed compatible with link conditions, and management of half-duplex connections so as to optimize throughput in both directions.

minimum-shift keying: digital modulation technique in which advantage is taken of phase continuity, and the waveforms representing the datastreams impressed on the quadrature carriers are sinusoidal (rather than NRZ pulses). Employing one-half of the frequency spacing used in (Sunde's) FSK, the main lobe of the PSD diagram spans a frequency range that is 25% less than ASK and 50% less than FSK. The power in the sidelobes drops off very quickly; more quickly than FSK and much more quickly than ASK.

mobile: a mobile station in a mobile telephone network.

mobile telephone: a telephone instrument in which part of the path between instrument and mobile telephone switching office is completed by radio. The radio link replaces the local loop so that the user (mobile) can roam freely.

mobile telephone switching office: in wireless networks, an electronic switching system programmed to provide call-processing and system fault detection and diagnostics. The MTSO is connected directly by wireline, or microwave radio, to a set of base stations that communicate by radio with the mobiles in their cells.

modem: data circuit terminating equipment that converts digital signals to analog signals that match the bandwidth available on voice-grade lines.

modified Huffman code: a combination of run-length coding and variable length coding designed for use with facsimile signals.

modified READ code: in the Modified Relative Element Address Designate

(READ) Code, run lengths are encoded by noting the position of change elements (i.e., where white changes to black, or black changes to white) in relation to a previous change element. Designed for use with facsimile machines.

modulating signal: see modulation.

modulated wave (signal): see modulation.

modulation: a signal processing technique in which one signal (the modulating signal) modifies a property of another signal (the carrier signal) so that a composite wave (the modulated wave) is formed.

monitor bit: in an IEEE 802.5 LAN, bit used to detect unclaimed frames.

multiplexer: equipment that combines digital signals (voice or data) into a TDM signal.

narrowband signal: an analog signal whose frequency components are contained within the nominal passband (4 kHz) of a voice channel.

National Institute for Standards and Testing (NIST): a government unit (formerly the National Bureau of Standards) charged with providing an orderly basis for the conduct of business by maintaining measurement standards. Coordinates voluntary product standards. Develops computer software and data communication standards.

network: set of links attached to nodes. The nodes serve to receive traffic from one link and send it on another.

network access point: switches that are not equipped with intelligent network switching capabilities may be equipped as network access points. NAP capabilities permit access to a limited set of IN services. The NAP routes the call to an IN switching system, identifies the user, and requests IN treatment.

network ACD: in intelligent networks, enables subscribers to control the destination to which 1-800...calls are routed in order to match peak loads with availability of attendants and provide relief for overflow from congested ACDs.

network addressable unit: entity that terminates the SNA network. Resident in peripheral and subarea nodes, and interconnected by the path control network, they provide services that permit end–users to send and receive data, and help perform network control and management functions.

network attendant service: in intelligent networks, routes calls requiring attendant assistance to a central location.

network control point: location in common channel signaling network that contains databases that are used to implement special services.

network control program: in SNA, resident in the front-end processor (communications controller), NCP provides network information to the TP access method, controls cluster controllers and attached terminal nodes, routes data to communications controllers in other subareas, performs as a primary station, polls secondary stations, dials and answers switched lines, detects and corrects errors, maintains operating records, and allocates buffers for data coming from and going to the host, and to other stations. NCP is a specific IBM product.

network header: in OSI reference model, provides destination address (in a format understood by the chosen network) and sequence number.

network hierarchy: the ways in which nodes (switching centers, multiplexers, etc.) are arranged to achieve efficient division and routing of different classes of communications traffic.

network information database: in intelligent networks, contains customer and network information.

network interface device: device installed by carrier at network extremity that performs testing, loopback, and powering functions.

network layer: layer 3 in OSI model. Provides end-to-end communications services to the transport layer. Makes use of the underlying data link layer services to establish, maintain, and terminate transmission across the network.

network management: the planning, organizing, monitoring, and controlling of activities and resources within a network to provide a level of service that is acceptable to the users, at a cost that is acceptable to the owner.

network protocol data unit: combination of network header and transport PDU.

network quality speech: contains the full range of frequencies without perceptible noise or distortion.

network resource manager: in IN, determines which module provides resources to continue call processing. May give routing instructions to establish connection to module.

network termination 1: digital interface point between ISDN and the customer's equipment. Facilitates functions principally related to the physical layer, i.e., physical and electrical termination of the network at the customer's premises, performance monitoring, power transfer, contention resolution, overvoltage protection, and testing.

network termination 2: digital interface point between ISDN and customer's equipment. Includes enhanced functions associated with the data link layer and the network layer. Facilitates protocol handling, switching, concentrating, and multiplexing functions that are associated with PBXs, cluster controllers, local networks, and multiplexers.

network topology: the ways in which nodes and links are interconnected.

(N)-layer managed object: in network management, object specifically contained within an individual layer (e.g., data link protocol),

node-network interface standard: defines the forms, formats, and sequences of messages that flow between nodes and communication networks across node-network interfaces.

noise: the sum of all the unwanted signals that are added to the message signal in the generation, transmission, reception, and detection processes employed in effecting telecommunication. The difference between the received (corrupted) signal and the intended (ideal) signal.

non-switched network: provides communication services to users through a common facility without resorting to switching.

normalization: process in which a signal is scaled so that its maximum magnitude is 1.

North American digital cellular standard: a wireless information system that provides three 16 kbits/s digitally encoded TDMA channels in a 30 kHz cellular phone channel.

North American numbering plan: numbering plan for North America and certain countries in Caribbean Basin. Within the NANP area geographic segmentation is accomplished by 3-digit area codes.

null compression: a data compression technique that reduces the number of zeros transmitted. A null-compression-indicating character is used followed by the number of all-zeros characters in the string.

null modem: cable that routes the transmit data circuit of one DTE to the receive data circuit of the other DTE, and *vice-versa*, to achieve direct connection of DTEs.

Nyquist rate: a sampling rate of $2W$ samples per second taken from a signal that is band-limited to W Hz.

Nyquist's theorem: the signaling rate (r) in bits/s is related to the passband (B) in Hertz by $r \leq 2B$.

octet-level multiplexing: octets from each input line are interleaved on the common bearer in time–slots reserved for them. After each sequence of octets (one from each input line) has been serviced, the multiplexer adds a bit to define the frame.

odd parity check: the parity bit makes (or leaves) the number of 1s odd.

offered traffic: the number of calls attempted, times the average holding time of the carried traffic.

office database: in a switching machine, a collection of data unique to a specific switching location that describes the office configuration and customer connections.

off-net call: call for which either the originating or the terminating end of the connection is within a private network, and the other end is in another network.

offset-QPSK: quaternary phase-shift keying in which one of the datastreams is delayed by one *bit* period. Phase changes are limited to 90°; this makes performance in a non-linear environment significantly better than QPSK.

on-net call: call for which the originating and terminating stations are within the same private network, and the traffic is carried over the facilities of the network.

open systems architecture: permits transparent communication between cooperating machines.

operating time: the time to establish and release a connection. It is the sum of call set-up time and time to disconnect.

operations, maintenance, and administration part: an SS7 protocol that provides the procedures required to monitor, coordinate and control all the network resources needed to make communication based on SS7 possible.

operations system: in IN, performs network operations functions such as resource administration, surveillance, testing, traffic management, and data collection.

operator services system: in an SS7 network, a facility that provides operator assistance, particularly directory assistance services.

optical carrier level 1: the optical equivalent of STS-1.

optical carrier level N: the optical equivalent of STS-N.

OSI reference model: model of communication between cooperating machines that incorporates OSA. Consists of seven layers defined by a service definition (i.e., what functions does the layer perform) and a protocol specification (i.e., what procedures are employed to execute the functions).

other common carrier: independent carrier offering network services in competition with AT&T.

out-of-band signaling: the associated signals are analog tones that do not occupy the same bandwidth as the information exchange.

out-of-frame event: in ESF error checking, when two out of four consecutive framing bits are incorrect.

out-of-slot signaling: the associated digital signals use a time-slot assigned exclusively for signaling.

outside plant: all facilities situated between the distribution frame of the serving office and the customers' premises.

packet: a sequence of bits that is divided into two parts—one contains the user's information, and the other contains control information.

packet assembler and disassembler: device that assembles packets from characters received from asynchronous DTEs and disassembles packets into characters for transmission to asynchronous DTEs. ITU-T Recommendation X.3 defines the PAD that is usually implemented in software resident at the boundary of the network.

packetized-ensemble-protocol: modem that employs a comb of 512 carrier tones within the voice band to send bits in parallel as QAM symbols.

packet medium access controller: in FDDI-II, supports data access.

packet switch: a network node that receives packets, performs error and congestion control, may store the packets briefly, and sends them on towards their destination over routes determined by information carried in the packet, by information stored at the node, and by the state of the network.

packet-switched network: network in which data messages are divided into segments that are inserted into packets. The packets are statistically multiplexed across links between store-and-forward nodes. Users employ X.25 protocols to connect synchronous DTE to the network. Within the network, the nodes route packets in accordance with information carried in individual packets, information stored in the node, and the state of the network.

pair-gain: in the local loop, concentration and multiplexing is added at appropriate points to replace a larger number of underutilized loops by a smaller number of loops that are in use more often.

passband: the range of frequencies transmitted without distortion by a bearer and associated equipment.

path overhead: in SONET, part of synchronous payload envelope that verifies connections, and monitors path errors, receiver status, and communication between path terminating equipment.

payload: in a frame, the data being transported between users.

PC Network: a bus-oriented network that can connect up to 72 stations (PCs) within a 1000 feet radius and support message exchanges at 2 Mbits/s.

peak busy hour: selected from a series of daily busy hours, the one that exhibits the highest traffic intensity.

peaked traffic: traffic in which many users demand service at the same time. The variance to mean ratio (α) is greater than 1.

peer polling: stations poll one another, in sequence, continually. If the polled station has something to send, it sends it; if it has nothing ready, it polls the next station in the sequence.

performance management: in network management, evaluate the performance of components, and the network as a whole, with regard to availability, response time, accuracy, throughput, and utilization.

periodic function: eternal function that repeats at regular intervals.

peripheral node: in SNA, terminal or cluster controller assigned to a specific subarea.

permanent virtual circuit: a virtual circuit that is permanently assigned between two stations. Fixed logical channel numbers are used by the sending DTE and the receiving DCE. Because it is a permanent (virtual) connection, no connect or disconnect packets are required to set up the circuit, and information packets may be transmitted at will by the originating DTE.

personal communication networks: first used in the United Kingdom, and then adopted in Europe, a term employed to identify forms of metropolitan–area, portable radio telephony networks.

personal communication services: a term employed by the FCC to mean a family of services provided over mobile or portable radio communications facilities for individuals and businesses on the move. These services will be provided in an integrated manner over a variety of competing networks.

personal services: voice, video, and document services used for the exchange of information among persons, and between persons and machines.

personal telecommunication: communication-at-a-distance that involves the distribution or exchange of messages among persons, or between persons and machines.

phase modulation: modulation produced by varying the phase angle of the carrier in sympathy with the amplitude of the message signal.

phase-reversal keying: phase-shift keying that employs two states. The presence of the carrier signal in one phase condition corresponds to a 1-state, and the presence of the carrier signal 180° out of phase with it corresponds to a 0-state. Also known as binary-PSK.

phase-shift keying: digital modulation technique in which the transmitted signal is a sinusoid of constant amplitude whose phase is varied by the modulating datastream. When two states are employed, the presence of the carrier signal in one phase condition corresponds to a 1-state, and the presence of the carrier signal 180° out of phase with it corresponds to a 0-state. This is known as *binary*-PSK (BPSK) and sometimes as *phase-reversal* keying (PRK).

phoneme: an elementary human sound uttered at rates up to 10 per second.

physical layer: layer 1 in OSI model. Makes the connection over which bit–streams flow between the users.

physical stage: with regard to using the telephone, establishing a path along which signals can flow between users.

physical unit: in SNA, a NAU that manages attached logical units and acts as a routing node for communication.

plesiochronous multiplexing: a multiplexing process in which two or more signals are generated at nominally the same digital rate and their significant instants occur at nominally the same time.

point-of-presence: the facility interface between LEC and IXC. A POP must be established on each trunk path that connects switches that belong to the local and long-distance carriers. On one side of the POP, the LEC is responsible for service; on the other side, the IXC is responsible.

polar orbit: orbit in which satellites travel around the earth passing over the poles.

Post, Telegraph, and Telephone Administration: agency responsible for national communication networks. Administers standards within the country and cooperates with the recommendations of ITU to achieve international communications.

power spectral density: plot of the variation of signal power (in watts per Hertz) with frequency.

presentation header: identifies the specific encoding, compression and encryption employed.

presentation layer: layer 6 of OSI model. Translates the information to be exchanged into terms that are understood by the end systems.

presentation protocol data unit: combination of presentation header and application PDU.

preventive cyclic retransmission: SS7 error control technique that uses forward error correction. It is employed on satellite links.

primary polling: in turn, the primary station asks each of the secondary stations if they have anything to send. If a secondary station has a frame (frames) to send, it transmits it (them) to the primary, when polled; if the secondary station has nothing ready to send, it indicates nothing to send and the primary station polls the next station.

primary rate interface: ISDN channel consisting of 23 B-channels + 1 D-channel. The PRI is supplied over a single connection that operates at 1.544 Mbits/s. This is the T-1 rate; it contains 1.536 Mbits/s of user and signaling data, and 8 kbits/s for framing the bitstream.

primary reference clock: in network synchronization, a timing source that establishes exact time. In AT&T's network, PRCs consists of an ensemble of primary or secondary atomic standards that achieve a frequency accuracy of better than 1×10^{-11}, and permit no more than two slips per year.

primary station: station that determines when secondary stations shall send, and when they shall receive.

priority bits: in an IEEE 802.5 LAN, they identify the level of priority a station must have to seize the token.

private branch exchange: a circuit switch that serves a community of stations. Usually, the community is located in a limited area that includes the PBX.

private network: one created by a corporate entity. It is used to complete some fraction of the voice, data, image, and video communications employed in running the business. Regardless of who owns them, the facilities employed in private networks provide communication services for the benefit of the corporate community.

procedural stage: with regard to using the telephone, ensuring that the information is in a common language and is exchanged between the right parties.

protocol data unit: defined by IEEE 802.2, the frames exchanged between data link layers in a local–area network. Also, the frames exchanged between peer entities in any protocol stack.

public network: provides voice, data, image, or video communications services to any person with the ability to pay the fee published in the tariff describing the service. The person paying the fee controls the purpose and content of the communication, and benefits from the exchange.

pulse amplitude–modulation: a stream of pulses whose amplitudes are varied according to some rule.

pulse code–modulation: in digital voice, a robust coding scheme that employs a sampling rate of 8000 samples/s, codes each sample in 8-bits, and employs µ-law or A-law companding to produce a signal with a bit-rate of 64,000 bits/s.

quadrature-carrier amplitude modulation: digital modulation technique that employs orthogonal carriers derived from the same source. By dividing the datastream in two streams (even-numbered bits and odd-numbered bits) that amplitude modulate orthogonal carriers ($\sin\omega_c t$ and $\cos\omega_c t$) symbols are transmitted at rate $2r$ in a bandwidth of r Hz.

quadrature phase-shift keying: digital modulation technique that employs orthogonal sinusoidal carriers derived from a common source and phase-shift keying. The datastream is divided in two streams (of rate r symbols/s) that phase modulate the orthogonal carriers ($\cos\omega_c t$ and $\sin\omega_c t$). By transmitting symbols at rate $2r$ symbols/s in a bandwidth of r Hz a spectral efficiency of 2 bits/s/Hz is achieved.

quadriphase PSK: see quadrature phase-shift keying.

quantizing: the process that segregates sample values into ranges and assigns a unique, discrete identifier to each range. Whenever a sample value falls within a range, the output is the discrete identifier assigned to the range.

quantizing noise signal: in digital voice, the difference between the signal at the output of the reconstruction filter, and the original band-limited signal, assuming no errors are introduced in transmission.

quaternary-PSK: see quadrature phase-shift keying.

Radiocommunications Sector: sector of ITU concerned with radio issues and the allocation of the electromagnetic spectrum.

raised-cosine pulse: signal consisting of one period of a cosine function ($-\pi$ to $+\pi$ radians) offset by its amplitude so that the values are always ≥ 0.

random function: function for which future values can be described only in probabilistic terms. Neither the value at a specific future time, nor the future time of a specific value, is known for sure.

random traffic: traffic in which the calling pattern is random and the variance-to-mean ratio (α) is 1.

real-time medium: an entity that makes it possible to communicate-at-a-distance, at the same time.

reconstruction filter: in digital voice, a low-pass filter that passes frequencies up to W Hz. The input is a train of pulses whose amplitudes correspond to sample values of a signal limited to W Hz, and whose repetition rate corresponds to the sampling rate ($\geq 2W$ samples/s). The output is a close likeness of the original band-limited signal ($0 < f \leq W$ Hz).

record medium: an entity that makes it possible to communicate at a later time, or at a distance, at a later time.

regenerator: see repeater.

Regional Bell Operating Company: an independent carrier that provides local exchange services within specific areas.

remote operations service element: in network management, general software support system that supports distributed, interactive applications processing between remote objects. In synchronous mode, ROSE requires the receiver to reply before the originator can invoke another operation. In asynchronous mode, ROSE permits the originator to continue invoking operations without waiting for a response.

repeater: in digital transmission, a device that reads the pulse stream before it degrades substantially, regenerates a new stream and passes it on to the next unit. Also, between LANs, if the differences are limited to the implementation of the physical layers, repeaters are used to connect them and amplify and regenerate the signals.

reservation-ALOHA: in satellite communication, the ALOHA process is modified to allow stations to reserve time-slots. The technique guarantees that messages will arrive one-at-a-time at the transponder and will be relayed without collisions.

reservation bits: in an IEEE 802.5 LAN, they provide a mechanism for devices to request the opportunity to transmit.

resistance design: in the local loop, a combination of wire sizes is chosen for each customer loop so that the total resistance does not exceed 1300 ohms, the value that allows sufficient current to flow from the serving office battery for ringing.

response: in SDLC, a frame that passes from a secondary station to the primary station.

retried traffic: blocked calls that are placed again at later times at the discretion of the callers.

return-to-zero: bits are represented by a positive or a negative current pulse that returns to zero before the next bit in the sequence is handled.

richness: a message is said to be richer than another when it includes more message dimensions. Thus, video messages are richer than audio messages, audio messages are richer than graphics messages, and graphics messages are richer than text messages.

ring network: each station is connected to two other stations so that there is a continuous, single-thread connection among all units.

router: device that modifies segment sizes, routes frames between users on non-identical, but the same type of networks, and connects at the network level.

run-length compression: a superset of null compression that is used to compress any string of repeating characters.

sampling: the process of determining the value of the amplitude of an analog signal at an instant in time. Usually, sampling is done in a regular repetitive manner at equal time intervals.

scoping: in network management, identifies the sub-tree of managed objects to which a particular filter is applied.

scrambling: a datastream is multiplied by a semi-random sequence to produce an acceptable ratio of 1s and 0s. After the scrambled stream is transmitted from sender to receiver, the receiver multiplies it by the inverse of the semi-random sequence to produce the original datastream.

section overhead: in SONET, part of the transport overhead that is concerned with framing, frame identification, section error monitoring, and communication between section terminating equipment.

security management: in network management, control access to, and use of, the network according to rules established by the network owner.

segmentation and reassembly sublayer: component of ATM adaptation layer. At the sending end, performs segmentation of large data packets into ATM cells, and the inverse operation at the receiving end.

selective dynamic overload: a protective routing scheme that relieves congestion at an overloaded switch by sending an instruction to all switches connected to it to reduce the volume of their requests for service for a fixed time interval.

selective-repeat: an ARQ technique used on duplex connections, the sender only repeats those frames for which negative acknowledgments are received from the receiver, or no acknowledgment is received.

selective trunk reservation: a protective routing scheme. When the number of idle trunks in a particular group drops below a specified threshold, the idle trunks remaining are used preferentially to complete single-link (first-routed) calls.

self-information: the self-information of a symbol is inversely proportional to the probability of the symbol being transmitted.

service access point: point within the sending or receiving device that acts as the logical port for the exchange of data.

service control point: in intelligent network, consists of processor-based service logic interpreter, network information databases, signaling network interface, and network resource manager. In an SS7 network, a facility that provides database and call processing procedures for special calls.

service logic interpreter: in intelligent network, executes service logic programs and handles exchanges between IN modules.

service logic program: in IN, program that defines the operation of specific network features or service application. SLPs reside in service control points adjunct processors, intelligent peripherals, and service nodes, or within the network switches themselves.

service node: in intelligent networks, provides service logic and implementing resources in a single network entity. Contains adjunct service point that responds to requests for service processing. Provides access to service logic programs that support a specific service, or group of services, or a full range of service functions (including attendant services).

service plane: describes IN services from a user's perspective.

services plane: a plane containing orthogonal axes that divide real-time media (right-hand side) from record media (left-hand side), and intercommunication (upper-half) from mass communication (lower-half). The axes are marked off in richness in both directions from the crossing point, and the horizontal axis represents the richness in the upstream direction, while the vertical axis represents the richness in the downstream direction

service switching point: in an SS7 network, a facility embedded in the switching point, that processes calls that require remote database translations. The SSP recognizes special call types, communicates with the service control point, and handles the calls according to SCP instructions.

serving area: contains all of the premises connected to a serving area interface. An area served by a group of distribution cables.

serving area interface: a cross–connect point connecting a feeder cable to the distribution cables that serve a serving area.

session: a temporary connection between machines or programs for an exchange of messages according to the rules established for such exchanges. In SNA, communication between network addressable units.

session header: added to presentation PDU to define line discipline, direct establishment, maintenance and termination of connection, and identify any specific markers employed.

session protocol data unit: combination of session header and presentation PDU.

setup channel: in CMRTS, radio channel that transmits and receives data used to set up calls.

setup channel: in mobile telephone networks, the channel over which the signaling messages used to set up all calls are transmitted.

severely-errored second: in ESF error checking, any second in which from 320 to 333 CRC-6 error events are detected, or one OOF event occurs.

shutting down: in network management, an administrative state in which the managed object will honor current users, but will not accept new users.

sights: static or dynamic relationships among objects arranged in space likely to present vistas, to witness events, or to assist with the visualization of ideas.

signal: manifestation of message exchanged-at-a-distance.

signal constellation: a polar diagram in which points represent the phase and amplitude of QAM signals.

signaling: the name given to the exchanges of messages that inform the users, and the telecommunications facilities, of what actions to take in order to set up, maintain, terminate, and bill a call.

signaling connection control part: an SS7 protocol that enhances the services of the MTP to provide full OSI network layer capabilities. In particular, SCCP extends MTP's addressing and routing capabilities and provides connectionless (CL-SCCP) and connection-oriented (CO-SCCP) classes of services.

signaling end-point: a SS7 node connected by high-speed data links to a network of signal transfer points. can be a switching point, a service switching point, a service control point, an ISDN service node, or an operator services system.

signaling rate: the number of independent symbols sent per second.

signaling system #7: a common-channel, network signaling system that supports normal message transfer activities as well as user-to-user signaling, closed-user groups, calling line identification, call forwarding, etc. The basic network architecture is a quad structure consisting of signaling end points and signal transfer points. To provide added reliability, each SEP is connected to the SS7 network by duplicate STPs.

signaling tone: in mobile telephone networks, an out-of-band tone burst (10 kHz). Among other things, the signaling tone is used to initiate ringing, to signal hand-off to another cell site (tune the mobile to another set of frequencies), and to disconnect.

signal-to-noise ratio: the ratio of the power in the message signal to the power in the noise signal. It is usual to express this ratio as a logarithm. The unit ratio is 1 decibel (db).

signal transfer point: facility that performs as a link concentrator and message switcher to interconnect signaling end points. Routes signaling messages directly to the terminating switch, or directly to the STP that serves the terminating switch.

signal unit error rate monitor: device that monitors error performance of SS7 links.

simplex: information flows in only one direction (a single channel is employed).

single-channel per carrier: a demand assignment, multiple access technique in which each carrier is used as one channel.

single-sideband AM: a modulation technique that produces a modulated wave in which the message energy is contained in one sideband. There is no carrier component. When the higher frequency sideband is employed, the modulated wave is described as upper-sideband (USB) SSB; when the lower frequency sideband is employed, the modulated wave is described as lower-sideband (LSB) SSB.

SLC-96: a digital loop carrier that multiplexes 96 individual voice or data channels over 4, DS-1 facilities or a DS-2 facility. Includes the option to concentrate 48 channels on each DS-1 facility.

sliding window flow control: process in which the transmitter can transmit up to W packets and must then wait for an acknowledgment from the receiving DTE. The value of W is negotiated between sending and receiving DTEs.

slip: in network synchronization, a slip occurs when a timing error causes a time-slot to be repeated or deleted.

slow associated control channel: in Global System for Mobile Communications, the signaling channel that handles call supervision and related activities.

slot: in an IEEE 802.6 MAN, consist of 53 octets (52–octet payload + 1–octet header). The number of slots depends on the data rate (13 slots at T-3, 45 slots at STS-3). When permitted, nodes write data to the slots, and read and copy data from them.

slotted-ALOHA: in satellite communication, the ALOHA process is modified so that all packets arrive at the transponder at the beginning of agreed upon time–slots.

smart terminal: device that contains memory and logic that can be set up to perform specific tasks.

SMDS interface protocol: a three level protocol that addresses, frames, and transmits PDUs using cell-relay techniques.

SMDS network interface: a broadband network access arrangement that supports switched, high-speed, connectionless data services. Through the use of a three-level, SMDS interface protocol, SMDS transfers information in variable length frames up to 9188 octets over connections that employ cell-relay techniques. Each PDU is routed independently to its destination. Error detection (but not correction) is performed and corrupted frames are discarded (without requesting retransmission).

smooth traffic: traffic in which calling is continuous. The variance-to-mean ratio (α) is zero.

source coding: in digital voice, speech sounds are identified and encoded.

space diversity: in satellite communication, when satellites are physically separated from each other, i.e., they occupy different orbit positions.

space diversity: to avoid signal interference, satellites are physically separated from each other. They occupy different orbit positions.

space-division network: one in which input and output are connected by physical links that provide a single channel. They are neither time-divided, nor frequency divided.

space-time-space switch: combination of space switching stages and time switching stage to serve a large customer community. The time stage interconnects the space stage serving the calling party with the space stage serving the called party.

specialized common carrier: carrier authorized to carry data signals over private facilities between major cities.

specific management function area: in network management, area of specialized technical responsibility.

spillover: in digital modulation, name given to power that extends beyond the transmission bandwidth.

spread spectrum signal: a very wideband signal whose energy density is low enough so as not to produce noticeable interference in conventional AM or FM signals using frequencies in the same spectrum space. Produced by direct sequence or frequency hopping spectrum spreading techniques.

stand-alone STP: in an SS7 network, a signaling point that provides only STP capability, or STP, and SCCP capabilities.

star network: all links radiate from a central node so that all user stations are connected to it, and all information exchanges must transit this node.

start-stop, DLC protocol: procedure that manages an asynchronous stream of characters framed by start and stop bits.

station signaling: the exchange of signaling messages over local loops that connect the users' stations to the serving switch; station signaling is limited to an exchange area.

statistical multiplexing: see asynchronous multiplexing.

stop-and-wait: an ARQ technique in which the sender sends a frame and waits for acknowledgment from the receiver.

store-and-forward: an operation in which messages are generated, sent to an intermediate location, stored, and forwarded to the recipient on demand.

Stratum 1: in AT&T's network, a group of 16, free-running, primary reference clocks.

Stratum 2: in AT&T's network, a group of clocks located in toll offices. Disciplined by PRCs, they achieve a frequency accuracy of better than 1.6×10^{-8}, and permit no more than ten slips per day.

Stratum 3: in AT&T's network, a group of clocks located in local networks. Disciplined by clocks in Stratum 2, they achieve a frequency accuracy of better than 4.6×10^{-6}, and permit no more than 130 slips per hour.

Stratum 4: a group of clocks located in private networks. Disciplined by clocks in Stratum 3, they achieve a frequency accuracy of better than 3.2×10^{-5}, and permit no more than 15 slips per minute.

subarea: in SNA, a network address associated with a subarea node and attached peripheral nodes.

subarea node: in SNA, communications controller or host.

subband coding: in digital voice, the input signal frequency band is divided into several subbands (typically 4 or 8). Each subband is encoded using an adaptive step size PCM (APCM) technique.

subnetwork access functions: a sublayer of ISO's network layer protocol, concerned with data transfer within individual networks.

subnetwork dependent convergence functions: a sublayer of ISO's network layer protocol, concerned with correcting network anomalies and deficiencies.

subnetwork independent convergence functions: a sublayer of ISO's network layer protocol, concerned with internetwork routing and relay functions, and global network protocol.

subrate data: data at 2.4, 4.8, 9.6, and 19.2 kbits/s.

subscriber code: identifies an individual station line that connects the subscriber to the serving switching facilities.

subscriber identity module: in GSM, the user can have a subscriber identity module that is inserted at the mobile station. SIM carries the personal number assigned to the user. GSM associates it with the registered number of the mobile station and thus can route calls to the subscriber.

subscriber loop carrier: digital loop carrier.

subscriber terminal 1: a terminal that is fully compliant with ISDN requirements. Connects directly to NT1.

subscriber terminal 2: a terminal that does not fully comply with ISDN requirements. Must connect to NT1 through a terminal adapter.

subscription teletext: specialized information service for the dissemination of business and other commercially valuable information that can be used only by those who pay a fee.

substitution: in encryption, an action in which new bit patterns, characters, words, or phrases are substituted for the originals without changing their order.

superframe format: also called D4 superframe format, 12 DS-1 frames form a superframe that contains 2,316 bits and takes 1.5 milliseconds to transmit at the DS-1 rate. The 12 framing bits form a pattern that is used for synchronization and other purposes.

supervisory audio tone: in mobile telephone networks, an out-of-band, continuous tone (either 5700, 6000 or 6300 Hz). The SAT is present whenever a call is in progress. Its absence indicates that one of the parties has gone on-hook, and the call is terminated.

SW56: 56 kbits/s switched digital service; this is switched DDS.

SW64: 64 kbits/s switched digital service; this is switched DS-0 or ISDN B channel.

SW384: 384 kbits/s switched digital service; this is ISDN H0.

SW1536: 1536 kbits/s switched digital service; this is ISDN H11.

switched multimegabit data service: a high-speed, public, packet–switched, connectionless data service. SMDS supports higher access speeds and greater throughputs than frame–relay networks, and has no distance limitations.

switched network: consists of transmission equipment that can be interconnected, on-demand to provide individual communication channels between many independent sets of users.

switch fabric: term used to describe distributed switching matrix in ATM switch.

switching matrix: makes connections so that a circuit is completed between the calling and called parties. The stored program control seeks out a suitable combination of time–slots and multiplexed highways to perform this task, and instructs the hardware elements to employ them for the duration of the communication session.

switching point: in an SS7 network, an end-office switch at which the customer gains access to the network. Using common-channel signaling, the switch provides normal call-switching functions—establish, maintain, and disconnect.

switching terminal exchange: DCE that implements X.75 procedures to connect to another packet network.

symbols: numbers, letters, shapes, and logical relationships that specify facts, define entities, and seek to inform.

synchronous data link control protocol: a bit-oriented DLC protocol.

synchronous demodulation: process that recovers the message stream from a modulated wave by using the carrier as a reference. The unmodulated carrier signal may be transmitted separately to the receiver, or the receiver may reconstruct the carrier signal from the modulated wave.

synchronous digital hierarchy: designed to transport BISDN signals, employs a basic speed of 155.52 Mbits/s. Higher levels of SDH are multiples of basic speed, i.e., *n* 155.52 Mbits/s. Compatible with ISDN basic and primary rate signals [i.e., 64 kbits/s (DS-0) and 1.544 Mbits/s (DS-1) or 2.048 Mbits/s (E-1)], and SONET STS-N signals.

synchronous multiplexing: to combine *n* lower-speed, digital signals onto a higher-speed, single bearer, time–slots are assigned to each signal in turn in a regular sequence paced by the system.

synchronous operation: data are sent or actions occur in response to a command from *external* equipment.

synchronous optical network: a synchronous digital network that employs optical fibers (exclusively). Each facility conforms to standards that include electrical and optical speeds, and frame formats.

synchronous payload envelope: in an STS-1 frame, 9 rows × 87 columns (783 octets) are called the STS-1 synchronous payload envelope (SPE). Of the 783, nine octets (1 column) are reserved for path overhead information.

synchronous transport module level 1: in SDH, a frame of 19,440 bits that is transported at 155.52 Mbits/s.

synchronous transport module level N: in SDH, a frame of $N \times 19{,}440$ bits that is transported at $N \times 155.52$ Mbits/s. STM-N frames are created by interleaving *N* STM-1 frames, octet by octet.

synchronous transport signal level 1: in SONET, a frame of 6480 bits that is transported at 51.84 Mbits/s. STS-1 signals are designed to carry DS-3 (synchronous or asynchronous) signals, or a combination of DS-1, DS-1C, and DS-2 signals that is equivalent to DS-3.

synchronous transport signal level N: in SONET, a frame of $N \times 6480$ bits that is transported at $N \times 51.84$ Mbits/s. STS-N signals are created by interleaving N STS-1 signals, octet-by-octet. For various reasons, the values N = 3, 12, 24, 48, and 96, are preferred.

synthetic quality speech: contains sounds generated by an electronic analog of the human speech production mechanism.

SYNTRAN DS-3: octets from 28, DS-1 channels of exactly 1.544 Mbits/s are multiplexed directly to DS-3. The format of the traditional DS-3 frame is restructured into a synchronous superframe so that DS-1 frames have fixed locations within the superframe. The purpose is to provide simple, reliable access to all DS-0 and DS-1 signals contained in the SYNTRAN DS-3 signal.

system managed object: in network management, object pertaining to more than a single layer (e.g., router or gateway),

system management application service element: in network management, creates and uses the protocol data units that are exchanged during the management process. It employs communications services supplied by specific applications service elements (ASEs) or by CMISE.

system management function: in network management, system function included in specific management function area.

systems network architecture: a proprietary network in which the communication functions are tightly integrated with the computing resources.

system services control point: in SNA, a NAU that is a part of the TP access method. Running on a host node, it instructs PUs to activate, control, and deactivate resources that have been defined as being in its domain.

T-1C: T-carrier multiplexer that operates at 3.152 Mbits/s. Interleaves two, DS-1 streams, bit by bit, and adds a control bit after every 52 bits (26 from each DS-1). To maintain the ones-density, one DS-1 stream is logically inverted before multiplexing.

T-1 multiplexer: employs a channel bank that processes 24, analog voice signals to become 24, 64 kbits/s, PCM signals, and octet multiplexes them to produce a bit stream of 1.544 Mbits/s that is carried on conditioned, four-wire circuits. The DS-1 frame contains 24 octets that represent samples from 24 voice signals and a single bit that is used for synchronizing (framing bit).

T-2: T-carrier multiplexer that operates at 6.312 Mbits/s. Interleaves four, DS-1 streams, bit by bit, and adds a control bit after every 48 payload bits. To maintain the ones-density, bits in DS-1 streams #2 and #4 are logically inverted before multiplexing.

T-3: T-carrier multiplexer that operates at 44.736 Mbits/s. Bit-interleaves groups of four DS-1s to create seven, DS-2 signals; in turn, the seven, DS-2 signals are bit-interleaved to create one DS-3 signal. In the conversion of seven, DS-2 to DS-3, a control bit is added for every 84 payload bits, and bits are stuffed in specific slots to equalize the speeds of the individual streams.

T-4: T-carrier multiplexer that operates at 274.176 Mbits/s. Bit–interleaves six DS-3 streams. Uses stuffing bits to achieve synchronization.

T-4A: T-carrier multiplexer that operates at 139.264 Mbits/s. Bit–interleaves three DS-3 streams Uses stuffing bits to achieve synchronization.

talkspurt: more or less connected period of vocal activity.

T-carrier hierarchy: multiplexing facilities that combine DS-0 signals into progressively higher–speed streams. The four levels are T-1, T-2, T-3, and T-4, with an intermediate level, T-1C between T-1 and T-2, and another intermediate level, T-4NA between T-3 and T-4. The output signals are DS-1C, DS-2, DS-3, DS-4NA, and DS-4.

T-carrier multiplexers: produce composite signals that contain 24, 48, 96, 672, 2016, and 4032 digital voice signals. Also, provided certain restrictions are observed, they transport data signals. The family of T-carrier equipment extends from 1.544 Mbits/s to 274.176 Mbits/s in four levels.

telecommunication: communication-from-afar; the action of communicating-at-a-distance.

telecommunications: the technology of communication-at-a-distance.

telecommunications access method: in SNA, a queued access method that transfers messages from one remote terminal to another, and between terminals and application programs. Manages a communications network for the exclusive use of the applications it supports.

Telecommunications Development Sector (ITU-D): sector of ITU responsible for promoting the development of technical facilities and services likely to improve the operation of public networks.

Telecommunications Standardization Sector (ITU-T): sector of ITU that develops and adopts Recommendations (consensus standards) designed to facilitate global telecommunication.

Telecommunication Technology Committee (TTC): established to develop and disseminate Japanese domestic standards for deregulated technical items and protocols.

telepoint: a single base station providing one-way wireless, payphone access into the telephone network. It provides cordless pay phone service to customer-owned handsets within limited range of public base stations in malls and downtown areas.

teleprocessing access method: in SNA, a software interface between the host main memory and its peripherals that assists in moving information between the network and the host over the I/O channel. TP access methods reside in a host and run as independent jobs under the operating system.

teleprocessing monitor: to implement real-time, on-line, transaction processing, some applications are separated from the access method by a TP monitor that supervises multiple transaction programs running in a single partition.

teletex: a simplex, record service [1]. A communication service that transports character-based messages to remote sites where they are printed out. Teletex provides direct electronic document distribution from electronic typewriters, word processors, or personal computers that are equipped with sending and receiving storage and communications capabilities.

teletext: an electronic magazine. A mass communication, information retrieval service that supplies a hundred or more pages in the vertical blanking interval (VBI) of a broadcast television signal without affecting the television service.

teletypewriter: device that accepts keyboard input, prints hard copy output, and stores and forwards messages.

telex: a text message service provided by communicating teletypewriters and teleprinters.

terminal adapter: provides the interface and protocol conversions needed for TE2 to connect to NT1.

text mail: an intercommunication, half-duplex, record service [1,1]. The text message is composed at, and delivered by, data terminals; it may be stored in a central system of electronic mailboxes, or stored at one or other of the terminals serving the persons involved. When the intended recipients command, the messages are forwarded and displayed on a video terminal, or printed.

throughput: the rate at which user's data bits are delivered, i.e., the number of user's data bits delivered in unit time.

tie line: non-switched line directly connecting two private network facilities (e.g., PBXs, data centers, etc.) for the exclusive, full-time use of those that lease it.

time division multiple access: a technique that assigns time-slots to stations in the network to create individual digital channels.

time division multiplexing: to each signal the entire bandwidth of the bearer is assigned for very short periods of time so that multiplexed signals exist in series, at different times, on the same bearer.

time-slot interchange process: the action of a time switching stage that, on instructions from the stored program control, processes octets in incoming frames to produce outgoing frames on the same highway that contain the same octets in a different order.

time-slot synchronization: state of transmission facility when receiver can identify time-slots within the frame and correctly allocate them to each channel.

time-space-time switch: combination of time switching stages and a space switching stage to serve a large customer community. The space stage interconnects the time stage serving the calling party with the time stage serving the called party.

time to dial-tone: the time between the moment the user goes off-hook, and the moment dial-tone is returned from the local switch.

time to disconnect: the time between the moment the user goes on-hook, and the moment the circuit is available to others.

token: a supervisory frame that grants access to the bearer so that a station may send.

token bit: in an IEEE 802.5 LAN, bit that gives the token status. If it is 0, the token may be seized by a station that has sufficient priority. If it is 1, the token has been seized by another station and the frame is in use.

token-ring LAN: LAN operating under IEEE 802.5 in which each station is connected to two others so as to form a single-thread loop that connects all of the stations. Data is transferred around the ring from station to station.

touchpad: number pad on telephone instrument that generates bursts of two frequencies when a button is pressed to enter a digit.

TP0, TP1 , TP2, TP3, TP4: ISO transport layer protocols.

traditional DS-3: the DS-3 signal is partitioned in frames containing 4,760 bits. Called M-frames, they consist of seven subframes of 680 bits each. The subframes are divided into eight blocks of 85 bits. In these blocks, the first bit is a control bit and the other 84 are payload bits. Each frame includes 54 overhead bits that are used to perform frame alignment, subframe alignment, performance monitoring, alarm and status reporting, and other functions.

traffic density: the number of calls in progress at a given moment.

traffic intensity: the traffic density averaged over a one hour period.

transaction capabilities part: an SS7 protocol that provides a set of tools in a connectionless environment that can be used by an application at one node to invoke execution of a procedure at another node, and exchange the results.

transaction terminal: device designed for capturing and processing specific on-line transactions. Examples are automatic teller machine and supermarket checkout terminal with bar-code reader.

transducer: device that converts messages from one medium to another.

transient function: function that exists for a limited time.

transmission control protocol: an Internet transmission layer procedure.

transmission design: in the local loop, inductance is added to long loops to reduce the voice band attenuation of the connection so that the parties can hear each other.

transponder: amplifying and frequency changing device carried in a satellite that operates over a portion of the frequency band assigned to the satellite.

transport header: contains the sequence number, and a frame check sequence (FCS) that checks the entire segment.

transport layer: layer 4 of OSI model. Maintains the integrity of end-to-end communication, independent of the performance and number of networks involved in the connection between end systems. Serves as interface between the three lower layers that are concerned with providing transparent connections between users, and the three upper layers that ensure that data are delivered in correct and understandable form.

transport overhead: in SONET, in an STS-1 frame, 9 rows × 3 columns (27 octets) are occupied by transport overhead. It is made up of nine octets (3 rows × 3 columns) of section overhead and 18 octets (6 rows × 3 columns) of line overhead.

transport protocol data unit: combination of transport header and session PDU.

transposition: in encryption, an action in which the order of the bit patterns, characters, words, or phrases is rearranged.

trellis code modulation: a special form of QAM that incorporates twice as many signal points in the constellation as are needed to represent the data. As a result, errors are reduced, and TCM modems are able to operate at higher speeds than their non-TCM counterparts.

tributary group: in SDH, several tributary units.

tributary signal: a lower–speed signal that consists of a synchronized stream of octets.

tributary unit: in SDH, the combination of virtual container and pointer information for a tributary signal.

twisted pair transmission system: consists of pairs of copper wires twisted together and packaged in plastic insulated cable.

two-wire connection: the user's terminal is connected to the serving central office by a pair of wires that constitutes a duplex circuit and carries information in both directions.

unacknowledged connectionless service: messages are not acknowledged by the receiving station.

unchannelized T-1: for transmission over a DS-1 line, the bit stream is divided into segments of 192 bits to which framing bits are attached. Neither the sending nor the receiving equipment makes use of the boundaries that normally define channels.

unipolar signal: for 1s, positive current flows, and for 0s, no current flows.

universal digital channel: 2B1Q digital loop carrier equipment that implements six digital circuits on a single pair of wires for circuit expansion purposes.

universal numbering plan: worldwide numbering plan in which customer stations are identified by subscriber codes, exchange codes, area codes, and country codes.

universal personal telecommunication: ubiquitous telecommunication capability with which the users can communicate across town, or around the world, from anywhere.

universal service requirement: within their operating areas, common carrier communication companies must serve all customers on an equal basis.

universal synchronous/asynchronous receiver transmitter: an integrated circuit chip that implements a start-stop, DLC protocol.

unlocked: in network management, an administrative state in which the managed object can be used.

upstream direction: the direction in which the less rich messages flow.

user agents: in X.400, entities that interact directly with users and message transfer agents. They prepare messages, submit messages for routing and transmission to destinations, accept messages, deliver messages to users, and provide filing, retrieving, forwarding, and other message support functions.

user datagram protocol: a connectionless Internet network layer procedure.

user-friendly: by most persons, the rules governing the operation of user-friendly services and equipment can be learned readily, and they cause little or no increase in anxiety level in users.

user-hostile: by most persons, the rules governing the operation of user-hostile services and equipment are difficult to learn, and they cause a significant increase in anxiety level in users.

user-network interface standard: defines the forms, formats, and sequences of messages that flow between users and communication networks across user-network interfaces.

user-seductive: rules governing operation of user-seductive services and equipment dispel anxiety, and produce pleasure in the user.

vertical redundancy check: parity bits are added to individual characters

vestigial-sideband AM: a modulation technique that produces most of one sideband of LCAM or DSBSC and a vestige of the other sideband.

video conference: an interactive, intercommunication, real-time service {12,12}, that may be augmented to {14,14}.

Video, audio, and text space is used to connect remote conference rooms so that the sights, sounds, and symbols employed by their occupants are available for all to experience.

videoseminar: an interactive, one-to-many, real-time service with unequal richness in the downstream and upstream directions {14,4}.

videotex: an electronic encyclopedia. A mass communication, information retrieval service that employs the telephone network in an interactive way to search through databanks containing thousands of pages of information.

virtual circuit: in a packet network, a path between DTEs that is defined by a virtual circuit identifier (VCI). A virtual path is established so that a sequence of packets that constitutes a block of user's information can use the same path and arrive at the terminating DTE in sequence.

virtual container: in SDH, the combination of container and path overhead for a tributary signal, or that part of the payload that contains several tributary groups.

virtual network: network created out of carrier equipment that is operated as a private network with the special capabilities and performance required by the corporation, but without the burden of investment, maintenance, and disaster recovery. It is configured to provide users with reports on calls, equipment usage, configuration status, and other information that is required to manage corporate telecommunications activities.

Virtual Telecommunications Access Method: in SNA, an access method that directly controls transmission of data to and from channel-attached devices. Manages a communications network for the exclusive use of the applications it supports.

virtual tributary: in SONET, a fixed number of octets that occupy 9 rows × *n* columns in the SPE.

visiting location register: in mobile telephone network, a database that maintains and updates visitor's location and service profile for local use.

vocoding: in digital voice, the speech signal is divided into several separate frequency bands, and the magnitude of the energy in each band is transmitted to the receiver together with information on the presence or absence of pitch (voiced or unvoiced sounds). At the receiver, voiced or unvoiced energy is applied to a set of receivers that produce recognizable speech.

voice mail: an intercommunication, half-duplex, record service [4,4]. Recorded using telephone connections to voice storage devices (tapes or solid-state memories), voice messages are delivered on commands from the intended recipients. Delivery may be by the storage devices directly (answering machines at the recipients' telephones, for instance), or over telephone connections initiated by the recipients.

VSAT network: consists of a large number of geographically dispersed microstations that are star-connected to a hub station through a geostationary satellite.

VT1.5: in SONET, virtual tributary for DS-1 It consists of 27 octets (9 rows × 3 columns, equivalent to 1.728 Mbits/s).

VT2: in SONET, virtual tributary for E-1. It consists of 36 octets (9 rows × 4 columns, equivalent to 2.304 Mbits/s).

VT3: in SONET, virtual tributary for DS-1C. It consists of 54 octets (9 rows × 6 columns).

VT6: in SONET, virtual tributary for DS-2 It consists of 108 octets (9 rows × 12 columns).

WATS line: connection that supports wide-area telephone service.

waveform coding: in digital voice, speech is reduced to a string of bits by operating on the sampled signal.

white noise: a random signal in which power is distributed uniformly at all frequencies. White noise has the property that different samples are uncorrelated, and that, if the probability density function of the amplitude spectrum is Gaussian, they are statistically independent. White, Gaussian noise is truly random.

wide–area network: a star or hierarchically connected network that interconnects widely scattered facilities, usually for a private purpose.

wide–area telephone service: service that offers bulk rates for station-to-station calls that are dialed directly. Calling categories are intrastate and interstate, and incoming or outgoing.

wideband signal: an analog signal with frequency components that exceed the nominal passband of a voice channel.

wide sense stationary process: a random process in which the expectation is constant, and the autocorrelation and covariance functions depend only on the time interval over which they are evaluated.

wire center: a physical location at which many cables come together to be served by a switching facility that spans several exchange codes.

wireless information network: a network that supports communication between parties, at least one of which is not restricted to remain in close proximity to a fixed terminal. A network to which access is gained through wireless means. A network that supports communication between mobiles or between a mobile and a fixed facility.

workstation: device that contains significant processing power in their own right, but nevertheless have need to access other processors.

World System Teletext: ITU-T standard that includes most national teletext systems.

X.3: ITU-T Recommendation that defines a PAD.

X.25: ITU-T Recommendation that defines a user-network interface procedure for a synchronous DTE to connect to a packet network.

X.28: ITU-T Recommendation that defines the procedures whereby an asynchronous DTE is connected to a PAD.

X.29: ITU-T Recommendation that defines the procedures that permit a packet-mode device to control the operation of the PAD.

X.75: ITU-Ts Recommendation that defines way in which packet networks are interconnected through STEs.

X.400: a family of Recommendations for message handling systems prepared by ITU-T.

zero-byte time–slot interchange: in a T-1 system carrying data, a technique used to provide 64 kbits/s clear channels to the user yet maintain ones–density requirements.

Index